T0318842

PERSONALIZED EPIGENETICS

PERSONALIZED EPIGENETICS

Trygve O. Tollefsbol

AMSTERDAM • BOSTON • HEIDELBERG • LONDON
NEW YORK • OXFORD • PARIS • SAN DIEGO
SAN FRANCISCO • SINGAPORE • SYDNEY • TOKYO
Academic Press is an imprint of Elsevier

ELSEVIER

Academic Press is an imprint of Elsevier
125 London Wall, London EC2Y 5AS, UK
525 B Street, Suite 1800, San Diego, CA 92101-4495, USA
225 Wyman Street, Waltham, MA 02451, USA
The Boulevard, Langford Lane, Kidlington, Oxford OX5 1GB, UK

Library of Congress Cataloging-in-Publication Data
A catalog record for this book is available from the Library of Congress

British Library Cataloging-in-Publication Data
A catalog record for this book is available from the British Library

ISBN: 978-0-12-420135-4

For information on all Academic Press publications
visit our website at http://store.elsevier.com/

Working together
to grow libraries in
developing countries

www.elsevier.com • www.bookaid.org

Acquisition Editor: Catherine Van Der Laan
Editorial Project Manager: Lisa Eppich
Production Project Manager: Karen East and Kirsty Halterman
Designer: Mark Rogers

Typeset by TNQ Books and Journals
www.tnq.co.in

Printed and bound in the United States of America

Contents

I

OVERVIEW

1. Epigenetics of Personalized Medicine

TRYGVE O. TOLLEFSBOL

II

EPIGENETIC VARIATIONS AMONG INDIVIDUALS

2. Interindividual Variability of DNA Methylation

LOUIS P. WATANABE AND NICOLE C. RIDDLE

3. Differences in Histone Modifications Between Individuals

CHRISTOPH A. ZIMMERMANN, ANKE HOFFMANN, ELISABETH B. BINDER
AND DIETMAR SPENGLER

4. Individual Noncoding RNA Variations: Their Role in Shaping and Maintaining the Epigenetic Landscape

EMILY MACHIELA, ANTHONY POPKIE AND LORENZO F. SEMPERE

5. Personalized Epigenetics: Analysis and Interpretation of DNA Methylation Variation

HEHUANG XIE

III

BIOINFORMATICS OF PERSONALIZED EPIGENETICS

6. Computational Methods in Epigenetics

VANESSA AGUIAR-PULIDO, VICTORIA SUAREZ-ULLOA, JOSE M. EIRIN-LOPEZ, JAVIER PEREIRA AND GIRI NARASIMHAN

IV

DIAGNOSTIC AND PROGNOSTIC EPIGENETIC APPROACHES TO PERSONALIZED MEDICINE

7. Epigenetic Biomarkers in Personalized Medicine

FABIO COPPEDÈ, ANGELA LOPOMO AND LUCIA MIGLIORE

8. Epigenetic Fingerprint

LEDA KOVATSI, ATHINA VIDAKI, DOMNIKI FRAGOU AND D. SYNDERCOMBE COURT

9. Epigenetics of Personalized Toxicology

ALEXANDRE F. AISSA AND LUSÂNIA M.G. ANTUNES

V

ENVIRONMENTAL PERSONALIZED EPIGENETICS

10. Environmental Contaminants and Their Relationship to the Epigenome

ANDREW E. YOSIM, MONICA D. NYE AND REBECCA C. FRY

11. *Nutriepigenomics*: Personalized Nutrition Meets Epigenetics

ANDERS M. LINDROTH, JOO H. PARK, YEONGRAN YOO AND YOON J. PARK

VI

PHARMACOLOGY AND DRUG DEVELOPMENT OF PERSONALIZED EPIGENETICS

12. Personalized Pharmacoepigenomics

JACOB PEEDICAYIL

13. Personalized Medicine and Epigenetic Drug Development

KENNETH LUNDSTROM

VII

PERSONALIZED EPIGENETICS OF DISORDERS AND DISEASE MANAGEMENT

14. Epigenetics and Personalized Pain Management

SEENA K. AJIT

15. Understanding Interindividual Epigenetic Variations in Obesity and Its Management

SONAL PATEL, ARPANKUMAR CHOKSI AND SAMIT CHATTOPADHYAY

16. Epigenetic Modifications of miRNAs in Cancer

AMMAD A. FAROOQI, MUHAMMAD Z. QURESHI AND MUHAMMAD ISMAIL

17. Managing Autoimmune Disorders through Personalized Epigenetic Approaches

CHRISTOPHER CHANG

18. Cardiovascular Diseases and Personalized Epigenetics

ADAM M. ZAWADA AND GUNNAR H. HEINE

VIII

CHALLENGES AND FUTURE DIRECTIONS

19. Future Challenges and Prospects for Personalized Epigenetics

PENG ZHANG, YING LIU, QIANJIN LU AND CHRISTOPHER CHANG

Contributors

Vanessa Aguiar-Pulido   School of Computing & Information Sciences, Florida International University, Miami, FL, USA; Department of Information & Communication Technologies, University of A Coruña, A Coruña, Spain

Alexandre F. Aissa   Department of Clinical Analyses, Toxicology and Food Sciences, School of Pharmaceutical Sciences of Ribeirão Preto, University of São Paulo (USP), Ribeirão Preto, São Paulo, Brazil

Seena K. Ajit   Department of Pharmacology & Physiology, Drexel University College of Medicine, Philadelphia, PA, USA

Lusânia M.G. Antunes   Department of Clinical Analyses, Toxicology and Food Sciences, School of Pharmaceutical Sciences of Ribeirão Preto, University of São Paulo (USP), Ribeirão Preto, São Paulo, Brazil

Elisabeth B. Binder   Max Planck Institute of Psychiatry, Translational Research, Munich, Germany

Christopher Chang   Division of Rheumatology, Allergy and Clinical Immunology, University of California at Davis, CA, USA

Samit Chattopadhyay   National Centre for Cell Science, Pune University Campus, Ganeshkhind, Pune, India

Arpankumar Choksi   National Centre for Cell Science, Pune University Campus, Ganeshkhind, Pune, India

Fabio Coppedè   Department of Translational Research and New Technologies in Medicine and Surgery, University of Pisa, Pisa, Italy

Jose M. Eirin-Lopez   Department of Biological Sciences, Florida International University, North Miami, FL, USA

Ammad A. Farooqi   Laboratory for Translational Oncology and Personalized Medicine, Rashid Latif Medical College, Lahore, Pakistan

Domniki Fragou   Laboratory of Forensic Medicine and Toxicology, School of Medicine, Aristotle University of Thessaloniki, Thessaloniki, Greece

Rebecca C. Fry   Department of Environmental Sciences and Engineering, University of North Carolina, Chapel Hill, NC, USA; Curriculum in Toxicology, School of Medicine, University of North Carolina, Chapel Hill, NC, USA

Gunnar H. Heine   Department of Internal Medicine IV, Nephrology and Hypertension, Saarland University Medical Center, Homburg, Germany

Anke Hoffmann   Max Planck Institute of Psychiatry, Translational Research, Munich, Germany

Muhammad Ismail   IBGE, Islamabad, Pakistan

Leda Kovatsi   Laboratory of Forensic Medicine and Toxicology, School of Medicine, Aristotle University of Thessaloniki, Thessaloniki, Greece

Anders M. Lindroth Graduate School of Cancer Science and Policy, National Cancer Center, Goyang-si, Republic of Korea

Ying Liu Department of Dermatology, Hunan Key Laboratory of Medical Epigenomics, Second Xiangya Hospital, Central South University, Hunan, China

Angela Lopomo Department of Translational Research and New Technologies in Medicine and Surgery, University of Pisa, Pisa, Italy; Doctoral School in Genetics, Oncology, and Clinical Medicine, University of Siena, Siena, Italy

Kenneth Lundstrom PanTherapeutics, Lutry, Switzerland

Qianjin Lu Department of Dermatology, Hunan Key Laboratory of Medical Epigenomics, Second Xiangya Hospital, Central South University, Hunan, China

Emily Machiela Laboratory of Aging and Neurodegenerative Disease, Van Andel Research Institute, Grand Rapids, MI, USA

Lucia Migliore Department of Translational Research and New Technologies in Medicine and Surgery, University of Pisa, Pisa, Italy

Giri Narasimhan School of Computing & Information Sciences, Florida International University, Miami, FL, USA

Monica D. Nye Department of Environmental Sciences and Engineering, University of North Carolina, Chapel Hill, NC, USA

Joo H. Park Department of Nutritional Science and Food Management, Ewha Womans University, Seoul, Republic of Korea

Yoon J. Park Department of Nutritional Science and Food Management, Ewha Womans University, Seoul, Republic of Korea

Sonal Patel National Centre for Cell Science, Pune University Campus, Ganeshkhind, Pune, India

Jacob Peedicayil Department of Pharmacology and Clinical Pharmacology, Christian Medical College, Vellore, India

Javier Pereira Department of Information & Communication Technologies, University of A Coruña, A Coruña, Spain

Anthony Popkie Laboratory of Cancer Epigenomics, Van Andel Research Institute, Grand Rapids, MI, USA

Muhammad Z. Qureshi Department of Chemistry, GCU, Lahore, Pakistan

Nicole C. Riddle Department of Biology, The University of Alabama at Birmingham, Birmingham, AL, USA

Lorenzo F. Sempere Laboratory of MicroRNA Diagnostics and Therapeutics, Van Andel Research Institute, Grand Rapids, MI, USA

Dietmar Spengler Max Planck Institute of Psychiatry, Translational Research, Munich, Germany

Victoria Suarez-Ulloa Department of Biological Sciences, Florida International University, North Miami, FL, USA

D. Syndercombe Court Faculty of Biological Sciences and Medicine, King's College London, London, UK

Trygve O. Tollefsbol   Department of Biology, University of Alabama at Birmingham, Birmingham, AL, USA; Comprehensive Cancer Center, University of Alabama at Birmingham, Birmingham, AL, USA; Comprehensive Center for Healthy Aging, University of Alabama at Birmingham, Birmingham, AL, USA; Nutrition Obesity Research Center, University of Alabama at Birmingham, Birmingham, AL, USA; Comprehensive Diabetes Center, University of Alabama at Birmingham, Birmingham, AL, USA

Athina Vidaki   Faculty of Biological Sciences and Medicine, King's College London, London, UK

Louis P. Watanabe   Department of Biology, The University of Alabama at Birmingham, Birmingham, AL, USA

Hehuang Xie   Department of Biological Sciences, Virginia Bioinformatics Institute, Virginia Tech, Blacksburg, VA, USA

Yeongran Yoo   Department of Nutritional Science and Food Management, Ewha Womans University, Seoul, Republic of Korea

Andrew E. Yosim   Department of Environmental Sciences and Engineering, University of North Carolina, Chapel Hill, NC, USA

Adam M. Zawada   Department of Internal Medicine IV, Nephrology and Hypertension, Saarland University Medical Center, Homburg, Germany

Peng Zhang   Department of Dermatology, Hunan Key Laboratory of Medical Epigenomics, Second Xiangya Hospital, Central South University, Hunan, China

Christoph A. Zimmermann   Max Planck Institute of Psychiatry, Translational Research, Munich, Germany

Preface

It is now apparent that epigenetics is a central component of medicine. In light of that fact, future diagnostic, prognostic, and therapeutic advances will almost certainly increasingly rely on personalized epigenetics for the optimal management of many, if not most, health-related conditions.

Personalized epigenetics notably stands out in that the emphasis is on the translatability of epigenetics to health management of individuals who have unique variations in their epigenetic signatures that can guide disorder or disease prevention or therapy. The goal of this book on personalized epigenetics is to provide a comprehensive analysis of interindividual variability of epigenetic markers of human disease and to illuminate the bench to bedside advances that are readily advancing in this field. More specifically, the purpose of this book focusing on personalized epigenetics is to facilitate understanding of the application of medical prevention and therapy based in part on the unique health and disease susceptible epigenetic profile of each individual.

This book is intended for those with interests ranging from basic molecular biology to clinical therapy and who have an interest in personalized medicine.

Trygve O. Tollefsbol

OVERVIEW

CHAPTER

1

Epigenetics of Personalized Medicine

Trygve O. Tollefsbol[1,2,3,4,5]

[1]Department of Biology, University of Alabama at Birmingham, Birmingham, AL, USA; [2]Comprehensive Cancer Center, University of Alabama at Birmingham, Birmingham, AL, USA; [3]Comprehensive Center for Healthy Aging, University of Alabama at Birmingham, Birmingham, AL, USA; [4]Nutrition Obesity Research Center, University of Alabama at Birmingham, Birmingham, AL, USA; [5]Comprehensive Diabetes Center, University of Alabama at Birmingham, Birmingham, AL, USA

OUTLINE

1. INTRODUCTION

While each species, cell, and system has a characteristic epigenetic profile, individuals also have an epigenetic profile that forms their unique epigenome. Epigenetic aberrations are known to play a key role in many human diseases, and the purpose of personalized epigenetics is to base medical prevention and therapeutics as well as diagnostics and prognostics on the distinctive health and disease susceptibility profile of each individual. In fact, interindividual differences in the epigenome are the basis of personalized epigenetics and these can manifest through epigenetic signatures that are characteristic of each individual. DNA methylation, for example, varies between individuals [1–2] as do the many different forms of epigenetic histone modifications [3–4]. In addition, noncoding RNA profiles are also variable from person to person [5–6]. These variations in epigenetic expression will undoubtedly become increasingly important as the potential for personalized epigenetics in medicine continues to grow.

Personalized epigenetics can serve as a guide not only for the therapy of epigenetic-based diseases, but also for many other aspects of medicine. Epigenetic biomarkers comprising epigenomic signatures unique to each individual have increasingly been shown to provide not only valuable information with respect to the diagnosis of disorders and diseases, but also prognostic information pertaining to the likely progression of diseases such as cancer [7–8]. There are also many applications of personalized epigenetics in forensics and in toxicology. Many toxic compounds leave distinct epigenomic signatures that could have significant utility in terms of diagnosing toxicity as well as treatment of patients that have been exposed to toxins [9–10]. Similar concepts apply to environmental contaminants [11–12] and nutrients [13–14] that also have an impact on the epigenome.

Equally exciting is the prospect of application of personalized epigenetics to the many disorders and diseases that have epigenetic aberrations as a component of their etiology or pathogenesis. For example, epigenetic alterations have been shown in a number of studies to be important in chronic pain, and studies are emerging that may lead to personalized management of pain based on the distinct epigenomic profile of the individual patient [15–16]. Approaches to patient management through personalized epigenetics are also rapidly developing for many other medical conditions such as obesity, cancer, autoimmune disorders, and cardiovascular diseases.

2. EPIGENETIC VARIATIONS AMONG INDIVIDUALS

One of the most important components of the collective epigenetic changes that occur in cells and tissues is DNA methylation. The role of DNA methylation in modulating gene expression has been known for quite some time, although it has been more recently appreciated that it can vary considerably between individuals within a species. As Watanabe and Riddle explain in Chapter 2, the variation in DNA methylation between individuals is a common feature ranging from plants to humans. There are a number of possible causes for extensive variance of DNA methylation among individuals, including gender, health status, age, and environmental exposures, as well as many other factors. The heterogeneity of DNA methylation patterns within individuals is of particular interest to those focused on personalized medicine. For example, interindividual differences in DNA methylation have considerable potential as a biomarker for a number of diseases as well as a means to guide therapy for diseases such as cancer. In fact, the optimal drug dosage that is administered to individual patients could be monitored through changes in the patient's DNA methylation profile, which serves as a classic example of the utility of knowledge of personalized epigenetics as applied to medicine, as reviewed in Chapter 2.

There are many different types of histone modifications that can modulate epigenetic gene expression of an organism and these modifications are also subject to interindividual variation. The importance of individual heterogeneity of histone modifications as well as techniques such as mapping of DNase I-hypersensitive sites and formaldehyde-assisted isolation of regulatory elements (FAIRE) and chromatin immunoprecipitation (ChIP), as well as their genomic counterparts (DNase-seq, FAIRE-seq, and ChIP-seqA), is highlighted in Chapter 3. Especially fascinating is the prospect of the potential use of induced pluripotent stem cells for elucidating chromatin modifications and transcription differences in healthy and disease states as a "bottom-up" approach (Chapter 3) to advance our knowledge of chromatin heterogeneity between individuals and the potential application of this knowledge to personalized medicine.

Noncoding RNA (ncRNA) variations have great importance to personalized epigenetics since ncRNAs such as microRNAs (miRNAs), trinucleotide repeats, and long noncoding RNAs not only regulate epigenetic processes, but are in turn regulated by epigenetic processes as well. Variation in ncRNA sequences can modulate not only RNA stability, but also its processing, which can affect the regulatory capacity of ncRNAs. As delineated in Chapter 4, these differences in ncRNAs between individuals can lead to major epigenetic changes in disease processes and may also have utility as biomarkers of various diseases. For example, there is some evidence that miRNA expression signatures may distinguish Alzheimer's disease and Parkinson's disease from normal controls (Chapter 4). Moreover, variations of ncRNAs between individuals could be employed for personalized medical therapy through monitoring of drug responses.

Despite great excitement in the potential uses of interindividual epigenetic variations in personalized medicine, there are some limitations that must be overcome. For instance, there can be challenges in detecting true epi-mutations, and epigenetic signatures vary not only among individuals, but also between the various cell types within individuals (Chapter 5). There can also be complications with respect to allele-specific or asymmetric DNA methylation changes that can add to the complexity of interpretations of epigenetic differences between individuals. Nevertheless, it is apparent that interindividual differences in epigenetic processes such as DNA methylation is substantial and the collective pattern differences in response to medical interventions have considerable promise in leading to many advances in medical diagnosis, prognosis, and therapy.

3. BIOINFORMATICS OF PERSONALIZED EPIGENETICS

Although the heterogeneity of epigenetic events presents a challenge in monitoring reliable changes in epigenetic patterns between individuals and in response to medical interventions, computational epigenetics provides many solutions to this challenge. Chapter 6 reviews the data types most often used in epigenetic studies and their use in medicine. Computational approaches to DNA methylation patterns, histone modifications, and ncRNAs as well as quantitative protein analyses are greatly enhancing the ability to identify specific differences in epigenetic profiles between individuals that can be readily applied to personalized medicine. Technologies such as grid computing, which involves coupling of networked and geographically dispersed computers that, in combination, will be able to perform the requisite computational tasks that will enable more accurate analyses of variations in epigenetic signatures, are discussed.

4. DIAGNOSTIC AND PROGNOSTIC EPIGENETIC APPROACHES TO PERSONALIZED MEDICINE

The further development of epigenetic biomarkers has considerable potential to significantly advance personalized medicine. Epigenetic signatures, for example, unique epigenetic features of DNA methylation or ncRNA, may be applied not only to diagnostic aspects of personalized care, but also to prognosticating the progression of diseases (Chapter 7). Moreover, epigenetic biomarkers are being developed to monitor the most efficacious approaches to therapy. The development of epigenetic biomarkers has been most readily apparent in the field of tumor biology. These biomarkers provide both diagnostic and prognostic information, since different epigenetic markers may be present at various stages of carcinogenesis, thereby conferring predictive potential of disease outcome in different individuals. Changes in the epigenetic signature during carcinogenesis could be very useful in illuminating the likely progression of the disease. Currently considerable efforts are aimed at identifying epigenetic biomarkers that may be individual-specific based on disease progression. The development of these biomarkers may be applied not only to personalized cancer care but also to a number of other diseases such as neurodegenerative diseases, psychiatric and behavioral disorders, autoimmune diseases, cardiovascular diseases, and obesity, as described in Chapter 7 by Migliore and colleagues.

Future developments in personalized epigenetics will almost certainly become a part of the criminal justice system through application of epigenetic signatures to forensics, as described in Chapter 8. What we are referring to as an epigenetic signature may eventually replace more conventional means of criminal identification. For instance, monozygotic twins have the same genomic sequence but vary in their epigenomes, especially as aging progresses and environmental factors cause the epigenome to become more unique with the passage of time and with life experiences. In each individual there is only one genome but many epigenomes, which could also provide additional information of use in forensics. However, to date personalized epigenetics has not yet been applied to the criminal justice system, in part because of limitations in the availability of material and validation methods (Chapter 8). Although the courts have not yet generally accepted personalized epigenetics as admissible evidence, further developments in this area will undoubtedly continue and it is hoped that personalized epigenetics will become a regular part of the justice system in the not too distant future.

Exposure to toxins has been well documented to confer epigenetic effects, and the fields of toxicoepigenetics and toxicoepigenomics are rapidly advancing as the knowledge of the epigenetic impact of toxic exposure has increased. At the heart of personalized epigenetics is interindividual variance in epigenetic signatures, and identification of alterations in epigenetic signatures after toxic exposure may enhance improved treatment decisions, as described by Aissa and Antunes in Chapter 9. Various toxins such as arsenic and lead can deregulate epigenetic mechanisms that can leave unique signatures on the epigenome, which may not only help guide therapeutic measures, but also could serve to diagnose the source of toxicity or to prognosticate the likely outcome from a toxic exposure.

Therefore it is apparent that epigenetic biomarkers, forensics, and toxicology are emerging fields in personalized medicine. Advances are rapidly indicating that personalized epigenetics may eventually play a major role in the legal system, monitoring of disease progression, diagnosis of specific diseases, and identification and treatment of toxic exposures. The key to these developments will be further advances in the detection of epigenetic changes that will permit reliable and reproducible monitoring of epigenetic signatures specific to each individual.

5. ENVIRONMENTAL PERSONALIZED EPIGENETICS

As mentioned earlier, epigenetic changes tend to increase with aging and these changes in epigenetic marks are likely to contribute to a number of age-associated diseases such as cancer, diabetes, and cardiovascular disease. Fry and colleagues emphasize in Chapter 10 that environmental exposures to contaminants, such as through occupational hazards, often lead to epigenetic changes within individuals that are likely to lead to a personalized risk for epigenetic diseases later in life. These epigenetic insults could even occur prenatally through diet or at other times in early life and lead to epigenetic aberrations that later manifest as a number of disorders or diseases. Moreover, exposure to factors such as air pollution and cigarette smoke can occur within specific gene targets (Chapter 10). These altered gene-specific epigenetic events may therefore serve as diagnostic or prognostic markers that can serve to guide personalized epigenetic health management.

Perhaps one of the most important factors that contribute to the environmental influence on personalized epigenetics is the diet. Many foods and beverages have been shown to modify the epigenome and to create changes in the epigenome as specific to individuals as their diets. These dietary bioactive components have been known to affect not only DNA methylation, but also many histone modifications and ncRNAs. As described in Chapter 11, health strategies based on nutriepigenomics have the potential to allow optimal design of dietary intervention in many different diseases. Analyses that identify the specific nutriepigenomic signatures characteristic of not only key epigenetic-modifying dietary components, but also combinations of these components, will have increasing application to personalized epigenetics and relevance to the many different disorders and diseases that can be prevented, managed, or treated through dietary means.

6. PHARMACOLOGY AND DRUG DEVELOPMENT OF PERSONALIZED EPIGENETICS

The many new epigenetic-modifying drugs such as DNA methyltransferase and histone deacetylase inhibitors that have been developed have given rise to the relatively new field of pharmacoepigenetics. While pharmacoepigenetics involves the epigenetic basis of drug-response variation, pharmacoepigenomics focuses on the genomic impact of these variations (Chapter 12). Pharmacoepigenomics is an important component of personalized medicine since epigenetic-modulating drugs can leave a unique signature on the genome that can be used for monitoring the efficacy of epigenetic drugs with respect to neutralizing the epigenetic aberrations that have given rise to disorders or diseases specific to each individual. Implicit in advances in pharmacoepigenomics will be continued epigenomic studies that characterize the normal state of the epigenome as well as the characteristic changes that occur in the epigenome in various disease states. Also important is knowledge of the specific epigenomic changes that are conferred by the epigenetic-modifying drugs that are administered. The potential for toxicity from epigenetic drugs is another area that needs further study and personalized epigenetics will undoubtedly benefit from inclusion of analyses of pharmacoepigenetics with attention to adverse drug effects as well.

As reviewed by Kenneth Lundstrom in Chapter 13, the field of epigenetics has considerable potential for new drug development and the application of these pharmacological agents to personalized medicine. One of the reasons personalized epigenetics is an especially attractive area in drug development is that individual differences are common in pharmacological therapy and prescreening the epigenome for disease-specific therapy could not only prevent adverse drug reactions, but also enhance the likelihood of accurate dosages and specificity of epigenetic-modifying drug interventions.

7. PERSONALIZED EPIGENETICS OF DISORDERS AND DISEASE MANAGEMENT

Development of personalized epigenetics for the management and therapy of many different disorders and diseases is rapidly leading to a major expansion of the tools available to clinicians in preventing and controlling the many medical conditions that have an epigenetic basis in their etiology and pathogenesis. For example, chronic pain is a component of numerous medical conditions and a number of studies have reported an important role for epigenetics in the development and/or maintenance of chronic pain, as reviewed by Seena Ajit in Chapter 14. The management of chronic pain is highly individualistic, and monitoring of the epigenomic signatures in patients suffering from chronic pain will have considerable utility in the choice of appropriate analgesics that can alleviate chronic pain in affected individuals.

Many genes that are under epigenetic control also influence the development of obesity and related metabolic disorders such as diabetes. Dietary factors are well known to elicit epigenetic changes in DNA methylation, histone modifications, and ncRNA and these dietary factors often have interindividual variability. This is also the case with the epigenetics of obesity, in which both the quality and the quantity of diet can alter the epigenetic signature of individuals leading to epigenetic aberrations that may be manageable through personalized medical therapy.

As indicated in Chapter 15, obesity is a medical challenge at all ages and personalized epigenetic approaches need to be considered for efficacious management of obesity in the young as well as adults. Preventive personalized epigenetics also has considerable potential in the management of obesity since environmental factors often play an important role in the development of obesity, and epigenetic aberrations can be readily reversible through changes in environmental factors that lead to disorders such as obesity.

It has been known for quite some time that epigenetic aberrations often contribute to cancer and, in fact, the field of tumor biology has led the way to much of what has been discovered about the role of epigenetics in disease prevention, progression, and therapy. In addition to DNA methylation and histone modifications, ncRNAs also have an important role in epigenetics as mentioned earlier, and this especially holds true in the case of cancer, in which ncRNAs such as miRNAs have been found to contribute to a number of cancers such as gastric, lung, and pancreatic cancer, to mention a few (Chapter 16). It is apparent that personalized approaches to the diagnosis and management of cancer will increasingly rely on interindividual differences in miRNAs in patients suffering from many different types of cancer. However, as reviewed in Chapter 16, very little is known about the role of miRNAs in individualized chemoresistance, and advances in personalized epigenetics are especially needed in this specific area of research.

Newer to the field of epigenetics is the broad area of autoimmune disorders, although it is clear that, like the case of cancer and many other diseases, autoimmune disorders have an important epigenetic basis and much of this can be attributed to the impact of the environment on the immune system. Chapter 16 consists of a fascinating account of not only the role of epigenetics in autoimmunity, but also how this fits into the field of personalized epigenetics. For example, Christopher Chang indicates in Chapter 16, "An understanding of the precise epigenetic changes that occur in a patient with an autoimmune disease may at least partially guide us in the knowledge of which gene may need to be regulated and by which epigenetic mechanisms." The author further explains that this is an example of how epigenetics shows promise in the field of personalized medicine. It is known that autoimmune disorders as well as cancer, diabetes, and many other diseases are often age-associated, and the increasing epigenetic drift that occurs with aging illustrates the importance of personalized epigenetics in the management of these many diseases that accompany the aging process.

Cardiovascular disease persists as a significant health problem and a major cause of death. Many studies have shown that epigenetic alterations contribute to the development and progression of cardiovascular diseases, and drugs that are commonly used for the management of cardiovascular diseases, such as the statins, can also have pleiotropic effects on epigenetics. As indicated in Chapter 18, knowledge of personalized epigenetics has a high translational potential to the application of personalized medicine in clinical practice. One aspect of this translational potential is the further development of epigenetic biomarkers for cardiovascular disease that would allow earlier intervention and lead to more effective preventive approaches to cardiovascular diseases before the development of overt disease. In addition, increased understanding of the underlying basic epigenetic aberrations in cardiovascular disease will not only facilitate the development of therapeutic targets, but also lead to individualized strategies for the unique cardiovascular epigenetic alterations in individual patients.

8. CHALLENGES AND FUTURE DIRECTIONS

Individual variance in disease pathogenesis and response to therapy are mainstays in personalized epigenetics, and Chapter 19 delineates some of the future prospects and challenges that face the implementation of personalized epigenetics in the clinic. Challenges consist of individual property rights, reimbursement policies, and patient privacy. Since the *individual* profile of a patient is central to personalized epigenetics, care must be taken to preserve individual rights and privacy. New policies may need to be mandated along with developments in personalized epigenetics to ensure complete protection of individual patients and against the risks that could be associated with greater focus on individual epigenetic signatures. Perhaps most exciting are the future prospects of personalized epigenetics that include advances in both epigenetic biomarkers and epigenetic therapy in personalized medicine. For example, Chapter 19 by Zhang et al. indicates that the development of panels of epigenetic biomarkers for specific disorders will be necessary to facilitate diagnosis and monitoring of epigenetic-based diseases for personalized management. The development of pharmacodynamic epigenetic biomarkers will also be important to ascertain individual responses to medical intervention.

9. CONCLUSION

It is clear that personalized epigenetics is now coming of age and few dispute the potential of this field to revolutionize the diagnosis, prognosis, and management of many medical conditions. At the basic core of personalized epigenetics is interindividual variability in a number of epigenetic signatures such as DNA methylation, histone modifications, and ncRNA. Although there are some limitations, such as variations in epigenetic signatures between cells and allele-specific or asymmetric DNA methylation changes within an individual, the characterization of individual differences in epigenetics has progressed significantly and is continuing to reveal the importance of interindividual epigenetic variability in medicine. Computational epigenetics is providing many opportunities to interpret and understand the inherent complexities in the use of epigenomic information and how it can be translated to individual diagnosis, prognosis, and therapy. Increasing knowledge of epigenetic biomarkers and personalized epigenetic responses to drugs and environmental toxicants will continue to be meaningful approaches to enhancing the application of epigenetics to personalized medicine. The role of personalized epigenetics in disease prevention and management is progressing rapidly. For example, advances have been made with respect to not only pain management and personalized epigenetics, but also the role of personalized epigenetics in obesity, cancer, autoimmune disorders, cardiovascular diseases, and many other medical conditions. The future of personalized epigenetics is very bright and I have little doubt that this area will continue to develop at a rapid pace and ultimately revolutionize approaches to clinical practice.

References

[1] Heyn H, Moran S, Hernando-Herraez I, Sayols S, Gomez A, Sandoval J, et al. DNA methylation contributes to natural variation. Genome Res 2013;23:1363–72.

[2] Chambwe N, Kormaksson M, Geng H, De S, Michor F, Johnson NA, et al. Variability in DNA methylation defines novel epigenetic subgroups of DLBCL associated with different clinical outcomes. Blood 2014;123:1699–708.

[3] Kadota M, Yang HH, Hu N, Wang C, Hu Y, Taylor PR, et al. Allele-specific chromatin immunoprecipitation studies show genetic influence on chromatin state in human genome. PLoS Genet 2007;3:e81.

[4] Kilpinen H, Waszak SM, Gschwind AR, Raghav SK, Witwicki RM, Orioli A, et al. Coordinated effects of sequence variation on DNA binding, chromatin structure, and transcription. Science 2013;342:744–7.

[5] Wang L, Wang J. MicroRNA-mediated breast cancer metastasis: from primary site to distant organs. Oncogene 2012;31:2499–511.

[6] Cittelly DM, Das PM, Spoelstra NS, Edgerton SM, Richer JK, Thor AD, et al. Downregulation of miR-342 is associated with tamoxifen resistant breast tumors. Mol Cancer 2010;9:317.

[7] Tänzer M, Balluff B, Distler J, Hale K, Leodolter A, Röcken C, et al. Performance of epigenetic markers SEPT9 and ALX4 in plasma for detection of colorectal precancerous lesions. PLoS One 2010;5:e9061.

[8] Gezer U, Holdenrieder S. Post-translational histone modifications in circulating nucleosomes as new biomarkers in colorectal cancer. In Vivo 2014;28:287–92.

[9] Dejeux E, Ronneberg J, Solvang H, Bukholm I, Geisler S, Aas T, et al. DNA methylation profiling in doxorubicin treated primary locally advanced breast tumours identifies novel genes associated with survival and treatment response. Mol Cancer 2010;9:68.

[10] Lombardi G, Rumiato E, Bertorelle R, Saggioro D, Farina P, Della Puppa A, et al. Clinical and genetic factors associated with severe hematological toxicity in glioblastoma patients during radiation plus temozolomide treatment: a prospective study. Am J Clin Oncol September 21, 2013. [Epub ahead of print] PMID:24064758.

[11] Bollati V, Marinelli B, Apostoli P, Bonzini M, Nordio F, Hoxha M, et al. Exposure to metal-rich particulate matter modifies the expression of candidate microRNAs in peripheral blood leukocytes. Environ Health Perspect 2010;118:763–8.

[12] Fu A, Leaderer BP, Gent JF, Leaderer D, Zhu Y. An environmental epigenetic study of ADRB2 5′-UTR methylation and childhood asthma severity. Clin Exp Allergy 2012;42:1575–81.

[13] Yoo JY, Lee S, Lee HA, Park H, Park YJ, Ha EH, et al. Can proopiomelanocortin methylation be used as an early predictor of metabolic syndrome? Diabetes Care 2014;37:734–9.

[14] Sohi G, Marchand K, Revesz A, Arany E, Hardy DB. Maternal protein restriction elevates cholesterol in adult rat offspring due to repressive changes in histone modifications at the cholesterol 7alpha-hydroxylase promoter. Mol Endocrinol 2011;25:785–98.

[15] Zhang Z, Cai YQ, Zou F, Bie B, Pan ZZ. Epigenetic suppression of GAD65 expression mediates persistent pain. Nat Med 2011;17:1448–55.

[16] Denk F, Huang W, Sidders B, Bithell A, Crow M, Grist J, et al. HDAC inhibitors attenuate the development of hypersensitivity in models of neuropathic pain. PAIN 2013;154:1668–79.

EPIGENETIC VARIATIONS AMONG INDIVIDUALS

2

Interindividual Variability of DNA Methylation

Louis P. Watanabe, Nicole C. Riddle

Department of Biology, The University of Alabama at Birmingham,
Birmingham, AL, USA

O U T L I N E

1. INTRODUCTION

Covalent DNA modifications occur in a large number of organisms and carry out a variety of functions, contributing, for example, to gene regulation and the defense against parasitic sequences. One of the most common DNA modifications is the addition of a methyl group to the 5 position of cytosine, leading to the formation of 5-methylcytosine (5mC). This modification, often referred to informally as "DNA methylation," is present in the majority of eukaryotic lineages (for example see ref. [1]). There are some exceptions, most notable among them the genetic model systems *Caenorhabditis elegans*, *Drosophila melanogaster*, and the yeasts *Saccharomyces cerevisiae* and *Schizosaccharomyces pombe* (but see ref. [2]), demonstrating that while DNA methylation is widely distributed among eukaryotes, it is not essential.

In organisms such as plants and mammals that have DNA methylation, it contributes to a variety of biological processes [3]. Its main function is in the control of gene regulation (for a review for mammals see ref. [4]). Often, DNA methylation is thought to be associated with gene silencing, but this commonly held belief is incorrect in many cases. The relationship between DNA methylation and gene expression is complex and depends on the genomic region in question, the sequence context, and the species [5,6]. In mammals, increased levels of 5mC are seen in the heterochromatin, along with methylation of lysine 9 on histone 3, in which it is associated with gene silencing. At CpG islands—regions in the genome enriched for CpG dinucleotides, often in the promoter region of genes—the presence of 5mC is also associated with silencing, as it often will prevent the binding of transcription factors. However, there are exceptions, by which DNA methylation can result in the recruitment of a transcription factor, thus leading to active transcription [7,8]. Increased levels of 5mC are seen also in the gene bodies of transcriptionally active genes in mammals [5,6], illustrating again the complex relationship between DNA methylation and a gene's activity.

There are several pathways that mediate the contribution of DNA methylation to gene regulation. The presence of DNA methylation can, for example, lead to the recruitment of silencing proteins and thus result in the formation of a chromatin structure inaccessible to the transcriptional machinery. Large protein complexes that include DNA methyltransferases as well as other chromatin-modifying enzymes, such as SUV39H1, a histone 3 lysine 9 methyltransferase, mediate these effects on chromatin [9,10]. In addition, some regulatory proteins contain methyl-binding domains (MBDs), which specifically recognize methylated DNA [11]. Thus, MBD proteins can be recruited to sequences such as methylated promoters where they can serve as transcriptional activators or repressors. The resulting diversity of interactions between DNA methyltransferases and 5mC with other gene-regulatory proteins leads to the complex relationship between 5mC levels and gene expression.

Given DNA methylation's important role in gene regulation, it is not surprising that DNA methylation is crucial for human health and well-being. Several well-studied genetic conditions are linked to DNA methylation defects. They include imprinting disorders such as the Prader–Willi, Angelman, and Beckwith–Wiedemann syndromes [12–14], and also Rett syndrome, which is due to a defect in *MECP2*, an MBD protein-encoding gene [15]. DNA methylation is also linked to several common complex diseases such as diabetes, heart disease, and various cancers (see below). In this chapter, we review the available data on interindividual variation in DNA methylation, its sources, and the implications of this variation for human health and personalized medicine.

2. BIOCHEMICAL PATHWAYS REGULATING DNA METHYLATION

Most of our knowledge about DNA methylation comes from studies carried out in plants, mammals, and the fungal model system *Neurospora crassa*. Work from many laboratories has elucidated the biochemical pathways that lead to the formation of 5mC, the initiation and maintenance of specific methylation patterns, and the removal of DNA methylation [16]. A simplified version of the current model for the regulation of DNA methylation is shown in Figure 1 [16]. DNA methyltransferases generate DNA methylation using S-adenosylmethionine as the methyl donor. Generally, two classes of DNA methyltransferases exist, de novo DNA methyltransferases (DNMT3 types), which can add a methyl group to any cytosine, and maintenance DNA methyltransferases (DNMT1 types),

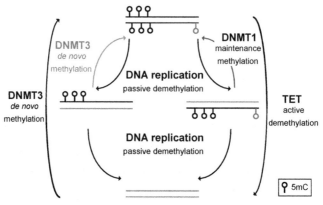

FIGURE 1   **Mechanisms of regulation of DNA methylation patterns.** DNA molecules carrying DNA methylation in both symmetrical (CpG and CpNpG) and asymmetrical (CpNpN) sequence contexts (top) will be hemimethylated after DNA replication (middle). If DNA methyltransferases are present, DNMT3-type enzymes can recognize hemimethylated DNA and produce fully methylated DNA at CpG and CpNpG sites. To regenerate DNA methylation in asymmetrical sequence contexts (red), DNMT3-class enzymes are required. If DNA replication continues in the absence of DNA methyltransferases, passive demethylation occurs, leading eventually to new double-stranded DNA without any DNA methylation (blue, bottom). This unmethylated state can also be achieved through active demethylation by TET demethylases or by glycosylases. From this unmethylated state, DNMT3-class enzymes are required to reestablish DNA methylation.

which recognize hemimethylated substrates, i.e., newly replicated DNA in which 5mC is present only on one strand. These maintenance DNA methyltransferases specifically recognize 5mC at so-called symmetrical sites, CpG and CpNpG, where a cytosine on one strand is adjacent to a cytosine on the opposing strand. The combined action of these two DNA methyltransferase classes establishes 5mC patterns and maintains them through DNA replication.

On the flip side, there are several pathways that can lead to the removal of DNA methylation [16–18]. These include the passive loss of 5mC if DNA is replicated in the absence of DNA methyltransferases that can maintain the DNA methylation. In this case, after the first cell division, the DNA in both daughter cells is hemimethylated. After a second cell division, two of the daughter cells will carry hemimethylated DNA, while the other two daughter cells will have no DNA methylation. Thus, the overall DNA methylation level is diluted over time. A second pathway to the removal of DNA methylation is the action of DNA glycosylases [18,19]. These enzymes can remove 5mC through a base excision repair mechanism and include several proteins that play important roles in plant development, for example. The most recently discovered pathway to the removal of DNA methylation involves the TET (ten–eleven translocation) family proteins. TET proteins can oxidate 5mC, leading to the formation of several intermediate products such as 5-hydroxymethylcytosine, 5-formylcytosine, and 5-carboxylcytosine [16–18]. Through this oxidation cascade, the action of TET proteins eventually leads to DNA demethylation [20]. Thus, the specific DNA methylation pattern found in any given cell depends on the coordinated action of DNA methyltransferases, glycosylases, and TET proteins [16–18]. While much of the initial work on the regulation of DNA methylation stems from research in model organisms, the basic biochemical pathways are conserved in humans.

3. CHALLENGES IN STUDYING DNA METHYLATION

While one field of study has focused on understanding the exact biochemical pathways involved in the generation and removal of 5mC, other researchers have concentrated their efforts on surveying the variation in DNA methylation and elucidating its impact on phenotypes. Given our understanding of the mechanisms for the propagation of DNA methylation patterns and their limitations, it was clear very early on that DNA methylation patterns have the potential to vary a great degree. In this aspect, DNA methylation is very different from DNA sequence. Genetic variation encoded in the DNA sequence is minimal within an individual (intraindividual), barring rare somatic mutations, and extensive between individuals (interindividual). In contrast, DNA methylation can vary significantly within individuals as well as between individuals [21]. This pattern of intra- and interindividual variation is typical of epigenetic systems, i.e., a system that is capable of altering gene expression—and ultimately organism-level phenotypes—by regulating access to the genetic information without affecting the DNA sequence.

The study of epigenotypes such as methylation patterns is further complicated by the fact that, in contrast to genotypes and DNA sequence, epigenomes are in constant flux. They reflect access to the DNA sequence and its expression status, properties that change depending on the assayed tissue, cellular composition of the tissue, time of assay, age of the subject, and environmental factors, to name a few. This flexibility of epigenetic information leads to their involvement in many important biological processes, but also makes their study—and even the simple cataloging of epigenotypes—much more complex than the study of DNA sequence variation.

The complexity in epigenotypes and DNA methylation patterns is now fully appreciated, owing to advances in the assay methods available. In the past, assays required large amounts of DNA, meaning that oftentimes different cell types or tissues from several individuals had to be pooled. Assays also were often limited in the fraction of the genome that could be evaluated; for example, DNA methylation assays used to rely on restriction enzyme assays, thus making only sequences matching the enzyme's recognition sequence available for studies. With the advent of next-generation sequencing and various array technologies (BS-seq, Me-DIP, RRBS, Illumina's Infinium and GoldenGate® methylation assays), it is now possible to assay DNA methylation genome-wide with single-base-pair resolution at a reasonable cost [22]. Despite the fact that even these new technologies only report aggregate measures from cell populations, these new methodologies have led to an enormous increase in our understanding of the variation in DNA methylation that occurs.

4. CLASSIFICATION OF EPIGENOTYPES

Based on the mechanisms leading up to the methylation of a specific cytosine residue, methylation patterns, just like other epigenotypes, can be classified into three groups based on their relationship to the genotype [23] (Figure 2). *Obligatory* epigenotypes are entirely dependent on the genotype. This dependence can be either *in cis* or *in trans*. An example of an obligatory epigenotype acting *in cis* would be the presence of a single-nucleotide polymorphism at a CpG site that changes the sequence to TpG. Without the cytosine, DNA methylation is absent, and so the epigenotype (methylation versus no methylation) is entirely dependent on the genotype (CpG versus TpG). An example of an obligatory epigenotype *in trans*

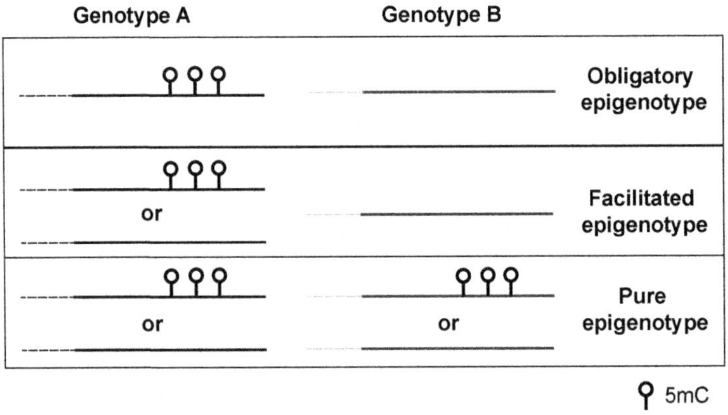

FIGURE 2 **Classification of epigenotypes.** Obligatory epigenotypes are entirely dependent on the genotype. If genotype A occurs, the methylated epigenotype is produced in all instances (top). Facilitated epigenotypes are somewhat dependent on the genotype. If genotype A occurs, the methylated epigenotype is produced in some instances. This means that the methylated epigenotype always occurs together with genotype A, but the occurrence of genotype A does not necessitate the methylated epigenotype (middle). Pure epigenotypes are independent of the genotype. Thus, the methylated epigenotype can occur with either genotype A or genotype B (bottom).

would be a loss of DNA methylation at a locus owing to a mutation in a DNA methyltransferase gene. In the absence of the DNA methyltransferase, no DNA methylation is produced at the locus, while if the DNA methyltransferase gene is intact, DNA methylation will be produced at the locus. Again, the epigenotype at the locus in question is entirely dependent on the genotype, this time the genotype of a locus elsewhere in the genome.

A second type of epigenotypes is classified as *facilitated*. Facilitated epigenotypes are dependent on the genotype but to a lesser degree. In these cases, the presence of a specific DNA sequence is required for a specific epigenotype, but not sufficient to always induce the epigenotype. Thus, there is a probabilistic element or randomness to this class of epigenotypes. An example of a facilitated epigenotype would be a transposable element that induces silencing of an adjacent gene in some cases. The silencing epigenotype occurs only if the transposable element is present, but the presence of the transposable element might or might not induce silencing.

The third class of epigenotypes is called *pure*. Pure epigenotypes have no link to the genotype, and these epigenotypes are thought to be generated by stochastic or environmental events. Thus, a single genotype can be associated with various epigenotypes, and the epigenotype cannot be predicted from the genotype. An example of a pure epigenotype would be DNA methylation that is simply lost or gained over time owing to the inaccuracies in replicating/maintaining the epigenotype. Thus, as this classification illustrates, the relationship between the epigenome, including DNA methylation, and the genome is highly complex.

Epimutation is another term that is sometimes used to refer to a specific epigenotype. As suggested by its relationship to the word "mutation," epimutation specifically refers to an epigenotype that is abnormal and that leads to abnormal gene expression/silencing and potentially abnormal phenotypes. Epimutations can be obligatory, facilitated, or pure epigenotypes, and their stability varies as well, from highly unstable, changing after a few cell divisions, to extremely stable and possibly transmitted transgenerationally.

5. EARLY EVIDENCE OF INTERINDIVIDUAL VARIATION IN DNA METHYLATION FROM DNA METHYLATION MUTANTS

Many of the early studies assessing variation in DNA methylation were carried out in several plant systems (for a review on epialleles and inheritance of 5mC in plants see refs [24,25]). Plants have high levels of DNA methylation in symmetric (CpG and CpNpG) as well as asymmetric

(CpNpN) sequence contexts and have served as important model systems for the study of DNA methylation [17]. In the earliest studies, variation in DNA methylation was not the focus of the work, but rather, the work aimed to progress the understanding of the mechanisms regulating DNA methylation. For example, the initial description of DNA methylation mutants (termed *ddm*, for deficient in DNA methylation) in *Arabidopsis* by Vongs and colleagues mostly reported aggregate/average DNA methylation data for the various mutant and wild-type genotypes that show reduced 5mC levels at the centromeres [26]. However, it also included some measurements from individual plants that illustrate variability between individuals in their levels of 5mC at the centromeres and rRNA loci [26]. Further characterization of one of these DNA methylation mutants, *ddm1*, also demonstrated that in this genetic background, DNA methylation at specific loci could be lost in some lineages and maintained in others, thus leading to increased interindividual variation in DNA methylation in these plants [27]. Some of this variation is retained even if a wild-type *DDM1* allele is crossed into the mutant background, leading again to interindividual variation in DNA methylation in the F1 and F2 generation [28]. Other examples of interindividual DNA methylation variation induced by mutations in the DNA methylation pathway include *FWA*, a gene involved in the regulation of flowering time [29,30]; *BAL*, a gene involved in disease resistance [31–33]; and transposable elements [34]. These examples illustrate that, while not the main focus of the work, information about differences in DNA methylation between individuals has been available for some time.

Similar to the above example from *Arabidopsis,* information about variation in DNA methylation has been available for a variety of species from studies of mutations in the pathways controlling the deposition, maintenance, and/or removal of DNA methylation. Mutations in the maintenance DNA methyltransferase Dnmt1 in mice in which they cause embryonic lethality were described in 1992/1993 [35,36], along with the alterations in 5mC patterns they induce. Other species with similar data sets from the 1990s include *Neurospora* and *Ascobolus*, in which, again, mutations affecting DNA methylation were isolated, demonstrating new variation in 5mC patterns induced by the mutations [37,38]. These examples illustrate that interindividual variation in DNA methylation could be experimentally induced.

The majority of the early studies that demonstrated variation in DNA methylation between individuals represent examples of obligatory epigenotypes, i.e., variation that is entirely dependent on the genotypes involved—a specific mutation in the genetic background will lead to an alteration in the DNA methylation pattern or levels. One exception is the variation in DNA methylation observed in crosses with *ddm1*, in which the genotype is identical but variation in 5mC patterns between individual plants is observed [26,28]. Given the large number of docu-mented cases of altered DNA methylation in response to mutations in the relevant pathways, this type of variation in DNA methylation is fairly well understood.

While these cases represent a very specific subset of variable 5mC epig-enotypes, they are nonetheless relevant to human health, as mutations in the DNA methylation biosynthetic pathways do occur in humans and lead to disease. For example, mutations in *DNMT1*, the human maintenance DNA methyltransferase, have been linked to autosomal dominant cerebel-lar ataxia, deafness, and narcolepsy [39] and to hereditary sensory neurop-athy with dementia and hearing loss (hereditary sensory and autonomic neuropathy) [40]. Mutations in *DNMT3B*, one of two human de novo DNA methyltransferases, are associated with immunodeficiency–centromere instability–facial anomalies syndrome 1 [41], and somatic mutations in several DNA methylation pathway components are prevalent in cancers [42]. Thus, these genetic disorders demonstrate that specific alleles at loci in the DNA methylation biosynthetic pathways such as *DNMT3B* can lead to obligatory DNA methylation variants in humans as well.

6. STUDIES OF NATURAL VARIATION IN DNA METHYLATION—MODEL SYSTEMS

6.1 Studies Focused on a Limited Number of Loci—Plants

The study of natural variation—in DNA sequence or epigenotypes—often complements experimental genetic studies. Again, several early studies come from *Arabidopsis* and other plant systems. In 1995, Bender and Fink describe variation in DNA methylation at the *PAI* gene, encoding the tryptophan biosynthesis enzyme phosphoribosylanthranilate isomer-ase [43]. When PAI enzyme levels are insufficient, plants are fluorescent under UV light. In the Wassilewskija accession of *Arabidopsis*, which has several *PAI* genes, two of the loci are silenced by DNA methylation, while in other accessions the same two loci are active and unmethylated. Crosses between accessions result in fluorescent and nonfluorescent plants owing to the segregation of methylated and unmethylated alleles [43], demon-strating naturally occurring variation in DNA methylation.

Another important study came from Enrico Coen's laboratory in 1999; they described a naturally occurring DNA methylation variant in toadflax *Linaria vulgaris*. The authors investigated the molecular basis of a vari-ant with altered floral morphology that changes the floral symmetry from bilateral to radial. This variant, which turned out to be due to altered DNA methylation, was originally described by Linneaus in the 1700s and thus might represent the oldest known epimutation [44]. This example is of interest because the phenotype affected by the interindividual DNA methylation variation—floral architecture—has clear fitness consequences for the organism and might thus influence the evolution of the species.

A larger scale study actually focused on natural variation in DNA methylation at the rDNA and five other sequences in 10 *Arabidopsis* accessions [45]. The survey revealed variation at the single-copy locus *MHC9.7/ MHC9.8* and methylation levels ranging from approximately 20% to more than 90% at the rDNA loci assayed by Southern blot analysis with methylation-sensitive restriction enzymes [45]. Within each accession, differences between individual plants were also detected, but these differences were relatively small compared to the differences observed between accessions/genotypes [45,46]. The presence of significant interindividual variation among *Arabidopsis* accessions has been confirmed since then for a variety of loci, including additional work at the rDNA locus [47], on transposable elements [48], at amplified fragment length polymorphism loci [49], and genome-wide [50].

More recently, the study of natural variation in DNA methylation has been extended to numerous other plant species. These include agriculturally important species such as maize [51,52], cotton [53], poplar [54,55], potato [56], soybean [57], and tobacco [58,59], but also species such as mangrove trees [60] and viola [61]. The focus of these studies varies widely. For example, the goal of the mangrove study was to understand the molecular mechanisms underlying phenotypic variation between trees at riversides and trees near salt marshes [60]. In this case, authors discovered that little genetic but extensive epigenetic variation in DNA methylation distinguishes the trees in different habitats [60]. Two of the studies, in tobacco [59] and poplar [55], focused on DNA methylation variation after callus culture and addressed the question of how different clones derived from callus culture are from their "parent" and how much they differ between clones. Both studies found significant variation in DNA methylation patterns between clones, and in tobacco, an interesting correlation was observed between the time spent in callus culture prior to plant regeneration and the amount of variation in DNA methylation between regenerated plants [59]. These findings are of great importance economically for species that have to be propagated by callus culture or similar to it, as they indicate that epigenetic pathways might make the reliable production of phenotypically identical clones difficult. Overall, studies of natural variation in DNA methylation in plants indicate that variation is extensive, relevant to agricultural traits, and potentially ecologically important.

6.2 Studies Focused on a Limited Number of Loci—Animals

Studies in animals other than humans are somewhat more limited. As noted above for plants, again, the earliest studies reporting natural variation in DNA methylation focused on select loci of interest. For example, a 1993 study investigated the promoter region of *Ha-ras* in three different mouse strains [62], while a 1995 study observed DNA hypermethylation close to the *Xist* locus on the active X chromosome in mouse [63], and a study in 1997 reported methylation differences at the *Pax3* and *Pax7* loci among mouse strains [64]. However, since generally there are fewer animal strains than plant strains maintained for research (seed storage is much easier than the continued culture required for most animal species), surveys of DNA methylation in animal models were not carried out as frequently.

One locus that has attracted a significant amount of interest is the *agouti* locus in mouse, owing to its metastable nature. The *agouti* locus controls coat color in mouse, and ectopic expression can lead to obesity, diabetes, and increased tumor susceptibility [65,66]. Several *agouti* alleles carry insertions of IAP (intracisternal A particle) retrotransposons, and for three of these alleles, variable DNA methylation levels at the IAP are reported that correlate with expression levels [67–70]. For example, the *agouti viable yellow* allele (a^{vy}) shows variable levels of DNA methylation in heterozygous sibling groups of isogenic backgrounds [69]. This variation in methylation level leads to variation in expression and ultimately results in variation in coat color and body weight/obesity [69]. The methylation levels at the *agouti* IAP can be altered by modulating the amount of methyl donor in the diet, providing an interesting precedent for affecting interindividual variation in DNA methylation by diet [71,72]. In addition, there is reported epigenetic inheritance of the methylation/expression state at a^{vy} [69]. There are a few other cases of metastable epialleles associated with interindividual variation in DNA methylation similar to the *agouti* locus, including *axin fused* ($axin^{fu}$) [73,74]. The findings from these mammalian epialleles, and especially the *agouti* locus, have significant implications for the application of personalized epigenetics to human medicine, as the interindividual variation in DNA methylation observed is affected by the environment (diet) and can be transmitted to subsequent generations.

Another area of interest in the study of animal epigenetics has been the study of imprinted loci and the variation of their DNA methylation patterns in response to reproductive cloning. Imprinting has to be reset during reproductive cloning—as it is during normal mammalian development—and thus, there are a significant number of studies examining the variation in DNA methylation among cloned animals. In 2001, Humpherys and colleagues described the variation in DNA methylation at imprinted loci in mice cloned from embryonic stem cells via nuclear transfer [75]. They found high levels of variation in DNA methylation at imprinted loci such as *H19* and also noted that this variation correlates with variation in gene expression [75]. Similar results have been reported for animals derived by nuclear transfer—and sometimes other forms of assisted reproductive technologies—from species such as sheep [76], pig [77], and cattle [78]. While this variation in DNA methylation—just as that observed in callus culture-regenerated plants—is not entirely natural, it is highly relevant to human health. Novel DNA methylation variation after somatic nuclear transfer and other assisted reproductive technologies is observed consistently among species and observations regarding increased frequencies of imprinting disorders in children conceived with assisted reproductive technologies [79] suggest that humans are not an exception.

Examining natural populations, surveys of DNA methylation variation have been carried out in bats [80], red grouse [81], house sparrows [82], and fish [83,84]. In all studies, significant amounts of epigenetic variation are found, even in the absence of DNA sequence variation. For example, Massicote and colleagues studied the clonally reproducing fish *Chrosomus eos–neogaeus*, in which DNA sequence variation is extremely low. Despite the almost complete absence of sequence variation, variation in DNA methylation is extensive, with approximately 15% of fragments analyzed by methylation-sensitive amplification polymorphism exhibiting polymorphisms [83]. However, the extent to which variation in DNA methylation is associated or correlated with DNA sequence variation appears to differ between species. In contrast to the previous example from fish, in bats, much of the variation in DNA methylation discovered is linked to sequence variation [80]. Carefully controlled studies will be required to determine what proportion of variation in DNA methylation is due to genetic variation—either obligatory or facultative—and what is independent of DNA sequence or pure.

6.3 Genome-Wide Studies of DNA Methylation

Tiling array methods for analyzing DNA methylation, and especially the advent of next-generation sequencing technologies, made a variety of DNA methylation studies possible that had been in the realm of dreams just a few years earlier. Rather than focusing on a few individual loci or sequences accessible by restriction enzymes, these methods made it feasible for the first time to examine DNA methylation genome-wide, one cytosine residue at a time. The community applied the new techniques quickly, and a variety of data sets from several species now exist that allow for the evaluation of interindividual variation in DNA methylation genome-wide.

These genome-wide studies demonstrate that variation in DNA methylation is widespread in a variety of species [25,51,80,83–85]. For example, comparing genome-wide DNA methylation of 20 inbred maize lines, Eichten and colleagues identified 1966 common and 1754 rare differentially methylated regions (DMRs). Interestingly, many of these DMRs were associated with local genetic variation such as single-nucleotide polymorphisms and also transposable elements [51], indicating that much of this variation in DNA methylation should be classified as obligatory or facilitated epigenotypes. In a follow-up study with near-isogenic lines focusing on just two maize lines, B73 and MO17, it was discovered that the 962 DMRs identified between these two strains showed four different inheritance patterns in the near-isogenic lines. Some DMRs showed an inheritance pattern consistent with local control, i.e., the impact of *cis*-genetic elements or direct inheritance of the DNA methylation (460 DMRs); some showed an inheritance pattern consistent with genetic control *in trans* (25 DMRs); some showed a complex pattern of inheritance (302 DMRs); and a small number of DMRs (13) showed a possible paramutation-like inheritance pattern [86], illustrating the complex nature of variation in DNA methylation.

These studies also show that variation in DNA methylation is not evenly distributed throughout the genome. In *Arabidopsis*, transposable elements were noted to show higher levels of DNA methylation in general compared to other regions of the genome, but this methylation was also more similar, i.e., less variable, when comparing different accessions [50]. In contrast, methylation at genes was more variable [50]. In a study examining approximately 40 million cytosines in the *Arabidopsis* genome with the methylC-seq method, 114,287 differentially methylated CpG sites were observed [87]. While this number is large, 91% of methylated CpG sites were invariantly methylated. Again, the majority of variable sites were in genes (60.5%), with 36.2% being found in repeats and 3.3% in intergenic regions (Figure 3). Non-CpG differentially methylated cytosines were detected as well (284), but these were localized mostly to intergenic regions (141) with a smaller number within genes (57) or transposons (86) [87]. In addition, differentially methylated cytosines were depleted

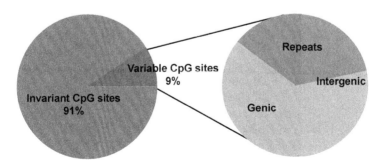

FIGURE 3   **Interindividual variation in DNA methylation is locus-specific.** Data from a methylC-seq study in *Arabidopsis* illustrate the nonrandom distribution of variable DNA methylation sites in the genome [87]. While a large majority of CpG sites are invariable in their DNA methylation levels, the fraction of variable sites is disproportionally enriched for genes and repeats, demonstrating locus specificity.

in heterochromatic regions of the genome and also from exons compared to introns [87]. Similar results were obtained in a study of epimutations in 10 *Arabidopsis* lines derived from a common ancestor 30 generations ago [85] and from a study of soybean recombinant inbred lines [57]. Thus, the consensus from the various studies in plant species is that variation in DNA methylation is extensive, occurs in all sequence contexts, and varies in level depending on the sequence context and/or genomic regions examined.

Studies from animals other than humans, which are discussed in detail below, are rare, but these studies are consistent with the observations from plants described above. For example, two mouse studies focus on the IAP retrotransposon, which is associated with two metastable methylation alleles, a^{vy} and $axin^{fu}$ (see Section 6.2). The first study sought to discover additional natural epialleles in the mouse genome by investigating retrotransposon-associated loci [88]. After identifying candidate loci by bioinformatics analysis of publicly available expression and histone 3 lysine 4 methylation data, the authors identified 143 candidate metastable epialleles. Further analyses of DNA methylation variation at these loci among littermates demonstrated that a subset of the loci shows interindividual variation in DNA methylation. The level of variation detected was locus-specific and also depended on the age of the animal and the tissue examined [88]. A second study of IAP elements in the mouse genome demonstrated that in addition, the methylation levels found at various IAP elements was dependent on their phylogenetic age [89]. Older, more divergent elements had more methylation than younger elements, indicating that the mouse's methylation machinery has developed efficient mechanisms to recognize and silence these transposable elements. In contrast, the younger elements appeared not to be as efficiently targeted by the DNA methylation machinery. This lower level of targeting resulted in more variation in DNA methylation levels at these elements, which were also more likely to be active. It is these younger elements with increased interindividual variation in DNA methylation that lead to the formation of metastable methylation alleles [89]. Thus, this study also confirms that the amount of interindividual variation in DNA methylation is highly dependent on the locus examined, even if the sequences involved are very similar.

7. STUDIES OF NATURAL VARIATION IN DNA METHYLATION—HUMANS

Not surprisingly, there is an extensive literature documenting DNA methylation patterns in the human genome. The goals of these studies are extremely varied, ranging from efforts to catalog tissue-specific DNA methylation patterns to efforts to observe methylation pattern changes associated with age or environmental exposures to efforts to identify DNA methylation biomarkers correlated with various disease states. While the sheer number of published studies makes it impossible to be comprehensive, in the following section, we highlight select case studies to illustrate the current state of knowledge about interindividual variation in DNA methylation in humans.

7.1 Natural Variation

Several large-scale, publicly funded projects are involved in mapping human epigenomes. These projects include the ENODE (Encyclopedia of DNA Elements) Consortium [90], the National Institutes of Health Roadmap Epigenomics Mapping Consortium [91] (both in the United States), and the Human Epigenome Project (http://www.epigenome.org/index.php; in Europe). In 2004, the Human Epigenome Project reported the results from their pilot experiments focusing on a 3.8-Mb region of the major histocompatibility complex (*MHC*) [92]. In addition to the analysis of variation in DNA methylation patterns among tissues, this study also included some analysis focused on interindividual variation. They found that within the *MHC*, most loci showed some variation in DNA methylation levels among individuals. However, only 118 of 253 amplicons analyzed showed high variability, defined as "a difference of greater than 50% between the lowest and highest median methylation values in at least one tissue" [92]. The variability observed is tissue-dependent, as not all loci showed the same level of variation in each tissue. While this study presents an important data set in the study of interindividual variation in DNA methylation, it also highlights some of its difficulties. Because the tissues analyzed were obtained from anonymous donors and not matched for age, sex, or genetic background, it is impossible to attribute the variation in DNA methylation observed to a specific mechanism, as the authors point out correctly, and it is also unclear what portion of the observed variation is due to technical variation rather than biological [92]. Unfortunately, this and similar problems affect many early studies of variation in DNA methylation, as it was not recognized that even subtle differences in tissue composition, such as variability in the contribution of the various blood cells to a blood sample, could lead to the observation of DNA methylation differences.

Widespread interindividual variation in DNA methylation is also reported in other studies with different tissues and sampling methodologies. DNA methylation variation, for example, occurs at imprinted loci (*IGF2/H19* and *PEG1/MEST*) [93–95]; it can be due to the sex of the tissue donor, especially for X-chromosome loci [96,97]; and age also is a significant contributor to variation in DNA methylation, as studies with more than one time point per individual report shifts in DNA methylation patterns over time [97–99]. For example, in a longitudinal study, Feinberg and colleagues report that half of the regions variable for methylation between individuals show stable methylation after an 11-year interval, while the other half have changed during this time [98]. Despite the difficulties associated with studies in humans (i.e., lack of precise matching for environmental exposures, age, etc.), several studies report that variation in DNA methylation patterns is higher between tissues—even from the same individual—than when comparing the same tissue from different individuals [100–103]. An exception to this observation might be sperm, as here a 2006 study reported more variable DNA methylation patterns in sperm cells obtained from one individual than when comparing between individuals [104].

Another observation emerging from these studies of DNA methylation in normal human samples is the importance of sequence context to the amount of 5mC variation detected. Generally speaking, DNA methylation levels in mammalian genomes tend to fall into two categories, fully methylated (almost everywhere) or unmethylated (CpG islands) [105]. However, some larger regions were discovered, where the average DNA methylation content was less than 70%. These regions, termed partially methylated domains (PMDs), showed intermediate levels of DNA methylation and constituted close to 40% of the human genome in fibroblasts [106]. These PMDs can also occur in human tissues (placenta, [107]), and their occurrence might reflect underlying sequence characteristics [108]. A small-scale study of 58 CpG sites in 30 mother–offspring pairs showed that loci with intermediate levels of DNA methylation (approximately 50%) had increased levels of interindividual variability in DNA methylation compared to loci with either high (close to 100%) or low (close to 0%) methylation levels [109], as did a small study of 16 candidate loci from blood samples [110]. This finding suggests that PMDs might contribute disproportionally to interindividual variation in DNA methylation. Shen and colleagues found, in a study of blood monocytes, that repetitive elements showed the highest levels of interindividual variation in DNA methylation, and genic regions were underrepresented among highly variable regions [111]. A bioinformatics analysis carried out by Bock and colleagues also suggests that interindividual DNA methylation variation is greater in sequence regions with lower CpG dinucleotide content [112]. Together, these observations suggest that interindividual variation in humans has characteristics similar to those in mouse and plants—levels of variation differ depending on the sequence context, and regions of intermediate methylation levels and repeated elements might contribute disproportionally to the variation (Figure 4).

7.2 DNA Methylation Variants Associated with Disease

A second important area of research in human epigenetics/epigenomics focuses on elucidating the links between variation in DNA methylation and various diseases. There is a large body of literature from this research,

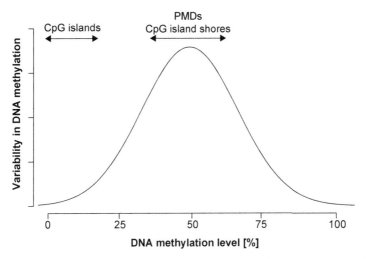

FIGURE 4   **Interindividual variation in DNA methylation is highest in regions of intermediate DNA methylation levels.** Interindividual variation in DNA methylation is low in regions of the genome that are mostly unmethylated and also in regions that are close to fully methylated. Interindividual variation in DNA methylation is observed most often in regions of intermediate levels of DNA methylation, which include CpG island shores and PMDs. This relationship is modeled by the bell curve, with DNA methylation on the x axis and interindividual variation in DNA methylation on the y axis.

and the diseases investigated include genetic disorders, cancers, diabetes, and also neurological and psychological disorders [113]. While the importance of DNA methylation in, for example, imprinting disorders such as Beckwith–Wiedemann and Prader–Willi/Angelman syndromes [14] and in cancer [114] is well understood at this point, for many other diseases the importance of DNA methylation and other epigenetic factors is less clear, as these studies are still very much in their infancy.

Genomic imprinting occurs when a gene is monoallelically expressed from one specific parental allele [115]. Imprinting disorders occur if this monoallelic expression is disturbed. Studies from several such disorders, but particularly from the Beckwith–Wiedemann, Prader–Willi, and Angelman syndromes, have shown that these disorders can have a variety of molecular causes (reviewed for the Beckwith–Wiedemann syndrome in [14]). These molecular causes include small local deletions/mutations of the gene in question, uniparental disomies of the imprinted region, and also deletions of the imprinting control region. For the Prader–Willi/Angelman genetic region, the imprint corresponds to a specific DNA methylation pattern that distinguishes the maternally derived allele from the paternally derived allele [13]. Loss of the DNA methylation imprint has been documented in some cases of Prader–Willi and Angelman syndrome [12], demonstrating that interindividual variation in DNA methylation at imprinted loci can result in the disease, essentially mimicking a genetic disorder. This example of imprinting disorders caused by abnormal DNA methylation illustrates the potential for DNA methylation variation to induce a significant, whole-organism disease phenotype via its impact on gene expression.

Abnormal DNA methylation patterns also are often observed in cancer, a group of diseases collectively characterized by unchecked cell proliferation. Simple models of cancer progression suggest that for the loss of proliferation control to occur, two alleles of a proliferation-limiting tumor suppressor gene have to be inactivated (two-hit model). Alternatively, a usually silent growth-promoting oncogene can be activated, leading again to unchecked cell proliferation. While the silencing of tumor suppressor genes or the activation of oncogenes can occur by genetic mechanisms, evidence from a variety of cancers demonstrates that the new gene expression state can be induced by altered DNA methylation. One well-studied example of this process are the epimutations in the DNA mismatch repair system genes found in a subset of patients with Lynch syndrome, an autosomal dominant cancer susceptibility syndrome. In these patients, *MLH1* or *MSH2* does not harbor genetic mutations but instead is silenced by aberrant DNA methylation (reviewed in [116]). This case illustrates how unusual DNA methylation variants can be found in tumor cells. Current models characterize cancer as a disease with both genetic and epigenetic components (for a review on this topic see [42]).

The extent of DNA methylation alterations is illustrated by a study of breast cancer from 2009. In this study, DNA methylation profiles for approximately 21,000 loci were obtained from nine breast cancer samples as well as nine healthy breast tissue samples from the same individuals [117]. Despite the small scale of the study, 220 regions with significantly altered DNA methylation were discovered. Both hypo- and hypermethylation were observed, and the biomarkers identified included genes previously reported to show differential methylation in breast cancer cells. In addition, a subset of the DNA methylation differences induced by cancer was validated in a larger panel of approximately 230 clinical samples, demonstrating their utility as biomarkers [117]. Similar studies have been carried out for other cancers, again characterizing the DNA methylation changes most often induced by any particular cancer type. Overall, these studies support the conclusion that, while there is variability between the methylation changes induced by the various cancer types, some cancer-induced methylation changes occur at a sufficiently high frequency to make them suitable as biomarkers [118].

As noted above, the investigations into the contribution of DNA methylation to many common complex human disease phenotypes are still in their infancy and tend to focus on identifying interindividual differences in DNA methylation that distinguish healthy from affected individuals. For example, a 2012 study looked at the potential contribution of DNA methylation to major depressive disorder [119]. They compared DNA samples taken postmortem from the frontal cortex of 39 patients with major depressive disorder to those of 26 sex- and age-matched controls. Among the 3.5 million CpG dinucleotides covered by the assay, 224 regions with methylation differences greater than 10% were identified between cases and controls. While 10 of 17 tested variable regions were validated within the data set, validation in a new set of samples was not successful, illustrating the difficulties with these types of studies.

Hypertension is another complex disease phenotype that has been investigated for a link to DNA methylation. Wang and colleagues generated genome-wide DNA methylation profiles from blood leukocytes for eight hypertensive individuals and eight age-matched control individuals, assaying 27,000 CpG dinucleotides [120]. While methylation differences between cases and controls were detected, because of the large number of tests conducted (27,000), no single methylation difference was statistically significant after multiple testing corrections [120]. They selected the two CpG sites with the largest methylation difference for validation in a second cohort of 36 cases and 60 controls and found that the methylation differences in *SULF1* (sulfatase 1) were statistically significant, indicating that DNA methylation variability at a subset of CpG sites within this gene might be associated with hypertension [120]. In addition, this study again illuminates the complexities involved in trying to relate DNA methylation differences to a disease phenotype. As large numbers of statistical tests are carried out, the sample sizes (number of cases and controls in the study) need to be large as well, which is often difficult with human subjects and also involves high costs for whole-genome DNA methylation analysis. In addition, similar to genome-wide association studies, validation of the candidate loci is difficult in independent cohorts, as many factors affect DNA methylation that often cannot be controlled for by matching cases and controls.

The contribution of DNA methylation to the disease process has been investigated for a large number of conditions, as illustrated by the selected studies described [113]. As noted above, the role of DNA methylation in cancer and some genetic disorders, such as the various imprinting disorders, is well understood. However, our understanding of the contribution of DNA methylation variation to other, complex diseases is fairly minimal. For most of these diseases, correlations between specific DNA methylation variants and the disease phenotypes have been observed, but functional links are generally lacking.

8. TWIN STUDIES

8.1 Overview

Twin studies are helping to shed some light on the correlations of DNA methylation variants with disease phenotypes that have been reported. They also help to improve our understanding of the nature of DNA methylation variation in general. Thus, twin studies represent an important subset of human DNA methylation studies as they provide experimental design advantages otherwise impossible to achieve with human subjects. Monozygotic twins in particular are an ideal system to study interindividual variation in DNA methylation for several reasons. First, there are no "genetic background effects," as the two individual twins share the same DNA sequence. Second, twin pairs also share other factors that usually must be controlled in DNA methylation studies, such as age and sex. Third, most twins are raised in similar environments, thus minimizing early life environmental effects and allowing us to potentially identify late life environmental differences that might explain variability among twins [121,122]. Given these characteristics, monozygotic twins present the best opportunity for well-controlled studies of DNA methylation in humans.

8.2 DNA Methylation Heritability in Twins

Over the past century, countless twin studies have been performed, beginning with simple comparisons of phenotypic variability between mono- and dizygotic twins [123]. For DNA methylation studies, especially with the rise of next-generation sequencing, twin studies offer a unique opportunity to understand epigenetic variability in humans. Traditionally, phenotypic differences between monozygotic twins have been attributed purely to environmental factors. More recent research, such as the pioneering studies by Fraga and colleagues [124], have indicated that epigenetic variation, for example in DNA methylation, may play a crucial role in the development of the distinct phenotypic characteristics [125].

Twin studies examining differences in DNA methylation in mono- versus dizygotic twins indicate that monozygotic twins on average have more similar DNA methylation profiles than their dizygotic counterparts. For example, the correlation coefficient for DNA methylation profiles was 0.200 in monozygotic twins compared to 0.109 in dizygotic twins in a study measuring DNA methylation across 417,069 probes in peripheral blood lymphocytes [126]. As monozygotic twins share 100% of their genome, while dizygotic twins on average share only 50% of their genomes, the higher correlation in DNA methylation in monozygotic twins suggests that, while the DNA methylation epigenotype might be inherited, it is likely largely regulated by the genetic background [121].

This finding is confirmed by other studies [126,127], with most research suggesting that a large portion of the heritability of DNA methylation profiles is due to genetic effects, i.e., they represent obligatory and/or facilitated epigenotypes.

With growing numbers of experiments studying the heritability of DNA methylation, it is becoming increasingly evident that a variety of factors are responsible for determining the heritability of epigenetic profiles (Table 1). For example, a study in 2007 that compared DMRs of the *H19* and *IGF2* genes in 98 adolescent and 88 middle-aged twin pairs found DNA methylation heritability at individual CpG sites to be 20–74% for *H19* and 57–97% for *IGF2* [127]. An investigation of DNA methylation at the MHR in 49 monozygotic and 40 dizygotic twin pairs found heritable CpG methylation at just 2–16% [128]. These data demonstrate that DNA methylation heritability is locus-specific and highly variable among different genomic regions.

Another major determinant of DNA methylation heritability is the cell type or tissue under study. Average DNA methylation heritability has been estimated in a variety of cell types including samples taken from adipose cells (19%) [129], blood (18% [130] to 23% [131]), cord-blood mononuclear cells (12%), human umbilical vascular endothelial cells (7%), placenta (5%) [130], and four brain regions (the cerebellum, frontal cortex, caudal pons,

TABLE 1   Heritability of DNA Methylation Estimated From Twin Studies Differs Depending on the Locus and Tissue Studied. (A) Heritability Estimates for Select Loci (Top). (B) Genome-Wide Heritability Estimates From Select Tissues (Bottom)

(A)		
Genetic locus	**Heritability estimates (%)**	**References**
H19	20–74	[127]
IGF2	57–97	[127]
MHC	2–16	[128]

(B)		
Cell type	**Heritability estimates (%)**	**References**
Adipose	19	[129]
Blood	18–23	[130,131]
Cord-blood mononuclear	12	[130]
Human umbilical vascular endothelial	7	[130]
Placenta	5	[130]
Brain regions	3–4	[132]

and temporal cortex; 3–4%) [132]. The estimates of DNA methylation heritability range from 3–4% [132] in the brain to approximately 20% in adipose cells [129] and blood [130,131]. These results illustrate that heritability of DNA methylation is a cell-type and tissue-specific measure. Further research is necessary to understand the mechanisms controlling this heritability as well as its variability.

8.3 Discordant Twin Studies

Although much of the variability in DNA methylation appears to result from genetic influences, it is evident that nongenetic factors influence DNA methylation as well. Monozygotic twins have identical DNA sequences and yet divergent epigenetic profiles, making their differences in DNA methylation profiles pure epigenetic changes as described in Section 4 [133]. We want to know what the nongenetic factors and mechanisms involved in the establishment and maintenance of these epigenetic profiles are. With crucial aspects such as age, sex, and many environmental factors controlled for, monozygotic twins are an excellent system for studying discordant phenotypes, for example, in DNA methylation profiles. Early monozygotic twin studies were primarily motivated by identification and research on similar traits, also known as concordant studies [134]. More recently, the focus has shifted to uncovering the mechanisms behind discordant, or variable, phenotypes among monozygotic twins, particularly in prominent diseases. Thus, in twin studies, concordant phenotypes are ideal for estimating heritability, while discordant phenotypes indicate an environmental factor and/or an environmentally driven epigenetic component [135].

An example of a disease studied in discordant twins is systemic lupus erythematosus (SLE), an inflammatory chronic autoimmune disorder that affects 1 in every 1000 females and 1 in every 9000 males [136]. A study from 2010 compared DNA methylation in DNA extracted from the white blood cells of monozygotic twin sets discordant for SLE, rheumatoid arthritis, or dermatomyositis ($n = 5$ twin sets for each disorder). The researchers concluded that of the three diseases, only SLE showed significant differences in levels of DNA methylation (49 genes of 807 tested) [136]. This finding suggests that variability in DNA methylation might contribute to the SLE disease phenotype.

Another prominent autoimmune disorder studied among discordant twin pairs is diabetes [122,137]. A genome-wide study in lymphocytes was performed in 2013, comparing DNA methylation levels between six concordant and three discordant twin pairs with type 1 diabetes [137]. More than 40 genes have been associated with type 1 diabetes in genome-wide association studies, 6 of which showed differentially methylated sites between the concordant and the discordant twin pairs in this DNA methylation study. A similar experiment was conducted with a total of

307 monozygotic twin pairs for the trait of insulin resistance, the primary pathway for type 2 diabetes [122]. Their results also showed that there was a significant difference in global DNA methylation between concordant and discordant twin pairs, thus indicating a significant association between insulin resistance and differential methylation. Overall, studies of discordant twins suggest that DNA methylation variation contributes to a number of complex human diseases.

Finally, twin studies have revealed novel epigenetic mechanisms in the development of certain cancers [118]. In the *BRCA1* gene, in which mutations are associated with breast and ovarian cancers, bisulfite sequencing from a twin study showed that 13% of the *BRCA1* allele was completely methylated in the cancerous twin, while the healthy twin had only individual CpG errors [138]. A second study used 15 monozygotic twin pairs and identified DNA hypermethylation of *DOK7* as a potential biomarker for early the diagnosis of breast cancer [139]. These studies, similar to the studies carried out without twins discussed above, indicate that there is extensive DNA methylation variation associated with cancers, even among twin pairs.

In summary, discordant twin studies allow us to observe what are probably pure epigenotypes and to relate the variation in DNA methylation to complex human disorders. These studies have great potential to improve our understanding of the molecular basis of complex diseases. While we have illustrated the power of discordant twin studies with select examples, it is important to note that the discordant twins model has also been used to define DMRs for disorders such as schizophrenia, bipolar disease, primary biliary cirrhosis, and multiple sclerosis [121,140]. Overall, monozygotic twin studies already have contributed greatly to our knowledge through identification of disease-associated DNA methylation profiles and by providing targets for novel treatments such as the epigenetic changes that may govern the onset of such diseases.

8.4 Limitations of Twin Studies and Future Opportunities

While twin studies are an accepted system for investigating heritability and discordant phenotypes, they also have some potential shortcomings that must be kept in mind when interpreting results. One assumption that many twin studies make is that of a shared and equal environment within the twin pair (equal environment assumption (EEA)). Specifically, the assumption is made that mono- and dizygotic twins are born and reared in very similar environments [121]. However, this assumption might not be correct. Recent studies have shown that the intrauterine environment may play a significant role in establishing the epigenome, including DNA methylation patterns [141]. Investigations of cell samples taken from monozygotic twins at birth show that interindividual variation exists even at these early time points and suggest that chorionicity and other intra-uterine factors might be involved in generating this variation [142,143]. Thus, there is evidence to suggest that the EEA might not be fulfilled even for monozygotic twins.

Another concern with many twin studies is sample size, which is often quite small, involving fewer than 10 twin pairs. As noted in Section 7.2, genome-wide DNA variation studies generally lead to a large number of tests being carried out and a requirement for large sample sizes to achieve statistical significance. Similar to the studies described in Section 7.2, twin studies often are limited in their power by the small sample size, a situation that is likely to continue until the cost of generating whole-genome DNA methylation profiles falls further. Until larger sample sizes can be routinely achieved, the small sample sizes necessitate caution in the interpretation of study results and independent validation of candidates of interest.

In addition, a study by Steve Horvath regarding DNA methylation and age points to another potential shortcoming that is shared by twin studies as well as other human studies [144]. It had been noted in the past by Feinberg and colleagues in a longitudinal study that some loci seemed to be unchanged in their DNA methylation levels over time, while others changed profoundly [98]. This finding has been supported in other studies as well [97,99], and it has been suggested that epigenotypes could drift over the course of a life span, simply because of imperfections in the maintenance systems [145,146]. In this new study by Horvath, 7844 publicly available, noncancer DNA methylation profiles were examined to determine if DNA methylation patterns can be used to predict chronological age. Horvath identified 353 informative CpG's (among 21,369 studied) that allowed him to predict chronological age from DNA methylation with high accuracy (age correlation 0.96, error = 3.6 years in the validation data set) [144]. This finding has potentially far-reaching consequences for DNA methylation studies in humans, as it indicates that age matching has to be very exact (within 3 years) not to introduce additional DNA methylation "noise" due to age into the study system. However, only ~1.7% of CpG sites investigated show this pattern and are included in the age prediction algorithm. In addition, Horvath notes that there are some tissues that do not follow this pattern, and there, DNA methylation levels do not correlate well with chronological age [144]. Furthermore, a twin study by Pirazzini and colleagues focusing on the *IGF2/H19* locus indicated that at this imprinted locus methylation differences are influenced by the aging process itself, more so than by environmental cues [147]. Together these findings indicate that age-matching individuals in DNA methylation studies might be more important than previously thought, but also point to the need for further study to determine if a "DNA methylation clock" is truly absent in some tissues or if it simply utilizes a different subset of CpG sites.

Despite these potential limitations, twin studies have offered substantial insights into the heritability of DNA methylation profiles and the factors generating interindividual variation in DNA methylation. Studies of discordant twins in particular have been useful for understanding the role of epigenetic mechanisms such as DNA methylation in disease processes and have suggested potential targets for novel treatments or biomarkers. The establishment of twin registries such as the MuTHER project (http://www.muther.ac.uk/), the United Kingdom Twin Registry (http://www.twinsuk.ac.uk/), and the NIEHS Twin Registry (http://www.niehs.nih.gov), with genomic profiles for thousands of twin pairs, will facilitate future twin studies with potentially larger sample sizes and more statistical power that will continue to contribute to our understanding of interindividual variation in DNA methylation and its role in human health and disease.

9. METHYLATION QUANTITATIVE TRAIT LOCI AND EPIGENETIC QUANTITATIVE TRAIT LOCI STUDIES

In addition to twin studies, two other more recently developed study types have contributed greatly to our understanding of the nature of interindividual DNA methylation, its causes, and its phenotypic consequences. These are the methylation quantitative trait loci (meQTL) studies [148–150] and epigenetic QTL (epiQTL) studies (also known as epigenome-wide association studies [113]). These study designs have their basis in classical QTL studies that relate a quantitative phenotype such as height or weight to the underlying genetic architecture and determine the location of genetic loci that contribute to the phenotype, the magnitude of their effects on the genotype, and their interactions. These classical quantitative genetics models are expanded to include epigenetic information such as DNA methylation.

MeQTL studies treat DNA methylation at a specific locus as a quantitative trait ranging from 0% (unmethylated) to 100% (fully methylated). Then, with DNA methylation levels as the trait, a regular QTL study is performed to determine if there are genetic determinants (meQTL) that contribute to the variation in DNA methylation observed. The meQTL approach is well illustrated by a study of soybeans published in 2013 [57]. This study used a set of 83 recombinant inbred lines derived from a cross between two different soybean strains to investigate the genetic basis of 1416 DMRs that they detected in non-CpG contexts. For 91% of the DMRs one or more meQTL were detected. Among these meQTL, the vast majority represent *cis*-acting QTL mapping back to the DMR locus (1260/1293), and only 33 meQTL represent *trans*-acting QTL located elsewhere in the genome [57]. Thus, it appears that much of the DNA methylation variation seen might link back to local genetic variation such as single-nucleotide polymorphisms.

Similar results have been obtained in other studies including several from human samples [148–152]. For example, an investigation of 210 human lung tissue samples identified 34,304 *cis*-acting genetic determinants affecting local DNA methylation variation [151]. In contrast, only 585 *trans*-acting meQTL were detected in this same study [151]. These findings show that *cis*-acting meQTL are much more prevalent than *trans*-acting meQTL, revealing a similar ratio of *cis*- to *trans*-acting meQTL as observed in the soybean study described above [57]. Overall, the available results indicate that obligatory and facilitated DNA methylation epigenotypes due to local, *cis* genetic variation are common.

EpiQTL studies take a slightly different approach. Rather than treating DNA methylation as the phenotype in the QTL analysis, epiQTL studies use DMRs instead of DNA sequence variations as physical markers along the chromosome to "map" a phenotype to the epigenome. One example of this approach is a study by Cortijo and colleagues investigating the contribution of DNA methylation to flowering time and root length in *Arabidopsis thaliana* [153]. Using 126 DMRs covering the majority of the *Arabidopsis* genome, they were able to define three epiQTL each for flowering time and root length, explaining 51% and 33% of the phenotypic variance, respectively. While the epiQTL regions in this case are too large to determine if a single DMR is responsible for the phenotypic effects [153], this study illustrates that the epiQTL approach can be used to identify candidate epialleles that contribute to complex phenotypes.

10. IMPACT OF VARIATION IN DNA METHYLATION ON PERSONALIZED MEDICINE

Variation in DNA methylation can be linked unequivocally to a variety of diseases, and for some of them, such as those illustrated above, we have a good understanding of how DNA methylation contributes to the disease process. DNA methylation is now considered an important candidate biomarker to guide cancer therapy and to predict treatment success (for example see ref. [154]). In addition, several papers demonstrate the DNA methylation variation at loci involved in drug metabolism regulates the repression of these proteins in the liver [155,156]. These findings demonstrate that it is not sufficient to genotype patients at loci such as *CYP3A4*, a major drug-metabolizing enzyme, to predict a patient's response to a medication. Similarly, a study of the levels of monoamine oxidase A (MAOA), an enzyme involved in the metabolism of neurotransmitters in the brain, demonstrated that sequence variation at the *MAOA* locus was not a good predictor of MAOA levels, but that DNA methylation levels at *MAOA* could be used to predict enzyme levels [157]. These studies suggest that knowledge of a patient's DNA methylation profile might be required to estimate an optimal, individualized drug dose or regimen for a patient.

Another pioneering study, by Rakyan and colleagues, explores DNA methylation links to type 1 diabetes and identifies several DMRs that can serve as predictive biomarkers that precede disease [113]. Prognostic DNA methylation biomarkers also have been identified for a variety of cancers (for a review focused on bladder cancer see [158]). While the use of most DNA methylation biomarkers has not advanced to the clinical setting, a colon cancer screening test that assays methylation at one genetic locus has been approved (August 2014) by the Food and Drug Administration for use in the United States [159]. These developments are very promising and again point to the importance of including DNA methylation analyses in personalized medicine initiatives, as DNA methylation analysis provides actionable information beyond the information derived from genome analysis.

While the data on interindividual variation in DNA methylation clearly demonstrate the utility of methylome analysis for personalized medicine, prior to the implementation of such analyses several issues have to be addressed. Given the constant flux of DNA methylation in its response to environment, health status, and age, it is unclear when a patient's methylome should be assayed. Should such an analysis be carried out at predefined intervals or only as a response, i.e., when a specific disease condition such as cancer is detected? If periodic monitoring is considered, what tissue is most informative? Many current studies use blood for convenience, but is this tissue the best choice for a personalized medicine monitoring program? Also, given the significant cost of genome-wide DNA methylation analyses, how can we evaluate cost effectiveness? Given the considerable interest in both personalized medicine and the contribution of epigenetic processes to human disease, research into these questions is undoubtedly ongoing.

11. CONCLUSION

In summary, the various studies discussed above demonstrate that interindividual variation in DNA methylation in humans and other species is extensive. This variation has many origins [23], including a surprisingly large effect of genetic variation as demonstrated, for example, by meQTL studies [57,148–150]. Other factors influencing DNA methylation patterns detected include gender and age of the DNA donor, health status of the donor, tissue type, and also the genomic loci under study. Studies of genetically identical individuals—inbred lines or twins—suggest that a variety of environmental factors also can play a role in shaping DNA methylation patterns. While the resulting variability in DNA methylation patterns makes its study—and that of other aspects of the epigenome—much more complex than genome studies, the large impact of DNA methylation variation on complex, ecologically and evolutionarily significant phenotypes illustrates why it is important to continue to research this challenging research field.

LIST OF ACRONYMS AND ABBREVIATIONS

5mC	5-Methylcytosine
a^{vy}	Agouti viable yellow
$axin^{fu}$	Axin fused
BRCA1	Breast cancer 1
BS-seq	Whole-genome shotgun bisulfite sequencing
CpG	Cytosine–phosphate–guanine
CpNpG	Cytosine–phosphate–(any base)–guanine
CpNpN	Cytosine–phosphate–(any base)–phosphate–(any base)
CYP3A4	Cytochrome P450, family 3, subfamily A, polypeptide 4
ddm	Deficient in DNA methylation
ddm1	Deficient in DNA methylation 1
DMR	Differentially methylated region
DNMT3	DNA (cytosine-5-) methyltransferase 3
DNMT3B	DNA (cytosine-5-) methyltransferase 3β
DNMT1	DNA (cytosine-5-) methyltransferase 1
DOK7	Docking protein 7
EEA	Equal environment assumption
epiQTL	Epigenetic QTL
F1	Filial generation 1
F2	Filial generation 2
Ha-ras	Harvey rat sarcoma viral oncogene homolog
IAP	Intracisternal A particle
IGF2	Insulin-like growth factor 2
MAOA	Monoamine oxidase A
MBD	Methyl-binding domain

meQTL	Methylation QTL
MHC	Major histocompatibility complex
MethylC-seq	Whole-genome shotgun bisulfite sequencing
MECP2	Methyl CpG binding protein 2
Me-DIP	Methylated DNA immunoprecipitation
MEST	Mesoderm-specific transcript
MuTHER	Multiple-tissue human expression resource
NIEHS	National Institute of Environmental Health Sciences
PAI	Phosphoribosylanthranilate isomerase
Pax3	Paired box 3
Pax7	Paired box 7
PEG1	Paternally expressed gene 1
PMD	Partially methylated domain
QTL	Quantitative trait loci
RRBS	Reduced representation bisulfite sequencing
rRNA	Ribosomal ribonucleic acid
rDNA	Ribosomal DNA
SLE	Systemic lupus erythematosus
SULF1	Sulfatase 1
SUV39H1	Suppressor of variegation 3-9 homolog 1 (*Drosophila*)
TET	Ten–eleven translocation
TpG	Thymine–phosphate–guanine
UV	Ultraviolet
Xist	X-inactive specific transcript

Acknowledgments

We thank the members of the Riddle lab and the anonymous reviewers for their helpful comments on the manuscript. In addition, we apologize to all our colleagues whose work we were unable to cite owing to space limitations.

References

[1] Jurkowski TP, Jeltsch A. On the evolutionary origin of eukaryotic DNA methyltransferases and Dnmt2. PloS One 2011;6(11):e28104.

[2] Capuano F, Mulleder M, Kok R, Blom HJ, Ralser M. Cytosine DNA methylation is found in *Drosophila melanogaster* but absent in *Saccharomyces cerevisiae, Schizosaccharomyces pombe*, and other yeast species. Anal Chem 2014;86(8):3697–702.
[3] Klose RJ, Bird AP. Genomic DNA methylation: the mark and its mediators. Trends Biochem Sci 2006;31(2):89–97.
[4] Guibert S, Weber M. Functions of DNA methylation and hydroxymethylation in mammalian development. Curr Top Dev Biol 2013;104:47–83.
[5] Feng S, Cokus SJ, Zhang X, Chen PY, Bostick M, Goll MG, et al. Conservation and divergence of methylation patterning in plants and animals. Proc Natl Acad Sci USA 2010;107(19):8689–94.
[6] Zemach A, McDaniel IE, Silva P, Zilberman D. Genome-wide evolutionary analysis of eukaryotic DNA methylation. Science 2010;328(5980):916–9.
[7] Gustems M, Woellmer A, Rothbauer U, Eck SH, Wieland T, Lutter D, et al. c-Jun/c-Fos heterodimers regulate cellular genes via a newly identified class of methylated DNA sequence motifs. Nucleic Acids Res 2014;42(5):3059–72.
[8] Niesen MI, Osborne AR, Yang H, Rastogi S, Chellappan S, Cheng JQ, et al. Activation of a methylated promoter mediated by a sequence-specific DNA-binding protein, RFX. J Biolo Chem 2005;280(47):38914–22.
[9] Fuks F, Hurd PJ, Deplus R, Kouzarides T. The DNA methyltransferases associate with HP1 and the SUV39H1 histone methyltransferase. Nucleic Acids Res 2003;31(9):2305–12.
[10] Geiman TM, Sankpal UT, Robertson AK, Zhao Y, Zhao Y, Robertson KD. DNMT3B interacts with hSNF2H chromatin remodeling enzyme, HDACs 1 and 2, and components of the histone methylation system. Biochem Biophys Res Commun 2004; 318(2):544–55.
[11] Buck-Koehntop BA, Defossez PA. On how mammalian transcription factors recognize methylated DNA. Epigenetics Official J DNA Methylation Soc 2013;8(2):131–7.
[12] Buiting K, Gross S, Lich C, Gillessen-Kaesbach G, el-Maarri O, Horsthemke B. Epimutations in Prader-Willi and Angelman syndromes: a molecular study of 136 patients with an imprinting defect. Am J Hum Genet 2003;72(3):571–7.
[13] Nicholls RD, Knepper JL. Genome organization, function, and imprinting in Prader-Willi and Angelman syndromes. Annu Rev Genomics Hum Genet 2001;2:153–75.
[14] Soejima H, Higashimoto K. Epigenetic and genetic alterations of the imprinting disorder Beckwith-Wiedemann syndrome and related disorders. J Hum Genet 2013;58(7):402–9.
[15] Amir RE, Van den Veyver IB, Wan M, Tran CQ, Francke U, Zoghbi HY. Rett syndrome is caused by mutations in X-linked MECP2, encoding methyl-CpG-binding protein 2. Nat Genet 1999;23(2):185–8.
[16] Jeltsch A, Jurkowska RZ. New concepts in DNA methylation. Trends Biochem Sci 2014;39(7):310–8.
[17] Furner IJ, Matzke M. Methylation and demethylation of the Arabidopsis genome. Curr Opin Plant Biol 2011;14(2):137–41.
[18] Kohli RM, Zhang Y. TET enzymes, TDG and the dynamics of DNA demethylation. Nature 2013;502(7472):472–9.
[19] Zhang H, Zhu JK. Active DNA demethylation in plants and animals. Cold Spring Harbor Symp Quant Biol 2012;77:161–73.
[20] Tan L, Shi YG. Tet family proteins and 5-hydroxymethylcytosine in development and disease. Development 2012;139(11):1895–902.
[21] Allis CD, Jenuwein T, Reinberg D. Overview and concepts. In: Allis CD, Jenuwein T, Reinberg D, editors. Epigenetics: official journal of the DNA methylation society. Cold Spring Harbor, NY: Cold Spring Harbor Laboratory Press; 2007. p. 23–61.
[22] Rivera CM, Ren B. Mapping human epigenomes. Cell 2013;155(1):39–55.
[23] Richards EJ. Inherited epigenetic variation–revisiting soft inheritance. Nat Rev Genet 2006;7(5):395–401.

[24] O'Malley RC, Ecker JR. Epiallelic variation in *Arabidopsis thaliana*. Cold Spring Harbor Symp Quant Biol 2012;77:135–45.

[25] Niederhuth CE, Schmitz RJ. Covering your bases: inheritance of DNA methylation in plant genomes. Mol Plant 2014;7(3):472–80.

[26] Vongs A, Kakutani T, Martienssen RA, Richards EJ. *Arabidopsis thaliana* DNA methylation mutants. Science 1993;260(5116):1926–8.

[27] Kakutani T, Jeddeloh JA, Flowers SK, Munakata K, Richards EJ. Developmental abnormalities and epimutations associated with DNA hypomethylation mutations. Proc Natl Acad Sci USA 1996;93(22):12406–11.

[28] Kakutani T, Munakata K, Richards EJ, Hirochika H. Meiotically and mitotically stable inheritance of DNA hypomethylation induced by ddm1 mutation of *Arabidopsis thaliana*. Genetics 1999;151(2):831–8.

[29] Kankel MW, Ramsey DE, Stokes TL, Flowers SK, Haag JR, Jeddeloh JA, et al. Arabidopsis MET1 cytosine methyltransferase mutants. Genetics 2003;163(3):1109–22.

[30] Soppe WJ, Jacobsen SE, Alonso-Blanco C, Jackson JP, Kakutani T, Koornneef M, et al. The late flowering phenotype of fwa mutants is caused by gain-of-function epigenetic alleles of a homeodomain gene. Mol Cell 2000;6(4):791–802.

[31] Stokes TL, Kunkel BN, Richards EJ. Epigenetic variation in Arabidopsis disease resistance. Genes Dev 2002;16(2):171–82.

[32] Stokes TL, Richards EJ. Induced instability of two Arabidopsis constitutive pathogen-response alleles. Proc Natl Acad Sci USA 2002;99(11):7792–6.

[33] Yi H, Richards EJ. Gene duplication and hypermutation of the pathogen resistance gene SNC1 in the arabidopsis bal variant. Genetics 2009;183(4):1227–34.

[34] Miura A, Yonebayashi S, Watanabe K, Toyama T, Shimada H, Kakutani T. Mobilization of transposons by a mutation abolishing full DNA methylation in Arabidopsis. Nature 2001;411(6834):212–4.

[35] Li E, Beard C, Jaenisch R. Role for DNA methylation in genomic imprinting. Nature 1993;366(6453):362–5.

[36] Li E, Bestor TH, Jaenisch R. Targeted mutation of the DNA methyltransferase gene results in embryonic lethality. Cell 1992;69(6):915–26.

[37] Foss HM, Roberts CJ, Claeys KM, Selker EU. Abnormal chromosome behavior in Neurospora mutants defective in DNA methylation. Science 1993;262(5140):1737–41.

[38] Malagnac F, Wendel B, Goyon C, Faugeron G, Zickler D, Rossignol JL, et al. A gene essential for de novo methylation and development in Ascobolus reveals a novel type of eukaryotic DNA methyltransferase structure. Cell 1997;91(2):281–90.

[39] Winkelmann J, Lin L, Schormair B, Kornum BR, Faraco J, Plazzi G, et al. Mutations in DNMT1 cause autosomal dominant cerebellar ataxia, deafness and narcolepsy. Hum Mol Genet 2012;21(10):2205–10.

[40] Klein CJ, Botuyan MV, Wu Y, Ward CJ, Nicholson GA, Hammans S, et al. Mutations in DNMT1 cause hereditary sensory neuropathy with dementia and hearing loss. Nat Genet 2011;43(6):595–600.

[41] Ehrlich M. The ICF syndrome, a DNA methyltransferase 3B deficiency and immunodeficiency disease. Clin Immunol 2003;109(1):17–28.

[42] Dawson MA, Kouzarides T. Cancer epigenetics: from mechanism to therapy. Cell 2012;150(1):12–27.

[43] Bender J, Fink GR. Epigenetic control of an endogenous gene family is revealed by a novel blue fluorescent mutant of Arabidopsis. Cell 1995;83(5):725–34.

[44] Cubas P, Vincent C, Coen E. An epigenetic mutation responsible for natural variation in floral symmetry. Nature 1999;401(6749):157–61.

[45] Riddle NC, Richards EJ. The control of natural variation in cytosine methylation in Arabidopsis. Genetics 2002;162(1):355–63.

[46] Riddle NC, Richards EJ. Genetic variation in epigenetic inheritance of ribosomal RNA gene methylation in Arabidopsis. Plant J Cell Mol Biol 2005;41(4):524–32.

[47] Woo HR, Richards EJ. Natural variation in DNA methylation in ribosomal RNA genes of *Arabidopsis thaliana*. BMC Plant Biol 2008;8:92.

[48] Rangwala SH, Elumalai R, Vanier C, Ozkan H, Galbraith DW, Richards EJ. Meiotically stable natural epialleles of Sadhu, a novel Arabidopsis retroposon. PLoS Genet 2006;2(3):e36.

[49] Cervera MT, Ruiz-Garcia L, Martinez-Zapater JM. Analysis of DNA methylation in *Arabidopsis thaliana* based on methylation-sensitive AFLP markers. Mol Genet Genomics 2002;268(4):543–52.

[50] Vaughn MW, Tanurdzic M, Lippman Z, Jiang H, Carrasquillo R, Rabinowicz PD, et al. Epigenetic natural variation in *Arabidopsis thaliana*. PLoS Biol 2007;5(7):e174.

[51] Eichten SR, Briskine R, Song J, Li Q, Swanson-Wagner R, Hermanson PJ, et al. Epigenetic and genetic influences on DNA methylation variation in maize populations. Plant Cell 2013;25(8):2783–97.

[52] Eichten SR, Swanson-Wagner RA, Schnable JC, Waters AJ, Hermanson PJ, Liu S, et al. Heritable epigenetic variation among maize inbreds. PLoS Genet 2011;7(11):e1002372.

[53] Keyte AL, Percifield R, Liu B, Wendel JF. Infraspecific DNA methylation polymorphism in cotton (*Gossypium hirsutum* L.). J Hered 2006;97(5):444–50.

[54] Ma K, Song Y, Yang X, Zhang Z, Zhang D. Variation in genomic methylation in natural populations of Chinese white poplar. PloS One 2013;8(5):e63977.

[55] Vining K, Pomraning KR, Wilhelm LJ, Ma C, Pellegrini M, Di Y, et al. Methylome reorganization during in vitro dedifferentiation and regeneration of *Populus trichocarpa*. BMC Plant Biol 2013;13:92.

[56] Cara N, Marfil CF, Masuelli RW. Epigenetic patterns newly established after interspecific hybridization in natural populations of Solanum. Ecol Evol 2013;3(11):3764–79.

[57] Schmitz RJ, He Y, Valdes-Lopez O, Khan SM, Joshi T, Urich MA, et al. Epigenome-wide inheritance of cytosine methylation variants in a recombinant inbred population. Genome Res 2013;23(10):1663–74.

[58] Jiao J, Jia Y, Lv Z, Sun C, Gao L, Yan X, et al. Analysis of methylated patterns and quality-related genes in tobacco (*Nicotiana tabacum*) cultivars. Biochem Genet 2014;52(7–8):372–86.

[59] Krizova K, Fojtova M, Depicker A, Kovarik A. Cell culture-induced gradual and frequent epigenetic reprogramming of invertedly repeated tobacco transgene epialleles. Plant Physiol 2009;149(3):1493–504.

[60] Lira-Medeiros CF, Parisod C, Fernandes RA, Mata CS, Cardoso MA, Ferreira PC. Epigenetic variation in mangrove plants occurring in contrasting natural environment. PloS One 2010;5(4):e10326.

[61] Schulz B, Eckstein RL, Durka W. Epigenetic variation reflects dynamic habitat conditions in a rare floodplain herb. Mol Ecol 2014;23(14):3523–37.

[62] Counts JL, Goodman JI. Comparative analysis of the methylation status of the 5′ flanking region of Ha-ras in B6C3F1, C3H/He and C57BL/6 mouse liver. Cancer Lett 1993;75(2):129–36.

[63] Courtier B, Heard E, Avner P. Xce haplotypes show modified methylation in a region of the active X chromosome lying 3′ to Xist. Proc Natl Acad Sci USA 1995;92(8):3531–5.

[64] Kay PH, Harmon D, Fletcher S, Ziman M, Jacobsen PF, Papadimitriou JM. Variation in the methylation profile and structure of Pax3 and Pax7 among different mouse strains and during expression. Gene 1997;184(1):45–53.

[65] Duhl DM, Stevens ME, Vrieling H, Saxon PJ, Miller MW, Epstein CJ, et al. Pleiotropic effects of the mouse lethal yellow (Ay) mutation explained by deletion of a maternally expressed gene and the simultaneous production of agouti fusion RNAs. Development 1994;120(6):1695–708.

[66] Duhl DM, Vrieling H, Miller KA, Wolff GL, Barsh GS. Neomorphic agouti mutations in obese yellow mice. Nat Genet 1994;8(1):59–65.

[67] Argeson AC, Nelson KK, Siracusa LD. Molecular basis of the pleiotropic phenotype of mice carrying the hypervariable yellow (Ahvy) mutation at the agouti locus. Genetics 1996;142(2):557–67.

[68] Michaud EJ, van Vugt MJ, Bultman SJ, Sweet HO, Davisson MT, Woychik RP. Differential expression of a new dominant agouti allele (Aiapy) is correlated with methylation state and is influenced by parental lineage. Genes Dev 1994;8(12):1463–72.

[69] Morgan HD, Sutherland HG, Martin DI, Whitelaw E. Epigenetic inheritance at the agouti locus in the mouse. Nat Genet 1999;23(3):314–8.

[70] Perry WL, Copeland NG, Jenkins NA. The molecular basis for dominant yellow agouti coat color mutations. BioEssays News Rev Mol Cell Dev Biol 1994;16(10):705–7.

[71] Cooney CA, Dave AA, Wolff GL. Maternal methyl supplements in mice affect epigenetic variation and DNA methylation of offspring. J Nutr 2002;132(Suppl. 8):2393S–400S.

[72] Wolff GL, Kodell RL, Moore SR, Cooney CA. Maternal epigenetics and methyl supplements affect agouti gene expression in Avy/a mice. FASEB J Official Publ Fed Am Soc Exp Biol 1998;12(11):949–57.

[73] Rakyan VK, Chong S, Champ ME, Cuthbert PC, Morgan HD, Luu KV, et al. Transgenerational inheritance of epigenetic states at the murine Axin(Fu) allele occurs after maternal and paternal transmission. Proc Natl Acad Sci USA 2003;100(5):2538–43.

[74] Waterland RA, Dolinoy DC, Lin JR, Smith CA, Shi X, Tahiliani KG. Maternal methyl supplements increase offspring DNA methylation at Axin Fused. Genesis 2006;44(9):401–6.

[75] Humpherys D, Eggan K, Akutsu H, Hochedlinger K, Rideout WM 3rd, Biniszkiewicz D, et al. Epigenetic instability in ES cells and cloned mice. Science 2001;293(5527):95–7.

[76] Young LE, Fernandes K, McEvoy TG, Butterwith SC, Gutierrez CG, Carolan C, et al. Epigenetic change in IGF2R is associated with fetal overgrowth after sheep embryo culture. Nat Genet 2001;27(2):153–4.

[77] Shen CJ, Cheng WT, Wu SC, Chen HL, Tsai TC, Yang SH, et al. Differential differences in methylation status of putative imprinted genes among cloned swine genomes. PloS One 2012;7(2):e32812.

[78] Smith LC, Suzuki Jr J, Goff AK, Filion F, Therrien J, Murphy BD, et al. Developmental and epigenetic anomalies in cloned cattle. Reprod Domest Animals Zuchthygiene 2012;47(Suppl. 4):107–14.

[79] Kochanski A, Merritt TA, Gadzinowski J, Jopek A. The impact of assisted reproductive technologies on the genome and epigenome of the newborn. J Neonatal-Perinatal Med 2013;6(2):101–8.

[80] Liu S, Sun K, Jiang T, Ho JP, Liu B, Feng J. Natural epigenetic variation in the female great roundleaf bat (*Hipposideros armiger*) populations. Mol Genet Genomics 2012;287(8):643–50.

[81] Wenzel MA, Piertney SB. Fine-scale population epigenetic structure in relation to gastro-intestinal parasite load in red grouse (*Lagopus lagopus scotica*). Mol Ecol 2014;23(17):4256–73.

[82] Schrey AW, Coon CA, Grispo MT, Awad M, Imboma T, McCoy ED, et al. Epigenetic variation may compensate for decreased genetic variation with introductions: a case study using house sparrows (*Passer domesticus*) on two continents. Genet Res Int 2012;2012:979751.

[83] Massicotte R, Whitelaw E, Angers B. DNA methylation: a source of random variation in natural populations. Epigenetics Official J DNA Methylation Soc 2011;6(4):421–7.

[84] Moran P, Perez-Figueroa A. Methylation changes associated with early maturation stages in the Atlantic salmon. BMC Genet 2011;12:86.

[85] Becker C, Hagmann J, Muller J, Koenig D, Stegle O, Borgwardt K, et al. Spontaneous epigenetic variation in the *Arabidopsis thaliana* methylome. Nature 2011;480(7376):245–9.

[86] Li Q, Eichten SR, Hermanson PJ, Springer NM. Inheritance patterns and stability of DNA methylation variation in maize near-isogenic lines. Genetics 2014;196(3):667–76.

[87] Schmitz RJ, Schultz MD, Lewsey MG, O'Malley RC, Urich MA, Libiger O, et al. Transgenerational epigenetic instability is a source of novel methylation variants. Science 2011;334(6054):369–73.
[88] Ekram MB, Kang K, Kim H, Kim J. Retrotransposons as a major source of epigenetic variations in the mammalian genome. Epigenetics Official J DNA Methylation Soc 2012;7(4):370–82.
[89] Faulk C, Barks A, Dolinoy DC. Phylogenetic and DNA methylation analysis reveal novel regions of variable methylation in the mouse IAP class of transposons. BMC Genomics 2013;14:48.
[90] Consortium EP. The ENCODE (Encyclopedia of DNA Elements) project. Science 2004;306(5696):636–40.
[91] Bernstein BE, Stamatoyannopoulos JA, Costello JF, Ren B, Milosavljevic A, Meissner A, et al. The NIH Roadmap Epigenomics Mapping Consortium. Nat Biotechnol 2010;28(10):1045–8.
[92] Rakyan VK, Hildmann T, Novik KL, Lewin J, Tost J, Cox AV, et al. DNA methylation profiling of the human major histocompatibility complex: a pilot study for the human epigenome project. PLoS Biol 2004;2(12):e405.
[93] Court F, Tayama C, Romanelli V, Martin-Trujillo A, Iglesias-Platas I, Okamura K, et al. Genome-wide parent-of-origin DNA methylation analysis reveals the intricacies of human imprinting and suggests a germline methylation-independent mechanism of establishment. Genome Res 2014;24(4):554–69.
[94] McMinn J, Wei M, Sadovsky Y, Thaker HM, Tycko B. Imprinting of PEG1/MEST isoform 2 in human placenta. Placenta 2006;27(2–3):119–26.
[95] Sandovici I, Leppert M, Hawk PR, Suarez A, Linares Y, Sapienza C. Familial aggregation of abnormal methylation of parental alleles at the IGF2/H19 and IGF2R differentially methylated regions. Hum Mol Genet 2003;12(13):1569–78.
[96] Illingworth R, Kerr A, Desousa D, Jørgensen H, Ellis P, Stalker J, et al. A novel CpG island set identifies tissue-specific methylation at developmental gene loci. PLoS Biol 2008;6(1):e22.
[97] Wang D, Liu X, Zhou Y, Xie H, Hong X, Tsai HJ, et al. Individual variation and longitudinal pattern of genome-wide DNA methylation from birth to the first two years of life. Epigenetics Official J DNA Methylation Soc 2012;7(6):594–605.
[98] Feinberg AP, Irizarry RA, Fradin D, Aryee MJ, Murakami P, Aspelund T, et al. Personalized epigenomic signatures that are stable over time and covary with body mass index. Sci Transl Med 2010;2(49):49ra67.
[99] Gautrey HE, van Otterdijk SD, Cordell HJ, Newcastle 85+ Study, Core T, Mathers JC, et al. DNA methylation abnormalities at gene promoters are extensive and variable in the elderly and phenocopy cancer cells. FASEB J Official Publ Fed Am Soc Exp Biol 2014;28(7):3261–72.
[100] Byun HM, Siegmund KD, Pan F, Weisenberger DJ, Kanel G, Laird PW, et al. Epigenetic profiling of somatic tissues from human autopsy specimens identifies tissue- and individual-specific DNA methylation patterns. Hum Mol Genet 2009;18(24):4808–17.
[101] Fernandez AF, Assenov Y, Martin-Subero JI, Balint B, Siebert R, Taniguchi H, et al. A DNA methylation fingerprint of 1628 human samples. Genome Res 2012;22(2):407–19.
[102] Ladd-Acosta C, Pevsner J, Sabunciyan S, Yolken RH, Webster MJ, Dinkins T, et al. DNA methylation signatures within the human brain. Am J Hum Genet 2007;81(6):1304–15.
[103] Slieker RC, Bos SD, Goeman JJ, Bovée JV, Talens RP, van der Breggen R, et al. Identification and systematic annotation of tissue-specific differentially methylated regions using the Illumina 450k array. Epigenetics Chromatin 2013;6(1):26.
[104] Flanagan JM, Popendikyte V, Pozdniakovaite N, Sobolev M, Assadzadeh A, Schumacher A, et al. Intra- and interindividual epigenetic variation in human germ cells. Am J Hum Genet 2006;79(1):67–84.

[105] Meissner A, Mikkelsen TS, Gu H, Wernig M, Hanna J, Sivachenko A, et al. Genome-scale DNA methylation maps of pluripotent and differentiated cells. Nature 2008;454(7205):766–70.

[106] Lister R, Pelizzola M, Dowen RH, Hawkins RD, Hon G, Tonti-Filippini J, et al. Human DNA methylomes at base resolution show widespread epigenomic differences. Nature 2009;462(7271):315–22.

[107] Schroeder DI, Blair JD, Lott P, Yu HO, Hong D, Crary F, et al. The human placenta methylome. Proc Natl Acad Sci USA 2013;110(15):6037–42.

[108] Gaidatzis D, Burger L, Murr R, Lerch , Dessus-Babus, Schübeler D, et al. DNA sequence explains seemingly disordered methylation levels in partially methylated domains of Mammalian genomes. PLoS Genet 2014;10(2):e1004143.

[109] Jacoby M, Gohrbandt S, Clausse V, Brons NH, Muller CP. Interindividual variability and co-regulation of DNA methylation differ among blood cell populations. Epigenetics Official J DNA Methylation Soc 2012;7(12):1421–34.

[110] Talens RP, Boomsma DI, Tobi EW, Kremer D, Jukema JW, Willemsen G, et al. Variation, patterns, and temporal stability of DNA methylation: considerations for epigenetic epidemiology. FASEB J Official Publ Fed Am Soc Exp Biol 2010;24(9):3135–44.

[111] Shen H, Qiu C, Li J, Tian Q, Deng HW. Characterization of the DNA methylome and its interindividual variation in human peripheral blood monocytes. Epigenomics 2013;5(3):255–69.

[112] Bock C, Walter J, Paulsen M, Lengauer T. Inter-individual variation of DNA methylation and its implications for large-scale epigenome mapping. Nucleic Acids Res 2008;36(10):e55.

[113] Rakyan VK, Down TA, Balding DJ, Beck S. Epigenome-wide association studies for common human diseases. Nat Rev Genet 2011;12(8):529–41.

[114] Plass C, Pfister SM, Lindroth AM, Bogatyrova O, Claus R, Lichter P. Mutations in regulators of the epigenome and their connections to global chromatin patterns in cancer. Nat Rev Genet 2013;14(11):765–80.

[115] Plasschaert RN, Bartolomei MS. Genomic imprinting in development, growth, behavior and stem cells. Development 2014;141(9):1805–13.

[116] Hitchins MP. The role of epigenetics in Lynch syndrome. Fam cancer 2013;12(2):189–205.

[117] Ordway JM, Budiman MA, Korshunova Y, Maloney RK, Bedell JA, Citek RW, et al. Identification of novel high-frequency DNA methylation changes in breast cancer. PloS One 2007;2(12):e1314.

[118] Heyn H, Mendez-Gonzalez J, Esteller M. Epigenetic profiling joins personalized cancer medicine. Expert Rev Mol Diagn 2013;13(5):473–9.

[119] Sabunciyan S, Aryee MJ, Irizarry RA, Rongione M, Webster MJ, Kaufman WE, et al. Genome-wide DNA methylation scan in major depressive disorder. PloS One 2012;7(4):e34451.

[120] Wang X, Falkner B, Zhu H, Shi H, Su S, Xu X, et al. A genome-wide methylation study on essential hypertension in young African American males. PloS One 2013;8(1):e53938.

[121] Nag A, Hammond CJ. Twin studies in inherited eye disease. Clin Exp Ophthalmol 2014;42(1):84–93.

[122] Zhao J, Goldberg J, Bremner JD, Vaccarino V. Global DNA methylation is associated with insulin resistance: a monozygotic twin study. Diabetes 2012;61(2):542–6.

[123] Steves CJ, Spector TD, Jackson SH. Ageing, genes, environment and epigenetics: what twin studies tell us now, and in the future. Age Ageing 2012;41(5):581–6.

[124] Fraga MF, Ballestar E, Paz MF, Ropero S, Setien F, Ballestar ML, et al. Epigenetic differences arise during the lifetime of monozygotic twins. Proc Natl Acad Sci USA 2005;102(30):10604–9.

[125] Bell JT, Spector TD. A twin approach to unraveling epigenetics. Trends Genet 2011;27(3):116–25.

[126] McRae AF, Powell JE, Henders AK, Bowdler L, Hemani G, Shah S, et al. Contribution of genetic variation to transgenerational inheritance of DNA methylation. Genome Biol 2014;15(5):R73.

[127] Heijmans BT, Kremer D, Tobi EW, Boomsma DI, Slagboom PE. Heritable rather than age-related environmental and stochastic factors dominate variation in DNA methylation of the human IGF2/H19 locus. Hum Mol Genet 2007;16(5):547–54.

[128] Gervin K, Hammero M, Akselsen HE, Moe R, Nygård H, Brandt I, et al. Extensive variation and low heritability of DNA methylation identified in a twin study. Genome Res 2011;21(11):1813–21.

[129] Grundberg E, Meduri E, Sandling JK, Hedman AK, Keildson S, Buil A, et al. Global analysis of DNA methylation variation in adipose tissue from twins reveals links to disease-associated variants in distal regulatory elements. Am J Hum Genet 2013;93(5):876–90.

[130] van Dongen J, Ehli EA, Slieker RC, Bartels M, Weber ZM, Davies GE, et al. Epigenetic variation in monozygotic twins: a genome-wide analysis of DNA methylation in buccal cells. Genes 2014;5(2):347–65.

[131] Boks MP, Derks EM, Weisenberger DJ, Strengman E, Janson E, Sommer IE, et al. The relationship of DNA methylation with age, gender and genotype in twins and healthy controls. PloS One 2009;4(8):e6767.

[132] Quon G, Lippert C, Heckerman D, Listgarten J. Patterns of methylation heritability in a genome-wide analysis of four brain regions. Nucleic Acids Res 2013;41(4):2095–104.

[133] Kaminsky ZA, Tang T, Wang SC, Ptak C, Oh GH, Wong AH, et al. DNA methylation profiles in monozygotic and dizygotic twins. Nat Genet 2009;41(2):240–5.

[134] Hrubec Z, Robinette CD. The study of human twins in medical research. N Engl J Med 1984;310(7):435–41.

[135] Javierre BM, Fernandez AF, Richter J, Al-Shahrour F, Martin-Subero JI, Rodriguez-Ubreva J, et al. Changes in the pattern of DNA methylation associate with twin discordance in systemic lupus erythematosus. Genome Res 2010;20(2):170–9.

[136] Sthoeger Z, Sharabi A, Mozes E. Novel approaches to the development of targeted therapeutic agents for systemic lupus erythematosus. J Autoimmun 2014;54:60–71.

[137] Stefan M, Zhang W, Concepcion E, Yi Z, Tomer Y. DNA methylation profiles in type 1 diabetes twins point to strong epigenetic effects on etiology. J Autoimmun 2014;50:33–7.

[138] Galetzka D, Hansmann T, El Hajj N, et al. Monozygotic twins discordant for constitutive BRCA1 promoter methylation, childhood cancer and secondary cancer. Epigenetics Official J DNA Methylation Soc 2012;7(1):47–54.

[139] Heyn H, Carmona FJ, Gomez A, Ferreira HJ, Bell JT, Sayols S, et al. DNA methylation profiling in breast cancer discordant identical twins identifies DOK7 as novel epigenetic biomarker. Carcinogenesis 2013;34(1):102–8.

[140] Selmi C, Cavaciocchi F, Lleo A, Cheroni C, De Francesco R, Lombardi SA, et al. Genome-wide analysis of DNA methylation, copy number variation, and gene expression in monozygotic twins discordant for primary biliary cirrhosis. Front Immunol 2014;5:128.

[141] Loke YJ, Novakovic B, Ollikainen M, Wallace EM, Umstad MP, Permezel M, et al. The Peri/postnatal epigenetic twins study (PETS). Twin Res Hum Genet Official J Int Soc Twin Stud 2013;16(1):13–20.

[142] Gordon L, Joo JE, Powell JE, Ollikainen M, Novakovic B, Li X, et al. Neonatal DNA methylation profile in human twins is specified by a complex interplay between intrauterine environmental and genetic factors, subject to tissue-specific influence. Genome Res 2012;22(8):1395–406.

[143] Teh AL, Pan H, Chen L, Ong ML, Dogra S, Wong J, et al. The effect of genotype and in utero environment on interindividual variation in neonate DNA methylomes. Genome Res 2014;24(7):1064–74.

[144] Horvath S. DNA methylation age of human tissues and cell types. Genome Biol 2013;14(10):R115.

[145] Issa JP. Aging and epigenetic drift: a vicious cycle. J Clin investigation 2014;124(1):24–9.

[146] Teschendorff AE, West J, Beck S. Age-associated epigenetic drift: implications, and a case of epigenetic thrift? Hum Mol Genet 2013;22(R1):R7–15.

[147] Pirazzini C, Giuliani C, Bacalini MG, Boattini A, Capri M, Fontanesi E, et al. Space/ population and time/age in DNA methylation variability in humans: a study on IGF2/ H19 locus in different Italian populations and in mono- and di-zygotic twins of different age. Aging 2012;4(7):509–20.

[148] Gibbs JR, van der Brug MP, Hernandez DG, Traynor BJ, Nalls MA, Lai SL, et al. Abundant quantitative trait loci exist for DNA methylation and gene expression in human brain. PLoS Genet 2010;6(5):e1000952.

[149] Shoemaker R, Deng J, Wang W, Zhang K. Allele-specific methylation is prevalent and is contributed by CpG-SNPs in the human genome. Genome Res 2010;20(7):883–9.

[150] Zhang D, Cheng L, Badner JA, Chen C, Chen Q, Luo W, et al. Genetic control of individual differences in gene-specific methylation in human brain. Am J Hum Genet 2010;86(3):411–9.

[151] Shi J, Marconett CN, Duan J, Hyland PL, Li P, Wang Z, et al. Characterizing the genetic basis of methylome diversity in histologically normal human lung tissue. Nat Commun 2014;5:3365.

[152] Smith AK, Kilaru V, Kocak M, Almli LM, Mercer KB, Ressler KJ, et al. Methylation quantitative trait loci (meQTLs) are consistently detected across ancestry, developmental stage, and tissue type. BMC Genomics 2014;15:145.

[153] Cortijo S, Wardenaar R, Colome-Tatche M, Gilly A, Etcheverry M, Labadie K, et al. Mapping the epigenetic basis of complex traits. Science 2014;343(6175):1145–8.

[154] Mikeska T, Bock C, Do H, Dobrovic A. DNA methylation biomarkers in cancer: progress towards clinical implementation. Expert Rev Mol Diagn 2012;12(5):473–87.

[155] Hammons GJ, Yan-Sanders Y, Jin B, Blann E, Kadlubar FF, Lyn-Cook BD. Specific site methylation in the 5'-flanking region of CYP1A2 interindividual differences in human livers. Life Sci 2001;69(7):839–45.

[156] Kacevska M, Ivanov M, Wyss A, Kasela S, Milani L, Rane A, et al. DNA methylation dynamics in the hepatic CYP3A4 gene promoter. Biochimie 2012;94(11):2338–44.

[157] Shumay E, Logan J, Volkow ND, Fowler JS. Evidence that the methylation state of the monoamine oxidase A (MAOA) gene predicts brain activity of MAO A enzyme in healthy men. Epigenetics Official J DNA Methylation Soc 2012;7(10):1151–60.

[158] Kandimalla R, van Tilborg AA, Zwarthoff EC. DNA methylation-based biomarkers in bladder cancer. Nat Rev Urol 2013;10(6):327–35.

[159] Imperiale TF, Ransohoff DF, Itzkowitz SH, Levin TR, Lavin P, Lidgard GP, et al. Multitarget stool DNA testing for colorectal-cancer screening. N Engl J Med 2014;370(14):1287–97.

3

Differences in Histone Modifications Between Individuals

Christoph A. Zimmermann, Anke Hoffmann,
Elisabeth B. Binder, Dietmar Spengler

Max Planck Institute of Psychiatry, Translational Research,
Munich, Germany

OUTLINE

1. INTRODUCTION

Modern-day research on the molecular etiology of common diseases has focused initially on the role of DNA sequence variation with some remarkable success [1]. Nonetheless, accumulating evidence for the multilayered biology of mammalian genomes has revived interest in the still poorly understood role of epigenetic variation in human health and disease [2]. As a result, the new field of epigenetic epidemiology aims to join rapidly evolving epigenomic technologies with traditional population-based epidemiological research to identify causes (i.e., genetic, environmental, or stochastic) and phenotypic outcomes (i.e., health and disease) of epigenetic variation [3].

Epigenome-wide association studies are under way to detect differences in DNA methylation associated with disease predisposition, course, and treatment outcomes [4]. This new research line promises to advance the classification and clinical diagnosis of a number of complex diseases and to catalyze personalized therapies [5]. In addition, epigenetic marks are assessed in large cohorts of individuals to determine the degree of variation among individuals across the life span [6] and in response to specific lifetime events [7,8]. Over the past years, rapid progress on sequencing- and array-based technologies has led to current high-throughput DNA methylation analyses [9]. While contemporary studies of epigenetic variation in humans relied mostly on changes in CpG dinucleotide methylation, evidence shows that the epigenome is certainly more complex, comprising additional mechanisms such as nucleosome positioning [10], noncoding RNAs [11], and multifaceted histone modifications [12], which influence one another in deposition [13]. Integrated analysis of various layers of the epigenome raises the prospect of substantially increasing our insight into the epigenetic architecture of common diseases and elucidating how DNA sequence polymorphisms can translate into functional outcomes underlying individual-to-individual variations and complex disease phenotypes.

This review focuses on such integrated approaches. We first briefly outline the role of chromatin structure and histone modifications in gene regulation and comment on advances in the analytical toolbox. Next, we elaborate on the rationale for investigating intra- and interindividual histone modifications by revisiting the series of ground-laying findings on DNA sequence variation and gene expression. Thereafter, we explore the question of how and to what degree epigenetic mechanisms—in particular histone modifications—can serve as the long-sought link between DNA sequence variation and transcriptional output. Last, we discuss pros and cons gained from the integrated analysis of sequence variation in the context of multilayered epigenomics and consider future implications for the clinic.

2. CHROMATIN STRUCTURE AND HISTONE MODIFICATIONS

The expression of genes in higher organisms is critically controlled by the DNA accessibility to regulatory factors, which are the driving force for recruiting the transcriptional machinery [14]. Eukaryotic DNA is several meters in length and must stay functional when deposited in the nucleus [15]. Since the length of DNA as a linear molecule would vastly exceed the capacity of the nuclear compartment, it must be tightly compressed to fit into the limited space supplied. In contrast to the popular visualization of DNA as a strung-out double helix, structural deformation of DNA to bend or fold it into a more compact form is the rule rather than the exception.

In higher eukaryotes, DNA is wound in its natural state around an octameric core comprising two copies each of histones H2A, H2B, H3, and H4 or their variants (Figure 1(A)) [16]. This structure constitutes the building block of chromatin around which 146 base pairs of DNA are wrapped in 1¾ superhelical turns to form single nucleosomes [17]. These are arranged like pearls on a string, the so-called 10-nm filament, and further compaction with the help of linker histone H1 and other nonhistone proteins generates condensed chromatin fibers, largely inaccessible to the transcriptional machinery [15] (Figure 1(A)).

Since 2005, it has become increasingly clear that nucleosomes fulfill a crucial role in gene expression in addition to packaging. The close proximity of nucleosomes can prevent assembly of transcription initiation complexes and interfere with progression of RNA polymerase II (PolII)-driven gene transcription [18]. Over the past years various categories of chromatin-remodeling enzymes have been isolated that act to configure the histone–DNA structure to favor or disfavor transcription. Some enzymes catalyze disentanglement of nucleosomes in an ATP-dependent fashion and mobilize the histone core relative to the DNA. Such nucleosome "sliding" can uncover DNA sequences that serve as an address code for various classes of DNA-binding transcription factors (TFs) [10].

According to this model, nucleosomes form a static, physical barrier to transcription. In contrast, a more dynamic role for nucleosomes has emerged from their capability to store information in the amino-terminal

tails of the four core histones. Their free termini protrude from the nucleosome surface and can serve as substrates for a variety of enzyme-dependent, posttranslational modifications [19] (Figure 1(B)). These enzymes
recognize select amino acids residing within the histone tails and catalyze
lysine acetylation, lysine and arginine methylation, serine phosphorylation,

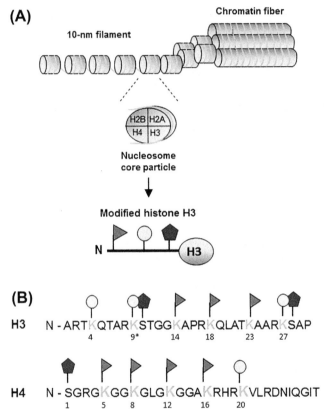

FIGURE 1   Scheme of chromatin structure and histone modifications. (A) Mammalian
DNA (blue string) is wound around a histone core (yellow cylinders) consisting of two copies
each of histones H2A, H2B, H3, and H4. This octamer constitutes the building block around
which 146 base pairs of DNA are wrapped to form the nucleosome (middle). These structures are arranged like pearls on a string in the chromatin filament and form, with the help
of the linker histone H1 and other nonhistone proteins, higher-order foldings resulting in a
condensed chromatin, inaccessible to the transcriptional machinery (top). The free amino
termini of the histones protrude from the nucleosome surface and can serve as substrate for a
variety of enzyme-dependent, posttranslational modifications, including acetylation (green
flag), methylation (yellow circle), and phosphorylation (blue pentagon), among others (bottom). (B) The amino termini of histone H3 and histone H4 are shown in single-letter amino
acid code. Exemplary, well-known posttranslational modifications are shown. Lysine (K),
marked in light gray and larger font, can be subject to site-specific acetylation or methylation.
Acetylation favors in general an open chromatin state, whereas methylation has two-sided
effects; it can either favor (H3K4me1, H3K4me3) or disfavor (H3K9me1-3, H3K27me1-3) an
open chromatin state. Lysine 9 (marked by an asterisk) can be either methylated or acetylated.

and covalent binding of the small peptide ubiquitin, among others [12,20] (Figure 1(B)). The diverse functions of these modifications have raised great interest, given the perspective that nucleosomes with their modified tails are more than a wrapper of DNA by capturing epigenetic information that controls how genes are expressed and how their expression patterns are transmitted from parental cells to progeny [13].

3. AN ANALYTICAL TOOLBOX FOR CHROMATIN STRUCTURE AND HISTONE MODIFICATIONS

Analytical tools to study chromatin status and epigenetic modifications in a locus-specific or genome-wide manner for single cells or whole tissues have evolved rapidly over the past years. Here, we focus on a limited set of methods that play an eminent role in the context of this review and refer readers interested in a comprehensive introduction to a series of monographs [21–23].

Two techniques, namely mapping of DNase I-hypersensitive sites (DHSs) [24,25] and formaldehyde-assisted isolation of regulatory elements (FAIRE) [26], have been developed to assess chromatin structure at a genome-wide scale and to map active regulatory elements (Figure 2).

Both techniques enable localization of active regions through detection of nucleosome-free regions where TF binding is thought to trigger

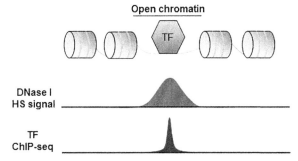

FIGURE 2 **Assays for chromatin structure and histone modifications.** The DNA (blue string) is wrapped around the histone core particles (yellow cylinders) to form the nucleosome. These structures localize tightly to each other in the case of a closed chromatin state. Binding of a TF (blue hexagon) to its DNA address code can reverse nucleosome formation and loosen the chromatin locally in favor of transcriptional activation. The green and blue peaks symbolize the genomic regions enriched by the DNase I and ChIP assays, respectively, which are mapped to a reference genomic sequence symbolized by the black straight line. The DNase I assay is based on the fact that regions of open chromatin show increased sensitivity to DNase I digestion. This method maps regulatory active elements via their impact on chromatin conformation without informing about the factors involved. In contrast, the ChIP assay relies on a specific antibody interaction to enrich for circumscribed DNA sequences serving as an address code for specific TFs or associated histone modifications. HS, hypersensitivity.

chromatin reorganization and nucleosome mobilization. In the case of DHSs, nucleosome-free sites are preferentially digested by DNase I and appear hypersensitive compared to compacted chromatin. By implication, hypersensitive sites are thought to correspond to "open," transcriptionally active chromatin, whereas hyposensitive sites represent "closed," transcriptionally inactive chromatin (Figure 2). DHS mapping can consistently detect all types of active elements comprising promoters, enhancers and silencers, insulators and boundary elements, control regions, and imprinting control centers. Although DHS mapping does not inform about the identity of any DNA-associated regulatory factor at the traced sites, this technique provides a valuable screening tool for functional regulatory elements in the genome and assessing open and closed chromatin states across cell types or within the same cell type between different individuals.

As a complementary method to DHS mapping, FAIRE relies on formaldehyde to cross-link native chromatin–DNA complexes in vivo, which are sheared by sonication. In a next step, phenol–chloroform extraction is used to separate chromatinized DNA from chromatin-free regions, which are recovered from the aqueous phase. Both DNA fractions are subsequently fluorescently labeled and hybridized to a DNA microarray [26]. Overall, FAIRE also enriches for open chromatin in a genome-wide manner irrespective of the underlying type of regulatory element or DNA-bound TF.

For chromatin immunoprecipitation experiments (ChIP), native chromatin–DNA complexes and associated proteins are similarly cross-linked by a brief treatment with formaldehyde. Cross-linked chromatin is subsequently immunoprecipitated with high-quality antibodies directed against proteins of interest (e.g., TFs, histone marks, etc.). Following reversal of the cross-link, DNA is purified for further study by gene-specific polymerase chain reaction, microarray (ChIP-chip), or sequencing (ChIP-seq) analyses [27]. In any case, ChIP techniques allow precise mapping of specific DNA–chromatin associated proteins, of core histones and their variants, as well as their posttranslational modifications throughout the genome (Figure 2).

All of these methods investigating chromatin structure and associated proteins have been customized for the needs of next-generation sequencing (NGS) enabling a genome-wide readout (DNase-seq, FAIRE-seq, and ChIP-seq). Sequencing libraries of DNA fragments enriched for open chromatin (DHS and FAIRE) or chromatin–DNA associated proteins (ChIP) contain fragments of varying size, which will be analyzed by short sequence reads and subsequently aligned to a reference genomic sequence. The short reading length and repetitive nature of mammalian genomes require careful interpretation of NGS data, in particular in the case of allele-specific data at heterozygous single-nucleotide polymorphisms (SNPs; see below). In further development, paired-end sequencing of both ends of enriched DNA fragments can help to dissolve inconsistent sequence reads [28].

4. THE DISCOVERY OF CIS-REGULATORY MUTATIONS

Genetic variation in gene expression has been recognized and exploited for a long time in animal breeding in the absence of molecular proof; and it was not before the first half of the twentieth century that Haldane postulated presciently the concept of variation in the timing of gene activity [29]. Remarkably, two decades before the conceptualization of the molecular gene, Haldane reasoned that genetic variation in a gene's activity follows from variation in the gene itself or in an unrelated locus.

The concept that noncoding mutations can fulfill a regulatory role for gene expression dates back to the discovery of regulatory sequences themselves. Soon after the appearance of their pioneering work on the *lac* operon in 1961, Jacob and Monod [30] hypothesized that mutations in operators (their term for *cis*-regulatory regions) might contribute to gene expression and the course of evolution. They further reasoned that proper function of every gene depends on the product it encodes and the conditions under which it is produced. As an example, a faultless TF can be worthless or even harmful if synthesized at the wrong place or time.

As far back as the 1940s, McClintock demonstrated in her genetic experiments the existence of special mobile elements in the maize genome, which were capable of transposing from one genetic locus to another and might regulate the activity of nearby genes [31]. Cloning of these elements showed that they encoded various classes of repetitive DNA sequences. This was followed by two influential publications from the 1970s, which provided further evidence for an important role of *cis*-acting mutations. First, Britten and Davidson [32] suggested that repetitive sequences are important to phenotypic evolution in addition to their role in gene regulation. Second, King and Wilson [33] noted that homologous proteins in humans and chimpanzees are nearly identical and cannot explain the profound differences in phenotypes between the two species. Alternatively, they hypothesized that phenotypic divergence among closely related species as well as between individuals within a population might be encoded primarily by regulatory mutations.

Compatible with this view, many aspects of organismal phenotype require dynamic changes in gene function (i.e., development, behavior, adaptation, immune responses). Transcription is a dynamic process that can be "fine-tuned" to swiftly fulfill contextual demands [30]. In this respect, phenotypic differences underlying traits associated with dynamic processes might evolve more readily through *cis*-regulatory changes that affect initiation, stability, and rate of transcription in an allele-specific manner [34]. On the other hand, mutations altering the activity or expression of factors that bind to cis-regulatory sequences to regulate transcription, also known as trans-regulatory mutations, appear generally less dynamic [35].

As a case in point, cis-regulatory regions are often organized in a modular manner, whereby mutations in a single module might influence one particular aspect of the overall transcription profile (e.g., altered temporospatial expression during development or an altered response to a specific signaling cue). On the other side, nonsynonymous coding mutations will produce an altered trans-regulatory protein with potential implications at any site of its expression.

On theoretical grounds, natural selection is predicted to operate differentially on mutations in cis-regulatory and trans-regulatory coding sequences [36]. Allele-specific analysis of transcript levels in diploid organisms supports that both alleles are transcribed to a major part independent of each other [37,38]. Mutations in cis-regulatory regions frequently act codominantly, whereas coding mutations are largely recessive. Consequently, codominant mutations are more likely to affect fitness in the heterozygous state and to provide a substrate for natural selection.

A number of studies support this hypothesis [39]. As an example, interspecific expression differences in Drosophila originate less from a few trans-regulatory changes associated with broad effects but preferentially from many cis-regulatory changes mapping throughout the fly genome [38].

The dissemination of microarray technologies in the mid-1990s drove simultaneous quantification of large numbers of transcripts between individuals within natural populations and provided further evidence for widespread variation in gene expression [40–42]. To contribute to evolutionary change, such variation must contain a heritable component. We will come back to this question later in this chapter.

5. AN ABUNDANCE OF CIS-REGULATORY MUTATIONS

Beyond a role as a selective force in species specification, cis-regulatory mutations are also thought to fulfill an important but possibly more subtle role in encoding individual traits and associated disease risks. Although the foregoing discussion implicitly assumed that noncoding mutations have a regulatory role, it remains an open question to what degree this applies at the genome-wide scale. Moreover, insight into the molecular nature of cis-acting mutations promises to illuminate their regulatory roles as well as their importance to epigenetic variation in humans (i.e., histone modifications between individuals).

In an endeavor to define the extent of genetic variation within the human population, the HapMap and 1000 Genomes projects detected about 40 million genetic variants at the genome-wide scale [43,44]. These variants consist of structural variants comprising insertions and deletions (so-called indels), copy-number variants (CNVs), inversions, and SNPs. The last represent, with 95%, the most frequent type of genetic variation, with more than 8900 SNPs showing an association with human traits or diseases in genome-wide association studies (GWAS) [44]. However, only about 16% of them map to coding sequences and strengthen the case for *cis*-regulatory mutations in individual-to-individual variation and associated disease.

The association of SNPs and CNVs with expression levels of transcripts detected in individual lymphoblastoid cell lines (LBCLs) from the Hap-Map project has been analyzed to assess the relevance of these different types of mutations for gene expression [45]. The findings show that about 84% and 18% of overall genetic variation in gene expression, respectively, is caused by SNPs and CNVs, with a minor overlap between these two variables.

Our understanding of how the *cis*-regulatory code operates dates back to the discovery that *cis*-regulatory regions typically comprise DNA-binding sites for TFs that recruit the transcriptional apparatus to confer gene regulation [14,46]. These docking sites consist of short address codes around 6 to 20 nucleotides in length in the vicinity of protein-coding DNA (but see also below). Computational studies on various species suggest that many TF-binding sites can evolve rather rapidly, leading to their existence in one species but not another [34,47]. The short and often degenerated sequence of TF-binding sites makes their conservation across species or individuals necessary to guarantee their consistent identification and limits the detection of those that are evolutionarily divergent or subject to individual polymorphisms. Moreover, DNA-associated regulatory proteins can interact transiently with nucleoprotein structures to loosen repressive states. This pioneer function is an early step, which precedes access to and occupation of any regulatory DNA motifs [48] (see also below). In this respect, computational analysis of sequence data per se is of limited use for the prediction of TF binding and associated biological functions.

6. CIS-REGULATORY MUTATIONS INFLUENCE CHROMATIN STRUCTURE, HISTONE MODIFICATIONS, AND TRANSCRIPTIONAL OUTPUT

The advent of chromatin-immunoprecipitation techniques (ChIP-chip and ChIP-seq) [27] opened up the possibility of globally mapping TF-binding sites and, furthermore, of approaching functional implications of allele-specific *cis*-regulatory mutations. As noted above, ChIP enables precise identification of TF-binding sites or other specific DNA-associated proteins, including histone modifications and histone variants within the genome (Figure 2).

For example, in the model system yeast, binding sites of the transcriptional regulators Ste12 and Tec1 were mapped in three species under low-nitrogen (pseudohyphal) conditions by means of ChIP-chip [49]. These experiments showed that most of the respective sites have diverged across species and exceeded by far interspecies variation in orthologous genes. Moreover, divergence among TF-binding sites evolved significantly faster than divergence among orthologous genes, indicating that diversity in TF-binding is a driving force in the formation of species [49]. Additionally, TF-binding sites might evolve faster than genes since TF-binding sites are less constrained.

As discussed in the foregoing section, current large-scale analysis of the genetics of gene expression originates from the advent of microarray technology in the mid-1990s. These studies evidenced that gene expression differs between various strains in *Drosophila* [38], yeast [49], and mice [50] and furthermore showed that such differences segregate in appropriate crosses [51].

In this context, the study by Kadota et al. [52] investigated the question of heritability in human gene expression at the genome-wide scale and was the first to provide preliminary insight into its relevance for the formation of the chromatin landscape and individual-to-individual differences in histone modifications.

Analysis of allele-specific protein–DNA interactions, termed ChIP-SNP, combines ChIP with genome-wide SNP genotyping via microarray detection (or NGS, see below) to determine allele-specific binding of regulatory proteins [53] or chromatin modifications [52].

Based on allele-specific ChIP-on-chip assays, Kadota et al. investigated 12 LBCLs representing two families (established by the Centre d'Etude du Polymorphisme Human; CEPH) using antibodies directed against PolII and five different posttranslational modifications of histone H3 protein (comprising the transcriptionally active marks H3 acetylated at lysines 9 and 14 (H3K9ac, K14ac) or dimethylated at lysine 4 (H3K4me2) and the transcriptionally repressive marks H3 dimethylated at lysine 9 or 27 (H3K9me2 or H3K27me2) or trimethylated at lysine 27 (H3K27me3)). Using principal component analysis the authors deduced similar global histone profiles in cell lines obtained from the same family, indicating that variations in the genetic blueprint couple to histone modifications in a genome-wide and heritable fashion [52].

One limitation inherent to this early work arises from incomplete genotyping data from these cell lines and a restricted number of sites at which allele-specific histone modifications were testable in the circumstances at the time. Moreover, only preselected heterozygous sites were present on the array for analysis and among these, only those that mapped to a binding site were informative. Consequently, only a minor proportion of allele-specific binding sites were identified. In contrast, ChIP-seq (see below) allows capturing allele-specific differences in TF binding or chromatin marking at all heterozygous sites in an individual through separate analysis of both alleles.

Finally, this early work did not answer how genetic variation links to differential histone modifications between individuals nor whether these marks actually correlate with changes in chromatin structure and gene expression—both events are commonly viewed as a major readout of genetic variation.

In anticipation of human studies to follow, shortly afterward a ChIP-seq study in mouse embryonic stem cells pioneered the single-molecule-based sequencing technology (also popularized as NGS) [54] and thereby was the first to explore the possibility of deducing allele-specific chromatin states from these data [55]. For this purpose, male embryonic stem cells were isolated from a cross between distantly related *129SvJae* (maternal) × *Mus musculus castaneus* (paternal) mice. Chromatin maps for the X chromosome versus autosomes and allele-specific distribution of histone marks at imprinted control regions served as positive controls in these experiments. Collectively, this study provided proof of principle that chromatin marks can be measured on a genome-wide scale in an allele-specific manner through assignment to SNPs.

Building on these findings, two complementary studies [56,57] aimed to infer the role of genetic variation for chromatin structure from variations in allele-specific binding of TFs at the genome-wide scale. Both studies benefited from the high percentage of allele-specific regulation in humans (10–20%) [58] and the progress in comprehensive allele-specific genotyping data from the 1000 Genomes project, which strongly expanded the number of sites associated with an informative genotype for a large panel of LBCLs.

First, McDaniell et al. addressed the impact of *cis*-regulatory mutations on variations in TF binding and chromatin structure as a proxy for gene expression [56]. Thereby, they assessed both allele-specific variations within the same individual and individual-to-individual variations using LBCLs derived from donors of geographically diverse ancestry in the 1000 Genomes project (one CEPH Utah reference family, residents with ancestry from Northern and Western Europe, both parents and their daughter; one Yoruba family from Ibadan, Nigeria, both parents and their daughter). This study design aimed to assess at the same time molecular mechanisms involved in allele-specific gene regulation and the connected but distinct question of phenotypic variation between individuals.

To explore individual genetic variation and its relationship to allele-specific chromatin structure, ChIP-seq analysis for the TF CTCF (CCTC-binding factor), a versatile regulator of transcription and chromatin structure [59], was conducted in parallel to DNase-seq as a direct readout of chromatin structure. This setup enabled the researchers to determine the effects of genetic variation on chromatin structure by two distinct approaches and showed that 10% of active chromatin sites were individual-specific, with a similar amount due to allele-specific regions. At sites of allele-specific CTCF binding, favored alleles had higher motif scores compared to disfavored ones, whereby highly conserved nucleotide positions within the address code contributed the most to allele-specific SNPs. In general, both individual-specific and allele-specific DHSs and CTCF-binding sites were transmitted from parents to children, indicating that they encode a heritable fraction of the human genome.

Overall, these experiments strongly support the hypothesis that heritable TF binding and chromatin status differ as a result of genetic variation and possibly underpin phenotypic variation in cells and, ultimately, in humans.

As a side note: age- and life event-dependent epigenetic differences in TF binding and chromatin structure [6,7] were largely undetectable in the investigated family trios. This finding might result from the need to actively maintain such marks across the life span [60].

Second, Kasowski et al. [57] mapped, by means of ChIP-seq, the genome-wide binding of nuclear factor κB (NF-κB, alias p65), a key regulator of immune responses, and of PolII, an integral component of the transcription machinery, in 10 LBCLs and a single chimpanzee. As a result, 7.5% and 25% of the NF-κB and PolII-binding sites differed between individuals and were frequently associated with SNPs and structural genomic variants such as deletions at NF-κB-binding motifs or CAAT elements. These differences correlated well with differences in gene expression, indicating that variation in TF binding translates into functional output. Moreover, individual-to-individual differences in PolII occupancy in humans were fewer than interspecies ones that amounted to up to 32% between humans and chimpanzees. This is consistent with the hypothesis that *cis*-regulatory variations account for species differences by encoding the selective recruitment of TFs.

Together, these numbers strongly surpass approximations for sequence variations in coding regions (expected to be 0.025% for the comparison of humans [61] and 0.71% for the comparison between humans and chimpanzees [62]) and attest to the importance of variation in TF binding for human diversity and specification among higher primates.

Overall, these two landmark studies suggest that individual-to-individual and allele-specific genetic variations are frequently encoded by TF-binding sites, associate with changes in chromatin status (open vs closed), and translate into downstream processes such as gene transcription. Moreover, a sizable fraction of these differences is heritable.

7. CIS-REGULATORY MUTATIONS CAN ACT IN A CELL-TYPE-DEPENDENT MANNER

In the section above, we discussed evidence for individual-to-individual and interspecies differences in chromatin structure, histone modifications, and transcriptional output. Owing to comprehensively analyzed genotypes, LBCLs have proved to be a valuable tool for these studies. However, it remains an open question to what degree genetically determined expression differences in LBCLs are informative for all cell lineages and their highly specialized progeny (Figure 3). This question is of particular importance to disease-related studies, for which it appears highly desirable to study disease-relevant tissues/cells.

As noted before, genetic variation in TF-binding sites has been proposed to act as a driving force in speciation and evolution. This hypothesis brings up the question whether the binding sites of those factors, which are conserved between species, escape from variation.

In this respect, an early study on a set of highly conserved liver-specific TFs revealed extensively varying binding sites between human and mouse [63]. In these experiments, binding of four TFs (FOXA2, HNF1A, HNF4A, and HNF6) was assessed for 4000 orthologous gene pairs (10 kb of sequence neighboring the transcriptional start site (TSS)) captured on a custom DNA microarray (ChIP-chip). Hepatocytes are functionally and structurally conserved among mammals; their gene expression programs are similar across species and largely unaffected by the isolation step. Yet, this set of core TFs occupied 41–89% of the orthologous promoters in one species but not in the other in a manner specific for each single factor and unpredictable from human–mouse sequence alignments alone. Indeed, analysis of genomic regions occupied by the same TF in either species revealed that about two-thirds of the actual binding sites diverged between the mouse and the human genomes. This result highlights the high flexibility in TF binding [63].

A limitation of this study arises from the ChIP-chip design, with a focus on core promoter regions (see below) and the lack of information on allele-specific expression differences in this cell type.

The work of Zhang et al. [58] addressed this shortcoming by using digital RNA allelotyping for quantitative interrogation of allele-specific gene expression. This method confines sequencing to the transcriptome fraction harboring SNPs and combines the sensitivity and accuracy of digital expression measurements (RNA-seq) with the advantage of site-specific

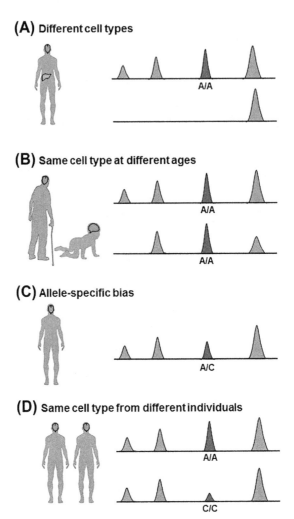

FIGURE 3 **Differences in chromatin state and histone modifications can result from multiple sources.** The straight black lines represent a genomic sequence investigated by chromatin-based assays such as DNase I or ChIP-seq. The gray and red peaks symbolize enriched genomic regions. (A) Intraindividual differences arise from different tissues such as liver and brain and probably extend to different cell types within the same tissue. (B) Intraindividual differences within the same tissue, for example, brain, can arise over a lifetime owing to different developmental time windows and in response to the cumulative effects of various environmental exposures. (C) Genetic variation at TF-binding sites confers allele-specific differences within the same tissue or cell type of an individual. (D) Individual-to-individual differences for the same tissue or cell type can result from genetic variation at TF-binding sites notwithstanding different environmental exposures and ages as noted in (B).

sequencing [64]. Four cell lines were studied; they consisted of LBCLs, primary fibroblasts, and primary keratinocytes, all from one male individual, and another primary fibroblast line from a female donor. About 11–22% of the heterozygous transcribed SNPs revealed allele-specific expression in each cell line, whereby 4.3–8.5% was tissue-specific, indicative of tissue-specific *cis* regulation. In accord with the studies from above on heritability of gene expression, allelotyping of human embryonic stem cells obtained from two pairs of siblings revealed greater similarities in allele-specific expression between siblings than genetically unrelated counterparts.

Overall, about 82% of global variation in allele-specific expression was encoded by the genome of this panel of eight cell lines representing various degrees of genetic similarity. Thereby, the variation in allelic ratios in gene expression among different cell lines was mainly due to genetic variation and less to tissue type.

These findings are supported by a second study, which investigated three cell types (primary fibroblasts obtained from umbilical cord, LBCLs, and T cells) in 75 individuals. In addition to determining allele-specific expression across different tissues, the size of this cohort allowed the combination of traditional genetic mapping methods with expression analysis in an approach termed "genetical genomics" [65]. The principal aims of this strategy were to define marker genotypes and to quantify gene expression in related or unrelated probands, to consider each gene expression level as a quantitative trait, and to correlate patterns of genetic variation with variation in expression. This procedure yields a set of expression quantitative trait loci (eQTL) that define genomic loci where genetic variation encodes one or more transcript levels.

Analysis of *cis*-regulatory effects across these three cell types in 75 individuals identified 1007 genes correlated with *cis* eQTL of which only 86 (8.5%) and 120 (12%) were shared among three and two cell types, respectively, and acted thereby in the same direction.

In contrast, 801 (79.5%) of these *cis* eQTL existed only in one cell type, similar to previous findings for tissue-specific eQTL [66]. Genes assigned to cell-type-specific eQTL displayed significantly larger variation in expression in the eQTL positive cell type compared to the negative one, indicating that a major part of this effect originates from the cell-type-specific activity of regulatory elements as opposed to differential gene expression levels between cell types.

Moreover, eQTL shared across cell types had a tendency to show larger effects and higher significance and to group closely around TSSs. On the other hand, cell-type-specific eQTL revealed smaller effect sizes and localized more broadly around TSSs. This result is consistent with the notion that enhancer elements, which map more remotely from genes, encode higher tissue specificity than core promoter elements [47].

Taken together, this study evidenced that eQTL operate preferentially in a cell-type-specific manner. This applies also to cell types with a common developmental origin such as lymphoblastoid and T cells, which share only a minor fraction of *cis* eQTL. Cell-type-specific eQTL confer smaller effects on transcription levels and localize remotely to TSSs, indicative of a role in modulating enhancer function. Thereby, eQTL seem to operate through differential use of regulatory elements of genes that are broadly expressed across different cell types.

8. EXPRESSION QUANTITATIVE TRAIT LOCI, CHROMATIN ACCESSIBILITY, AND DISEASE PATHWAYS

As in the case of SNPs, identification of eQTL raises the question of by which mechanisms genetic variation translates into changes in gene regulation. To explore this topic, Degner et al. [67] applied DNase-seq to investigate chromatin accessibility in 70 Yoruba LBCLs for which genome-wide genotypes and gene expression levels had been previously established. Genome-wide maps of chromatin accessibility for each individual were based on a total of 2.7 billion uniquely mapped DNase-seq reads and enabled inference of 8902 locations at which DNase-seq read depth correlated significantly with genetic variation. These sites, termed "DNase I-sensitive quantitative trait loci" (dsQTL), were strongly enriched within predicted TF-binding sites and frequently associated with allele-specific changes in TF binding. Consistent with these data, 16% of the dsQTL associated with variation in expression levels of at least one nearby gene. On the other side, of 1271 eQTL previously established by RNA-seq in these cell lines, 23% of the most significant SNPs also encoded dsQTL and most of these joint dsQTL–eQTL were associated with increased gene expression levels indicative of their functional role.

As a whole, this study supports that common genetic variants affect chromatin accessibility at a genome-wide scale and map within or near to DHSs. The underlying sequence variants are predicted to change the DNA-binding affinity of TFs at their recognition sites. Since a substantial fraction of eQTL concur with dsQTL, it appears likely that changes in chromatin accessibility serve to translate genetic variation into changes in gene regulation, and ultimately cellular phenotypes.

This hypothesis has received further support from a study published at the same time [68], which showed that noncoding variants associated with common diseases and traits are enriched at regulatory regions marked by DHSs. Genome-wide DNase I mapping of 349 cell and tissue samples comprising 85 cell types investigated under the ENCODE project [69] and 264 samples investigated under the Roadmap Epigenomics program [70], as well as 233 diverse fetal tissues, led to the identification of an average of 198,180 DHSs per cell type. Of 5654 noncoding genome-wide associations—inferred from 207 diseases and 447 quantitative traits [71]—a collective 40% of GWAS SNPs were enriched at DHSs. Consistent with a tissue-specific regulatory role for many of these common variants, multiple independently associated SNPs were broadly distributed across the genome and showed a cell-specific localization within pathogenically or physiologically important cell types as evidenced by DNase I mapping.

Compatible with the critical role of early development for adult common diseases [72], 88% of 2931 noncoding disease- and trait-associated SNPs within global DHSs localize also within DHSs present in fetal cells and tissues. Moreover, by focusing on those DHSs that harbor disease-associated variations, about 58% emerge first in fetal cells and tissues and endure in adult cells.

GWAS SNPs residing at DHSs mapped at about 93% to a TF-binding site, whereby common variants associated with specific diseases or traits showed an enrichment of motifs recognized by factors that regulate physiological processes important to the same kind of disease [68]. Genetic variants that change TF recognition sequences are in turn thought to modify local chromatin structure via histone modifications (see previous section).

In anticipation of our discussion on the pros and cons gained from integrated analysis of sequence variations, we note that this study [68] also found that disease-associated variants are enriched in specific transcriptional regulatory pathways. As an example, noncoding variants predisposing to deregulation of glucose homeostasis disturb the same regulatory network affected by monogenic forms of type 2 diabetes.

Remarkably, disease-associated variants from the same or related disorders and traits recurrently map within recognition sequences of TFs that coordinate the interplay between complex regulatory networks, indicating that groups of such TFs might shape common regulatory architectures. For example, about 24% of GWAS SNPs associated with autoimmune disorders such as Crohn disease, rheumatoid arthritis, lupus, and type 1 diabetes map to DHSs bound by TFs that interact with interferon regulatory factor 9 and thus bookmark a segment of a regulatory network relevant to autoimmune disorders.

Together, these findings open up a new perspective on the genetic architecture and heritability of common human diseases and phenotypic traits. Integration of disease-associated DNA variation with more dynamic molecular data such as DHSs can allow the construction of disease-associated regulatory networks. Chromatin states seem to play a critical role in modulating disease risk, progression, and severity. Therefore, integration of DHS data with those from molecular profiling (transcriptomic, proteomic, and metabolomics), imaging, and clinical data promises to advance our insight into various diseases.

9. CHROMATIN TRANSLATES GENETIC VARIATION INTO TRANSCRIPTIONAL PHENOTYPES

So far, we have considered data on genetic variation, TF binding, chromatin modifications, and gene expression with the implicit assumption that these different genomic layers correlate in one way or the other with one another and between individuals. However, apart from numerous examples of single genes, the actual direction of such interactions has not been investigated systematically at the genome-wide scale. The key question in this context is causality [73]—are histone modifications driving differences between chromatin states or, alternatively, are differences in histone modifications mainly the result of sequence-dependent dynamic processes such as nucleosome remodeling and transcription?

Although epigenetic mechanisms direct heritable gene expression and are commonly thought to operate independent of underlying DNA sequences [74], specific combinations of histone modifications associate with TFs that increase or decrease gene expression through recruitment of the transcriptional apparatus [75,76]. Since noncoding *cis*-regulatory variations occur frequently at TF-binding sites, histone marks can be hypothesized to mirror to varying degrees underlying genetic variations. If yes, individuals of different genetic makeups are expected also to differ in a systematic manner in their histone profiles. Here, we focus on three studies aimed to investigate this hypothesis on a genome-wide scale.

McVicker et al. [77] conducted ChIP-seq for PolII and four posttranslational modifications of histone H3 (H3K4me1, H3K4me3, H3K27ac, and H3K27me3) in 10 unrelated Yoruba LBCLs that have been extensively genotyped by the International HapMap project. The marks H3K4me1 and H3K4me3 are preferentially associated with active enhancers and promoters, respectively, while H3K27ac is found at either active site. In contrast, H3K27me3 is a characteristic of repressive polycomb complexes [78]. Correlation with previously established polymorphic sites resulted in identification of more than 1200 histone marks and PolII QTL (hmQTL and PolIIQTL) comprising a total of 27 distinct QTL for H3K4me1, 469 for H3K3me3, 730 for H3K27ac, 118 for PolII, and 2 for H3K27me3. Moreover, a large fraction of hmQTL localized to previously identified dsQTL [67]. Thereby, genotypes encoding low nucleosome occupancy within DHSs (so-called high-sensitive dsQTL) displayed higher levels of TF binding [67] and higher levels of the active marks H3K4me1, H3K4me3, H3K27ac, and PolII. In support of a functional role, overall patterns of dsQTL were similar to previously identified eQTL in these cell lines [79] and correlated strongly with the allele-specific modus of these active marks.

Computational analysis of the 10 Yoruba genomes predicted 11,437 high-confidence polymorphic TF-binding sites, whereby increased binding activity was associated with proximity to activating histone marks. Further categorization into diverse groups of TFs based on underlying sequence motifs showed that most groups associated with activating, but not with repressive, marks and that their DNA binding most likely sets off a chain of events that converge on histone modifications. Thus, single DNA variants can influence histone modifications at multiple functionally related regions in the genome such as distal enhancers and proximal promoters, possibly via chromatin looping [80].

The results of this study, taken together, demonstrate that genetic variation in humans underlies variation in chromatin states. QTL associated with histone modifications and PolII binding are enriched at both dsQTL and eQTL, whereby a single genetic variant can affect multiple features of chromatin formation such as nucleosome positioning, DNase I sensitivity, and histone modifications. The effects of genetic variants on histone modifications can be encoded through polymorphic TF-binding sites directing differential TF binding and tethering of enzymatic complexes.

A study by Kilpinen et al. [81] complemented these findings by incorporating additional layers of analysis and by studying the transmission of histone marks from parents to offspring. Five posttranslational modifications (H3K4me1, H3K4me3, H3K27ac, H3K27me3, and H4K20me1; the last a well-established repressive mark at enhancers and promoters [82]), three TFs (TFIIB, PU.1, and MYC), and the second largest RNA polymerase subunit RPB2 (POLR2B) were investigated by ChIP-seq in LBCLs originating from two parent–offspring trios [44]. All samples were RNA profiled for expression analysis and one of the trios was further analyzed by global run-on sequencing (GRO-seq) to determine incipient transcription.

Genome-wide mapping identified allele-specific TF binding (reaching from 11 to 12%) and posttranslational histone modifications (reaching from 6 to 30%) at heterozygous sites. Allele-specific effects were less pronounced for mRNA-seq (5%) compared to GRO-seq (27–28%), pointing to compensatory posttranscriptional mechanisms.

Although alignment of posttranslational histone modifications with specific DNA sequences remains in general a thorny task [73], allele-specific effects at TF-binding sites offer the opportunity to assess the impact of motif-disruptive polymorphisms. In this respect, SNP-dependent disruption of TF binding accounted for most cases of allele-specific effects (PU.1, 70%; MYC, 97%) although the respective consensus motifs were largely conserved. This at first sight counterintuitive outcome can be explained at least in part by the presence of additional nearby TF-binding sites that support cooperative allele-specific binding.

On the other hand, a majority of motif-disruptive SNPs lack significant allelic effects owing to the presence of multiple motifs for the same TF (i.e., homotypic motifs), which counterbalance the outcome from single motif-disruptive SNPs and buffer as a whole genetic variation.

Among the various epigenetic and transcriptional layers affected by genetic variation, mRNA expression showed the highest degree of consistency in allelic directions within the study group, followed by TF binding and histone modifications, respectively.

For the two parent–offspring trios the degree of transmission was similar for RNA expression levels and TF binding, with both parameters surpassing posttranslational histone modifications. Transmission of active histone marks (i.e., H3K4me1, H3K4me3, and H3K27ac) was more efficient close to known eQTL [83] and dsQTL [67] than on a genome-wide scale, indicating that transmission of composite chromatin states is facilitated in the presence of underlying regulatory DNA sequences.

Consistent with these findings, pairwise comparison of allele-specific effects and distinct genomic regulatory layers showed a strong allele-specific interconnectivity among specific and general TFs, histone modifications, and transcription. In addition to a role as local organizer, allele-specific effects of a single or a few variants underpinned long-distance coordination of regulatory layers acting on a larger genomic region.

Together, these data suggest integrated, genetically determined directionality between TF binding and histone modifications, with the former acting as a primary source of the regulatory interplay.

Overall, the studies of Kilpinen et al. [81] and McVicker et al. [77] demonstrate that genetic variations establish epigenetic effects on gene regulation in a heritable fashion. TF binding mediates these effects and translates them into different molecular layers at a local and genome-wide scale, whereby individual-to-individual histone modifications show the lowest degree of consistency.

The third relevant study in this context is the work of Kasowski et al. [84], with a focus on variation of chromatin states between human individuals. Similar to the experimental design from above, 19 LBCLs (5 European (CEPH), 7 Yoruban (YRI), and 2 Asian individuals from the 1000 Genomes project) were investigated by RNA-seq for expression analysis, ChIP-seq for mapping of histone modifications (H3K4me1, H3K4me3, H3K27ac, H3K27me3, and H3K36me3, the last with a role in DNA mismatch control [85]), and two general regulatory factors (CTCF and SA1, the latter a subunit of cohesin).

While chromatin marks were in general more variable than RNA expression levels (see above), enhancer regions showed the highest variability across individuals. In this category, bivalent enhancers (i.e., enhancers harboring simultaneously active and repressive histone marks) comprised the highest fraction of individual-specific regions, followed by weak and strong enhancers, respectively.

The active marks H3K27ac and H3K4me3 revealed significantly higher variability at enhancers compared with promoters, indicating that underlying regulatory elements play a critical role in the variability of histone marks. In support of this assumption, the repressive mark H3K27me3 was more variable when present with other marks in the context of bivalent domains characteristic of proactive (so-called "poised") enhancers and promoters than at domains stably silenced owing to polycomb occupancy. The most dynamic changes in chromatin states among individuals took place between switches from highly active to weakly active or repressed states.

Genes with or without one variable enhancer showed, however, no differences in expression variability, whereas expression variability increased significantly when more than 60% of the enhancers at a gene varied. This behavior suggests that multiple enhancers are needed to alter gene expression. About 74% of nonvariable genes and 99% of variable genes harbor at least one variable enhancer, and correlations between enhancers and gene expression are higher for genes containing a single enhancer than for those containing multiple ones. In consequence, a major fraction of the enhancers showing variability among individuals does not confer corresponding changes in gene expression, pointing to the existence of unknown compensatory mechanisms or enhancer redundancy. Alternatively, the effects of enhancer variability on gene expression might be masked under resting and/or in vitro conditions and manifest only in certain contexts.

Variable regions were enriched in SNPs compared to nonvariable ones, suggesting a genetic basis to enhancer variability and showing a higher correlation between genotype and allele-specific histone marks.

Consistent with the foregoing studies, allele-specific CTCF or SA1 binding and enhancer or promoter histone marks were heritable between parents and daughters. Moreover, an individual's ancestry can also influence to some degree (some 20% for most regions) which genomic regions exhibit genetically driven variability in chromatin marks.

In conclusion, this study shows that histone marks at enhancers are highly variable and may underlie differences in cellular phenotype between individuals and ancestry on the basis of SNPs overlying TF-binding sites.

The overarching picture emerging from these key studies [77,81,84] is that genetic variation at TF-binding sites is the primary source of allele-specific differences in TF occupancy, histone modifications, and enhancer variability. At the same time, all three studies agree that many of the DNA variants affecting these interconnected layers do not translate into corresponding changes in gene expression, indicating an abundance of unproductive regulatory variation and/or the need for specific environmental conditions to unmask such genetic variability.

10. PROSPECTS FOR THE ANALYSIS OF HISTONE MODIFICATIONS IN HUMANS

Chromatin signatures differ strongly between cell types within a single individual as well as between individuals (Figure 3). To exploit the potential of histone marks as a functional proxy to genetic variation and its impact on health and disease, particular attention has to be paid to the question of which kinds of cells are isolated from an individual and how to obtain access to disease-relevant tissues in the case of clinical studies. Given the vast number of different cells within intact whole organs (e.g., brain), anatomical regions (e.g., cortex, cerebellum, etc.), or even single cell layers (e.g., anterior vs posterior layers in the frontal cortex), qualitative and quantitative histone measurements represent an ongoing challenge. In addition to spatial differences among cells and tissues, any individual ages over a certain period; this factor might additionally contribute to intraindividual and individual-to-individual variation if specimens are sampled at different time points. In this respect, many environmental exposures can leave their lasting marks and modify effects from genetic variation in terms of magnitude and directionality in the same individual or between individuals. Conceivably, genetic variation unfolds in interactions with the environment and over time [86,87]—conditions that can be only partly recapitulated in vitro.

Moreover, isolation of almost any cell population from living humans depends on invasive procedures that might affect the specimens themselves. As an example, enrichment by fluorescence-activated cell sorting can result in activation of isolated cell populations and may confound their analysis in the case of immune cells.

Collectively, spatiotemporal complexity of histone marks across tissues and cells, intra- and interindividual influences of environmental exposures, and technical inaccessibility of many cell types require well-defined inclusion criteria for studying histone modifications in healthy or diseased human populations despite remarkable technical progress over the past years.

For the present, the largest collection of cells from different individuals is available from Epstein–Barr virus-transformed white blood cells known as LBCLs, which have been comprehensively genotyped and whole-genome sequenced by the HapMap and 1000 Genomes projects. As we have learned in the previous chapters, many effects of genetic variation and allele-specific gene expression—buffered by local redundancy of regulatory elements and/or integrated regulatory networks—seem to be tissue-specific and/or context-dependent and do not translate into measurable functional outputs (i.e., mature RNA and protein).

A promising approach for studying effects from genetic variation in various cell types has emerged from the availability of induced pluripotent stem cells (iPSCs). Cells from various sources, such as skin or lymphocytes, can be reprogrammed into pluripotent stem cells [88]. Such self-renewing cells are thought to retain all of the donor's germ line and somatic DNA variations, whereas the process of reprogramming erases most, if not all, lifetime epigenetic information. Importantly, iPSCs can be differentiated into different lineages to give rise to postmitotic terminally differentiated cell types closely resembling their in vivo counterparts. This approach opens the prospect of assessing the functional output of otherwise masked genetic variations on a permissive cellular background. For example, a genetic variation in neurotransmitter expression, undetectable on a lymphoblastoid background, might manifest only in a defined neuronal context under resting or stimulated conditions. Cellular phenotypic data from iPSCs, including histone modifications as an essential relay between genetic variation, chromatin, and transcription, can increase the insight into tissue-specific effects and render them easier to identify.

Such a bottom-up approach from cellular to organismal phenotypic analysis is needed to advance our insight into the biology of complex traits and diseases. The use of iPSC-derived cellular resources and cellular phenotyping also offers a realistic alternative to the use of conventional model organisms for answering specific effects of human genetic variation on different genomic layers spanning the nucleotide sequence, to chromatin modifications and transcription, to cellular functions in healthy and diseased individuals.

LIST OF ACRONYMS AND ABBREVIATIONS

ChIP	Chromatin immunoprecipitation
CNV	Copy number variants
DHS	DNase I-hypersensitive site
dsQTL	DNase I-hypersensitive quantitative trait loci
eQTL	Expression quantitative trait loci
FAIRE	Formaldehyde-assisted isolation of regulatory elements
GWAS	Genome-wide association studies
hmQTL	Histone mark quantitative trait loci
iPSC	Induced pluripotent stem cells

LBCL	Lymphoblastoid cell line
NGS	Next-generation sequencing
PolII	RNA polymerase II
PolIIQTL	RNA polymerase II expression quantitative trait loci
SNP	Single-nucleotide polymorphism
TF	Transcription factor
TSS	Transcriptional start site

References

[1] Visscher PM, Brown MA, McCarthy MI, Yang J. Five years of GWAS discovery. Am J Hum Genet 2012;90:7–24.

[2] Bernstein BE, Birney E, Dunham I, Green ED, Gunter C, Snyder M. An integrated encyclopedia of DNA elements in the human genome. Nature 2012;489:57–74.

[3] Mill J, Heijmans BT. From promises to practical strategies in epigenetic epidemiology. Nat Rev Genet 2013;14:585–94.

[4] Rakyan VK, Down TA, Balding DJ, Beck S. Epigenome-wide association studies for common human diseases. Nat Rev Genet 2011;12:529–41.

[5] Heyn H, Esteller M. DNA methylation profiling in the clinic: applications and challenges. Nat Rev Genet 2012;13:679–92.

[6] Heyn H, Li N, Ferreira HJ, Moran S, Pisano DG, Gomez A, et al. Distinct DNA methylomes of newborns and centenarians. Proc Natl Acad Sci USA 2012;109:10522–7.

[7] Mehta D, Klengel T, Conneely KN, Smith AK, Altmann A, Pace TW, et al. Childhood maltreatment is associated with distinct genomic and epigenetic profiles in posttraumatic stress disorder. Proc Natl Acad Sci USA 2013;110:8302–7.

[8] Yehuda R, Daskalakis NP, Lehrner A, Desarnaud F, Bader HN, Makotkine I, et al. Influences of maternal and paternal PTSD on epigenetic regulation of the glucocorticoid receptor gene in Holocaust survivor offspring. Am J Psychiatry 2014;171: 872–80.

[9] Zillner K, Németh A. Single-molecule, genome-scale analyses of DNA modifications: exposing the epigenome with next-generation technologies. Epigenomics 2012;4:403–14.

[10] Cairns BR. Chromatin remodeling: insights and intrigue from single-molecule studies. Nat Struct Mol Biol 2007;14:989–96.

[11] Zaratiegui M, Irvine DV, Martienssen RA. Noncoding RNAs and gene silencing. Cell 2007;128:763–76.

[12] Kouzarides T. Chromatin modifications and their function. Cell 2007;128:693–705.

[13] Bernstein BE, Meissner A, Lander ES. The mammalian epigenome. Cell 2007;128:669–81.

[14] Ptashne M. The chemistry of regulation of genes and other things. J Biol Chem 2014;289:5417–35.

[15] Misteli T. Beyond the sequence: cellular organization of genome function. Cell 2007;128:787–800.

[16] Strahl BD, Allis CD. The language of covalent histone modifications. Nature 2000;403:41–5.

[17] Luger K, Mäder AW, Richmond RK, Sargent DF, Richmond TJ. Crystal structure of the nucleosome core particle at 2.8 A resolution. Nature 1997;389:251–60.

[18] Li B, Carey M, Workman JL. The role of chromatin during transcription. Cell 2007;128:707–19.

[19] Berger SL. The complex language of chromatin regulation during transcription. Nature 2007;447:407–12.

[20] Bhaumik SR, Smith E, Shilatifard A. Covalent modifications of histones during development and disease pathogenesis. Nat Struct Mol Biol 2007;14:1008–16.

[21] Tollefsbol TO. Epigenetics protocols. Methods in molecular biology, vol. 287. Totowa, NJ: Humana Press; 2004.

[22] Tollefsbol TO. Epigenetics protocols. 2nd ed. vol. 791. Totowa, NJ: Springer; 2011.

[23] Vancura A. Transcriptional regulation: methods and protocols, vol. 809. New York, NY: Springer; 2012.

[24] Boyle AP, Davis S, Shulha HP, Meltzer P, Margulies EH, Weng Z, et al. High-resolution mapping and characterization of open chromatin across the genome. Cell 2008;132:311–22.

[25] Hesselberth JR, Chen X, Zhang Z, Sabo PJ, Sandstrom R, Reynolds AP, et al. Global mapping of protein-DNA interactions in vivo by digital genomic footprinting. Nat Methods 2009;6:283–9.

[26] Giresi PG, Kim J, McDaniell RM, Iyer VR, Lieb JD. FAIRE (Formaldehyde-assisted isolation of regulatory elements) isolates active regulatory elements from human chromatin. Genome Res 2007;17:877–85.

[27] Park PJ. ChIP-seq: advantages and challenges of a maturing technology. Nat Rev Genet 2009;10:669–80.

[28] Henikoff JG, Belsky JA, Krassovsky K, MacAlpine DM, Henikoff S. Epigenome characterization at single base-pair resolution. Proc Natl Acad Sci USA 2011;108:18318–23.

[29] Haldane JBS. The time of action of genes and its bearing on some evolutionary problems. Am Nat 1932;66:5–24.

[30] Jacob FMP. Genetic regulatory mechanisms in the synthesis of proteins. J Mol Biol 1961;3:318–56.

[31] McClintock B. Chromosome organization and genic expression. Cold Spring Harbor Symp Quant Biol 1951;16:13–47.

[32] Britten RJ, Davidson EH. Repetitive and non-repetitive DNA sequences and a speculation on the origins of evolutionary novelty. Q Rev Biol 1971;46:111–38.

[33] King MC, Wilson AC. Evolution at two levels in humans and chimpanzees. Science 1975;188:107–16.

[34] Doniger SW, Fay JC, Davidson EH. Frequent gain and loss of functional transcription factor binding sites. The regulatory genome: gene regulatory networks in development and evolution, vol. 3. Burlington MA, San Diego: Academic; 2006.

[35] Carroll SB, Grenier JK, Weatherbee SD. From DNA to diversity: molecular genetics and the evolution of animal design. 2nd ed. Malden MA: Blackwell Pub; 2005.

[36] Davidson EH. The regulatory genome: gene regulatory networks in development and evolution. Burlington MA, San Diego: Academic; 2006.

[37] Pastinen T, Sladek R, Gurd S, Sammak A, Ge B, Lepage P, et al. A survey of genetic and epigenetic variation affecting human gene expression. Physiol Genomics 2004;16:184–93.

[38] Wittkopp PJ, Haerum BK, Clark AG. Evolutionary changes in cis and trans gene regulation. Nature 2004;430:85–8.

[39] Wray GA. The evolutionary significance of cis-regulatory mutations. Nat Rev Genet 2007;8:206–16.

[40] Cheung VG, Conlin LK, Weber TM, Arcaro M, Jen KY, Morley M, et al. Natural variation in human gene expression assessed in lymphoblastoid cells. Nat Genet 2003;33:422–5.

[41] Oleksiak MF, Churchill GA, Crawford DL. Variation in gene expression within and among natural populations. Nat Genet 2002;32:261–6.

[42] Storey JD, Madeoy J, Strout JL, Wurfel M, Ronald J, Akey JM. Gene-expression variation within and among human populations. Am J Hum Genet 2007;80:502–9.

[43] International HapMap 3 Consortium. Integrating common and rare genetic variation in diverse human populations. Nature 2010;467:52–8.

[44] 1000 Genomes Project Consortium, Abecasis GR, Auton A, Brooks LD, DePristo MA, Durbin RM, Handsaker RE, et al. An integrated map of genetic variation from 1092 human genomes. Nature 2012;491:56–65.

[45] Stranger BE, Forrest MS, Dunning M, Ingle CE, Beazley C, Thorne N, et al. Relative impact of nucleotide and copy number variation on gene expression phenotypes. Science 2007;315:848–53.

[46] Latchman DS. Eukaryotic transcription factors. 5th ed. Amsterdam: Elsevier Acad Press; 2008.

[47] ENCODE Project Consortium. Identification and analysis of functional elements in 1% of the human genome by the ENCODE pilot project. Nature 2007;447:799–816.

[48] Voss TC, Hager GL. Dynamic regulation of transcriptional states by chromatin and transcription factors. Nat Rev Genet 2014;15:69–81.

[49] Borneman AR, Gianoulis TA, Zhang ZD, Yu H, Rozowsky J, Seringhaus MR, et al. Divergence of transcription factor binding sites across related yeast species. Science 2007;317:815–9.

[50] Sandberg R, Yasuda R, Pankratz DG, Carter TA, Del Rio JA, Wodicka L, et al. Regional and strain-specific gene expression mapping in the adult mouse brain. Proc Natl Acad Sci USA 2000;97:11038–43.

[51] Jin W, Riley RM, Wolfinger RD, White KP, Passador-Gurgel G, Gibson G. The contributions of sex, genotype and age to transcriptional variance in *Drosophila melanogaster*. Nat Genet 2001;29:389–95.

[52] Kadota M, Yang HH, Hu N, Wang C, Hu Y, Taylor PR, et al. Allele-specific chromatin immunoprecipitation studies show genetic influence on chromatin state in human genome. PLoS Genet 2007;3:e81.

[53] Maynard ND, Chen J, Stuart RK, Fan J, Ren B. Genome-wide mapping of allele-specific protein-DNA interactions in human cells. Nat Methods 2008;5:307–9.

[54] Service RF. Gene sequencing. The race for the $1000 genome. Science 2006;311:1544–6.

[55] Mikkelsen TS, Ku M, Jaffe DB, Issac B, Lieberman E, Giannoukos G, et al. Genome-wide maps of chromatin state in pluripotent and lineage-committed cells. Nature 2007;448:553–60.

[56] McDaniell R, Lee BK, Song L, Liu Z, Boyle AP, Erdos MR, et al. Heritable individual-specific and allele-specific chromatin signatures in humans. Science 2010;328:235–9.

[57] Kasowski M, Kyriazopoulou-Panagiotopoulou S, Grubert F, Zaugg JB, Kundaje A, Liu Y, et al. Variation in transcription factor binding among humans. Science 2010;328:232–5.

[58] Zhang K, Li JB, Gao Y, Egli D, Xie B, Deng J, et al. Digital RNA allelotyping reveals tissue-specific and allele-specific gene expression in human. Nat Methods 2009;6:613–8.

[59] Phillips JE, Corces VG. CTCF: master weaver of the genome. Cell 2009;137:1194–211.

[60] Hoffmann A, Spengler D. DNA memories of early social life. Neuroscience 2014;264:64–75.

[61] Levy S, Sutton G, Ng PC, Feuk L, Halpern AL, Walenz BP, et al. The diploid genome sequence of an individual human. PLoS Biol 2007;5:e254.

[62] Watanabe H, Fujiyama A, Hattori M, Taylor TD, Toyoda A, Kuroki Y, et al. DNA sequence and comparative analysis of chimpanzee chromosome 22. Nature 2004;429:382–8.

[63] Odom DT, Dowell RD, Jacobsen ES, Gordon W, Danford TW, MacIsaac KD, et al. Tissue-specific transcriptional regulation has diverged significantly between human and mouse. Nat Genet 2007;39:730–2.

[64] Nilsson M, Malmgren H, Samiotaki M, Kwiatkowski M, Chowdhary BP, Landegren U. Padlock probes: circularizing oligonucleotides for localized DNA detection. Science 1994;265:2085–8.

[65] Jansen RC, Nap JP. Genetical genomics: the added value from segregation. Trends Genet 2001;17:388–91.

[66] Heinzen EL, Ge D, Cronin KD, Maia JM, Shianna KV, Gabriel WN, et al. Tissue-specific genetic control of splicing: implications for the study of complex traits. PLoS Biol 2008;6:e1.

[67] Degner JF, Pai AA, Pique-Regi R, Veyrieras JB, Gaffney DJ, Pickrell JK, et al. DNase I sensitivity QTLs are a major determinant of human expression variation. Nature 2012;482:390–4.

[68] Maurano MT, Humbert R, Rynes E, Thurman RE, Haugen E, Wang H, et al. Systematic localization of common disease-associated variation in regulatory DNA. Science 2012;337:1190–5.

[69] Thurman RE, Rynes E, Humbert R, Vierstra J, Maurano MT, Haugen E, et al. The accessible chromatin landscape of the human genome. Nature 2012;489:75–82.

[70] Bernstein BE, Stamatoyannopoulos JA, Costello JF, Ren B, Milosavljevic A, Meissner A, et al. The NIH roadmap epigenomics mapping consortium. Nat Biotechnol 2010;28:1045–8.

[71] Welter D, MacArthur J, Morales J, Burdett T, Hall P, Junkins H, et al. The NHGRI GWAS Catalog, a curated resource of SNP-trait associations. Nucleic Acids Res 2014;42:D1001–6.

[72] Gluckman PD, Hanson MA, Bateson P, Beedle AS, Law CM, Bhutta ZA, et al. Towards a new developmental synthesis: adaptive developmental plasticity and human disease. Lancet 2009;373:1654–7.

[73] Henikoff S, Shilatifard A. Histone modification: cause or cog? Trends Genet 2011;27: 389–96.

[74] Allis CD, Jenuwein T, Reinberg D, Caparros M. Epigenetics. 1st ed. New York: Cold Spring Harbor; 2007.

[75] Ptashne M. On the use of the word 'epigenetic'. Curr Biol 2007;17:R233–6.

[76] Ptashne M. Epigenetics: core misconcept. Proc Natl Acad Sci USA 2013;110:7101–3.

[77] McVicker G, van de Geijn B, Degner JF, Cain CE, Banovich NE, Raj A, et al. Identification of genetic variants that affect histone modifications in human cells. Science 2013;342:747–9.

[78] Margueron R, Reinberg D. The polycomb complex PRC2 and its mark in life. Nature 2011;469:343–9.

[79] Pickrell JK, Marioni JC, Pai AA, Degner JF, Engelhardt BE, Nkadori E, et al. Understanding mechanisms underlying human gene expression variation with RNA sequencing. Nature 2010;464:768–72.

[80] Sanyal A, Lajoie BR, Jain G, Dekker J. The long-range interaction landscape of gene promoters. Nature 2012;489:109–13.

[81] Kilpinen H, Waszak SM, Gschwind AR, Raghav SK, Witwicki RM, Orioli A, et al. Coordinated effects of sequence variation on DNA binding, chromatin structure, and transcription. Science 2013;342:744–7.

[82] Cheng J, Blum R, Bowman C, Hu D, Shilatifard A, Shen S, et al. A role for H3K4 monomethylation in gene repression and partitioning of chromatin readers. Mol Cell 2014;53:979–92.

[83] Lappalainen T, Sammeth M, Friedländer MR, Hoen PA, Monlong J, Rivas MA, et al. Transcriptome and genome sequencing uncovers functional variation in humans. Nature 2013;501:506–11.

[84] Kasowski M, Kyriazopoulou-Panagiotopoulou S, Grubert F, Zaugg JB, Kundaje A, Liu Y, et al. Extensive variation in chromatin states across humans. Science 2013;342:750–2.

[85] Li G. Decoding the histone code: role of H3K36me3 in mismatch repair and implications for cancer susceptibility and therapy. Cancer Res 2013;73:6379–83.

[86] Lahiri DK, Maloney B. Gene × environment interaction by a longitudinal epigenome-wide association study (LEWAS) overcomes limitations of genome-wide association study (GWAS). Epigenomics 2012;4:685–99.

[87] Zannas AS, Binder EB. Gene-environment interactions at the FKBP5 locus: sensitive periods, mechanisms and pleiotropism. Genes Brain Behav 2014;13:25–37.

[88] McKernan R, Watt FM. What is the point of large-scale collections of human induced pluripotent stem cells? Nat Biotechnol 2013;31:875–7.

Individual Noncoding RNA Variations: Their Role in Shaping and Maintaining the Epigenetic Landscape

Emily Machiela[1], Anthony Popkie[2], Lorenzo F. Sempere[3]

[1]Laboratory of Aging and Neurodegenerative Disease, Van Andel Research Institute, Grand Rapids, MI, USA; [2]Laboratory of Cancer Epigenomics, Van Andel Research Institute, Grand Rapids, MI, USA; [3]Laboratory of MicroRNA Diagnostics and Therapeutics, Van Andel Research Institute, Grand Rapids, MI, USA

OUTLINE

1. INTRODUCTION

When C.H. Waddington first put forward the concept of epigenetics in 1942, DNA and chromatin structures were unknown, and noncoding regulatory RNAs (ncRNAs) were, of course, unimaginable. While studies have assigned only 3% of the human genome sequence to protein-coding genes, we know that 76% of the genome is actually transcribed [1]. There is growing evidence for the biological significance of the ncRNAs resulting from these transcripts. Noncoding RNAs encompass a variety of regulatory RNA classes having different characteristics, biogenesis, and functions. Short ncRNA classes (17–31 nucleotides) include microRNAs (miRNAs) and PIWI-interacting RNAs, midsize classes (22–200 nucleotides) include small nucleolar RNAs and the 5′ regions of protein-coding genes, and long classes (over 200 nucleotides) include long intergenic ncRNAs and transcribed ultraconserved regions [2]. Members of different regulatory ncRNA classes have been linked to epigenetic processes. The dynamic and multilayered regulatory interactions between ncRNAs and the epigenetic machinery are important for normal development and differentiation; disruption of these interactions contributes to initiation and progression of human disease [2].

Originally, epigenetics was an abstract concept used to describe the emerging properties of interacting genes that unfold during development to form and maintain specialized cell types, tissues, and organs [3]. Over the years, the term "epigenetics" has been used in different contexts to describe a wide variety of nongenetic (nonmutational) processes that affect global gene expression (the transcriptome) or regulation of gene products. These different uses of the term can create confusion [3]. For the purpose of this chapter, we define epigenetics as a chromatin event that can be inherited mitotically (somatic) or meiotically (germ line). Within this framework, (1) enzymes that catalyze a chemical modification such as methylation or acetylation in the chromatin will be referred as "effectors" of epigenetic machinery; (2) ncRNAs that mediate interactions of epigenetic machinery with the chromatin will be referred to as "guides"; (3) ncRNAs that regulate expression or activity of epigenetic machinery (e.g., posttranscriptional regulation) will be referred to as "regulators"; and (4) ncRNAs that interfere with the function of guides and/or regulators (e.g., RNA decoys) and thereby indirectly affect the function of epigenetic effectors will be referred as "interferers."

In addition to cardiovascular disease, neurodegenerative conditions and cancer are the main causes of death in developed countries. We discuss the clinical implications for inherited, congenital, and sporadic neurodegenerative conditions, as well as for complex somatic diseases such as breast, bladder, and lung cancers. These diseases provide illustrative examples of the close interplay and reciprocal regulatory feedback loops between ncRNAs and epigenetic processes. These regulatory interactions shape and remodel the epigenetic landscape and dictate transcriptional programs that can affect ncRNA expression and activity (Figure 1). This chapter has three main sections, which highlight specific ncRNA classes and the interactions of their members with epigenetic processes. First, we describe regulatory roles of miRNAs as posttranscriptional inhibitors of epigenetic effectors and epigenetic regulation of miRNA expression. Second, we describe the pathogenic roles of trinucleotide repeat (TNR) expansion, a form of sense-strand lncRNA, as a driver of neurotoxicity and neurodegeneration. TNRs can act as guides of the epigenetic

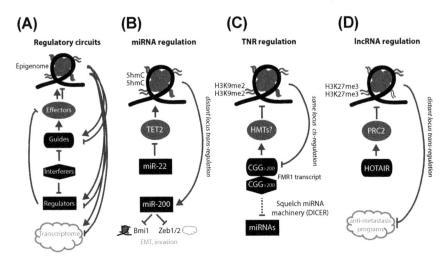

FIGURE 1 **Regulatory circuits and interplay between ncRNAs and epigenetic processes.** (A) ncRNAs (solid black shapes) can act as guides to direct epigenetic effectors (gray ovals) to particular chromosomal regions; as regulators of effector expression, reducing or increasing their activity; or as interferers that affect how other ncRNAs engage in the regulation of epigenetic effectors. Collectively, this interplay shapes the local and global epigenetic landscape and affects the transcriptional output of the cell (the transcriptome). (B) miR-22 posttranscriptionally inhibits the expression of TET2. Reduced TET2 activity causes increased DNA methylation of the *MIR200* promoter and thus loss of miR-200b expression. Zeb1/2 and Bmi1 expression is elevated in the absence of miR-200 posttranscriptional regulation, promoting invasive and prometastatic programs in breast cancer. (C) Expanded TNR repeats in *FMR1* mRNA can act as ncRNA guides for chromatin modifiers, inducing gene silencing on *cis* or *trans* configurations. These expanded CGG-containing hairpins interact and sequester Dicer away from the pre-miRNA hairpin. Decreased processing of pre-miRNAs into mature miRNAs reduces miRNA activity in the cell. (D) The HOTAIR lncRNA acts as a guide to direct polycomb repressive complex 2 (PRC2) to specific genomic regions, where it modifies histone tails and thus affects metastatic transcriptional programs in breast cancer.

machinery by promoting local epigenetic silencing changes (same-gene *cis* regulation), but they can also interfere with proper miRNA processing and expression in some disease settings (e.g., fragile X syndrome). Third, we describe regulatory roles of lncRNAs as guides of the epigenetic machinery. We conclude the chapter with a discussion of the potential applications of ncRNAs for personalized diagnostics based on epigenetic changes during disease initiation and progression.

2. REGULATORY INTERACTIONS BETWEEN MICRORNAS AND EPIGENETIC MACHINERY

MicroRNAs are evolutionarily conserved short ncRNAs. MicroRNA biogenesis and maturation is a stepwise process [4,5] that begins with miRNA gene transcription by RNA polymerase II, followed by cleavage and processing of the primary transcript into precursor miRNA (pre-miRNA) hairpins (~70 nt) by a microprocessor in the nucleus. DGCR8 (DiGeorge critical region gene 8) is the protein in the microprocessor complex that recognizes secondary structures in the pre-miRNA and clamps the pre-miRNA hairpin for the RNase type III Drosha to effectuate the actual cleavage [6]. The next step is the cleavage and processing of the pre-miRNA into mature and biologically active miRNA (19–24 nt) by the RNase type III Dicer-containing multiprotein complex in the cytoplasm. Finally, the mature miRNA is loaded into the Argonaute-containing miRNA-induced silencing complex (miRISC). MicroRNAs generally bind to partially complementary sites in the 3′ UTR of target mRNAs, serving as a guide for the miRISC to trigger mRNA degradation and/or translational repression of the target genes [5,7]. Via this mechanism, a single miRNA can modulate the expression of hundreds of target mRNAs to different extents of repression. In some contexts and specific cell types, the miRNA-mediated regulation of a single (or a few key) target(s) may significantly influence a biological process; in other situations, collective modulation of a larger set of targets may be more influential.

This RNA–RNA interaction is the most widespread regulatory mechanism by which miRNAs negatively control gene expression at the posttranscriptional level. In this chapter, we discuss only examples of miRNA-mediating regulation via binding to the 3′ UTR of target mRNAs. However, miRNAs can act via other mechanisms, some of which have been uncovered only recently and need further characterization. For example, miRNAs can bind to 5′ UTRs and coding sequences and cause mRNA stabilization, mRNA degradation, or translational inhibition. They can also bind to DNA promoter regions and affect transcriptional firing and output rates, or they can be secreted in exosomes from one cell type and act like a hormone in another cell type via Toll-like receptor signaling [8–11].

Most miRNAs are transcribed by RNA polymerase II and as such are subject to epigenetic and transcription regulation similar to that for protein-encoding gene transcripts. Epigenetic changes that affect the expression of a small number of miRNAs can have overt phenotypic consequences because each individual miRNA can inhibit expression of a large number of target genes. Epigenetic reprogramming of key miRNA genes can greatly increase the capability of cells to initiate or accelerate the progression of a pathological process. The involvement of miRNAs and their interplay with epigenetic efforts is being extensively studied in human diseases such as neurodegenerative conditions and cancer, with the hope that this research can lead to the discovery of useful biomarkers and actionable clinical information. There are excellent reviews that cover in detail the role of miRNAs in Alzheimer's disease, Parkinson's disease, Huntington's disease, and amyotrophic lateral sclerosis [12–15] and in hematologic and solid tumors (including chronic lymphocytic leukemia; acute myeloid lymphoma; and cancers of the bladder, breast, liver, lung, and prostate) [16–20]. We provide here a few examples to illustrate this regulatory paradigm in such diseases, and when possible we comment on parallels and differences among the involved miRNAs, comparing disease factors such as relative risk, etiology, and clinical management.

2.1 Epigenetic Modifications That Affect the Expression of MicroRNAs under Neurodegenerative Conditions

Most miRNAs that regulate the expression of genes associated with neurodegenerative diseases were identified through in vitro studies. To further study their clinical relevance, changes in expression of selected miRNAs have been analyzed in diseased tissues, and epigenetic marks have been characterized in some of these miRNA loci to determine if those marks were a major contributor to altered miRNA expression.

2.1.1 MicroRNAs in Alzheimer's Disease

Alzheimer's disease (AD) is the most prevalent form of dementia in older individuals (>65 years) and a major public health concern in aging populations. Intracellular Tau neurofibrillary tangles and the formation of extracellular amyloid-β (Aβ) plaques from amyloid precursor protein (APP) are molecular hallmarks of AD. Processes that limit the expression of these proteins or the activity of enzymes required for their pathological misfolding could protect against disease development and lessen or delay disease progression. MicroRNAs have been identified as potential regulators of such processes. Age-associated epigenetic changes in miRNA genes can reduce miRNA expression levels and disrupt their ability to regulate AD-associated target genes. Examples of these regulatory interactions include the following cases. miR-20a family members (miR-20a, miR-17-5p, and miR-106b) have been shown to interact with sites on the 3′ UTR of the *APP* mRNA [21]. The expression of miR-106b was detected at lower levels in a patient cohort of sporadic AD cases, suggesting that decreased activity of this miRNA may increase the total amount of APP and facilitate accumulation of pathogenic APP aggregates [21]. BACE1 (β-site amyloid precursor protein-cleaving enzyme 1) cleaves aggregation-prone αβ1–42 peptide from APP. miR-29a, miR-29b, miR-107, and miR-124, among other miRNAs, have been shown to interact with sites on the 3′ UTR of the *BACE1* mRNA [22,23]. Decreased expression levels of these miRNAs correlated with higher expression of BACE1 in brain tissues from AD cases [22,23]. Serine palmitoyltransferase (SPT) is a rate-limiting enzyme in the de novo production of ceramides, which can increase the expression of BACE1 protein by stimulation of interleukin-1β and other cytokines. miR-137 and miR-181c have been shown to interact with sites on the 3′ UTR of the *SPTLC1* mRNA, and miR-9, miR-29a, and miR-29b with the *SPTLC2* mRNA, which together encode the subunits of SPT [24]. miR-9 and miR-181c expression is downregulated in AD patients, and exposure to Aβ peptides inhibits miR-9 and miR-181c expression in mouse models of AD [25]. Together, these observations suggest that decreased activity of these miRNAs, collectively or individually, may increase the total amount and/or activity of BACE1 and skew processing of APP to aggregation-prone αβ1–42 peptides. Expression of miR-137 is epigenetically regulated by DNA methyl-CpG-binding protein-2, which binds to methylated promoter regions in adult murine neural stem cells [26]. In this context, miR-137 inhibits posttranscriptionally the expression of *EZH2* (enhancer of Zeste homolog 2), decreasing the protein output of its gene product, a histone methyltransferase component of polycomb repressive complex 2 (PRC2). Lower amounts of fully assembled PRC2 result in global epigenetic changes owing to decreased histone H3 trimethyllysine 27 modifications [26]. It will be important to determine how this epigenetic regulatory interplay is at work in AD.

2.1.2 MicroRNAs in Parkinson's Disease

Parkinson's disease (PD) is the second most prevalent neurodegenerative disease. Loss of dopaminergic neurons in the substantia nigra is a hallmark of PD. Mutations in several genes have been casually linked to familial and sporadic PD, including mutations in the *SNCA* gene encoding α-synuclein. Mutant *SNCA* can produce too much or an abnormally folded α-synuclein protein, which is central to disease pathology in PD. miR-7 and miR-153 have been shown to interact with sites on the 3′ UTR of the *SNCA* mRNA [27], suggesting that these miRNAs may lower α-synuclein levels and pathology in neurons. The genetic variants of *LRRK2* (leucine-rich repeat kinase 2) are known as the most common cause of hereditary PD. miR-205 has been shown to interact with sites on the 3′ UTR of the *LRRK2* mRNA [28]. Single-nucleotide polymorphisms (SNPs) on the 3′ UTR of the *LRRK2* mRNA were associated with increased risk for PD in a cohort of Spanish patients [29]. The high-risk rs66737902 C allele was predicted to disrupt a binding site for miR-138-3p, which may explain the elevated expression of LRRK2 protein [29]. Together, these observations suggest that loss of miRNA regulation of *LRRK2* expression could increase the total protein amount of LRRK2 and lead to a higher LRRK2 kinase activity and/or pathogenic interactions with α-synuclein, Tau, and other PD-associated proteins. Last, SNPs on the 3′ UTR of the *FGF20* mRNA were shown to be associated with increased risk for PD in a multi-institutional patient cohort. The high-risk rs12720208 T allele disrupted a binding site for miR-433, which resulted in the higher output of FGF20 protein both in vitro and in vivo. Increased FGF20 protein levels correlate positively with higher expression levels of α-synuclein [23], which may provide a large cellular pool of pathology-related proteins.

2.1.3 MicroRNAs in Huntington's Disease

Huntington's disease (HD) is an autosomal-dominant genetic disease caused by the expansion of CAG trinucleotide repeats in the huntingtin gene (*HTT*). The mutant protein resulting from the translation of the *HTT* mRNA with the CAG expansion is neurotoxic. miR-125b, miR-137, miR-146a, miR-148a, miR-150, miR-196a, and miR-214 have been shown to interact with sites on the 3′ UTR of the *HTT* mRNA [30–32]. Mutant HTT has been shown to cause changes in the expression of these miRNAs and others in cell lines and mouse models. Various mechanisms can account for altered miRNA expression, including interference by mutant HTT in the interaction of wild-type HTT with Ago2 and increased nuclear localization of the transcriptional repressor RE1-silencing transcription factor [33,34]. In addition to the mutant protein effects, the transcript having pathogenic CAG TNR expansion can also contribute to altered miRNA expression and function.

2.2 Epigenetic Modifications That Affect MicroRNA Expression in Cancer

Germ line genetic and epigenetic variants can increase the risk of developing some cancer types, but most cancers are sporadic in origin. Somatic genetic and epigenetic variants primarily dictate tumor evolution and the presence of aggressive features. Some cancers have well-known driver mutations, but many do not have dominant driver mutations. Epigenetic variants in cancer cells may allow for the selection of aggressive transcriptional programs in conjunction with dominant driver mutations or enhancement of weaker tumorigenic alterations. Epigenetic drugs have already shown promising results in clinical trials for various cancer types, mainly of hematological origin. Derepression of tumor-suppressive miRNAs may account for some of the beneficial effects observed after treatment with epigenetic drugs.

Well-characterized cell lines and tissue specimens are more readily available for most cancer sites than for neurodegenerative conditions. These have enabled researchers to collect integrated data sets of miRNA expression changes and epigenetic modifications on miRNA gene loci from large cohorts of cancer patients representing many different tumor types. EpimiR is a comprehensive database of studies that document epigenetic regulation of miRNA expression and miRNA-mediated regulation of epigenetic effectors [35]. The EpimiR database can be queried by miRNA name or epigenetic effector name; the search results indicate if data were generated in cell lines (in vitro studies) or in patient samples (in vivo studies). EpimiR already contains more than 3000 references. We provide here some general observations about this body of literature and highlight a few miRNAs that have been studied in more detail and that have been associated with multiple cancer types.

2.2.1 Epigenetic Drugs and Regulation of MicroRNA Expression in Cancer Cell Lines

A general strategy for identifying miRNA loci subject to epigenetic regulation has been to treat cancer cell lines with inhibitors of epigenetic effectors and analyze changes in miRNA expression [18,20]. Most approaches include an inhibitor of DNA methyltransferases (DNMTs) such as 5-azacytidine (5-Aza-CR), 5-aza-2'-deoxycytidine (5-Aza-CdR), or 1-(β-D-ribofuranosyl)-1,2-dihydropyrimidin-2-one (zebularine), as well as an inhibitor of histone deacetylases (HDACs) such as trichostatin A (TSA), 4-phenylbutyric acid (PBA), or suberoylanilide hydroxamic acid (SAHA). The genomic region of the differentially expressed miRNA is then thoroughly characterized for epigenetic modifications of DNA or histones.

2.2.1.1 Reexpression of miR-127 in Bladder Cancer Cell Lines

Pioneering work by Peter Jones's group identified epigenetic regulation of miR-127 in the T24 bladder cancer cell line [36]. miR-127 can be transcribed as part of a polycystronic miRNA primary transcript from a distal promoter or individually from a proximal promoter. Treatment with both a DNMT inhibitor (5-Aza-CR) and an HDAC inhibitor (PBA) derepressed miR-127 expression [36]. Demethylation of CpG sites (not a CpG island) around the proximal promoter increased miR-127 transcription (Figure 2(A)). Consistent with this finding, increases in acetylated histone 3 (H3ac) and in histone 3 trimethylated on lysine 4 (H3K4me3), which are epigenetic marks of an open chromatin state, are observed only around the proximal promoter region. Higher expression of miR-127 in T24 cells causes apoptosis, at least in part, via the downregulation of B-cell lymphoma-6 (BCL-6). Several studies using DNMT and/or HDAC inhibitors to treat cancer cell lines, as well as those using cell lines that lack activity of DNMT1 and DNMT3B, have uncovered the epigenetic regulation of multiple miRNAs [18,20]. Briefly, altered expression (mainly upregulation) of let-7 family members, miR-9 family members, miR-34 family members, and miR-124 family members after treatment with epigenetic drugs has been reported for more than five cancer types, including both hematologic and solid tumors.

2.2.1.2 Reexpression of miR-9 and miR-124 in Cell Lines from Cancer Models

miR-124 expression is epigenetically silenced in breast, colorectal, gastric, liver, and cervical cancer, among others. MiR-9 expression is also epigenetically silenced in breast and colorectal cancer as well as in melanoma and head and neck cancer. These findings are interesting because these miRNAs are well known for their brain-specific roles [37,38]. Both miR-9 and miR-124 are engaged in double-negative feedback regulatory loops with the RE1-silencing transcription factor (REST) [39,40]. These regulatory loops occur during neuronal development and differentiation,

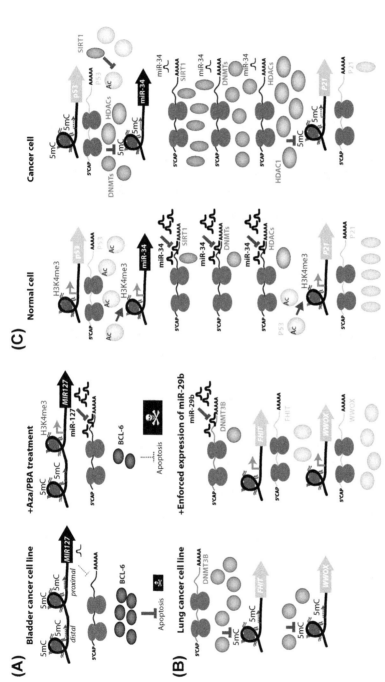

FIGURE 2 Multiregulatory layers of interaction between microRNAs and epigenetic machinery. (A) A distal promoter transcribes mir-127 in an miRNA polycystronic transcript in normal cells (not shown). That distal promoter is silenced by DNA hypermethylation in bladder cancer cells. After epigenetic treatment, a proximal promoter transcribes mir-127 as a single miRNA transcript. Increased amounts of miR-127 can now direct miRNAs to binding sites on the 3′ UTR of BCL-6 and decrease expression of this antiapoptotic protein. (B) Loss of miR-29 expression during lung carcinogenesis contributes to increased DNMT3B expression. Increased DNMT3B activity results in hypermethylation of many chromosomal regions. Enforced expression of miR-29 restores the inhibition of DNMT3B. miR-29 indirectly reshapes the epigenetic landscape, which includes the reactivation of tumor-suppressive genes such as FHIT and WWOX. (C) In this composite scenario, miR-34 and p53 reinforce each other's expression and activity via epigenetic, transcriptional, and posttranscriptional mechanisms. Normal epithelial cells express wild-type p53 and miR-34; in cancer cells, p53 and

reinforcing the specification and maintenance of neuronal transcription programs in neurons and of nonneuronal transcription programs in other cell types [39,40]. miR-124, along with miR-9* (coexpressed on the same pre-miRNA as miR-9), inhibits the expression of BAF53a—a subunit of Swi/Snf-like neural-progenitor-specific BAF complexes—via binding sites on the 3′ UTR of its mRNA [41]. miRNA-mediated regulation triggered an exchange of BAF53a for its homologous BAF53b subunit and the consequent formation of neuron-specific BAF complexes [41].

There are also many studies that suggest roles for miR-9 and miR-124 in the suppression of multiple solid tumor types and of central nervous system (CNS) malignancies. Expression of miRNAs in the *MIR106b–MIR25* gene cluster is induced by hypoxia and inhibits posttranscriptional expression of REST in cell line models of prostate cancer [42]. Loss of REST relieves repression of miR-9 and other neuron-expressed miRNAs and correlates with neuroendocrine differentiation in advanced prostate cancer tumors [42]. The dysregulation of REST expression or activity may be a recurrent mechanism for ectopic expression in cancer types of nonneuronal origin, which may lead to different molecular phenotypes.

2.2.1.3 Reexpression of miR-34 Family Members in Cell Lines from Cancer Models

miR-34 family members (miR-34a, miR-34b, and miR-34c) are transcriptionally upregulated by p53. Regulatory input from p53 is relayed to these miRNAs, which exert tumor-suppressive functions by posttranscriptionally regulating the expression of protein networks involved in proliferation and survival pathways, including CDK4/6, cyclin D1, cyclin E2, MET, Myc, and BCL-2 [43]. Because of their potent tumor-suppressive functions, restoration therapies based on miR-34a mimetics are being actively pursued [43]. miR-34a is located on chromosome 1 and is transcribed as a single miRNA transcript. miR-34b and miR-34c are both located on chromosome 11, form part of the same miRNA gene cluster, and are cotranscribed as a polycystronic unit (miR-34b–miR-34c transcript). In the *MIR34A* locus, there is a well-defined CpG island within 2.5 kb upstream of the transcription start site, and the p53-binding site is located about 200 bp upstream of that site. miR-34a expression is silenced by promoter hypermethylation at a high frequency (>40% of examined cases) in prostate cancer and melanoma and at a lower frequency (<30% of examined cases) in other cancer types, including breast, bladder, colon, and lung cancer [44]. Other studies report a higher frequency (>55% of examined cases) of DNA hypermethylation on the promoter region of *MIR34A* and even higher frequency (>70%) on the promoter region of the *MIR34B* gene cluster in breast, colon, pancreatic, and ovarian cancer and others [45,46]. miR-34a expression is frequently silenced by promoter hypermethylation in 110 cancer cell lines established from these and other cancer types [44].

The treatment of cell lines from breast cancer (Hs578T), melanoma (IGR-39), and prostate cancer (LAPC-4 and PC3) with 5-Aza-CdR and TSA reactivated miR-34a expression [44]. Enforced expression of miR-34a in the PC3 line caused cellular senescence [44], but there was no direct test of whether reexpression of miR-34a with epigenetic drugs was sufficient to cause the same molecular phenotype. Silencing of miR-34b/c expression by promoter hypermethylation has been found in each of nine colorectal cancer cell lines [46]. Treatment with 5-Aza-CdR increased miR-34b/c expression in these cancer cell lines. Enforced expression of miR-34b or miR-34c in the HCT116 cancer cell line inhibited the expression of bona fide target genes, including CDK4, cyclin E2, and MET. A similar decrease in these target genes was induced by treatment with 5-Aza-CdR. These results suggest that reactivation of these miRNAs by epigenetic reprogramming strongly contributes to the molecular phenotype.

2.2.1.4 Epigenetic Modifications That Increase Expression of Tumorigenic MicroRNAs

A few studies have reported increased expression of cancer-associated miRNAs in normal cells after treatment with epigenetic drugs. These findings suggest that epigenetic changes can select for overexpression of tumorigenic miRNAs in normal or premalignant cells that contribute to cellular transformation. miR-224 is highly expressed in hepatocarcinoma cell lines and liver tumor tissues [47]. Treatment with HDAC inhibitors (TSA or SAHA) increases the expression of miR-224 and that of neighboring genes in immortalized liver cells [47]. Consistent with the effects of these drugs, there is an increase in acetylated H3K9 and H3K14 around the *MIR224* locus. This observation was extended by a clinical study of 100 hepatocarcinoma tumor specimens, in which increased expression of miR-224 was caused by preferential binding of histone acetyl transference EP300, displacing HDAC1 binding in the *MIR224* chromosomal region [47].

Expression of miRNAs in the *MIR17–MIR92* gene cluster, also known as Oncomir-1 [48], was downregulated by several HDAC inhibitors (i.e., butyrate, SAHA, and TSA) in HT29 and HCT116 colorectal cancer cell lines [49]. Increased expression of PTEN, BCL2L11, and CDKN11A—known targets of the miRNAs in this cluster (miR-17, miR-18, miR-19a, miR-19b, miR-20a, and miR-92)—at least partially explains the decreased proliferation and increased apoptosis of these cell lines induced by HDAC inhibitors. Increased expression of other targets of individual miRNAs in the cluster also contributed to the observed molecular phenotype. This study did not characterize the epigenetic modification of the genomic region around the *MIR17–MIR92* gene cluster. It will be important to study whether treatment with HDAC inhibitors causes either local remodeling of the miR-17–miR-92 genomic cluster regions or altered expression of positive (e.g., c-Myc) or negative transcriptional regulators.

2.2.2 Epigenetic Modifications on Genomic Regions around MicroRNA Loci in Tumor Tissues

In several of the aforementioned studies, epigenetic characterization of highlighted miRNA loci has been extended to clinical specimens. These locus-specific studies have, in addition to genome-wide discovery studies, found prognostic value in the epigenetic signatures of several cancer types. These observations suggest that, in most contexts, there is a selection and evolution of altered epigenomes that globally silence expression of miRNAs with tumor-suppressive attributes.

Technical challenges have limited genome-wide studies of the epigenetic regulation of miRNA loci in cell lines and tissue specimens. One of these challenges is the identification of a transcription start site for an miRNA locus. The start site for a primary miRNA transcript can be up to 50 kb upstream of the pre-miRNA hairpin, and therefore the epigenetic status of the closest CpG island to the pre-miRNA may not influence transcription, or the presence of CpG island may not be required for promoter firing of some miRNA transcripts [50–52]. A second challenge is that a single miRNA may be transcribed from multiple promoters depending on cell type or physiological condition. The epigenetic status of the chromosomal regions surrounding a pre-miRNA may be relevant in certain circumstances [36,53]. Strategies to overcome these challenges include the use of RNA polymerase II binding regions, H3K4me3 open chromatin mark, cap analysis of gene expression, and the transcription start site sequencing library to identify likely miRNA promoter regions [19,51,52]. Because the detection of histone modifications requires fresh samples and a large number of cells for chromatin immunoprecipitation and downstream analysis, the characterization of histone modifications in miRNA promoter regions has been carried out in cancer cell lines, in cells isolated from hematological malignancies, or in normal counterparts such as B cells. In contrast, the characterization of DNA methylation changes requires a relatively small amount of starting material, so more is known about altered methylation patterns of miRNA loci.

2.2.2.1 Transcriptional and Epigenetic Regulation of miR-200 Family Members in Breast Cancer

miR-200 family members (miR-141, miR-200a, miR-200b, miR-200c, miR-429) are expressed in epithelial cells and help maintain an epithelial phenotype by inhibiting the expression of zinc finger transcriptional factors Zeb1/2, which in turn can bind to E boxes on promoter regions of *MIR200* gene loci and E-cadherin, repressing their transcription [54,55]. Loss of miR-200b expression (and that of other family members) facilitates the epithelial-to-mesenchymal transition (EMT), enhances the migratory and invasive features of cancer cells, and correlates with metastatic disease and the expansion of tumor-initiating cells [54,55]. In normal mammary gland epithelia, DNA is unmethylated in the promoter region of the *MIR200C* gene cluster (miR-200c–miR-141 transcript) and these miRNAs are thus expressed. Conversely, DNA is hypermethylated in this promoter region in mammary gland fibroblasts and these miRNAs are not expressed [56,57]. Similarly, the repressive chromatin mark of H3K27me3 in the promoter region of the *MIR200B* gene cluster (miR-200b–miR-200a–miR-429 transcript) results in no expression of these miRNAs in mammary fibroblasts.

Characterization of DNA methylation of the promoter regions of 93 miRNAs associated with breast cancer identified to two regions for the *MIR200B* cluster that were differentially methylated. DNA hypermethylation of either promoter was negatively correlated with miR-200b expression. DNA hypermethylation of promoter 1 (about 4 kb upstream of the miR-200b pre-miRNA) in metastatic lymph node sites was compared to matching primary tumors, whereas DNA hypermethylation of promoter 2 (about 2 kb of miR-200b pre-miRNA) in primary tumor tissues was associated with loss of the estrogen receptor or progesterone receptor. There was no prognostic association between the DNA methylation status of *MIR200B* promoter regions and the clinical outcome, which may not be unexpected in a small cohort of patients ($n = 26$). Another study identified an enhancer element 5.1 kb upstream of the transcription start site of the *MIR200B* gene cluster that is differentially methylated in breast epithelial and mesenchymal cells and may influence expression of the miR-200b–miR-200a–miR-429 transcript [58]. Together, these findings suggest that epigenetic silencing contributes to the regulation of miR-200b expression in breast cancer. Indeed, 5-Aza-CdR treatment in breast cancer cell lines reactivates expression of the miR-200c cluster [57]. It will be interesting to determine whether negative regulation by Zeb1/2 on the promoter regions of *MIR200B* and/or *MIR200B* (miR-200c–miR-141 transcript) clusters is also reinforced by repressive epigenetic marks such as DNA methylation, trimethylation or deacetylation of certain lysines in histone tails.

2.2.2.2 Epigenetic Signatures for Prognosis in Solid and Hematological Tumors

An epigenetic signature consisting of the promoter hypermethylation status of the *MIR9-3*, *MIR34B* cluster, and *MIR148* loci is associated with lymph node metastasis in breast, colorectal, and head and neck cancer, as well as in melanoma [59]. This was the first demonstration of a potential clinical application based on epigenetic modifications of miRNA loci. The promoter hypermethylation status of the *MIR9-1* and *MIR9-3* loci has been associated with shorter disease-free survival and risk of metastatic recurrence in clear cell renal cell carcinoma [60]. Studies on acute lymphoblastic leukemia (ALL) and lung cancer have associated other epigenetic signatures of miRNA loci with disease outcome.

Cancer cells can be collected at high purity from ALL cases and they can be used as input for chromatin immunoprecipitation analysis, which is still a challenge for most solid tumors. Using this approach, an epigenetic signature consisting of high numbers of H3K9me2 (a repressive mark) and/or low numbers of H3K4me3 (an active mark) in the CpG islands around 13 miRNA genomic regions was identified in ALL-derived cell lines [61]. The miRNA loci that constitute this epigenetic signature are *MIR9-1*, *MIR9-2*, *MIR9-3*, *MIR10B*, *MIR34B* cluster, *MIR124A-1*, *MIR124A-2*, *MIR124A-3*, *MIR132* cluster, *MIR196B*, and *MIR203*. These histone marks correlated with DNA methylation on the same chromosomal regions. Using DNA methylation as a proxy for overall chromatin status in a patient cohort of 353 ALL cases, DNA hypermethylation on any of these 13 miRNA loci was associated with shorter disease-free and overall survival [61]. A more detailed study of epigenetic modifications around all three independent *MIR124A* loci was conducted on the same ALL patient cohort [62]. There was a strong correlation between DNA hypermethylation status, decreased numbers of active marks (H3K4me3 and H3ac), and increased numbers of repressive marks (H3K9me2, H3K9me3, H3K27me3) in *MIR124A* genomic regions and the loss of miR-124a expression in ALL cancer cells [62]. Alone, DNA hypermethylation of the *MIR124A* loci was still an independent prognostic indicator of poor clinical outcome. The DNA hypermethylation status of all three *MIR124A* loci clearly separates normal healthy gastric tissues from gastric cancer tissues, noncancerous gastric tissues from cancer patients, and *Helicobacter pylori*-infected gastric tissues [63]. These results suggest that aberrant DNA methylation could be a useful marker for field defect and early disease detection.

CpG islands associated with the *MIR9-3* and *MIR193A* genomic regions were hypermethylated in non-small-cell lung cancer (NSCLC) tissues in a cohort of 101 cases [64]. DNA hypermethylation of a CpG island near the *MIR9-3* locus was associated with shorter overall survival only for the squamous cell carcinoma subtype [64]. Promoter hypermethylation of the *MIR34B* gene cluster correlated with shorter recurrence-free and overall survival in a cohort of 161 stage I NSCLC cases [65]. The expression levels of miR-34b and miR-34c did not correlate with promoter methylation status and had no prognostic value [65]. An independent study validated the prognostic value of promoter hypermethylation in the *MIR34B* gene cluster in a cohort of 140 early stage (stage I/II) lung adenocarcinoma cases [66], and another study found that promoter hypermethylation of that cluster correlated with metastatic disease in lymph nodes in a cohort of 99 NSCLC cases [67]. The latter study did not report survival analysis based on methylation changes, but association with lymph node involvement was a prognostic factor for poor clinical outcome. DNA hypermethylation around the *MIR127* locus was associated with shorter overall survival in a cohort of 97 NSCLC cases [68].

Somewhat puzzling results were reported for let-7a-3 in ovarian cancer and NSCLC patient cohorts [69,70]. let-7a and other family members are known to act as tumor suppressors by inhibiting the expression of Ras, c-Myc, HMGA2, and other oncogenic proteins [71,72]. Expression profiling of let-7 in lung and other cancers found that low levels of let-7 expression correlated with poor outcome [73,74]. Nonetheless, these two studies found that DNA hypermethylation around the *LET7A-3* locus was associated with better clinical outcome, suggesting repression of an oncogenic activity. In the ovarian cancer study, this hypermethylation did not correlate with loss of let-7 expression but rather with decreased expression of insulin-like growth factor-II at the mRNA and protein levels [70]. Thus, the characterized genomic region may not contain a promoter for let-7, and the methylation status of this region may not be linked to let-7a expression. By extension, caution should be exercised when interpreting changes in DNA methylation status in CpG islands or chromatin features in genomic regions around miRNA loci. Characterization of epigenetic modification should be accompanied by expression analysis to determine the functional link between them.

2.3 MicroRNA-Mediated Regulation of Epigenetic Machinery

miR-29 family members were the first miRNAs described as regulating DNMT3A/B, and they established a precedent for a functional class of miRNAs called epi-miRNAs [75], which regulate expression of epigenetic effectors. Epi-miRNAs can regulate one or more epigenetic efforts and can indirectly alter the epigenetic landscape (Figure 2(B) and (C)). In some instances, these epigenetic changes can lead to expression reactivation or to the silencing of key ncRNAs or tumor suppressor genes that contribute significantly to observed molecular phenotypes.

2.3.1 *miR-29 Family Members Regulate DNMT3 Expression*

In the original study by Fabbri and colleagues, an inverse correlation of miR-29 family members (miR-29a, miR-29b, miR-29c) and DNMT3A expression in lung cancer tissues led them to investigate the mechanistic link involving the DNMT enzymes DNMT1, DNMT3A, and DNMT3B [76]. Enforced expression of any of the miR-29 family members was sufficient to interact with binding sites on the 3' UTRs of *DNMT3A* and *DNMT3B* mRNAs in lung cancer cell lines. Consistent with these regulatory interactions, enforced expression of a miR-29 family member reduced genome-wide DNA methylation patterns. The most robust effect was observed with enforced expression of miR-29b, which reduced genome-wide DNA methylation to levels comparable to that after treatment with the DNMT inhibitor 5-Aza-CdR. Expression of the tumor suppressor genes *FHIT* and *WWOX* was increased after enforced expression of miR-29b or other miR-29s. These tumor suppressor genes are frequently silenced by promoter methylation. Thus, miR-29-mediated regulation of DNMT3A and DNMT3B appears to normalize the epigenetic landscape in lung cancer, including reactivation of tumor suppressor genes (Figure 2(B)). Enforced expression of miR-29s inhibits the growth of in vitro and in vivo models of lung cancer, presumably by engaging DNMT3A and DNMT3B, but probably also via other targets having oncogenic functions.

Several follow-up papers reinforced the idea that the miR-29/DNMT3 axis is engaged in other cancer types and has important phenotypic consequences via local and global epigenetic alterations. Enforced expression of miR-29b inhibited DNMT3A and DNMT3B expression via binding sites on the 3′ UTR of their respective mRNAs in cell lines and primary blasts from acute myeloid leukemia (AML) cases [77]. miR-29b also affected the expression of DNMT1 indirectly by inhibiting the expression of Sp1, a zinc finger transcription factor that activates DNMT1 expression. Thus, miR-29b modulates expression of all DNMTs in AML, reducing genome-wide DNA methylation and triggering reactivation of tumor-suppressive p15INK4b and estrogen receptor-1, which were silenced by promoter hypermethylation. The methylation status of these genes correlated with poor clinical outcome in AML patient cohorts [78]. Similar regulatory effects were uncovered for miR-29a in hepatocarcinoma cell lines, in which enforced expression of miR-29a resulted in reexpression of the lncRNA maternally expressed gene 3, which has tumor-suppressive properties [79]. The expression of miR-29c and DNMT3B was inversely correlated during melanoma progression [80]. Higher levels of miR-29c were associated with better disease outcome in stage III melanoma cases, and virtually identical survival curves were obtained for cases having low expression of DNMT3B [80]. In multiple myeloma cell lines and xenograft mouse models, synthetic miR-29b mimetic compounds that replenish miR-29 activity reduce protein output of DNMT3A/B and induce global DNA demethylation changes [81]. In xenograft models, miR-29 synthetic compound has a potent antitumoral activity [81]. In a follow-up study, miR-29b-induced promoter demethylation and derepression of suppressor of cytokine signaling-1 (SOCS-1) gene expression was identified as an important molecular consequence of miR-29 mimetic therapy [82]. SOCS-1 activity inhibited migration and invasiveness of multiple myeloma cancer cells [82].

Other targets of miR-29b have also been described that can be more important than DNMT3A/B or contribute in parallel to the indirect effects of epigenetic reprogramming triggered by miR-29-mediated regulation of DNMT3A/B. Antiapoptotic myeloid cell leukemia-1 and BCL-2 have been identified as gene targets of miR-29s in different disease contexts, including AML and liver cancer [83,84]. As mentioned above, miR-29s can regulate the expression of BACE1 and SPTLC2, which are etiologically relevant proteins in Alzheimer's disease. In cell lines and mouse models of breast cancer, miR-29b inhibits a metastatic program by targeting expression of a protein network involved in angiogenesis and extracellular matrix remodeling, including VEGFA, ANGPTL4, MMP9, and TGFβ [85]. It is an intriguing possibility that epigenetic changes reinforce posttranscriptional changes imposed by miR-29, or vice versa. It will be important to investigate this potential regulatory mechanism in different disease-specific settings.

2.3.2 miR-22 Regulates Ten–Eleven Translocation-2 Expression

Myelodysplastic syndrome (MDS) and other hematopoietic malignancies, including leukemia, are characterized by epigenetic changes, and specific aberrations have been linked to poor clinical outcome. Mutations in epigenetic effectors, including ten–eleven translocation gene 2 (TET2), have been casually linked to malignancy. TET2 catalyzes the oxidation of 5-methylcytosine (5mC) to 5-hydroxymethylcytosine (5hmC), an intermediate in active DNA demethylation opposing the roles of DNMTs. miR-22-mediated regulation of TET2 expression largely explained why overexpression of miR-22 in the mouse hematopoietic compartment triggered an MDS-like phenotype [86]. miR-22 expression was upregulated in MDS and leukemias [86], and inhibition of miR-22 activity in leukemia cell lines reduced their proliferative capacity. This effect was associated with increased expression of TET2 and of some of its targets, such as absent in melanoma-2 and SP140 nuclear body protein, presumably by decreasing DNA methylation in their promoter regions [86].

The same group investigated the miR-22/TET2 axis in breast cancer. Overexpression of miR-22 in the mammary epithelia compartment caused enhanced cancer cell aggressiveness and increased metastasis to the lungs in mouse models [87]. In these models, miR-22 targeting of TET2 caused promoter hypermethylation (assayed only for the *MIR200C* gene cluster) and silencing of miR-200a and miR-200c expression, as well as a concomitant increase in Zeb1/2. These altered epigenetic and transcriptional states promoted EMT, cancer stem cell expansion, and prometastatic programs (Figure 1(B)). Treatment with 5-Aza-CR opposed miR-22-induced epigenetic silencing and was able to reactivate miR-200 expression. In clinical samples, overexpression of miR-22 correlated with poor clinical outcome and with loss of TET2 and miR-200 expression [87].

Other studies in breast cancer and gastric cancer have suggested a tumor-suppressive role for miR-22 [88,89]. Curiously, miR-22 can inhibit expression of Sp1, a transactivator of DNMT1 expression, in MDA-MB-231 and BT-549 breast cancer cell lines [88]. miR-22 repressed expression of prometastatic CD147 directly posttranscriptionally and indirectly by reducing the protein output of Sp1, which is required for increased

CD147 expression [88]. While this study did not directly measure DNMT1 expression, it seems plausible that in different contexts miR-22 could affect the epigenetic landscape via modulation of TET2 and DNTM1 to create environments that are more prone to cellular transformation and invasion. In the MGC-803 gastric cancer cell line, miR-22 and miR-200b cooperatively exert tumor-suppressive activities via inhibition of Wnt-1 expression [89].

2.3.3 miR-34 Regulates HDAC1 Expression

p53-induced transcriptional activation of miR-34 expression is an important regulatory cue for miR-34 tumor-suppressive activities (Figure 2(C)). Upregulation of miR-34 expression in turn can amplify p53 expression by targeting negative regulators of p53 expression or activity [90–92], which include effectors of the epigenetic machinery. miR-34a inhibited expression of SIRT1 (silent information regulator 1) posttranscriptionally in HCT116 colorectal cancer cells [91]. SIRT1 deacetylates histones and DNMTs. SIRT1 also deacetylates p53, lessening its ability to activate transcription of its target genes, p21 and Puma, among others [91]. This study did not assess whether miR-34a-mediated loss of SIRT1 had an effect on histone markers or DNMT activity. In another study, enforced expression of miR-34b in PC3 and LNCaP prostate cancer cells caused downregulation of the expression of DNMT1, DNMT3b, HDAC1, HDAC2, HDAC3, and HDAC4 proteins [92]. miR-34b binding sites were identified on the 3′ UTRs of all these epigenetic effectors' mRNAs, suggesting that miR-34b could directly inhibit their expression. This collective targeting of the epigenetic machinery reshaped the miR-34b genomic region, moving from a hypermethylated DNA state to a partially demethylated DNA state and histone modifications consistent with open chromatin conformation. Therefore, it appears that miR-34b and these epigenetic effectors reciprocally inhibited each other's expression.

Other studies have suggested that such regulatory feedback loops need not be fully engaged for p53 to function independent of miR-34 [93], or vice versa [94]. Replenishing the expression of individual miR-34 family members with mimetic compounds inhibited proliferation of isogenic breast cancer and colorectal cancer cells with or without functional p53 activity [94]. In the absence of p53, expression of p21 was required for miR-34-mediated effects in these cell line models. Interestingly, miR-34 inhibited expression of HDAC1, which led to p21 reexpression. Treatment with TSA (an HDAC inhibitor) recapitulated miR-34-mediated induction of p21 expression [94]. Local chromatin remodeling of the p21 genomic regions appears to be key for miR-34-mediated regulation; lowering of p21 expression by siRNA knockdown negated the antiproliferative effects induced by miR-34 mimetics.

3. EPIGENETIC ALTERATIONS AND REGULATION OF TRINUCLEOTIDE REPEATS IN NEURODEGENERATIVE DISEASES

Neurodegenerative diseases are caused by a vast array of genetic, epigenetic, and environmental factors. Within this field, ncRNAs are becoming increasingly interesting. ncRNAs are highly expressed in the brain; many neurological diseases involve expansions of TNRs that can act as ncRNAs in pathogenesis. In fact, more than 30 neurological/neuromuscular diseases are caused by TNR expansions, which induce gain- or loss-of-function mutations within the CNS. The pathogenic expansions can occur within or outside of the protein-coding region of a gene. Pathogenic RNAs can cause disease through both pathogenic protein translation and aberrant ncRNA transcription, which can induce autorepression, recruit epigenetic machinery such as HDACs, or recruit chromatin modifiers (Figure 3). In this section we discuss the proposed mechanisms by which ncRNA-mediated pathogenesis causes disease in TNR expansion-driven diseases.

3.1 Evidence for Toxic Noncoding RNAs as Drivers in Trinucleotide Repeat Expansion Disease

The first evidence of pathogenic ncRNAs in TNR-driven disease was found in myotonic dystrophy (DM). DM is an autosomal-dominant neurological disease caused by a CTG expansion in the 3′ UTR of the myotonic dystrophy protein kinase gene (*DMPK*), which produces long CUG RNA repeats. DM is characterized by both cerebral neurodegeneration and muscle atrophy. Subsequently, other autosomal-dominant TNR-driven diseases, such as HD and spinocerebellar ataxia type 8 (SCA8), have been shown to produce an ncRNA that interferes with cellular processes. HD is an autosomal-dominant inherited disease caused by abnormal expansion of a CAG repeat in *HTT*. This pathogenic TNR expansion produces a mutant gain-of-function HTT protein that interacts with multiple processes of the cell, leading to neurodegeneration of the striatum (Figure 3(A)). SCA8 is an autosomal-dominant degenerative disease of the cerebellum caused by bidirectional expansion of CAG/CTG repeats in the *ATXN8OS* gene, which also produce pathogenic CUG RNA repeats.

Autosomal-recessive diseases such as fragile X syndrome (FXS) and Friedreich's ataxia (FRDA) also exhibit pathogenesis at the RNA level. FRDA is the most common autosomal-recessive ataxia, caused by an expansion of the GAA repeat in intron 1 of the frataxin gene. This expansion reduces the level of the frataxin protein, which leads to degeneration of central and peripheral neurons and of the heart. Similarly, FXS is an

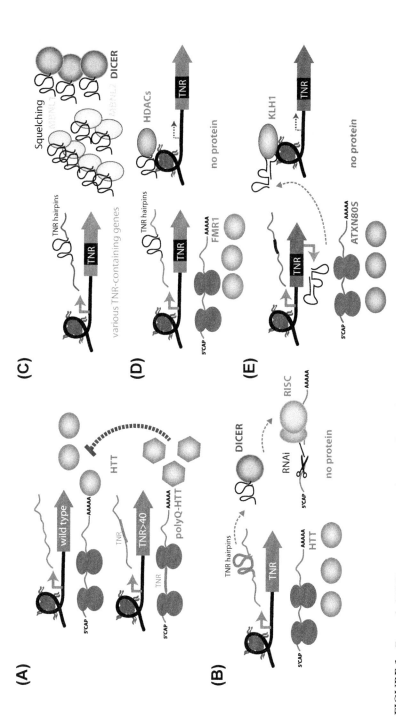

FIGURE 3 Expanded TNRs can cause neurodegeneration at the protein and RNA levels. (A) In some diseases such as Huntington's disease, the TNR codes for amino acids and can be translated to form misfolded, aggregate-prone proteins that interfere with cellular functions. (B) However, the TNR in *HTT* mRNA can also function as ncRNA hairpins that are cleaved by Dicer and enter the RNAi pathway to knock down its own transcripts. (C) TNR ncRNA hairpins can also bind to and squelch proteins such as MBNL1 and MBNL2 (in the nucleus) as well as Dicer, in the latter case indirectly reducing global miRNA processing and activity. (D) TNR ncRNA can duplex with DNA once transcribed, acting as a guide for HDACs to silence gene promoters. (E) Antisense transcription of the TNRs can bind to chromatin modifiers and scaffolding proteins such as KLH1 and CBP.

autosomal-recessive disease caused by an abnormal expansion of the CGG trinucleotide on the fragile X mental retardation 1 (*FMR1*) gene. Silencing of *FMR1* causes little or no FMR1 protein to be made, which disrupts the nervous system and causes physical and cognitive abnormalities. While the genetic cause of these diseases is known, the mechanism of silencing is still debated.

3.2 Mechanisms of Trinucleotide Repeat Expansion Pathogenicity

The first proposed mechanism for ncRNA effects came from research into myotonic dystrophy RNA [95]. Because the CTG expansion is outside the translated region of the gene, researchers were curious about the mechanism of pathogenesis and discovered aggregated RNA foci containing the CUG expansion inside the nucleus of affected cells [96]. These RNA-expansion foci were subsequently found in SCA8, FXS, HD, and other TNR-driven diseases [97]. Models of TNR-driven diseases in invertebrate model organisms such as *Caenorhabditis elegans* display toxicity at the RNA level [98]. Additionally, an antisense transcript was found in affected nuclei of Friedreich's ataxia cells, suggesting an alternative mechanism [99].

3.2.1 RNA Hairpins

Several studies have found that the RNAs harboring TNR expansion in DM and SCA8 are capable of forming extended hairpin structures that interact with proteins and DNA sequences [100–102]. Dicer can also interact with these structures (Figure 3(C)). In addition to cleaving mature miRNA from pre-miRNA hairpins, Dicer recognizes longer double-stranded RNA sequences and cleaves the structure into small interfering RNAs (siRNAs), inducing the RNA interference pathway [102–104]. SCA8-derived TNR-containing hairpins can be processed by Dicer and serve as siRNA guides to cleave SCA8 mRNA, decreasing wild-type as well as mutant SCA8 mRNA and protein levels. The hairpins formed as a result of CUG-repeat RNAs have been shown to sequester protein kinase R (PKR) in a repeat-length-dependent manner [101]. PKR is activated by the binding of double-stranded RNA and functions in the stress response pathway [105]. Loss of this protein may suppress the inflammatory stress response pathway in affected neurons, contributing to pathogenesis.

3.2.2 Protein Sequestration

ncRNAs may function in the sequestration of gene-regulatory proteins in the nucleus (Figure 3(C)). This has been shown most prominently with the dysregulation of MBNL1 and MBNL2 (muscleblind-like RNA-binding proteins). MBNL proteins were first discovered as factors that bind to

CUG-repeat RNAs and these RNA–protein complexes accumulate in nuclear foci of patients with DM [106]. Via this interaction, the expanded TNR RNAs sequester MBNL proteins, leading to loss of function of MBNL1 and MBNL2 and deregulated splicing of neuronal proteins in DM patients [107,108]. There is also evidence for aberrant splicing and protein sequestration in HD and FXS. A study has found aberrant splicing of the *HTT* in HD, possibly due to the binding of SRSF6 (splicing protein serine/arginine-rich splicing factor 6) to CAG/CUG repeats [109]. In FXS, mass spectrometry analysis identified MBNL1, hnRNP A2/B1, and lamin A/C proteins in RNA aggregates. In addition, RNA affinity pull down analysis identified binding factors such as Sam68, Purα, DGCR8, and Drosha [110–112]. DGCR8 and Drosha are proteins involved in the processing of miR-NAs. Drosha and DGCR8 are sequestered by the pathogenic RNA hairpin structures within the nucleus, which decreases the processing of pri-miR-NAs to pre-miRNAs and thus reduces the amount of mature miRNAs in cells [113]. In HD, decreased Drosha activity and subsequent miRNA dysregulation leads to the expression of genes normally repressed by miR-NAs [114]. Additionally, CAG repeats can be recognized and cleaved by Dicer. Association of Dicer with expanded TNR RNAs squelches Dicer activity required for other RNA substrates, including miRNA processing [104].

3.2.3 RNA/DNA Duplexes

An alternative mechanism to ncRNA toxic fragmentation is ncRNA/DNA duplexing. Noncoding RNA can bind to complementary sequences on DNA and silence gene transcription (Figure 3(D)). The structure of CUG-repeat ncRNA can form short duplexes on itself and on CAG DNA in vitro [115]. This structure demonstrates the possibility that ncRNAs with CUG expansion can spontaneously bind and silence CAG DNA in vivo. In addition, work has provided evidence for autosilencing of the *FMR1* gene by the complementary sequence on RNA CGG repeats. *FMR1* ncRNA has been shown to fold back onto and interact with the adjacent complementary DNA sequence to induce methylation and silencing of the *FMR1* gene promoter region, which ultimately leads to decreased protein production [116].

3.2.4 Antisense Transcription

The initial studies of SCA8 led to the discovery of antisense transcription in disease pathogenesis [117]. In SCA8, the sense strand of the *ATXN8OS* gene is transcribed and translated into the pathogenic protein ATXN8, while the antisense strand is transcribed and produces CUG-expanded ncRNA, which is also pathogenic [118]. This antisense transcript has the ability to bind nuclear proteins, squelching protein activity in the cell. An antisense transcript for HD (*HTTAS*) has been described, suggesting an additional source for pathogenic RNA repeats [119]. In FXS, two antisense transcripts have been described at the *FMR1* locus. Antisense *FMR1* (*ASFMR1*) encodes two ncRNAs that may function independently to silence and signal antiapoptotic pathways [120,121]. The mechanism by which these antisense strands avoid degradation is not yet known.

4. LONG NONCODING RNAs GUIDE THE EPIGENETIC MACHINERY AND INTERFERE WITH OTHER NONCODING RNA FUNCTIONS TO REGULATE GENE EXPRESSION

Another class of ncRNAs consists of transcripts longer than 200 nucleotides, termed long noncoding RNAs. Thousands of lncRNAs have been characterized in the human genome [122]. Most lncRNAs do not exhibit high sequence conservation relative to protein-coding sequences; the majority of identified lncRNAs appear to have evolved recently [123]. Although the lack of strong evolutionary sequence conservation may raise doubts about their biological significance, lncRNAs have a striking tendency to be composed of two exons [122], indicating evolutionary conservation in their structure. A large body of evidence demonstrates a functional role for a number of lncRNAs in the regulation of gene expression, including X-chromosome inactivation (Xist, Tsix, Jpx) [124–126], genomic imprinting [127,128], directing chromatin modifiers to specific target loci [129], regulation of DNA methylation [130], and as a decoy for miRNAs [131,132]. Furthermore, identification of cancer-risk loci by genome-wide association studies indicates that some 80% of cancer-risk loci are located within noncoding sequences [133]. Genetic and expression variants of the lncRNAs are increasingly found to contribute to human disease. The lncRNAs, ANRIL in particular, were found in a hot spot for gliomas and basal cell carcinomas [134]. Genetic variants of ANRIL are also linked to the risk of coronary artery disease and diabetes [135]. A subclass of lncRNAs is encoded in ultraconserved regions (UCRs) that are 100% identical in humans, rats, and mice [136] and are termed transcribed UCRs (T-UCRs). These T-UCR lncRNAs have altered expression in leukemia and certain cancers [137]. Furthermore, particular T-UCRs can function as tumor suppressors and may be aberrantly silenced via epigenetic mechanisms such as DNA hypermethylation [138].

4.1 Paradigms of Long Noncoding RNA Function in the Regulation of Gene Expression

There is growing evidence that certain risk loci are transcribed into lncRNAs; these lncRNAs function via a variety of mechanisms to regulate the expression of target genes both *in cis* and *trans* (Figure 4(A)).

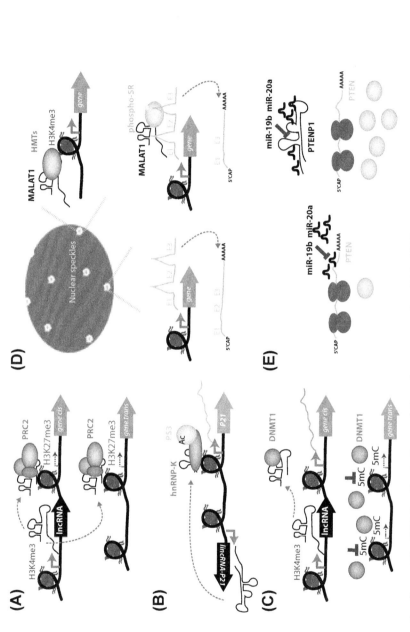

FIGURE 4 Paradigms of lncRNA function in the regulation of gene expression. (A) lncRNAs can act as guides, recruiting epigenetic modifiers such as PRC2 to target genes either in cis or trans. (B) lncRNA can function as a transcriptional coactivator (or corepressor, not shown). (C) LncRNAs sequester DNMT1, preventing methylation at actively transcribed loci. (D) lncRNAs are involved in localization of genetic loci to specific regions of the nucleus (nuclear speckles) where activating epigenetic marks are enforced. lncRNAs can also affect the distribution of splicing factors, phosphorylation of SR protein splicing factors, and selection of alternative splice site. (E) PTENP1 expression regulates PTEN expression by sequestration of PTEN-targeting miRNAs.

4.1.1 Long Noncoding RNAs as Guides for the Epigenetic Machinery

A number of lncRNAs have been found to act as guides to recruit chromatin-modifying complexes to target genes and thereby regulate epigenetic control of gene expression. The archetypical lncRNA that employs this mechanism is HOTAIR (Figure 1(D)), which functions *in trans* to silence HOXD expression via the interaction and recruitment of PRC2, an epigenetic effector that directs methylation of lysine residues on histone H3 [129]. Other lncRNAs have also been shown to interact and recruit chromatin-modifying complexes [139], including recruitment of PRC2 by Xist [140] and PRC2 and PRC1 by ANRIL [141,142].

Large intergenic noncoding RNAs (lincRNAs) are a subclass of lncRNAs defined by their genomic location between two protein-encoding genes. Most lincRNAs bind to chromatin modifying complexes, with an estimated 20% of lincRNAs binding to PRC2 [139]. Unlike most lincRNAs, murine lincRNA-p21 regulates a set of PRC2 target genes without directly interacting with PRC2. The proapoptotic lincRNA-p21 regulates the expression of genes in response to p53 by functioning as a transcriptional corepressor [143] or coactivator [144]. p53 and lincRNA-p21 coactivate p21 expression *in cis* at the p21 promoter. This activation is also facilitated by the RNA-binding protein hnRNP-K (Figure 4(B)) and results in global downstream gene expression changes in a number of genes involved in response to DNA damage-induced p53 activation, G1/S checkpoint programs, and cellular differentiation programs [144]. Even though these downstream genes are targeted by PRC2 in other contexts, lincRNA-p21 is not functioning as a guide and has no direct involvement in PRC2 recruitment to these loci.

lncRNAs can also function as regulators of other epigenetic modifications such as DNA methylation. A same-sense nonpolyadenylated lncRNA is transcribed concomitant with the protein-coding message from the CEPBA locus, which binds and blocks the epigenetic effector DNMT1 such that the act of transcription blocks promoter DNA methylation (Figure 4(C)). It was also found that many other loci also share an lncRNA-mediated mechanism of DNMT1 regulation [130].

The lncRNA MALAT1 has a role as a guide in the localization of growth control genes to nuclear speckles (interchromatin granules or sites of active transcription), where they can receive activating epigenetic marks [145] (Figure 4(D)). Because nuclear speckles are also sites enriched in pre-RNA splicing factors, MALAT1 could regulate alternative splicing of a number of pre-mRNAs by regulating the localization of splicing factors and the phospho-dependent activity of SR protein splicing factors [146]. Therefore, lncRNAs are involved not only in transcriptional regulation but also in posttranscriptional processes.

4.1.2 Long Noncoding RNA Interferes with Other Noncoding RNA Functions

lncRNAs can also function as interferers to regulate other ncRNAs such as miRNAs. Some lncRNAs can function as competing endogenous RNAs (ceRNAs), acting as "sponges" or decoys for miRNAs, preventing them from hitting their target mRNAs (Figure 4(E)). PTEN expression is regulated by a lncRNA encoded by its paralog, the pseudogene *PTENP1*, which is located on a different chromosome. This lncRNA serves as a decoy to prevent miRNA targeting of the PTEN transcript [132]. Therefore, epigenetic dysregulation at the distant *PTENP1* locus can affect PTEN expression. Aberrant DNA methylation at the *PTENP1* locus has been proposed to result in downregulation of PTEN in cases of endometrial cancer even though the *PTEN* locus displayed no aberrant methylation [147]. ceRNAs from antisense transcripts have also been described [148] and may be subjected to epigenetic regulation. These examples help to establish the biological relevance of pseudogenes and antisense transcripts, which at one time were considered junk or transcriptional noise. In certain contexts, such as an in vivo liver model of a miR-122 target, it seems unlikely that ceRNAs would achieve high enough levels to function as an effective decoy [149].

4.2 Long Noncoding RNAs are Associated with Disease Risk

There is growing evidence for the involvement of lncRNAs in a number of human diseases including cancer [2]. Variants in the nucleotide sequence at lncRNA loci and aberrant expression of lncRNAs have been reported in human cancers. These findings further dispel the notion of junk DNA and substantiate that regions of the genome that do not code for proteins make significant contributions to the risk, development, and progression of human diseases. While genetic variants have also been associated with miRNA loci and cancer, we focused here exclusively on the genetic variants of lncRNAs that directly interfere with their role of guides of epigenetic machinery.

Allele variants associated with a higher risk of developing cancer have been identified in several lncRNAs: ENST0000051508 in breast cancer [150], PTCSC3 in papillary thyroid carcinoma [151], CCAT2 in colorectal cancer [152], and RERT in hepatocellular carcinoma [153]. The chromosome 9p21 locus is a hot spot for genetic variants associated with diseases including cardiovascular disease; type 2 diabetes [135]; and cancers including glioma, basal cell carcinoma, nasopharyngeal carcinoma, and breast cancer [134]. Risk loci for cardiovascular disease, diabetes, melanoma, and glioma are all associated with allelic expression of ANRIL [154]. ANRIL is transcribed from the antisense strand of the $p15^{INK4b}$–$p14^{ARF}$–$p16^{INK4a}$

gene cluster on the 9p21 chromosomal region and it recruits polycomb repressive complexes to this locus to silence p16[INK4A] via PRC1 [142] and p15[INK4B] via PRC2 [141]. ANRIL may represent the archetypical *cis*-acting lncRNA in regulating the expression of cell cycle regulators by acting as a guide to recruit epigenetic effectors.

Elevated expression of ANRIL is associated with prostate cancer [142] and gastric cancer [150], reinforcing the negative regulatory role of ANRIL in tumor suppressor gene expression. ANRIL may regulate other target genes *in trans* since ANRIL can silence p15[INK4B] via heterochromatin formation and DNA methylation when overexpressed from a distant site [155]. Altered expression of other lncRNAs in cancer has been frequently reported. Aberrant upregulation of HOTAIR has been implicated in gastric cancer [156], non-small-cell lung cancer [157], metastasis in estrogen receptor-positive primary breast cancer [158], colorectal cancer [159], and renal carcinoma [160]. HOTAIR is negatively regulated by miR-141, a miR-200 family member, in renal carcinoma [160]. Aberrant expression of MALAT1 promotes lung cancer metastasis [161].

Xist, one of the first lncRNAs to be discovered, has an essential role in X-chromosome inactivation and functions as a guide for polycomb repressive complex recruitment [162]. Xist has been postulated to have a role in cancer, as the Barr body is frequently lost in breast cancer, and it is proposed that reactivation of genes on the inactive X chromosome may contribute to tumorigenesis. Downregulation of Xist expression has been observed in ovarian cancer cell lines and deletion of Xist in mice leads to highly aggressive hematologic cancer [163].

5. CONCLUSION

Mutated or dysregulated ncRNAs are emerging as major players in human disease. There is mounting evidence that the main function of many ncRNAs is linked to epigenetic processes, as regulators of epigenetic machinery and/or as targets of epigenetically regulated processes. DNA variants within ncRNA sequences can affect ncRNA functions by diminishing their processing, RNA stability, or binding affinity to their mRNA, DNA, or protein targets. DNA variants that result in disruption of binding sites on the target mRNA, DNA, or protein would similarly reduce the regulatory capacity of ncRNAs. Expression variants of ncRNA can increase their regulatory potency when overexpressed or decrease their regulatory potential when downregulated or silenced. Locus-specific and global epigenetic modifications can affect the expression of ncRNAs and their functions. In some instances, epigenetic modifications can be the main regulatory mechanism for silencing ncRNA expression, while in other instances they can reinforce signaling and transcriptional inputs. Therefore, detailed epigenetic characterizations of ncRNA loci can be used to understand their regulatory context during disease and may be used as biomarkers to interrogate disease status. For certain applications, epigenetic signatures may provide more specific information on the status of diseased cells and aberrant molecular pathways than ncRNA expression alone. These signatures may also highlight cases that would be more responsive to treatment with epigenetic drugs. In this context, reexpression of key ncRNAs could be useful as a pharmacodynamic marker to monitor whether treatment is working.

5.1 MicroRNA Biomarkers

There have been numerous studies in recent years characterizing miRNA expression in blood and other fluids as a means of providing new diagnostic tools. Blood or cerebrospinal fluid samples can be obtained by relatively noninvasive means from patients with neurodegenerative conditions or individuals at high risk for developing AD, PD, or HD. Early studies suggest that miRNA expression signatures can separate AD and PD cases from normal controls. However, these signatures vary between studies and there is overlap with the miRNA signatures from brain tumors when using similar experimental approaches. Thus, some of these changes may represent more general processes such as cellular stress, neurotoxicity, or neural cell death rather than disease-associated molecular changes. Moreover, the cellular source of the miRNAs in circulation is not known; confounding biological processes such as reactive stroma and inflammation can have similar manifestations in advanced stages of neurodegenerative conditions and cancer. In this context, epigenetic signatures that reflect changes within specific cell types could help in interpreting miRNA expression changes.

Tissue studies of neurodegenerative conditions are mainly limited to postmortem examination, but they are readily available for most cancer types, so that pretreatment and posttreatment samples can be compared. Epigenetic signatures can predict risk of metastasis and clinical outcome for various solid tumors, including breast, bladder, colorectal, and lung cancers [59,60,64–68]. Intratumoral and intertumoral tissue heterogeneity within cancer cells, reactive stroma, and immune cell infiltrate is still an important confounding factor when interpreting miRNA expression or epigenetic signatures. Interestingly, some of these studies also suggest that epigenetic changes may be more informative than expression changes of the involved miRNAs. This raises the intriguing possibility that epigenetic signatures based on miRNA may also be useful in understanding the clonal selection and evolution of cancer cells better fitted to specific microenvironments and challenges. Characterization of epigenetic and expression miRNA signatures in tumor tissues before and after chemotherapy or other treatments could better suggest which cases would be likely to respond to a specific treatment.

5.2 Trinucleotide Repeat Biomarkers

It is likely that multiple processes contribute to TNR expansion diseases, including both pathogenic protein translation and dysregulation of cellular processes through toxic RNAs. While these diseases all have the TNR expansion in common, the mechanisms of gain and loss of function appear to differ. For example, in FXS, an antisense lncRNA is transcribed from the *FMR1* gene if it has fewer than 200 TNRs (premutation carriers), but it is not transcribed from the gene if it has more than 200 TNRs (full-mutation stage) [120]. This suggests that the mechanism of aberrant gene regulation by the TNR expansion differs between the pre- and the full-mutation disease stage and that therapeutics could be developed targeting these separate mechanisms. In addition, miRNA profiling in brain tissues from HD patients has revealed differential expression between stages of the disease [164]. This may represent a greater squelching of Dicer or an increase in toxic fragmentation of the expanded transcript. In these instances, simply increasing Dicer levels would be a good target for the first proposed mechanism but not for the latter. This may, in part, explain why current experimental treatments have failed. In the latter case, more precise therapies could be applied to increase the expression of miRNAs that show lower expression as the disease progresses. Further understanding of such mechanisms could lead to therapeutics designed at both the RNA and the protein levels.

5.3 Long Noncoding RNA Biomarkers

Both genetic and expression variants of lncRNAs not only elucidate potential therapeutic targets in cancer but also may serve as effective biomarkers of risk and prognosis. HOTAIR and ANRIL are upregulated in a significant number of cancers and have value as biomarkers, because circulating RNA molecules can be readily and noninvasively isolated from blood and other biological fluids. MALAT1 is currently a candidate as a blood-based biomarker for the diagnosis of non-small-cell lung cancer [165] and prostate cancer [166]. As lncRNAs are expressed in a tissue-specific manner—more so than protein-coding genes—they may be useful in identifying specific subpopulations within a tumor such as cancer stem cells [133].

Glossary

Chromatin A dynamic complex of nucleic acid and protein in the nuclei of eukaryotic cells.
cis **interaction** The case in which a regulatory factor is encoded at the same locus as the gene it regulates.
CpG site A cytidine 5′ to a guanosine on a strand of DNA; the primary site of DNA methylation in mammals.

CpG island A region of >200 bp with a high frequency of CpG sites relative to the rest of the genome.

Dicer An RNA endonuclease that cleaves pre-miRNA hairpins into mature miRNAs and double-stranded RNA into shorter double strands.

DNA methylation An epigenetic mark involving addition of a methyl group to the 5-carbon of cytidine by DNA methyltransferases to produce 5-methylcytosine.

DNA methyltransferase A family of enzymes that catalyze the addition of a methyl group to the 5-carbon of cytidine.

Enforced expression Ectopic expression or overexpression of an miRNA minigene in cells after plasmid transfection or viral transduction using a heterologous system.

Epigenetic Heritable information within chromatin that is not encoded by the DNA sequence.

Genome-wide association studies Genome-wide studies of genetic variants and their relation to disease.

Histone acetyltransferase An enzyme that catalyzes the addition of an acetyl group to lysine residues on histones.

Histone deacetylase An enzyme that catalyzes the removal of an acetyl group from acetylated lysine residues on histones.

Long noncoding RNAs Noncoding RNAs greater than 200 nt in length.

MicroRNAs Small noncoding RNAs that generally guide (mi)RISC to mRNA for RNA degradation or translational repression of target genes.

Noncoding RNAs RNA molecules that are transcribed from a gene but are not translated into protein.

Polycomb repressive complex A multiprotein complex that facilitates repression of target loci by catalyzing posttranslational modification of histone amino acids.

RNA-induced silencing complex A multiprotein complex that facilitates the silencing effect of small RNAs such as siRNA and miRNA.

Ten–eleven translocation A family of enzymes that catalyze the sequential oxidation of the methyl group of 5-methylcytosine (5mC) first to 5-hydroxymethylcytosine and then 5-formylcytosine and 5-carboxylcytosine.

Trans **interaction** The case in which a regulatory factor is encoded at a locus separate from the gene it regulates.

Trinucleotide repeats Repeats of a 3-nt sequence that undergo aberrant expansion; found in both polyglutamine and nonpolyglutamine disorders.

X chromosome inactivation The process by which one X chromosome is inactivated in female mammals.

LIST OF ACRONYMS AND ABBREVIATIONS

AD	Alzheimer's disease
CNS	Central nervous system
HD	Huntington's disease
miRNA	MicroRNA
kb	kilobase (pairs)
mRNA	Messenger RNA
ncRNA	Noncoding RNA
Nt	Nucleotide

lncRNA	Long noncoding RNA
PD	Parkinson's disease
siRNA	Small interfering RNA
TNR	Trinucleotide repeat
UTR	Untranslated region

References

[1] Pennisi E. Genomics. ENCODE project writes eulogy for junk DNA. Science 2012;337(6099):1159–61.

[2] Esteller M. Non-coding RNAs in human disease. Nat Rev Genet 2011;12(12):861–74.

[3] Haig D. The (dual) origin of epigenetics. Cold Spring Harb Symp Quant Biol 2004;69:67–70.

[4] Ketting RF. MicroRNA biogenesis and function: an overview. Adv Exp Med Biol 2011;700:1–14.

[5] Krol J, Loedige I, Filipowicz W. The widespread regulation of microRNA biogenesis, function and decay. Nat Rev Genet 2010;11(9):597–610.

[6] Quick-Cleveland J, Jacob JP, Weitz SH, Shoffner G, Senturia R, Guo F. The DGCR8 RNA-binding heme domain recognizes primary MicroRNAs by clamping the hairpin. Cell Rep 2014;7(6):1994–2005.

[7] Bartel DP, Chen CZ. Micromanagers of gene expression: the potentially widespread influence of metazoan microRNAs. Nat Rev Genet 2004;5(5):396–400.

[8] Vasudevan S, Tong Y, Steitz JA. Switching from repression to activation: microRNAs can up-regulate translation. Science 2007;318(5858):1931–4.

[9] Fabbri M, Paone A, Calore F, Galli R, Gaudio E, Santhanam R, et al. MicroRNAs bind to toll-like receptors to induce prometastatic inflammatory response. Proc Natl Acad Sci USA 2012;109(31):E2110–6.

[10] Vatolin S, Navaratne K, Weil RJ. A novel method to detect functional microRNA targets. J Mol Biol 2006;358(4):983–96.

[11] Eiring AM, Harb JG, Neviani P, Garton C, Oaks JJ, Spizzo R, et al. miR-328 functions as an RNA decoy to modulate hnRNP E2 regulation of mRNA translation in leukemic blasts. Cell 2010;140(5):652–65.

[12] Tan L, Yu JT, Tan L. Causes and consequences of MicroRNA dysregulation in neurodegenerative diseases. Mol Neurobiol Jun 29, 2014. http://www.ncbi.nlm.nih.gov/pubmed/?term=24973986.

[13] Goodall EF, Heath PR, Bandmann O, Kirby J, Shaw PJ. Neuronal dark matter: the emerging role of microRNAs in neurodegeneration. Front Cell Neurosci 2013;7:178.

[14] Van den Hove DL, Kompotis K, Lardenoije R, Kenis G, Mill J, Steinbusch HW, et al. Epigenetically regulated microRNAs in Alzheimer's disease. Neurobiol Aging 2014;35(4):731–45.

[15] Bicchi I, Morena F, Montesano S, Polidoro M, Martino S. MicroRNAs and molecular mechanisms of neurodegeneration. Genes (Basel) 2013;4(2):244–63.

[16] Kita Y, Vincent K, Natsugoe S, Berindan-Neagoe I, Calin GA. Epigenetically regulated microRNAs and their prospect in cancer diagnosis. Expert Rev Mol Diagn 2014;14(6):673–83.

[17] Fabbri M, Calore F, Paone A, Galli R, Calin GA. Epigenetic regulation of miRNAs in cancer. Adv Exp Med Biol 2013;754:137–48.

[18] Liu X, Chen X, Yu X, Tao Y, Bode AM, Dong Z, et al. Regulation of microRNAs by epigenetics and their interplay involved in cancer. J Exp Clin Cancer Res 2013;32:96.

[19] Baer C, Claus R, Plass C. Genome-wide epigenetic regulation of miRNAs in cancer. Cancer Res 2013;73(2):473–7.

[20] Kunej T, Godnic I, Ferdin J, Horvat S, Dovc P, Calin GA. Epigenetic regulation of microRNAs in cancer: an integrated review of literature. Mutat Res 2011;717(1–2):77–84.

[21] Hebert SS, Horre K, Nicolai L, Bergmans B, Papadopoulou AS, Delacourte A, et al. MicroRNA regulation of Alzheimer's amyloid precursor protein expression. Neurobiol Dis 2009;33(3):422–8.

[22] Fang M, Wang J, Zhang X, Geng Y, Hu Z, Rudd JA, et al. The miR-124 regulates the expression of BACE1/beta-secretase correlated with cell death in Alzheimer's disease. Toxicol Lett 2012;209(1):94–105.

[23] Wang WX, Rajeev BW, Stromberg AJ, Ren N, Tang G, Huang Q, et al. The expression of microRNA miR-107 decreases early in Alzheimer's disease and may accelerate disease progression through regulation of beta-site amyloid precursor protein-cleaving enzyme 1. J Neurosci 2008;28(5):1213–23.

[24] Geekiyanage H, Chan C. MicroRNA-137/181c regulates serine palmitoyltransferase and in turn amyloid beta, novel targets in sporadic Alzheimer's disease. J Neurosci 2011;31(41):14820–30.

[25] Schonrock N, Humphreys DT, Preiss T, Gotz J. Target gene repression mediated by miRNAs miR-181c and miR-9 both of which are down-regulated by amyloid-beta. J Mol Neurosci 2012;46(2):324–35.

[26] Szulwach KE, Li X, Smrt RD, Li Y, Luo Y, Lin L, et al. Cross talk between microRNA and epigenetic regulation in adult neurogenesis. J Cell Biol 2010;189(1):127–41.

[27] Doxakis E. Post-transcriptional regulation of alpha-synuclein expression by mir-7 and mir-153. J Biol Chem 2010;285(17):12726–34.

[28] Cho HJ, Liu G, Jin SM, Parisiadou L, Xie C, Yu J, et al. MicroRNA-205 regulates the expression of Parkinson's disease-related leucine-rich repeat kinase 2 protein. Hum Mol Genet 2013;22(3):608–20.

[29] Cardo LF, Coto E, Ribacoba R, Mata IF, Moris G, Menendez M, et al. The screening of the 3'UTR sequence of LRRK2 identified an association between the rs66737902 polymorphism and Parkinson's disease. J Hum Genet 2014;59(6):346–8.

[30] Packer AN, Xing Y, Harper SQ, Jones L, Davidson BL. The bifunctional microRNA miR-9/miR-9* regulates REST and CoREST and is downregulated in Huntington's disease. J Neurosci 2008;28(53):14341–6.

[31] Marti E, Pantano L, Banez-Coronel M, Llorens F, Minones-Moyano E, Porta S, et al. A myriad of miRNA variants in control and Huntington's disease brain regions detected by massively parallel sequencing. Nucleic Acids Res 2010;38(20):7219–35.

[32] Sinha M, Mukhopadhyay S, Bhattacharyya NP. Mechanism(s) of alteration of micro RNA expressions in Huntington's disease and their possible contributions to the observed cellular and molecular dysfunctions in the disease. Neuromolecular Med 2012;14(4):221–43.

[33] Savas JN, Makusky A, Ottosen S, Baillat D, Then F, Krainc D, et al. Huntington's disease protein contributes to RNA-mediated gene silencing through association with Argonaute and P bodies. Proc Natl Acad Sci USA 2008;105(31):10820–5.

[34] Johnson R, Buckley NJ. Gene dysregulation in Huntington's disease: REST, microRNAs and beyond. Neuromolecular Med 2009;11(3):183–99.

[35] Dai E, Yu X, Zhang Y, Meng F, Wang S, Liu X, et al. EpimiR: a database of curated mutual regulation between miRNAs and epigenetic modifications. Database (Oxford) 2014;2014(bau023).

[36] Saito Y, Liang G, Egger G, Friedman JM, Chuang JC, Coetzee GA, et al. Specific activation of microRNA-127 with downregulation of the proto-oncogene BCL6 by chromatin-modifying drugs in human cancer cells. Cancer Cell 2006;9(6):435–43.

[37] Lagos-Quintana M, Rauhut R, Yalcin A, Meyer J, Lendeckel W, Tuschl T. Identification of tissue-specific microRNAs from mouse. Curr Biol 2002;12(9):735–9.

[38] Sempere LF, Freemantle S, Pitha-Rowe I, Moss E, Dmitrovsky E, Ambros V. Expression profiling of mammalian microRNAs uncovers a subset of brain-expressed microRNAs with possible roles in murine and human neuronal differentiation. Genome Biol 2004;5(3):R13.

[39] Conaco C, Otto S, Han JJ, Mandel G. Reciprocal actions of REST and a microRNA promote neuronal identity. Proc Natl Acad Sci USA 2006;103(7):2422–7.

[40] Laneve P, Gioia U, Andriotto A, Moretti F, Bozzoni I, Caffarelli E. A minicircuitry involving REST and CREB controls miR-9-2 expression during human neuronal differentiation. Nucleic Acids Res 2010;38(20):6895–905.

[41] Yoo AS, Staahl BT, Chen L, Crabtree GR. MicroRNA-mediated switching of chromatin-remodelling complexes in neural development. Nature 2009;460(7255):642–6.

[42] Liang H, Studach L, Hullinger RL, Xie J, Andrisani OM. Down-regulation of RE-1 silencing transcription factor (REST) in advanced prostate cancer by hypoxia-induced miR-106b~25. Exp Cell Res 2014;320(2):188–99.

[43] Agostini M, Knight RA. miR-34: from bench to bedside. Oncotarget 2014;5(4):872–81.

[44] Lodygin D, Tarasov V, Epanchintsev A, Berking C, Knyazeva T, Korner H, et al. Inactivation of miR-34a by aberrant CpG methylation in multiple types of cancer. Cell Cycle 2008;7(16):2591–600.

[45] Vogt M, Munding J, Gruner M, Liffers ST, Verdoodt B, Hauk J, et al. Frequent concomitant inactivation of miR-34a and miR-34b/c by CpG methylation in colorectal, pancreatic, mammary, ovarian, urothelial, and renal cell carcinomas and soft tissue sarcomas. Virchows Arch 2011;458(3):313–22.

[46] Toyota M, Suzuki H, Sasaki Y, Maruyama R, Imai K, Shinomura Y, et al. Epigenetic silencing of microRNA-34b/c and B-cell translocation gene 4 is associated with CpG island methylation in colorectal cancer. Cancer Res 2008;68(11):4123–32.

[47] Wang Y, Toh HC, Chow P, Chung AY, Meyers DJ, Cole PA, et al. MicroRNA-224 is up-regulated in hepatocellular carcinoma through epigenetic mechanisms. FASEB J 2012;26(7):3032–41.

[48] Mendell JT. miRiad roles for the miR-17-92 cluster in development and disease. Cell 2008;133(2):217–22.

[49] Humphreys KJ, Cobiac L, Le Leu RK, Van der Hoek MB, Michael MZ. Histone deacetylase inhibition in colorectal cancer cells reveals competing roles for members of the oncogenic miR-17-92 cluster. Mol Carcinog 2013;52(6):459–74.

[50] Chang TC, Yu D, Lee YS, Wentzel EA, Arking DE, West KM, et al. Widespread microRNA repression by Myc contributes to tumorigenesis. Nat Genet 2008;40(1):43–50.

[51] Ozsolak F, Poling LL, Wang Z, Liu H, Liu XS, Roeder RG, et al. Chromatin structure analyses identify miRNA promoters. Genes Dev 2008;22(22):3172–83.

[52] Chien CH, Sun YM, Chang WC, Chiang-Hsieh PY, Lee TY, Tsai WC, et al. Identifying transcriptional start sites of human microRNAs based on high-throughput sequencing data. Nucleic Acids Res 2011;39(21):9345–56.

[53] Wee EJ, Peters K, Nair SS, Hulf T, Stein S, Wagner S, et al. Mapping the regulatory sequences controlling 93 breast cancer-associated miRNA genes leads to the identification of two functional promoters of the Hsa-mir-200b cluster, methylation of which is associated with metastasis or hormone receptor status in advanced breast cancer. Oncogene 2012;31(38):4182–95.

[54] Hill L, Browne G, Tulchinsky E. ZEB/miR-200 feedback loop: at the crossroads of signal transduction in cancer. Int J Cancer 2013;132(4):745–54.

[55] Feng X, Wang Z, Fillmore R, Xi Y. MiR-200, a new star miRNA in human cancer. Cancer Lett 2014;344(2):166–73.

[56] Vrba L, Garbe JC, Stampfer MR, Futscher BW. Epigenetic regulation of normal human mammary cell type-specific miRNAs. Genome Res 2011;21(12):2026–37.

[57] Vrba L, Jensen TJ, Garbe JC, Heimark RL, Cress AE, Dickinson S, et al. Role for DNA methylation in the regulation of miR-200c and miR-141 expression in normal and cancer cells. PLoS One 2010;5(1):e8697.

[58] Attema JL, Bert AG, Lim YY, Kolesnikoff N, Lawrence DM, Pillman KA, et al. Identification of an enhancer that increases miR-200b~200a~429 gene expression in breast cancer cells. PLoS One 2013;8(9):e75517.

[59] Lujambio A, Calin GA, Villanueva A, Ropero S, Sanchez-Cespedes M, Blanco D, et al. A microRNA DNA methylation signature for human cancer metastasis. Proc Natl Acad Sci USA 2008;105(36):13556–61.

[60] Hildebrandt MA, Gu J, Lin J, Ye Y, Tan W, Tamboli P, et al. Hsa-miR-9 methylation status is associated with cancer development and metastatic recurrence in patients with clear cell renal cell carcinoma. Oncogene 2010;29(42):5724–8.

[61] Roman-Gomez J, Agirre X, Jimenez-Velasco A, Arqueros V, Vilas-Zornoza A, Rodriguez-Otero P, et al. Epigenetic regulation of microRNAs in acute lymphoblastic leukemia. J Clin Oncol 2009;27(8):1316–22.

[62] Agirre X, Vilas-Zornoza A, Jimenez-Velasco A, Martin-Subero JI, Cordeu L, Garate L, et al. Epigenetic silencing of the tumor suppressor microRNA Hsa-miR-124a regulates CDK6 expression and confers a poor prognosis in acute lymphoblastic leukemia. Cancer Res 2009;69(10):4443–53.

[63] Ando T, Yoshida T, Enomoto S, Asada K, Tatematsu M, Ichinose M, et al. DNA methylation of microRNA genes in gastric mucosae of gastric cancer patients: its possible involvement in the formation of epigenetic field defect. Int J Cancer 2009;124(10):2367–74.

[64] Heller G, Weinzierl M, Noll C, Babinsky V, Ziegler B, Altenberger C, et al. Genome-wide miRNA expression profiling identifies miR-9-3 and miR-193a as targets for DNA methylation in non-small cell lung cancers. Clin Cancer Res 2012;18(6):1619–29.

[65] Wang Z, Chen Z, Gao Y, Li N, Li B, Tan F, et al. DNA hypermethylation of microRNA-34b/c has prognostic value for stage non-small cell lung cancer. Cancer Biol Ther 2011;11(5):490–6.

[66] Nadal E, Chen G, Gallegos M, Lin L, Ferrer-Torres D, Truini A, et al. Epigenetic inactivation of microRNA-34b/c predicts poor disease-free survival in early-stage lung adenocarcinoma. Clin Cancer Res 2013;19(24):6842–52.

[67] Watanabe K, Emoto N, Hamano E, Sunohara M, Kawakami M, Kage H, et al. Genome structure-based screening identified epigenetically silenced microRNA associated with invasiveness in non-small-cell lung cancer. Int J Cancer 2012;130(11):2580–90.

[68] Tan W, Gu J, Huang M, Wu X, Hildebrandt MA. Epigenetic analysis of microRNA genes in tumors from surgically resected lung cancer patients and association with survival. Mol Carcinog Mar 24, 2014. http://www.ncbi.nlm.nih.gov/pubmed/?term=24665010.

[69] Brueckner B, Stresemann C, Kuner R, Mund C, Musch T, Meister M, et al. The human let-7a-3 locus contains an epigenetically regulated microRNA gene with oncogenic function. Cancer Res 2007;67(4):1419–23.

[70] Lu L, Katsaros D, de la Longrais IA, Sochirca O, Yu H. Hypermethylation of let-7a-3 in epithelial ovarian cancer is associated with low insulin-like growth factor-II expression and favorable prognosis. Cancer Res 2007;67(21):10117–22.

[71] Johnson SM, Grosshans H, Shingara J, Byrom M, Jarvis R, Cheng A, et al. RAS is regulated by the let-7 microRNA family. Cell 2005;120(5):635–47.

[72] Kumar MS, Erkeland SJ, Pester RE, Chen CY, Ebert MS, Sharp PA, et al. Suppression of non-small cell lung tumor development by the let-7 microRNA family. Proc Natl Acad Sci USA 2008;105(10):3903–8.

[73] Yanaihara N, Caplen N, Bowman E, Seike M, Kumamoto K, Yi M, et al. Unique microRNA molecular profiles in lung cancer diagnosis and prognosis. Cancer Cell 2006;9(3):189–98.

[74] Takamizawa J, Konishi H, Yanagisawa K, Tomida S, Osada H, Endoh H, et al. Reduced expression of the let-7 microRNAs in human lung cancers in association with shortened postoperative survival. Cancer Res 2004;64(11):3753–6.

[75] Iorio MV, Piovan C, Croce CM. Interplay between microRNAs and the epigenetic machinery: an intricate network. Biochim Biophys Acta 2010;1799(10–12):694–701.

[76] Fabbri M, Garzon R, Cimmino A, Liu Z, Zanesi N, Callegari E, et al. MicroRNA-29 family reverts aberrant methylation in lung cancer by targeting DNA methyltransferases 3A and 3B. Proc Natl Acad Sci USA 2007;104(40):15805–10.

[77] Garzon R, Liu S, Fabbri M, Liu Z, Heaphy CE, Callegari E, et al. MicroRNA-29b induces global DNA hypomethylation and tumor suppressor gene reexpression in acute myeloid leukemia by targeting directly DNMT3A and 3B and indirectly DNMT1. Blood 2009;113(25):6411–8.

[78] Hess CJ, Errami A, Berkhof J, Denkers F, Ossenkoppele GJ, Nygren AO, et al. Concurrent methylation of promoters from tumor associated genes predicts outcome in acute myeloid leukemia. Leuk Lymphoma 2008;49(6):1132–41.

[79] Braconi C, Kogure T, Valeri N, Huang N, Nuovo G, Costinean S, et al. MicroRNA-29 can regulate expression of the long non-coding RNA gene MEG3 in hepatocellular cancer. Oncogene 2011;30(47):4750–6.

[80] Nguyen T, Kuo C, Nicholl MB, Sim MS, Turner RR, Morton DL, et al. Downregulation of microRNA-29c is associated with hypermethylation of tumor-related genes and disease outcome in cutaneous melanoma. Epigenetics 2011;6(3):388–94.

[81] Amodio N, Leotta M, Bellizzi D, Di Martino MT, D'Aquila P, Lionetti M, et al. DNA-demethylating and anti-tumor activity of synthetic miR-29b mimics in multiple myeloma. Oncotarget 2012;3(10):1246–58.

[82] Amodio N, Bellizzi D, Leotta M, Raimondi L, Biamonte L, D'Aquila P, et al. miR-29b induces SOCS-1 expression by promoter demethylation and negatively regulates migration of multiple myeloma and endothelial cells. Cell Cycle 2013;12(23): 3650–62.

[83] Xiong Y, Fang JH, Yun JP, Yang J, Zhang Y, Jia WH, et al. Effects of microRNA-29 on apoptosis, tumorigenicity, and prognosis of hepatocellular carcinoma. Hepatology 2010;51(3):836–45.

[84] Mott JL, Kobayashi S, Bronk SF, Gores GJ. mir-29 regulates Mcl-1 protein expression and apoptosis. Oncogene 2007;26(42):6133–40.

[85] Chou J, Lin JH, Brenot A, Kim JW, Provot S, Werb Z. GATA3 suppresses metastasis and modulates the tumour microenvironment by regulating microRNA-29b expression. Nat Cell Biol 2013;15(2):201–13.

[86] Song SJ, Ito K, Ala U, Kats L, Webster K, Sun SM, et al. The oncogenic microRNA miR-22 targets the TET2 tumor suppressor to promote hematopoietic stem cell self-renewal and transformation. Cell Stem Cell 2013;13(1):87–101.

[87] Song SJ, Poliseno L, Song MS, Ala U, Webster K, Ng C, et al. MicroRNA-antagonism regulates breast cancer stemness and metastasis via TET-family-dependent chromatin remodeling. Cell 2013;154(2):311–24.

[88] Kong LM, Liao CG, Zhang Y, Xu J, Li Y, Huang W, et al. A regulatory loop involving miR-22, Sp1, and c-Myc modulates CD147 expression in breast cancer invasion and metastasis. Cancer Res 2014;74(14):3764–78.

[89] Tang H, Kong Y, Guo J, Tang Y, Xie X, Yang L, et al. Diallyl disulfide suppresses proliferation and induces apoptosis in human gastric cancer through Wnt-1 signaling pathway by up-regulation of miR-200b and miR-22. Cancer Lett 2013;340(1):72–81.

[90] Okada N, Lin CP, Ribeiro MC, Biton A, Lai G, He X, et al. A positive feedback between p53 and miR-34 miRNAs mediates tumor suppression. Genes Dev 2014;28(5): 438–50.

[91] Yamakuchi M, Lowenstein CJ. MiR-34, SIRT1 and p53: the feedback loop. Cell Cycle 2009;8(5):712–5.

[92] Majid S, Dar AA, Saini S, Shahryari V, Arora S, Zaman MS, et al. miRNA-34b inhibits prostate cancer through demethylation, active chromatin modifications, and AKT pathways. Clin Cancer Res 2013;19(1):73–84.

[93] Concepcion CP, Han YC, Mu P, Bonetti C, Yao E, D'Andrea A, et al. Intact p53-dependent responses in miR-34-deficient mice. PLoS Genet 2012;8(7):e1002797.

[94] Zhao J, Lammers P, Torrance CJ, Bader AG. TP53-independent function of miR-34a via HDAC1 and p21(CIP1/WAF1.). Mol Ther 2013;21(9):1678–86.

[95] Ranum LP, Cooper TA. RNA-mediated neuromuscular disorders. Annu Rev Neurosci 2006;29:259–77.

[96] Mankodi A, Lin X, Blaxall BC, Swanson MS, Thornton CA. Nuclear RNA foci in the heart in myotonic dystrophy. Circ Res 2005;97(11):1152–5.

[97] Wojciechowska M, Krzyzosiak WJ. CAG repeat RNA as an auxiliary toxic agent in polyglutamine disorders. RNA Biol 2011;8(4):565–71.

[98] Wang LC, Chen KY, Pan H, Wu CC, Chen PH, Liao YT, et al. Muscleblind participates in RNA toxicity of expanded CAG and CUG repeats in *Caenorhabditis elegans*. Cell Mol Life Sci 2011;68(7):1255–67.

[99] De Biase I, Chutake YK, Rindler PM, Bidichandani SI. Epigenetic silencing in Friedreich ataxia is associated with depletion of CTCF (CCCTC-binding factor) and antisense transcription. PLoS One 2009;4(11):e7914.

[100] Koch KS, Leffert HL. Giant hairpins formed by CUG repeats in myotonic dystrophy messenger RNAs might sterically block RNA export through nuclear pores. J Theor Biol 1998;192(4):505–14.

[101] Tian B, White RJ, Xia T, Welle S, Turner DH, Mathews MB, et al. Expanded CUG repeat RNAs form hairpins that activate the double-stranded RNA-dependent protein kinase PKR. RNA 2000;6(1):79–87.

[102] Krol J, Fiszer A, Mykowska A, Sobczak K, de Mezer M, Krzyzosiak WJ. Ribonuclease dicer cleaves triplet repeat hairpins into shorter repeats that silence specific targets. Mol Cell 2007;25(4):575–86.

[103] Handa V, Saha T, Usdin K. The fragile X syndrome repeats form RNA hairpins that do not activate the interferon-inducible protein kinase, PKR, but are cut by Dicer. Nucleic Acids Res 2003;31(21):6243–8.

[104] Banez-Coronel M, Porta S, Kagerbauer B, Mateu-Huertas E, Pantano L, Ferrer I, et al. A pathogenic mechanism in Huntington's disease involves small CAG-repeated RNAs with neurotoxic activity. PLoS Genet 2012;8(2):e1002481.

[105] Garcia MA, Gil J, Ventoso I, Guerra S, Domingo E, Rivas C, et al. Impact of protein kinase PKR in cell biology: from antiviral to antiproliferative action. Microbiol Mol Biol Rev 2006;70(4):1032–60.

[106] Miller JW, Urbinati CR, Teng-Umnuay P, Stenberg MG, Byrne BJ, Thornton CA, et al. Recruitment of human muscleblind proteins to (CUG)(n) expansions associated with myotonic dystrophy. EMBO J 2000;19(17):4439–48.

[107] Jiang H, Mankodi A, Swanson MS, Moxley RT, Thornton CA. Myotonic dystrophy type 1 is associated with nuclear foci of mutant RNA, sequestration of muscleblind proteins and deregulated alternative splicing in neurons. Hum Mol Genet 2004;13(24): 3079–88.

[108] Ho TH, Savkur RS, Poulos MG, Mancini MA, Swanson MS, Cooper TA. Colocalization of muscleblind with RNA foci is separable from mis-regulation of alternative splicing in myotonic dystrophy. J Cell Sci 2005;118(Pt 13):2923–33.

[109] Sathasivam K, Neueder A, Gipson TA, Landles C, Benjamin AC, Bondulich MK, et al. Aberrant splicing of HTT generates the pathogenic exon 1 protein in Huntington disease. Proc Natl Acad Sci USA 2013;110(6):2366–70.

[110] Hagerman P. Fragile X-associated tremor/ataxia syndrome (FXTAS): pathology and mechanisms. Acta Neuropathol 2013;126(1):1–19.

[111] Jin P, Duan R, Qurashi A, Qin Y, Tian D, Rosser TC, et al. Pur alpha binds to rCGG repeats and modulates repeat-mediated neurodegeneration in a Drosophila model of fragile X tremor/ataxia syndrome. Neuron 2007;55(4):556–64.

[112] Sellier C, Rau F, Liu Y, Tassone F, Hukema RK, Gattoni R, et al. Sam68 sequestration and partial loss of function are associated with splicing alterations in FXTAS patients. EMBO J 2010;29(7):1248–61.

[113] Sellier C, Freyermuth F, Tabet R, Tran T, He F, Ruffenach F, et al. Sequestration of DROSHA and DGCR8 by expanded CGG RNA repeats alters microRNA processing in fragile X-associated tremor/ataxia syndrome. Cell Rep 2013;3(3):869–80.

[114] Lee ST, Chu K, Im WS, Yoon HJ, Im JY, Park JE, et al. Altered microRNA regulation in Huntington's disease models. Exp Neurol 2011;227(1):172–9.

[115] Kiliszek A, Kierzek R, Krzyzosiak WJ, Rypniewski W. Atomic resolution structure of CAG RNA repeats: structural insights and implications for the trinucleotide repeat expansion diseases. Nucleic Acids Res 2010;38(22):8370–6.

[116] Colak D, Zaninovic N, Cohen MS, Rosenwaks Z, Yang WY, Gerhardt J, et al. Promoter-bound trinucleotide repeat mRNA drives epigenetic silencing in fragile X syndrome. Science 2014;343(6174):1002–5.

[117] Koob MD, Moseley ML, Schut LJ, Benzow KA, Bird TD, Day JW, et al. An untranslated CTG expansion causes a novel form of spinocerebellar ataxia (SCA8). Nat Genet 1999;21(4):379–84.

[118] Ikeda Y, Daughters RS, Ranum LP. Bidirectional expression of the SCA8 expansion mutation: one mutation, two genes. Cerebellum 2008;7(2):150–8.

[119] Chung DW, Rudnicki DD, Yu L, Margolis RL. A natural antisense transcript at the Huntington's disease repeat locus regulates HTT expression. Hum Mol Genet 2011;20(17):3467–77.

[120] Ladd PD, Smith LE, Rabaia NA, Moore JM, Georges SA, Hansen RS, et al. An antisense transcript spanning the CGG repeat region of FMR1 is upregulated in premutation carriers but silenced in full mutation individuals. Hum Mol Genet 2007;16(24): 3174–87.

[121] Sopher BL, Ladd PD, Pineda VV, Libby RT, Sunkin SM, Hurley JB, et al. CTCF regulates ataxin-7 expression through promotion of a convergently transcribed, antisense noncoding RNA. Neuron 2011;70(6):1071–84.

[122] Derrien T, Johnson R, Bussotti G, Tanzer A, Djebali S, Tilgner H, et al. The GENCODE v7 catalog of human long noncoding RNAs: analysis of their gene structure, evolution, and expression. Genome Res 2012;22(9):1775–89.

[123] Necsulea A, Soumillon M, Warnefors M, Liechti A, Daish T, Zeller U, et al. The evolution of lncRNA repertoires and expression patterns in tetrapods. Nature 2014;505(7485):635–40.

[124] Brown CJ, Hendrich BD, Rupert JL, Lafrenière RG, Xing Y, Lawrence J, et al. The human XIST gene: analysis of a 17 kb inactive X-specific RNA that contains conserved repeats and is highly localized within the nucleus. Cell 1992;71(3):527–42.

[125] Lee JT, Davidow LS, Warshawsky D. Tsix, a gene antisense to Xist at the X-inactivation centre. Nat Genet 1999;21(4):400–4.

[126] Tian D, Sun S, Lee JT. The long noncoding RNA, Jpx, is a molecular switch for X chromosome inactivation. Cell 2010;143(3):390–403.

[127] Sleutels F, Zwart R, Barlow DP. The non-coding air RNA is required for silencing autosomal imprinted genes. Nature 2002;415(6873):810–3.

[128] Thakur N, Tiwari VK, Thomassin H, Pandey RR, Kanduri M, Göndör A, et al. An antisense RNA regulates the bidirectional silencing property of the Kcnq1 imprinting control region. Mol Cell Biol 2004;24(18):7855–62.

[129] Rinn JL, Kertesz M, Wang JK, Squazzo SL, Xu X, Brugmann SA, et al. Functional demarcation of active and silent chromatin domains in human HOX loci by noncoding RNAs. Cell 2007;129(7):1311–23.

[130] Di Ruscio A, Ebralidze AK, Benoukraf T, Amabile G, Goff LA, Terragni J, et al. DNMT1-interacting RNAs block gene-specific DNA methylation. Nature 2013;503(7476):371–6.

[131] Cesana M, Cacchiarelli D, Legnini I, Santini T, Sthandier O, Chinappi M, et al. A long noncoding RNA controls muscle differentiation by functioning as a competing endogenous RNA. Cell 2011;147(2):358–69.

[132] Poliseno L, Salmena L, Zhang J, Carver B, Haveman WJ, Pandolfi PP. A coding-independent function of gene and pseudogene mRNAs regulates tumour biology. Nature 2010;465(7301):1033–8.

[133] Cheetham SW, Gruhl F, Mattick JS, Dinger ME. Long noncoding RNAs and the genetics of cancer. Br J Cancer 2013;108(12):2419–25.

[134] Pasmant E, Sabbagh A, Vidaud M, Bièche I. ANRIL, a long, noncoding RNA, is an unexpected major hotspot in GWAS. FASEB J 2011;25(2):444–8.

[135] Broadbent HM, Peden JF, Lorkowski S, Goel A, Ongen H, Green F, et al. Susceptibility to coronary artery disease and diabetes is encoded by distinct, tightly linked SNPs in the ANRIL locus on chromosome 9p. Hum Mol Genet 2008;17(6):806–14.

[136] Bejerano G, Pheasant M, Makunin I, Stephen S, Kent WJ, Mattick JS, et al. Ultraconserved elements in the human genome. Science 2004;304(5675):1321–5.

[137] Calin GA, Liu CG, Ferracin M, Hyslop T, Spizzo R, Sevignani C, et al. Ultraconserved regions encoding ncRNAs are altered in human leukemias and carcinomas. Cancer Cell 2007;12(3):215–29.

[138] Lujambio A, Portela A, Liz J, Melo SA, Rossi S, Spizzo R, et al. CpG island hypermethylation-associated silencing of non-coding RNAs transcribed from ultraconserved regions in human cancer. Oncogene 2010;29(48):6390–401.

[139] Khalil AM, Guttman M, Huarte M, Garber M, Raj A, Morales DR, et al. Many human large intergenic noncoding RNAs associate with chromatin-modifying complexes and affect gene expression. Proc Natl Acad Sci USA 2009;106(28):11667–72.

[140] Zhao JJ, Lin J, Yang H, Kong W, He L, Ma X, et al. MicroRNA-221/222 negatively regulates estrogen receptor alpha and is associated with tamoxifen resistance in breast cancer. J Biol Chem 2008;283(45):31079–86.

[141] Kotake Y, Nakagawa T, Kitagawa K, Suzuki S, Liu N, Kitagawa M, et al. Long non-coding RNA ANRIL is required for the PRC2 recruitment to and silencing of p15(INK4B) tumor suppressor gene. Oncogene 2011;30(16):1956–62.

[142] Yap KL, Li S, Muñoz-Cabello AM, Raguz S, Zeng L, Mujtaba S, et al. Molecular interplay of the noncoding RNA ANRIL and methylated histone H3 lysine 27 by polycomb CBX7 in transcriptional silencing of INK4a. Mol Cell 2010;38(5):662–74.

[143] Huarte M, Guttman M, Feldser D, Garber M, Koziol MJ, Kenzelmann-Broz D, et al. A large intergenic noncoding RNA induced by p53 mediates global gene repression in the p53 response. Cell 2010;142(3):409–19.

[144] Dimitrova N, Zamudio JR, Jong RM, Soukup D, Resnick R, Sarma K, et al. LincRNA-p21 activates p21 in cis to promote polycomb target gene expression and to enforce the G1/S checkpoint. Mol Cell 2014;54(5):777–90.

[145] Yang L, Lin C, Liu W, Zhang J, Ohgi KA, Grinstein JD, et al. ncRNA- and Pc2 methylation-dependent gene relocation between nuclear structures mediates gene activation programs. Cell 2011;147(4):773–88.

[146] Tripathi V, Ellis JD, Shen Z, Song DY, Pan Q, Watt AT, et al. The nuclear-retained noncoding RNA MALAT1 regulates alternative splicing by modulating SR splicing factor phosphorylation. Mol Cell 2010;39(6):925–38.

[147] Kovalenko TF, Sorokina AV, Ozolinia LA, Patrushev LI. [Pseudogene PTENP1 5'-region methylation in endometrial cancer and hyperplasias]. Bioorg Khim 2013;39(4):445–53.

[148] Memczak S, Jens M, Elefsinioti A, Torti F, Krueger J, Rybak A, et al. Circular RNAs are a large class of animal RNAs with regulatory potency. Nature 2013;495(7441):333–8.

[149] Denzler R, Agarwal V, Stefano J, Bartel DP, Stoffel M. Assessing the ceRNA hypothesis with quantitative measurements of miRNA and target abundance. Mol Cell 2014;54(5):766–76.

[150] Li N, Zhou P, Zheng J, Deng J, Wu H, Li W, et al. A polymorphism rs12325489C>T in the LincRNA-ENST00000515084 exon was found to modulate breast Cancer risk via GWAS-based association analyses. PloS One 2014;9(5).

[151] Jendrzejewski J, He H, Radomska HS, Li W, Tomsic J, Liyanarachchi S, et al. The polymorphism rs944289 predisposes to papillary thyroid carcinoma through a large intergenic noncoding RNA gene of tumor suppressor type. Proc Natl Acad Sci USA 2012;109(22):8646–51.

[152] Ling H, Spizzo R, Atlasi Y, Nicoloso M, Shimizu M, Redis R, et al. CCAT2, a novel non-coding RNA mapping to 8q24, underlies metastatic progression and chromosomal instability in colon cancer. Genome Res 2013;23(9):1446–61.

[153] Zhu Z, Gao X, He Y, Zhao H, Yu Q, Jiang D, et al. An insertion/deletion polymorphism within RERT-lncRNA modulates hepatocellular carcinoma risk. Cancer Res 2012;72(23):6163–72.

[154] Cunnington MS, Santibanez Koref M, Mayosi BM, Burn J, Keavney B. Chromosome 9p21 SNPs associated with multiple disease phenotypes correlate with ANRIL expression. PLoS Genet 2010;6(4).

[155] Yu W, Gius D, Onyango P, Muldoon-Jacobs K, Karp J, Feinberg AP, et al. Epigenetic silencing of tumour suppressor gene p15 by its antisense RNA. Nature 2008;451(7175):202–6.

[156] Endo H, Shiroki T, Nakagawa T, Yokoyama M, Tamai K, Yamanami H, et al. Enhanced expression of long non-coding RNA HOTAIR is associated with the development of gastric cancer. PloS One 2013;8(10).

[157] X-h L, Z-l L, Sun M, Liu J, Wang Z-x, De W. The long non-coding RNA HOTAIR indicates a poor prognosis and promotes metastasis in non-small cell lung cancer. BMC Cancer 2013;13.

[158] Sørensen KP, Thomassen M, Tan Q, Bak M, Cold S, Burton M, et al. Long non-coding RNA HOTAIR is an independent prognostic marker of metastasis in estrogen receptor-positive primary breast cancer. Breast Cancer Res Treat 2013;142(3):529–36.

[159] Svoboda M, Slyskova J, Schneiderova M, Makovicky P, Bielik L, Levy M, et al. HOTAIR long non-coding RNA is a negative prognostic factor not only in primary tumors, but also in the blood of colorectal cancer patients. Carcinogenesis 2014;35(7):1510–5.

[160] Chiyomaru T, Fukuhara S, Saini S, Majid S, Deng G, Shahryari V, et al. Long non-coding RNA HOTAIR is targeted and regulated by miR-141 in human cancer cells. J Biol Chem 2014;289(18):12550–65.

[161] Gutschner T, Hämmerle M, Eissmann M, Hsu J, Kim Y, Hung G, et al. The noncoding RNA MALAT1 is a critical regulator of the metastasis phenotype of lung cancer cells. Cancer Res 2013;73(3):1180–9.

[162] Maclary E, Hinten M, Harris C, Kalantry S. Long nonoding RNAs in the X-inactivation center. Chromosome Res 2013;21(6–7):601–14.

[163] Chaligné R, Heard E. X-chromosome inactivation in development and cancer. FEBS Lett 2014;588(15):2514–22.

[164] Hoss AG, Kartha VK, Dong X, Latourelle JC, Dumitriu A, Hadzi TC, et al. MicroRNAs located in the Hox gene clusters are implicated in Huntington's disease pathogenesis. PLoS Genet 2014;10(2):e1004188.

[165] Weber DG, Johnen G, Casjens S, Bryk O, Pesch B, Jockel K-H, et al. Evaluation of long noncoding RNA MALAT1 as a candidate blood-based biomarker for the diagnosis of non-small cell lung cancer. BMC Res Notes 2013;6.

[166] Ren S, Wang F, Shen J, Sun Y, Xu W, Lu J, et al. Long non-coding RNA metastasis associated in lung adenocarcinoma transcript 1 derived miniRNA as a novel plasma-based biomarker for diagnosing prostate cancer. Eur J Cancer (Oxford, Engl 1990) 2013;49(13):2949–59.

Personalized Epigenetics: Analysis and Interpretation of DNA Methylation Variation

Hehuang Xie

Department of Biological Sciences, Virginia Bioinformatics Institute,
Virginia Tech, Blacksburg, VA, USA

OUTLINE

1. INTRODUCTION

The term "epigenetic" mainly refers to histone modifications and DNA methylation, which are alternative ways to control gene expression while maintaining the nucleotide sequence of the genome. The eukaryotic DNA molecules are frequently packaged into nucleosomes consisting of a DNA segment 146 base pairs long wound around an octamer of histone core proteins, including histone 2A, histone 2B, histone 3 (H3), and histone 4 [1]. The N-terminal tails of histone proteins protruding from the nucleosome can be methylated, acetylated, phosphorylated, etc. Functional genomic regulatory elements are associated with different types of histone modifications on specific amino acids, which have diverse effects on the regulation of gene expression. For instance, histone acetylation at lysine residues on the amino-terminal tails alters nucleosomal conformation to free DNA molecules by neutralizing the positive charge of histone tails. Histone H3 lysine residue 27 acetylation (H3K27ac) and lysine residue 4 monomethylation (H3K4me1) are markers for enhancers, and lysine residue 4 trimethylation is a marker for the promoters of actively transcribed genes [2,3].

Mammalian DNA methylation mainly occurs at the cytosine of CpG dinucleotides. There are approximately 28 million CpG dinucleotides present in the human genome and roughly 60–80% of these cytosines are methylated in differentiated cells. The distribution of CpG sites in the human genome is nonrandom and their methylation probabilities depend highly on their position within the genome. Genomic regions with high GC content and high frequency of CpG dinucleotides are defined as CpG islands (CGI), which are prevalent near promoters, including the majority of the transcription start sites of housekeeping genes. Active promoters and CpG islands are largely resistant to DNA methylation. In contrast, repetitive DNA elements, including short interspersed nuclear elements (SINEs) and long interspersed nuclear elements (LINEs), are often heavily methylated. The methylated CpG-binding domain (MBD) proteins recognize methylated cytosines and form protein complexes with histone modification enzymes including histone deacetylases and histone methyltransferases. Methylated DNA regions are tightly packed into nucleosomes with reduced accessibility to transcription machinery. On the other hand, unmethylated promoters are often organized as transcriptionally active chromatin associated with histone acetylation.

The comparisons of interindividual epigenetic variation lead to the identification of valuable epigenetic biomarkers. A number of clinical trials have demonstrated that some disease-associated epigenetic markers open new avenues for early diagnosis and prognosis and individualize clinical decision-making of certain diseases. For instance, the promoter methylation of O(6)-methylguanine-DNA methyltransferase (MGMT) is an effective predictor of the responsiveness of patients with glioblastoma to alkylating agent chemotherapy [4]. Temozolomide (TMZ) is one of the most widely used and effective alkylating agents for glioblastoma patients [5]. It induces DNA methylation at

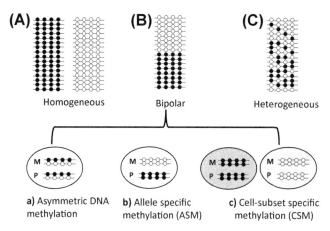

(A) Homogeneous **(B)** Bipolar **(C)** Heterogeneous

a) Asymmetric DNA methylation — M / P

b) Allele specific methylation (ASM) — M / P

c) Cell-subset specific methylation (CSM) — M / P , M / P

FIGURE 1 **Diverse methylation patterns reflect distinct epigenetic mechanisms.** Each line represents a sequence read or a DNA strand; open circles represent unmethylated cytosine; filled circles represent methylated cytosine; "M" and "P" indicate maternal and paternal alleles, respectively.

the O6-position of guanine to have the cytotoxic effect on cancer cells. MGMT is a DNA repair enzyme that removes alkyl adducts from the O6-position of guanine and thus antagonizes the genotoxic effects of TMZ. For glioblastoma patients, the silencing of the MGMT gene via promoter methylation indicates a favorable clinical outcome of alkylating agent chemotherapy.

Personalized epigenetic studies are complicated by the epigenetic variation resulting from the complexity of genome–environmental interactions. It has been documented that DNA methylation is influenced by the heterogeneity of the human condition, including demographics, lifestyle, and socioeconomic factors [6]. In response to external environmental stimuli and internal cellular metabolisms, changes in histone modifications and DNA methylation patterns may occur in postmitotic cells, at the regulatory elements in particular. In addition, genetic variations and cellular heterogeneity within tissues are anticipated to create greater challenges for epigenetic data analysis. Heterozygous genetic variations may result in allele-specific DNA methylation, and cell differentiation in mammals is accompanied by cell-subset-specific methylation. In normal tissues, genetic mutations occur at a low frequency but epigenetic modifications are relatively more dynamic. During cell division, genomic DNA methylation patterns may be transmitted from one generation to another. The fidelity of methylation inheritance varies considerably across genome regions and cell types [7]. The failure of DNA methylation inheritance leads to asymmetric DNA methylation in daughter cells. Altogether, within a given tissue, methylation variation may be observed at three levels: between DNA double strands (asymmetric DNA methylation), between two alleles (allele-specific DNA methylation), and among cells (cell-subset-specific DNA methylation) (Figure 1). This chapter begins with an introduction to DNA methylation and demethylation mechanisms and then discusses the possible sources of DNA methylation variation within tissues and the methods for quantitative assessment of methylation heterogeneity.

2. DNA METHYLATION AND DEMETHYLATION MECHANISMS

In mammalian genomes, DNA methylation patterns are established and maintained by three catalytically active DNA methyltransferases (DNMTs): DNMT1, DNMT3A, and DNMT3B [8]. DNMTs transfer the methyl group from S-adenosylmethionine to the C-5 carbon of cytosines. Depending on their target preference, the three DNMTs may be classified into two types: DNMT1 as a maintenance DNMT and DNMT3A/3B as de novo DNMTs. DNMT1 preferentially recognizes hemimethylated CpG sites, whereas DNMT3A/3B can methylate cytosines on both strands at previously unmethylated sites. DNMT1 has interaction partners, proliferating cell nuclear antigen (PCNA) and the ubiquitin-like plant homeodomain and RING finger domain 1 (UHRF1, also known as NP95). During DNA replication, PCNA and UHRF1 recruit DNMT1 to DNA replication forks to ensure the proper inheritance of DNA methylation patterns from mother to daughter strand [9]. DNMTs may continue working on the DNA, skipping sites at linker or nucleosome-free DNA regions during G2/M phase [10,11]. Thus, on both mother and daughter DNA strands, the cytosines at CpG/CpG dyads are frequently found to be either symmetrically methylated or completely unmethylated. Without DNMT1, DNA methylation will be gradually diluted through DNA replication.

DNMT1 alone, however, is incapable of successful maintenance of DNA methylation [12]. With the assistance of DNMT3L, DNMT3A/3B can catalyze de novo DNA methylation [13]. During early embryonic development, dramatic changes in genomic methylation occur. Shortly after fertilization, considerable loss of methylation is observed for both of the parental genomic DNAs and such demethylation is largely completed at the two-cell stage [14]. The level of DNA methylation continues to drop to around 29% at the blastocyst stage in the inner cell mass but rapidly increases following implantation. DNMT3A and DNMT3B are responsible for the establishment of the tissue-specific DNA methylation patterns following implantation. Both enzymes are highly expressed in undifferentiated embryonic stem cells (ESCs) but downregulated after differentiation [15,16]. The activities of these de novo methylation enzymes were also shown to be indispensable to keep some genomic regions including repeats fully methylated [12]. Genome-wide maps of DNMTs revealed that the DNMTs have specific but overlapping binding profiles [17], and the distribution of DNMTs has been associated with histone modifications and nucleosome structure [18].

A significant understanding of the DNA demethylation pathway has also been achieved. The three TET enzymes (TET1, TET2, and TET3) were discovered as mediators of DNA demethylation via the addition of hydroxyl groups onto cytosines and subsequent oxidation of 5-hydroxymethylcytosine (5hmC) to 5-formylcytosine and then 5-carboxylcytosine [19–21]. Both TET1 and TET2 are abundantly expressed and catalyze the conversion of 5-methylcytosine (5mC) to 5hmC in mouse ESCs [22,23]. However, the depletions of TET1 or TET2 are not equivalent. ESCs with the loss of TET1 show skewed differentiation toward trophectoderm in vitro [22] but mice with TET2 loss show defects in the differentiation of hematopoietic stem cells [23]. Two kinase inhibitors (2i), PD0325901 and CHIR99021, were found to block differentiation by targeting mitogen-activated protein kinase and glycogen synthase kinase-3 and enable self-renewal by stimulating Wnt signaling [24]. Cultured in 2i medium supplement with the cytokine leukemia inhibitory factor (LIF), the morphologically homogeneous ESCs represent a ground state of pluripotency with uniform Nanog protein expression [25]. Transcriptome comparison revealed that "2i ESCs" have higher TET2 expression and lower DNMT3A/3B expression than "serum ESCs" but with no significant alteration in the expression levels of TET1 and DNMT126. Accompanied by significantly decreased DNMT3A/3B protein levels and a substantial loss of 5mC, the level of 5hmC increased and reached a peak at 72 h after 2i addition. The knockdown of DNMT3A/3B did not lead to an increase in 5hmC in "serum ESCs" but TET1/2 is indispensable for 5hmC acquisition after 2i addition. Interestingly, revealed by locus-specific hairpin-bisulfite sequencing and hairpin oxidative bisulfite sequencing, substantial variations in the dynamics of 5mC/5hmC turnover were observed among different genomic loci [26]. In contrast to TET1 and TET2, TET3 is not or is hardly expressed in mouse ESCs but enriched in oocytes. Growing evidence indicates that TET enzymes play nonoverlapping functional roles in renewal and differentiation of ESCs [22, 27]. Importantly, 5hmC, the product of TET enzymes, is not randomly distributed but enriched within gene bodies and at transcriptional start sites bivalently marked with both active and repressive histone marks [28].

Despite the tightly controlled DNA methylation mechanisms, within a cell population, methylation patterns often show molecule-to-molecule variation, and a substantial portion of hemimethylated CpG dyads has been speculated [29]. Asymmetric DNA methylation present at certain genomic loci is associated with stochastic methylation changes in normal tissues [30] and increases in tumor cells with an unstable epigenome [31]. In our previous study [32], extremely high-fidelity DNA methylation transmission was found for approximately 30% of the genomic segments derived from CpG islands (CGIs) and 10% of the segments from Alu repeats, while only a small subset of CGIs and Alus are with highly variable methylation patterns. To determine the symmetry of CpG methylation, Laird and colleagues developed a hairpin-bisulfite PCR technique to generate methylation data of both complementary strands simultaneously [30]. They examined two fragile X mental retardation 1 alleles in human lymphocytes and estimated 83% as the fidelity of inheritance for the unmethylated cytosine in a hypermethylated CGI but 99% fidelity in the hypomethylated CGI. With the hairpin-bisulfite PCR technique, we demonstrated that the highly variable methylation patterns resulted from low fidelity of DNA methylation inheritance [32]. Notably, the methylation fidelities of CGIs and repetitive elements decrease in tumor tissues [32–34]. The hairpin-bisulfite sequencing technique has been applied to four single-copy genes and several repetitive elements [7]. Despite the lack of genome-wide data, Arand et al. demonstrated that DNMT1 and DNMT3A/3B may work cooperatively on some specific loci and the percentage of hemimethylated CpG dyads varies considerably among these genomic elements and different cell types.

We have modified the method of Laird and colleagues to implement a genome-wide hairpin-bisulfite sequencing approach (Figure 2). Briefly, after sonication or restriction enzyme cleavage, genomic DNA fragments are ligated to hairpin linkers and Illumina sequencing adaptors simultaneously. Following the streptavidin capture and bisulfite PCR, only the fragments linked to both the hairpin adaptor and the Illumina sequencing adaptor will be amplified for high-throughput pair-end sequencing. According to the methylation statuses of cytosines on two DNA strands, the symmetrical methylation status of CpG dyads (either methylated or

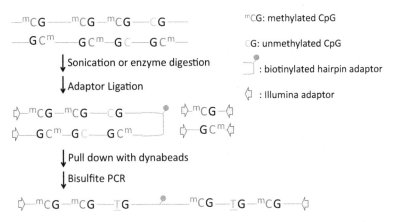

FIGURE 2 **A schematic diagram for genome-wide hairpin-bisulfite sequencing.** After sonication or restriction enzyme cleavage, genomic DNA fragments are ligated to hairpin linkers and Illumina sequencing adaptors simultaneously. Following the streptavidin capture and bisulfite PCR, only the fragments linked to both the hairpin adaptor and the Illumina sequencing adaptor will be amplified for high-throughput pair-end sequencing.

unmethylated) indicates successful methylation inheritance, while an asymmetrical methylation status (hemimethylated CpG dyads) indicates gain or loss of the methylation pattern. We defined the methylation fidelity for a given CpG dyads as 100% subtracted by the percentage of asymmetrically methylated CpGs. Thus, this approach may be used to determine the fidelity of methylation transmission on a genome-wide scale. In mouse ESCs, we found that the fidelity of 5mC inheritance is bimodally distributed and the majority of intermediately (40–60%) methylated CpG dinucleotides are hemimethylated and with low methylation fidelity, particularly in the differentiating ESCs [35]. Thus, the methylation differences between two DNA strands are important sources of DNA methylation variation.

3. GENETIC AND EPIGENETIC VARIATIONS

Genetic variations are sequence differences identified in a population, including single-nucleotide polymorphisms (SNPs), insertions/deletions, and other structural variations. An endeavor to provide a comprehensive resource on genetic variants in humans, the 1000 Genomes Project, was initiated in the year 2008 to determine the whole-genome sequences from a large number of people. The June 2014 release (http://www.1000genomes.org) delivers over 79 million human genetic variations based on the genomic sequences from 2535 individuals from 26 different populations around the world. The profiles of genetic variations for individuals were unique for geographic populations [36]. Many genetic variations identified do not affect human survivability, but a significant number of variations have been associated with various kinds of human disorders. The effects of genetic variations on gene expression are highly diverse. Genetic variations within genes may alter transcript structures and generate malfunctioning proteins. Genetic variations in the intergenic regions may disrupt the motifs in transcription factor-binding sites and result in changes to mRNA or noncoding RNA expression levels. Integrative annotation has been provided for a large collection of variants and nearly 100 cancer-related genetic variations have been revealed [37]. These disease-associated genetic variations could become molecular markers for disease classification and important determinants of drug response.

Deleterious genetic mutations have been found in genes encoding critical components of the epigenetic machineries and resulting in global methylation changes. In cancer genomes, various kinds of genetic disruptions have been documented for over 40 epigenetic genes, including DNA methyltransferases, chromatin remodeling complexes, and enzymes controlling histone modifications [38]. For instance, a total of

62 of 281 (22.1%) patients with acute myeloid leukemia carried somatic mutations in the DNMT3A gene [39]. These mutations, highly recurrent in patients, disrupt DNMT3A catalytic activity and were associated with several disease phenotypes, prognosis, and response of patients to chemotherapy [40]. Although homozygous germ-line variants have been identified in chromatin remodeling complexes or histone modifiers, such as SMARCC1 and SETD2, most deleterious genetic variants were heterozygous in normal tissues [41]. Apparently, homozygous germ-line variants affecting protein functions are often lethal for epigenetic genes playing vital roles in normal development and lack of family members with the overlapping functions.

Genetic variants have also been found for another de novo DNA methyltransferase, DNMT3B. The mutated DNMT3B gene leads to the immunodeficiency, centromeric region instability, and facial anomalies (ICF) syndrome. Deleterious mutations are frequently observed in the catalytic domain of the DNMT3B proteins and such DNMT3B mutants were believed to have residual activity [42]. Accordingly, DNMT3B mutations may be present homozygously. One characteristic of ICF syndrome is a decrease in the level of satellite DNA methylation, which might result in chromosome anomalies and expression of noncoding RNAs [43]. The mutations in methyl CpG binding protein 2 (MeCP2) have been widely studied because MeCP2 gene mutations are the cause of Rett syndrome. The MeCP2 gene is located on the X chromosome and codes for a protein highly expressed in the brain as an important component of neuronal chromatin [42]. Because of the genome location and expression profile of MeCP2, Rett syndrome is a severe neurological disorder and a prominent cause of mental retardation in females. Although MeCP2 mutations are also found in mentally handicapped males, without a backup copy to compensate for the defective one, males carrying germ-line mutations in the MeCP2 gene often die shortly after birth [42]. Although targets for therapeutic intervention are clear for ICF or Rett syndrome, specific drugs effective against these disorders have not yet been developed.

According to genome annotation (http://genome.ucsc.edu/), approximately half of the sequence content of the reference human genome is annotated as repetitive elements. At the epigenome level, SINEs, LINEs, and long terminal repeats are three major repeat families and contribute 25%, 13%, and 8% of CpG dinucleotides to the human genome, respectively. De novo retrotransposition of LINEs and Alu elements has been recognized as a major driver for insertion polymorphisms. Single-neuron genome sequencing revealed that LINEs may actively retrotranspose during neurogenesis, thereby contributing to genome diversity between neurons [44]. The influence of LINE transposition on the neuron epigenome landscape remains largely unexplored. We have exploited Alu-anchored bisulfite PCR libraries to identify evolutionarily recent Alu element insertions in the human brain. As the principal SINEs, Alu elements contribute over

7.1 million (23% of the total) CpG dinucleotides. We identified a total of 327 putatively recent Alu insertions in the human brains from eight Alu methylomes [45]. The majority of these Alu element insertions remained hemizygous in individuals and heavily methylated. Methylation analyses of three genomic loci revealed that the Alu insertions did not result in methylation changes in flanking CpG dinucleotides.

SNPs are the most common genetic variations in the human genome. According to the SNP database (http://www.ncbi.nlm.nih.gov/projects/ SNP/snp_summary.cgi), over 62 million SNPs have been identified, 43 million of which have been validated. According to the 1000 Genomes Project, many more genetic mutations with low frequency are present in populations [36]. Although some knowledge has been gained for the functional relevance of SNPs within the aforementioned epigenetic genes, numerous SNPs fall into intergenic regions, and their influences on local or genome-wide DNA methylation are largely unexplored. A large number of genome-wide association studies conducted since 2005 have led to the identification of SNPs enriched in patients with a given disease. The National Human Genome Research Institute provides a catalog for disease-associated SNPs identified to date. Approximately 8229 variants have been reported for 807 diseases/traits (http://www.genome.gov/gwastudies/). It has been shown that intergenic disease-associated SNPs are enriched in functional elements, enhancers in particular. Enhancers are known to be frequently associated with tissue-specific methylation patterns. Integrative analysis with genetic variation and epigenetic profiles may shed new light on the epigenetic importance of these disease-associated genetic variants.

4. ALLELIC-SPECIFIC DNA METHYLATION

Allele-specific methylation (ASM) has been well recognized in X-chromosome inactivation in females to achieve dosage balance and genomic imprinting essential for embryogenesis. Mechanistically, ASM events can be either parent-of-origin-dependent (imprinted) or sequence-dependent. The parent-of-origin-dependent ASM (P-ASM) may arise during gametogenesis for gametic imprints or in postimplantation embryos for somatic imprints [46,47]. During early embryogenesis, maternal factor PGC7 (also known as Dppa3, Stella) protects imprinted loci marked with dimethylated histone H3 lysine 9 from TET3-mediated conversion to 5hmC [48]. Although some imprinted domains span over megabases, the number of such regions has been found to be very limited [47]. These regions are well documented in imprinted gene databases such as Geneimprint (http://www.geneimprint.com) or the catalog of Imprinted Genes and Parent-of-Origin Effects [49]. Genome-wide studies have provided mouse [50] and human [51] P-ASM maps. A total of 55 discrete genomic loci with 1952 CG dinucleotides were identified as imprinted in the mouse genome and 51 imprinted in the human genome. Over a dozen human P-ASM loci were found to be placenta specific and the majority of them are unmethylated in sperm and all human embryonic stem cells [51].

DNA methylation shifts at P-ASM loci have been observed in a number of human diseases, which are characterized by complex phenotypes that affected development, viability, and neurological functions. A total of nine human imprinted syndromes and corresponding mouse models have been summarized [52]. For individuals suspected of having any of these imprinting related disorders, the first-tier diagnostic test in the clinic is DNA methylation analysis of P-ASM loci. Since only one of the two parental alleles is active at imprinted loci, genetic mutations in the active allele result in function defects that cannot be compensated for by the other inactive one. On the other hand, copy number variations leading to extra copies of an active allele can also be harmful. For genetic mutations on chromosome X in females, the X-inactivation process is random during cell differentiation and thus results in somatic mosaicism. Defects in proteins controlling the erasure, establishment, or maintenance of imprinted loci may lead to aberrant DNA methylation at multiple loci. For instance, a transcriptional repressor, ZFP57 zinc finger protein, recognizes the methylated CpG within a "TGCCGC" consensus sequence. A single base pair deletion identified in the ZFP57 gene in a Transient Neonatal Diabetes Mellitus (TNDM) patient leads to methylation defects at multiple maternal P-ASM loci [53].

Human somatic cells normally contain two sets of chromosomes, and genetic variations within an individual are frequently heterozygous. These heterozygous genetic variants at nonimprinted loci may lead to methylation differences between the parental alleles. Such sequence-dependent allele-specific methylation (S-ASM) events were found to be prevalent in the human genome and approximately 23–37% of heterozygous SNPs were associated with S-ASM [54]. Over 65% of the human SNPs documented in the NCBI SNP database are C/T (or G/A) transitions, 21.7% of which occur at CpG dinucleotides [55]. These SNPs directly result in the gain or loss of methylation target sites, the CpG dinucleotides. Accordingly, a heterozygous SNP in the CpG dinucleotide is the primary source of sequence-dependent ASM [54]. In contrast to P-ASM, the influence of S-ASM is highly localized and frequently restricted to a very small number of neighboring CpG sites [50,54].

Experimentally, both P-ASM and S-ASM identifications rely on methylation analyses of SNP alleles. The high-density SNP microarrays have been exploited intensively to quantitatively assess ASM with DNA from people of different populations [56–58]. Prior to array hybridization, genomic DNA is first digested with methylation-sensitive enzymes or enriched for methylated DNA with antibodies against methylated cytosine or MBD proteins.

Therefore, differentially methylated alleles will show allele-skewed hybridization signals. Rapid advances in high-throughput sequencing technologies now offer alternatives to identifying ASM events. High-throughput sequencing with bisulfite-treated genomic DNA provides the methylation information for CpG sites on the two alleles together with sequence information for genetic variations. Based on adjacent SNP information, these CpG sites may be assigned to each of the two alleles. The methylation status of the two alleles will be subjected to standard statistical tests such as the Fisher's exact test for ASM determination [50,54]. However, it is a challenging task to identify SNPs from bisulfite-converted sequence reads. Most of the current high-throughput methylation data sets are with the genomic coverage at 10×, but it was estimated that an average 30× sequence read depth would be required to call 96% of SNPs accurately from bisulfite sequencing data [59]. In addition, the true C/T SNPs and the C to T substitutions resulting from bisulfite conversion cannot be distinguished with bisulfite sequencing data, since many human SNPs are C/T (or G/A) transitions and unmethylated cytosines will be converted to uracils during bisulfite treatment and substituted by thymines in the following PCR step.

There is a growing interest in the understanding of the impact of local sequence content on DNA methylation patterns. A 2012 study performed reciprocal crosses between two distantly related inbred mouse strains (129x1/SvJ and Cast/EiJ) and determined the methylomes for F1 offspring at a single-base resolution [50]. In the genomes of the two strains, over 20 million SNPs provide a high-density SNP map (one SNP in every 133 bp) powerful enough for the identification of allele-associated methylation events. A total of 1952 and 131,765 CpG dinucleotides were identified as P-ASM and S-ASM, respectively. Of 131,765 S-ASM sites, 9030 CpG dinucleotides were clustered into 1051 regions. The rest of the S-ASM sites were found to have very restricted influence on local DNA methylation and affected only one or a few CpG sites adjacent to genetic variants. The identification of such a large number of S-ASM sites provides a unique opportunity to study the relationship between genetic and epigenetic variations. Compared to scattered S-ASM, clustered S-ASM CpG sites are more frequently localized within gene promoter regions and around 20% of genes associated with clustered S-ASM show allele-specific transcription [50]. The methylation analysis of scattered S-ASM revealed a significant sequence preference for S-ASM events. The allele carrying a sequence variant G at the −1 or C at the +1 position of a CpG dinucleotide tends to be hypermethylated. As mentioned in the previous section, over 43 million SNPs have been validated in the human genome. The determination of SNP influence on DNA methylation in the context of methylation comparisons between disease patients and the normal population will certainly provide a number of epigenetic markers for future development of novel strategies to improve disease management.

5. BLOOD CELLULAR HETEROGENEITY AND METHYLATION VARIATION

Human peripheral blood cells consist of many functionally distinct cell subsets in varying proportions, including erythrocytes, leukocytes, and thrombocytes. According to the presence of granules, leukocytes could be divided into two groups: granular and agranular cells. With the assistance of molecular tools, lymphocytes could further be classified into a growing list of cell subsets, many more than are shown in Figure 3. The most widely used molecular markers to distinguish blood cell subsets are cell surface markers. For instance, the surface expression of CD14, CD16, and CCR2 molecules could unequivocally classify monocytes into three groups [60]. A number of transcription factors have been recognized as master regulators of blood cell differentiation. In addition, the immune cells could also be classified by the effector molecules that they produce, including cytokines [61].

Apparently, the distortion in the normal distribution of blood cell subsets could result in severe health problems. Various blood tests have been established to measure the number of cells in each subset. The determination of the white blood cell count is a classic marker of the immune or the inflammatory response and the absolute CD4+ T cell count is a powerful way to monitor progression of human immunodeficiency virus disease [62]. To accurately distinguish subtypes of immune cells, many antibodies

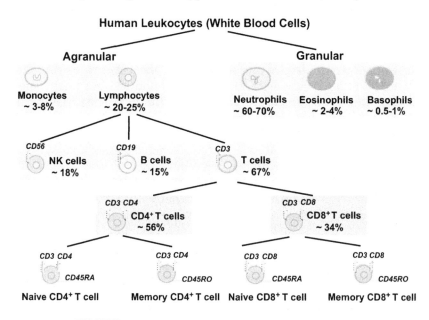

FIGURE 3   The cellular composition of human leukocytes.

against cell surface markers have been developed and extensively used in flow cytometric measurements. However, the analysis of cell subsets with flow cytometry depends on the availability and quality of antibodies. It also requires fresh cells with an intact cellular membrane. Most importantly, it remains a question whether cell surface markers alone are sufficient to precisely distinguish the cell subsets with subtle differences [63]. The plasticity in cellular phenotypes greatly enhances the challenges. Especially, the potential lineages that cells can differentiate into (i.e., CD4$^+$ T helper) far exceed expectation.

DNA methylation maintains homeostasis of hematopoietic stem cells and controls proper hematopoietic differentiation. Aberrant DNA methylation would lead to the premature activation of hematopoietic stem cells, and eventually, the development of leukemia [64]. Mouse hematopoietic stem cells with DNMT1 deficiency may differentiate into myeloerythroid, but not lymphoid, progeny [64]. As a relatively easily accessible tissue, human peripheral blood is the most widely studied system in clinical research. In the years 2012 to 2014, over 1000 manuscripts were published on the study of DNA methylation with human blood. In addition to blood disorders, the methylation status of peripheral blood samples may be used as an indicator of other diseases. For instance, to identify schizophrenia DNA methylation biomarkers in blood, a study examined methylomes for 759 cases of schizophrenia and 738 controls [65]. A total of 139 highly significant methylation sites were identified and linked to gene networks related to neuronal differentiation and associated with schizophrenia development.

With the development of the next-generation sequencing technique, the first methylome for human peripheral blood mononuclear cells at single-base resolution was revealed in 2010 [66]. In 2011, the entire methylomes for a number of human blood cell subsets became available [67]. A number of genome-wide DNA methylation studies have been done to examine the methylation differences between cell subsets directly sorted from human blood [67–69]. Various array-based or sequencing-based techniques have been explored to further uncover the methylation changes during hematopoiesis. These studies demonstrated that DNA methylation stabilizes cell fate decisions and controls the tissue-specific gene regulation within the hematopoietic compartment. Not surprisingly, genomic loci associated with hematopoietic-specific genes are frequently differentially methylated among cell subsets. Differentially methylated regions associated with hematopoiesis could be found in the promoters of lineage-specific genes, intragenic CpG islands, CpG island shores, and transcription factor-binding sites. High-resolution DNA methylation maps of purified cell populations of blood lineages have been generated for mouse models [70]. The cellular differentiation hierarchy of the blood lineage could be inferred from the methylation patterns of the genomic loci with lineage-specific DNA methylation. These studies paved the way for generating a collection of epigenetic fingerprints for distinct blood cell subsets. A 2012 review has summarized the immune genes regulated by DNA methylation and used CD4$^+$ T cells to illustrate how DNA methylation dynamics may contribute to the plasticity of cell subsets [71].

The heterogeneity in blood cell composition has gradually been recognized as an important confounding factor that could compromise the result interpretations for blood methylation comparisons. With a cohort of whole blood samples and cell fractions isolated using density gradient centrifugation, Adalsteinsson et al. found that blood cellular heterogeneity largely explained the interindividual variation observed in the whole-blood DNA methylation levels [72]. Therefore, the highly dynamic cell composition of peripheral blood poses a challenge to blood methylation studies. On the other hand, the clinical importance of blood cellular composition motivates the development of epigenetic-based methods to accurately categorize the blood cell subtypes. Human blood composition provides much information for prognosis, disease diagnosis, immune status, health state, and even aging rates. Currently, the ability to quantitatively categorize every kind of blood cell subset is still limited. Methylation profile-based methods to quantitatively assess the percentage of blood cell subsets and to evaluate the association of blood composition and biological processes (immune state, aging) are highly desired.

6. BRAIN CELLULAR HETEROGENEITY AND METHYLATION VARIATION

Human brains can roughly be divided into three functionally distinct regions: cerebrum, cerebellum, and brain stem. These parts consist of terminally differentiated glial cells and neurons, which can no longer divide. Approximately 86 billion neurons and a roughly equal number of glial cells are present in a mature human brain [73]. Glial cells consist of macroglial, microglial, and ependymal cells. The macroglial cells, including astrocytes and oligodendrocytes, are a heterogeneous population of cells with distinct molecular composition, structure, and activity. The microglial cells are mesodermal in origin and derived from blood monocytes that mediate the local immune response in the central nervous system [74]. As the principal brain functional units, neurons can be classified into two major groups: projection neurons and interneurons. With molecular, morphological, and physiological analyses, hundreds of distinct subtypes can be characterized for the neuronal cells. Within a neuronal subtype, substantial cellular heterogeneity has been documented for diverse molecular characteristics, firing patterns, and connections. The systematic and accurate classification of these cells is far beyond a challenging task.

For instance, to classify interneurons, it took the effort of many neuroscientists to develop a complex taxonomical system equipped with supervised classification models [75].

The understanding of the complexity of brain cellular composition is an important field of basic and clinical neuroscience. From the induction of the neural plate to the formation of synapses, normal brain development is accompanied by extreme dynamics in a cell population. Postnatal brains show continuing dramatic changes in the number of different cell subsets [76]. The proper functioning of a normal brain depends on the establishment of such diverse repertoire of neuronal subtypes and glial cells. Apparently, distortion in the normal distribution of brain cell subsets may lead to severe health problems. Brain tumors are associated with overproliferation of neurons or glial cells, while progressive dopamine neuron loss is a characteristic of Parkinson disease. Various attempts, such as design-based stereology and flow cytometry, have been taken to measure the number of brain cells in each subset [73,77,78]. Unfortunately, the molecular description of each cell subset in the human brain is an extremely challenging task. Despite several immune-labeling strategies that have been developed for nuclei sorting [79,80], a large number of molecular markers would be needed to precisely distinguish brain cell subsets.

Since 2005, great efforts have been made to generate cell-subset-specific gene expression profiles by isolating the small populations of fluorescently labeled neurons [81]. Although it remains impossible to digitally represent every single cell for the whole brain, the Allen Human Brain Atlas provides a cellular-resolution gene expression profile for ~1000 genes important for neural functions by in situ hybridization (ISH) [82]. The ISH patterns were used to define a set of genes as markers for specific cell populations. The same group has released a comprehensive map of the human brain transcriptome for approximately 900 anatomically defined brain regions in two males of similar age and ethnicity [83]. It was found that the two individuals' brain gene expression signatures were highly conserved but varied among anatomical regions. This reflects the fact that different brain regions contain diverse cell populations and have distinct functions.

The importance of DNA methylation has been firmly established for neural plasticity and brain functions. During neuronal differentiation, de novo DNA methylation occurs at the promoters of germ-line-specific genes to repress pluripotency in progenitor cells, while the methylation loss at the promoters activates neuron-specific genes [84]. Even in postmitotic neuronal cells, active methylation changes were observed before and after synchronous neuronal activation [85,86]. On the other hand, aberrant DNA methylation would lead to the premature activation of neuronal progenitor cells and, potentially, the development of brain tumors

[87]. Over a dozen neurological diseases have been linked to disorders in epigenetic machinery [88]. As mentioned in a previous section, mutations in DNMT3B lead to mental retardation and defective brain development, and mutations in MeCP2 have been linked to Rett symptoms. Notably, 5-hydroxymethylcytosine, an intermediate of the active DNA demethylation process, was found to be abundant in brains and enriched within the gene bodies of neuron function-related genes [89]. A global loss in both methylation and hydroxymethylation has been observed in the hippocampus of Alzheimer's disease patients [90].

Owing to the lack of an effective cell dissociation technique and the antibodies against cell-subset-specific markers, the majority of brain methylation studies have been performed with tissues derived from regions of interest. However, diverse subpopulations of neurons and/or glial cells can often be found even within one given brain region. Despite efforts to compare methylation profiles of neurons and glial cells with the separation of nuclei by fluorescence-activated cell sorting [91], the current understanding of the methylation heterogeneity of the human brain is still very limited. With the development of high-throughput sequencing technologies, it becomes feasible and affordable to generate many sequence reads for a given genomic locus to survey genomes from a significant number of cells. With the combination of advances in sequencing technologies and algorithms for methylation data analysis, the decoding of the epigenetic language of brain heterogeneity is within reach.

7. STEM CELLULAR HETEROGENEITY AND METHYLATION VARIATION

Stem cells have self-renewal abilities and may differentiate into tissue-specific cells with specialized functions. According to their origins, stem cells can be classified into embryonic stem cells, somatic stem cells, and induced pluripotent stem cells (iPSCs). Embryonic stem cells are derived from the undifferentiated inner cell mass of blastocysts and somatic stem cells are undifferentiated cells in tissues that frequently stay quiescent and do not actively divide. In 2006, Takahashi and Yamanaka demonstrated that iPSCs can be generated by introducing four transcription factors, Oct3/4, Sox2, c-Myc, and Klf4, into fully differentiated fibroblasts [92]. Induced PSCs resemble ESCs in morphology, differentiation potential, and gene expression profiles. All these stem cells have a wide range of applications throughout the fields of basic and clinical research. The self-renewal capacity of stem cells implies that an unlimited number of cells with isogenic nature can be obtained. In addition, iPSCs from patients with genetic mutations will provide invaluable disease models and tools for drug screening. The remarkable potential of stem cells that can give rise to various types of cells is the foundation for cell replacement strategies critical for organs with little self-repair capacity. Many innovations have also been made in the generation of pluripotent stem cells and several publications present significant progress on stem cell culture technology [93]. However, the tumorigenicity and cellular heterogeneity of stem cells have become the highest hurdles for translational applications.

Stem cells derived from the same founder display extensive phenotypic and functional heterogeneity [94,95]. There has been increasing attention paid to the causes and consequences of such clonal diversity. At the level of gene expression, substantial information has been gathered from single-cell gene expression profiling regarding the cell-to-cell variability within clones. A variety of transcriptional factors demonstrate significant variation in expression, including both pluripotency factors and lineage-specific factors. Apparently, such a prominent phenomenon indicates that clonal diversity can significantly influence cell fate decisions [96]. For instance, ESCs expressing low levels of Hes1 tend to differentiate into neural cells, while ESCs with high Hes1 expression favor a mesodermal fate. Cell cycle progression has been linked to the expression dynamics of transcriptional factors [97]. Such heterogeneity at the single-cell level was shown to be present in both pluripotent stem cells and their derivatives.

In addition to the stochastic fluctuations in gene transcription, varying external stimuli in the microenvironment and oscillations in cell activity (e.g., cell cycling) may trigger the population diversity of stem cells. The in vitro culture environment plays a critical role in sustaining the pluripotency of ESCs [98]. The cytokine LIF activates signal transducer and activator of transcription 3 and suppresses the spontaneous differentiation of ESCs in the absence of a feeder layer of fibroblasts [99,100]. However, mouse ESCs cultured in serum show significant heterogeneity and may be classified into subpopulations based on pluripotency markers such as SSEA-1 [101] or Nanog [26]. As mentioned in the previous section, 2i, PD0325901 and CHIR99021, may block stem cell differentiation and result in morphologically homogeneous ESCs [24]. ESCs cultured under different conditions ("2i ESCs" vs "serum ESCs") [102] show remarkable differences in gene expression, histone modifications, and DNA methylation. While serum ESCs with a heavily methylated genome resemble E6.5 blastocysts, the 2i ESCs with a hypomethylated methylome are comparable to E3.5 blastocysts and migratory primordial germ cells [26]. Interestingly, these two distinct methylomes are interconvertible with a shift in culture conditions [103]. These results suggest that specific signaling pathways and regulatory factors play essential roles in maintaining the "dynamic equilibrium" of a stem cell population.

Despite the advances made with single-cell analyses, the epigenetic basis for such cellular heterogeneity remains largely unexplored. DNA methylation has been recognized as an essential mechanism guiding stem cell self-renewal and differentiation [104]. During development and cellular differentiation, the establishment of tissue-specific patterns of DNA methylation enables cells with the same genetic composition to exhibit distinct phenotypes [105]. On the other hand, fully differentiated cells could be reprogrammed into pluripotent cells through different approaches, including nuclear transfer, cell fusion, and transcription factor transduction [106]. Epigenome remodeling is key to allowing cells to reacquire pluripotency [107]. Great efforts have also been made to compare the clones of human ESCs and iPSCs. Consequently, we gained a substantial understanding of how cellular sources, donor phenotype, diverse reprogramming, and culture methods can affect the functional variability, including lineage and even the malignant potential of stem cells. Studies have suggested that the epigenetic inheritance mechanism in stem cells may be incompetent and the stem cell epigenome can be extremely unstable [108]. In addition to the frequent loss of heterozygosity in multiple human ESCs [109], prolonged culture-induced epigenetic heterogeneity has also been observed. Abnormal epigenetic aberrations could directly result in carcinogenesis [64] and iPSC-derived chimeric mice were found to develop tumors, in particular those induced with c-Myc [110]. Although the industrial production of human ESCs and iPSCs is within reach, the challenges of obtaining functionally homogeneous and genetic/epigenetic stable stem cells remain to be resolved.

The methylation variations among cell lines or between ESCs and iPSCs have been extensively examined. Important findings have been achieved, including iPSCs maintaining an epigenetic memory of source cells [111,112] and aberrant epigenetic reprogramming events during reprogramming of iPSCs [113]. However, little is known about the methylation heterogeneity within a pluripotent stem cell line [114]. We have observed partially methylated (e.g., 20–80% methylated) genomic loci showing variation in DNA methylation patterns [32]. Such variations constitute epigenetic heterogeneity among a cell population. In particular, some genomic loci display strikingly bipolar DNA methylation patterns: the presence of both hypomethylated and hypermethylated patterns within a mixed cell population. In mouse embryonic stem cells, we found that some bipolar methylated genomic loci host symmetrically methylated CpG dyads with high fidelity in DNA methylation inheritance [35]. More interestingly, the symmetrically methylated CpG dyads are enriched in transcription factor-binding sites and regions with active histone modifications, especially H3K27ac and H3K4me1. Since H3K27ac marks distinguished active and cell-type-specific enhancers from poised ones with H3K4me1 alone [115,116], this indicates that the partially methylated status of these CpG sites may have resulted from cell-subset-specific methylation events, and bipolar methylated genomic loci may have important biological implications.

In summary, stem cells and patient-specific iPSCs in particular are invaluable for personalized medicine, and methods for generating homogeneous cell populations with stable genomes and epigenomes are urgently required.

8. QUANTITATIVE ASSESSMENT OF DNA METHYLATION VARIATION

The most frequently used techniques to distinguish differentially methylated DNA are bisulfite treatment, methylation-sensitive enzyme digestion, and immunoprecipitation with antibodies specific for methylated cytosines or MBD proteins. The popular platforms for the generation of genome-wide DNA methylation data are either array-based or sequencing-based. The combinations of different techniques and platforms produce various types of DNA methylation data. Currently, the high-throughput sequencing of bisulfite-treated DNA is still considered the gold standard for DNA methylation profiling. With bisulfite treatment, unmethylated cytosines will be converted to uracils and substituted by thymines in the following PCR step, whereas methylated cytosines remain unchanged during bisulfite treatment. The comparisons of sequences derived from bisulfite-treated DNA with a reference genomic sequence enable the determination of the methylation status of cytosines in the input DNA. Only bisulfite sequencing data can provide DNA methylation information for a DNA molecule at single-base resolution. In contrast, array-based approaches, such as the Illumina methylation chip, may provide methylation information for a single CpG dinucleotide but at the average methylation level (the percentage of methylated CpGs for a given locus in a cell population). Thus, array-based methylation data cannot be used to assess methylation variation within unsorted tissue samples.

We have introduced the concept of "methylation entropy" to quantitatively assess DNA methylation variation within a given DNA sample. A DNA methylation status is coded in binary: methylated versus unmethylated. DNA methylation patterns are herein defined as the combination of methylation statuses of neighboring CpG dinucleotides in a DNA strand. Thus, methylation variation of a given genomic locus in a cell population refers to the variability that might be observed in the pattern of DNA methylation [32]. Methylation entropy is calculated with the following parameters (Figure 4(A)): (1) number of CpG sites in a given genomic locus; (2) number of sequence reads generated for a genomic locus; and (3) frequency of each distinct DNA methylation pattern observed in a genomic locus, calculated based upon the sequence reads that were generated for the locus. The methylation entropy is minimal at the value of 0

(A) $$ME = \frac{e}{b}\sum(-\frac{n_i}{N}Log\frac{n_i}{N})$$

ME: Methylation Entropy
e: Entropy for code bit
b: Number of CpG sites
n_i: Observed occurrence of methylation pattern i
N: Total number of sequence reads generated

(B)

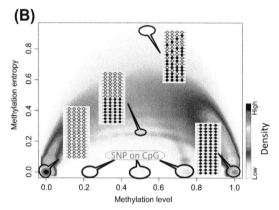

FIGURE 4 (A) The formula for methylation entropy and (B) the heat-map plot of methylation level versus methylation entropy. The heat-map plot was generated with human prefrontal cortex methylome data. Genomic segments with four neighboring CpG dinucleotides and at least 16-read coverage were included in the analysis.

when DNA molecules in all cells share the same methylation pattern and is maximal at the value of 1 when all possible DNA methylation patterns are equally represented in a population of cells. Thus, genomic loci with the same methylation entropy may have different methylation levels on average. However, genomic loci with different methylation entropies may share the same average level of methylation. Therefore, methylation entropy analysis differs significantly from conventional methylation level-based analyses in that it enables the assessment of epigenetic homogeneity of a cell population.

As discussed in the previous section, phenotypic diversity and functional variation have been commonly observed in human tissues, including blood, brain, and even stem cells. However, the epigenetic sources of such population variations are still largely unknown. Traditional DNA methylation analyses focus on the determination and comparison of the average methylation levels and, thus, cannot reveal the methylation variation within tissues. The analysis of DNA methylation patterns with bisulfite sequencing data can provide important clues for the understanding of epigenetic heterogeneity. Stochastic methylation events may lead to highly variable methylation patterns at some loci; but deterministic methylation events may be associated with well-structured methylation patterns, e.g., bipolar DNA methylation patterns. Stochastic methylation events may be associated with the failure in methylation inheritance (e.g., the skipping of sites by DNMT1), random de novo methylation by DNMT3A/3B, and demethylation by TET enzymes. Deterministic methylation events may be linked to allele-specific DNA methylation and cell-subset-specific DNA methylation.

To illustrate the differences between methylation level and entropy analyses, we reanalyzed a published methylation map for the human prefrontal cortex [117]. Over 1.3 billion methylation sequence reads were progressively scanned to determine the methylation levels and entropies for all possible genomic segments containing four neighboring CpG dinucleotides. A total of 850,763 segments covering 1,812,803 CpG sites were identified. The average DNA methylation level for these segments was determined to be 54.7%, and the mean methylation entropy was 0.15. Several regions are highlighted in Figure 4(B) to indicate the potential distributions of methylation patterns. A significant number of genomic segments are associated with deterministic methylation events showing well-structured methylation patterns. Among all the fragments, 95,928 (11.3%) are completely methylated, and 227,213 (26.7%) are completely unmethylated. For intermediately methylated genomic regions (methylation level in the range of 40–60%), a small number of genomic segments are associated with stochastic methylation events showing highly variable methylation patterns, and the others are with bipolar DNA methylation patterns. Some fragments with methylation levels near 0.25, 0.5, or 0.75 show a significant decrease in methylation entropy. These genomic fragments are likely to harbor an SNP on the CpG site. Such SNPs at one or both alleles eliminate the CpG sites, prevent methylation from occurring, and result in allele-specific methylation and a decrease in methylation variation. The genome-wide analysis of epigenetic heterogeneity in the human prefrontal cortex revealed the diverse origins of DNA methylation variation. More sophisticated algorithms and advanced computational tools would be needed to further dissect the mechanisms underlying epigenetic heterogeneity and quantitate the contributions of asymmetric DNA methylation, allele-specific DNA methylation, and cell-subset-specific DNA methylation.

9. CLOSING REMARKS

Decoding the epigenome is a new and growing field of postgenomic research. The analysis of epigenetic heterogeneity can provide target loci for precise genome and epigenome editing in specific types of cell and lead to the personalized understanding of the origins and processes of human diseases and, eventually, the development of noninvasive diagnostic procedures and personalized therapeutic protocols. With the advancements in high-throughput sequencing technologies, large-scale "omics" data including methylomes have accumulated rapidly. Continuing efforts are needed to develop powerful computational tools to meet the rising challenges in today's bioinformatics for Big Data. The integration of genome/epigenome, transcriptome, and clinical–pathological information will provide a more comprehensive understanding of diseases and invaluable biomarkers for diagnosis and prognosis in the context of personalized medicine.

LIST OF ACRONYMS AND ABBREVIATIONS

2i	Two kinase inhibitors, PD0325901 and CHIR99021
5hmC	5-hydroxymethylcytosine
ASM	Allele-specific methylation
DNMT	DNA methyltransferase
ESC	Embryonic stem cell
H3	Histone 3
H3K4me	Histone 3 lysine residue 4 methylated
ICF	Immunodeficiency, centromeric region instability, and facial anomalies syndrome
ISH	In situ hybridization ⁻
LIF	Leukemia inhibitory factor
LINE	Long interspersed nuclear element
MBD	Methylated CpG-binding domain
MeCP2	Methylated CpG-binding protein 2
MGMT	$O(6)$-methylguanine-DNA methyltransferase
NCBI	National Center for Biotechnology Information
P-ASM	Parent-of-origin-dependent allele-specific methylation
S-ASM	Sequence-dependent allele-specific methylation
SINE	Short interspersed nuclear element
SNP	Single-nucleotide polymorphism
TET	Ten–eleven translocation family of dioxygenases
TMZ	Temozolomide
TNDM	Transient Neonatal Diabetes Mellitus

References

[1] Karlić R, Chung H-R, Lasserre J, Vlahoviček K, Vingron M. Histone modification levels are predictive for gene expression. Proc Natl Acad Sci 2010;107(7):2926–31.

[2] Suganuma T, Workman JL. Signals and combinatorial functions of histone modifications. Annu Rev Biochem 2011;80:473–99.

[3] Barski A, Cuddapah S, Cui K, Roh TY, Schones DE, Wang Z, et al. High-resolution profiling of histone methylations in the human genome. Cell 2007;129(4):823–37.

[4] Hegi ME, Diserens AC, Gorlia T, Hamou MF, de Tribolet N, Weller M, et al. MGMT gene silencing and benefit from temozolomide in glioblastoma. N Engl J Med 2005;352(10):997–1003.

[5] Yung WK. Temozolomide in malignant gliomas. Seminars Oncol 2000;27(3 Suppl. 6):27–34.

[6] Lam LL, Emberly E, Fraser HB, Neumann SM, Chen E, Miller GE, et al. Factors underlying variable DNA methylation in a human community cohort. Proc Natl Acad Sci USA 2012;109(Suppl. 2):17253–60.

[7] Arand J, Spieler D, Karius T, Branco MR, Meilinger D, Meissner A, et al. In vivo control of CpG and non-CpG DNA methylation by DNA methyltransferases. PLoS Genet 2012;8(6):e1002750.

[8] Law JA, Jacobsen SE. Establishing, maintaining and modifying DNA methylation patterns in plants and animals. Nat Rev Genet 2010;11(3):204–20.

[9] Bostick M, Kim JK, Esteve PO, Clark A, Pradhan S, Jacobsen SE. UHRF1 plays a role in maintaining DNA methylation in mammalian cells. Science (New York, NY) 2007;317(5845):1760–4.

[10] Jeong S, Liang G, Sharma S, Lin JC, Choi SH, Han H, et al. Selective anchoring of DNA methyltransferases 3A and 3B to nucleosomes containing methylated DNA. Mol Cell Biol 2009;29(19):5366–76.

[11] Sharma S, De Carvalho DD, Jeong S, Jones PA, Liang G. Nucleosomes containing methylated DNA stabilize DNA methyltransferases 3A/3B and ensure faithful epigenetic inheritance. PLoS Genet 2011;7(2):e1001286.

[12] Jones PA, Liang G. Rethinking how DNA methylation patterns are maintained. Nat Rev Genet 2009;10(11):805–11.

[13] Cheng X, Blumenthal RM. Mammalian DNA methyltransferases: a structural perspective. Structure 2008;16(3):341–50.

[14] Guo H, Zhu P, Yan L, Li R, Hu B, Lian Y, et al. The DNA methylation landscape of human early embryos. Nature 2014;511(7511):606–10.

[15] Okano M, Xie S, Li E. Cloning and characterization of a family of novel mammalian DNA (cytosine-5) methyltransferases. Nat Genet 1998;19(3):219–20.

[16] Okano M, Bell DW, Haber DA, Li E. DNA methyltransferases Dnmt3a and Dnmt3b are essential for de novo methylation and mammalian development. Cell 1999;99(3):247–57.

[17] Jin B, Ernst J, Tiedemann RL, Xu H, Sureshchandra S, Kellis M, et al. Linking DNA methyltransferases to epigenetic marks and nucleosome structure genome-wide in human tumor cells. Cell Rep 2012;2(5):1411–24.

[18] Choi SH, Heo K, Byun HM, An W, Lu W, Yang AS. Identification of preferential target sites for human DNA methyltransferases. Nucleic Acids Res 2011;39(1):104–18.

[19] Tahiliani M, Koh KP, Shen Y, Pastor WA, Bandukwala H, Brudno Y, et al. Conversion of 5-methylcytosine to 5-hydroxymethylcytosine in mammalian DNA by MLL partner TET1. Science (New York, NY) 2009;324(5929):930–5.

[20] Hackett JA, Sengupta R, Zylicz JJ, Murakami K, Lee C, Down TA, et al. Germline DNA demethylation dynamics and imprint erasure through 5-hydroxymethylcytosine. Science (New York, NY) 2013;339(6118):448–52.

[21] Ito S, Shen L, Dai Q, Wu SC, Collins LB, Swenberg JA, et al. Tet proteins can convert 5-methylcytosine to 5-formylcytosine and 5-carboxylcytosine. Science (New York, NY) 2011;333(6047):1300–3.

[22] Dawlaty MM, Ganz K, Powell BE, Hu YC, Markoulaki S, Cheng AW, et al. Tet1 is dispensable for maintaining pluripotency and its loss is compatible with embryonic and postnatal development. Cell Stem Cell 2011;9(2):166–75.

[23] Ko M, Bandukwala HS, An J, Lamperti ED, Thompson EC, Hastie R, et al. Ten-Eleven-Translocation 2 (TET2) negatively regulates homeostasis and differentiation of hematopoietic stem cells in mice. Proc Natl Acad Sci USA 2011;108(35):14566–71.

[24] Ying QL, Wray J, Nichols J, Batlle-Morera L, Doble B, Woodgett J, et al. The ground state of embryonic stem cell self-renewal. Nature 2008;453(7194):519–23.

[25] Wray J, Kalkan T, Smith AG. The ground state of pluripotency. Biochem Soc Trans 2010;38(4):1027–32.

[26] Ficz G, Hore TA, Santos F, Lee HJ, Dean W, Arand J, et al. FGF signaling inhibition in ESCs drives rapid genome-wide demethylation to the epigenetic ground state of pluripotency. Cell Stem Cell 2013;13(3):351–9.

[27] Koh KP, Yabuuchi A, Rao S, Huang Y, Cunniff K, Nardone J, et al. Tet1 and Tet2 regulate 5-hydroxymethylcytosine production and cell lineage specification in mouse embryonic stem cells. Cell Stem Cell 2011;8(2):200–13.

[28] Yu M, Hon GC, Szulwach KE, Song CX, Zhang L, Kim A, et al. Base-resolution analysis of 5-hydroxymethylcytosine in the mammalian genome. Cell 2012;149(6):1368–80.

[29] Bird A. DNA methylation patterns and epigenetic memory. Genes Dev 2002;16(1):6–21.

[30] Laird CD, Pleasant ND, Clark AD, Sneeden JL, Hassan KM, Manley NC, et al. Hairpin-bisulfite PCR: assessing epigenetic methylation patterns on complementary strands of individual DNA molecules. Proc Natl Acad Sci USA 2004;101(1):204–9.

[31] Shao C, Lacey M, Dubeau L, Ehrlich M. Hemimethylation footprints of DNA demethylation in cancer. Epigenetics 2009;4(3):165–75.

[32] Xie H, Wang M, de Andrade A, Bonaldo Mde F, Galat V, Arndt K, et al. Genome-wide quantitative assessment of variation in DNA methylation patterns. Nucleic Acids Res 2011;39(10):4099–108.

[33] Watanabe N, Okochi-Takada E, Yagi Y, Furuta JI, Ushijima T. Decreased fidelity in replicating DNA methylation patterns in cancer cells leads to dense methylation of a CpG island. Curr Top Microbiol Immunol 2006;310:199–210.

[34] Ushijima T, Watanabe N, Shimizu K, Miyamoto K, Sugimura T, Kaneda A. Decreased fidelity in replicating CpG methylation patterns in cancer cells. Cancer Res 2005;65(1):11–7.

[35] Zhao L, Sun MA, Li Z, Bai X, Yu M, Wang M, et al. The dynamics of DNA methylation fidelity during mouse embryonic stem cell self-renewal and differentiation. Genome Res 2014;24(8):1296–307.

[36] Abecasis GR, Auton A, Brooks LD, DePristo MA, Durbin RM, Handsaker RE, et al. An integrated map of genetic variation from 1,092 human genomes. Nature 2012;491(7422):56–65.

[37] Khurana E, Fu Y, Colonna V, Mu XJ, Kang HM, Lappalainen T, et al. Integrative annotation of variants from 1092 humans: application to cancer genomics. Science (New York, NY) 2013;342(6154):1235587.

[38] Simo-Riudalbas L, Esteller M. Cancer genomics identifies disrupted epigenetic genes. Hum Genet 2014;133(6):713–25.

[39] Ley TJ, Ding L, Walter MJ, McLellan MD, Lamprecht T, Larson DE, et al. DNMT3A mutations in acute myeloid leukemia. N Engl J Med 2010;363(25):2424–33.

[40] Li Y, Zhu B. Acute myeloid leukemia with DNMT3A mutations. Leuk Lymphoma 2014.

[41] Yoshikawa Y, Sato A, Tsujimura T, Otsuki T, Fukuoka K, Hasegawa S, et al. Biallelic germline and somatic mutations in malignant mesothelioma: multiple mutations in transcription regulators including mSWI/SNF genes. Int J Cancer 2014.

[42] Matarazzo MR, De Bonis ML, Vacca M, Della Ragione F, D'Esposito M. Lessons from two human chromatin diseases, ICF syndrome and Rett syndrome. Int J Biochem Cell Biol 2009;41(1):117–26.

[43] Ehrlich M, Sanchez C, Shao C, Nishiyama R, Kehrl J, Kuick R, et al. ICF, an immunodeficiency syndrome: DNA methyltransferase 3B involvement, chromosome anomalies, and gene dysregulation. Autoimmunity 2008;41(4):253–71.

[44] Evrony GD, Cai X, Lee E, Hills LB, Elhosary PC, Lehmann HS, et al. Single-neuron sequencing analysis of L1 retrotransposition and somatic mutation in the human brain. Cell 2012;151(3):483–96.

[45] de Andrade A, Wang M, Bonaldo MF, Xie H, Soares MB. Genetic and epigenetic variations contributed by Alu retrotransposition. BMC Genomics 2011;12:617.

[46] Reik W, Walter J. Genomic imprinting: parental influence on the genome. Nat Rev Genet 2001;2(1):21–32.

[47] John RM, Lefebvre L. Developmental regulation of somatic imprints. Differ Res Biol Diversity 2011;81(5):270–80.

[48] Nakamura T, Liu YJ, Nakashima H, Umehara H, Inoue K, Matoba S, et al. PGC7 binds histone H3K9me2 to protect against conversion of 5mC to 5hmC in early embryos. Nature 2012;486(7403):415–9.

[49] Glaser RL, Ramsay JP, Morison IM. The imprinted gene and parent-of-origin effect database now includes parental origin of de novo mutations. Nucleic Acids Res 2006;34(Database issue):D29–31.

[50] Xie W, Barr CL, Kim A, Yue F, Lee AY, Eubanks J, et al. Base-resolution analyses of sequence and parent-of-origin dependent DNA methylation in the mouse genome. Cell 2012;148(4):816–31.

[51] Court F, Tayama C, Romanelli V, Martin-Trujillo A, Iglesias-Platas I, Okamura K, et al. Genome-wide parent-of-origin DNA methylation analysis reveals the intricacies of human imprinting and suggests a germline methylation-independent mechanism of establishment. Genome Res 2014;24(4):554–69.

[52] Peters J. The role of genomic imprinting in biology and disease: an expanding view. Nat Rev Genet 2014;15(8):517–30.

[53] Court F, Martin-Trujillo A, Romanelli V, Garin I, Iglesias-Platas I, Salafsky I, et al. Genome-wide allelic methylation analysis reveals disease-specific susceptibility to multiple methylation defects in imprinting syndromes. Hum Mutat 2013;34(4):595–602.

[54] Shoemaker R, Deng J, Wang W, Zhang K. Allele-specific methylation is prevalent and is contributed by CpG-SNPs in the human genome. Genome Res 2010;20(7):883–9.

[55] Xie H, Wang M, Bischof J, Bonaldo MF, Soares MB. SNP-based prediction of the human germ cell methylation landscape. Genomics 2009;93(5):434–40.

[56] Schalkwyk LC, Meaburn EL, Smith R, Dempster EL, Jeffries AR, Davies MN, et al. Allelic skewing of DNA methylation is widespread across the genome. Am J Hum Genet 2010;86(2):196–212.

[57] Paliwal A, Temkin AM, Kerkel K, Yale A, Yotova I, Drost N, et al. Comparative anatomy of chromosomal domains with imprinted and non-imprinted allele-specific DNA methylation. PLoS Genet 2013;9(8):e1003622.

[58] Kerkel K, Spadola A, Yuan E, Kosek J, Jiang L, Hod E, et al. Genomic surveys by methylation-sensitive SNP analysis identify sequence-dependent allele-specific DNA methylation. Nat Genet 2008;40(7):904–8.

[59] Liu YP, Siegmund KD, Laird PW, Berman BP. Bis-SNP: Combined DNA methylation and SNP calling for Bisulfite-seq data. Genome Biol 2012;13(7).

[60] Shantsila E, Wrigley B, Tapp L, Apostolakis S, Montoro-Garcia S, Drayson MT, et al. Immunophenotypic characterization of human monocyte subsets: possible implications for cardiovascular disease pathophysiology. J Thromb Haemost 2011;9(5):1056–66.

[61] Sarrazin S, Sieweke M. Integration of cytokine and transcription factor signals in hematopoietic stem cell commitment. Semin Immunol 2011;23(5):326–34.

[62] Soghoian DZ, Jessen H, Flanders M, Sierra-Davidson K, Cutler S, Pertel T, et al. HIV-specific cytolytic CD4 T cell responses during acute HIV infection predict disease outcome. Sci Transl Med 2012;4(123):123ra25.

[63] Schatteman GC, Dunnwald M, Jiao C. Biology of bone marrow-derived endothelial cell precursors. Am J Physiol Heart Circ Physiol 2007;292(1):H1–18.

[64] Broske AM, Vockentanz L, Kharazi S, Huska MR, Mancini E, Scheller M, et al. DNA methylation protects hematopoietic stem cell multipotency from myeloerythroid restriction. Nat Genet 2009;41(11):1207–15.

[65] Aberg KA, McClay JL, Nerella S, Clark S, Kumar G, Chen W, et al. Methylome-wide association study of schizophrenia: identifying blood biomarker signatures of environmental insults. JAMA Psychiatry 2014;71(3):255–64.

[66] Li Y, Zhu J, Tian G, Li N, Li Q, Ye M, et al. The DNA methylome of human peripheral blood mononuclear cells. PLoS Biol 2010;8(11):e1000533.

[67] Hodges E, Molaro A, Dos Santos CO, Thekkat P, Song Q, Uren PJ, et al. Directional DNA methylation changes and complex intermediate states accompany lineage specificity in the adult hematopoietic compartment. Mol Cell 2011;44(1):17–28.

[68] Bocker MT, Hellwig I, Breiling A, Eckstein V, Ho AD, Lyko F. Genome-wide promoter DNA methylation dynamics of human hematopoietic progenitor cells during differentiation and aging. Blood 2011;117(19):e182–9.

[69] Reinius LE, Acevedo N, Joerink M, Pershagen G, Dahlen SE, Greco D, et al. Differential DNA methylation in purified human blood cells: implications for cell lineage and studies on disease susceptibility. PloS one 2012;7(7):e41361.

[70] Bock C, Beerman I, Lien WH, Smith ZD, Gu H, Boyle P, et al. DNA methylation dynamics during in vivo differentiation of blood and skin stem cells. Mol Cell 2012;47(4):633–47.

[71] Suarez-Alvarez B, Rodriguez RM, Fraga MF, Lopez-Larrea C. DNA methylation: a promising landscape for immune system-related diseases. Trends Genet 2012;28(10):506–14.

[72] Adalsteinsson BT, Gudnason H, Aspelund T, Harris TB, Launer LJ, Eiriksdottir G, et al. Heterogeneity in white blood cells has potential to confound DNA methylation measurements. PloS one 2012;7(10):e46705.

[73] Lent R, Azevedo FA, Andrade-Moraes CH, Pinto AV. How many neurons do you have? some dogmas of quantitative neuroscience under revision. Eur J Neurosci 2012;35(1):1–9.

[74] Prinz M, Priller J, Sisodia SS, Ransohoff RM. Heterogeneity of CNS myeloid cells and their roles in neurodegeneration. Nat Neurosci 2011;14(10):1227–35.

[75] DeFelipe J, Lopez-Cruz PL, Benavides-Piccione R, Bielza C, Larranaga P, Anderson S, et al. New insights into the classification and nomenclature of cortical GABAergic interneurons. Nat Rev Neurosci 2013;14(3):202–16.

[76] Bandeira F, Lent R, Herculano-Houzel S. Changing numbers of neuronal and non-neuronal cells underlie postnatal brain growth in the rat. Proc Natl Acad Sci USA 2009;106(33):14108–13.

[77] Collins CE, Young NA, Flaherty DK, Airey DC, Kaas JH. A rapid and reliable method of counting neurons and other cells in brain tissue: a comparison of flow cytometry and manual counting methods. Front Neuroanat 2010;4:5.

[78] Schmitz C, Hof PR. Design-based stereology in neuroscience. Neuroscience 2005;130(4):813–31.

[79] Okada S, Saiwai H, Kumamaru H, Kubota K, Harada A, Yamaguchi M, et al. Flow cytometric sorting of neuronal and glial nuclei from central nervous system tissue. J Cell Physiol 2011;226(2):552–8.

[80] Jiang Y, Matevossian A, Huang HS, Straubhaar J, Akbarian S. Isolation of neuronal chromatin from brain tissue. BMC Neurosci 2008;9:42.

[81] Nelson SB, Hempel C, Sugino K. Probing the transcriptome of neuronal cell types. Curr Opin Neurobiol 2006;16(5):571–6.

[82] Zeng H, Shen EH, Hohmann JG, Oh SW, Bernard A, Royall JJ, et al. Large-scale cellular-resolution gene profiling in human neocortex reveals species-specific molecular signatures. Cell 2012;149(2):483–96.

[83] Hawrylycz MJ, Lein ES, Guillozet-Bongaarts AL, Shen EH, Ng L, Miller JA, et al. An anatomically comprehensive atlas of the adult human brain transcriptome. Nature 2012;489(7416):391–9.

[84] Mohn F, Weber M, Rebhan M, Roloff TC, Richter J, Stadler MB, et al. Lineage-specific polycomb targets and de novo DNA methylation define restriction and potential of neuronal progenitors. Mol Cell 2008;30(6):755–66.

[85] Guo JU, Ma DK, Mo H, Ball MP, Jang MH, Bonaguidi MA, et al. Neuronal activity modifies the DNA methylation landscape in the adult brain. Nat Neurosci 2011;14(10):1345–51.

[86] Moore LD, Le T, Fan G. DNA methylation and its basic function. Neuropsychopharmacology 2013;38(1):23–38.

[87] Fanelli M, Caprodossi S, Ricci-Vitiani L, Porcellini A, Tomassoni-Ardori F, Amatori S, et al. Loss of pericentromeric DNA methylation pattern in human glioblastoma is associated with altered DNA methyltransferases expression and involves the stem cell compartment. Oncogene 2008;27(3):358–65.

[88] Jakovcevski M, Akbarian S. Epigenetic mechanisms in neurological disease. Nat Med 2012;18(8):1194–204.

[89] Hahn MA, Qiu R, Wu X, Li AX, Zhang H, Wang J, et al. Dynamics of 5-hydroxymethylcytosine and chromatin marks in Mammalian neurogenesis. Cell Rep 2013;3(2):291–300.

[90] Chouliaras L, Mastroeni D, Delvaux E, Grover A, Kenis G, Hof PR, et al. Consistent decrease in global DNA methylation and hydroxymethylation in the hippocampus of Alzheimer's disease patients. Neurobiol Aging 2013.

[91] Iwamoto K, Bundo M, Ueda J, Oldham MC, Ukai W, Hashimoto E, et al. Neurons show distinctive DNA methylation profile and higher interindividual variations compared with non-neurons. Genome Res 2011;21(5):688–96.

[92] Takahashi K, Yamanaka S. Induction of pluripotent stem cells from mouse embryonic and adult fibroblast cultures by defined factors. Cell 2006;126(4):663–76.

[93] Couture LA. Scalable pluripotent stem cell culture. Nat Biotechnol 2010;28(6):562–3.

[94] Enver T, Pera M, Peterson C, Andrews PW. Stem cell states, fates, and the rules of attraction. Cell Stem Cell 2009;4(5):387–97.

[95] Graf T, Stadtfeld M. Heterogeneity of embryonic and adult stem cells. Cell Stem Cell 2008;3(5):480–3.

[96] Nakai-Futatsugi Y, Niwa H. Transcription factor network in embryonic stem cells: heterogeneity under the stringency. Biol Pharm Bull 2013;36(2):166–70.

[97] Singh AM, Chappell J, Trost R, Lin L, Wang T, Tang J, et al. Cell-cycle control of developmentally regulated transcription factors accounts for heterogeneity in human pluripotent cells. Stem Cell Rep 2013;1(6):532–44.

[98] Chou YF, Chen HH, Eijpe M, Yabuuchi A, Chenoweth JG, Tesar P, et al. The growth factor environment defines distinct pluripotent ground states in novel blastocyst-derived stem cells. Cell 2008;135(3):449–61.

[99] Smith AG, Heath JK, Donaldson DD, Wong GG, Moreau J, Stahl M, et al. Inhibition of pluripotential embryonic stem cell differentiation by purified polypeptides. Nature 1988;336(6200):688–90.

[100] Williams RL, Hilton DJ, Pease S, Willson TA, Stewart CL, Gearing DP, et al. Myeloid leukaemia inhibitory factor maintains the developmental potential of embryonic stem cells. Nature 1988;336(6200):684–7.

[101] Canham MA, Sharov AA, Ko MS, Brickman JM. Functional heterogeneity of embryonic stem cells revealed through translational amplification of an early endodermal transcript. PLoS Biol 2010;8(5):e1000379.

[102] Marks H, Kalkan T, Menafra R, Denissov S, Jones K, Hofemeister H, et al. The transcriptional and epigenomic foundations of ground state pluripotency. Cell 2012;149(3):590–604.

[103] Habibi E, Brinkman AB, Arand J, Kroeze LI, Kerstens HH, Matarese F, et al. Whole-genome bisulfite sequencing of two distinct interconvertible DNA methylomes of mouse embryonic stem cells. Cell Stem Cell 2013;13(3):360–9.

[104] Gereige LM, Mikkola HK. DNA methylation is a guardian of stem cell self-renewal and multipotency. Nat Genet 2009;41(11):1164–6.

[105] Mohn F, Schubeler D. Genetics and epigenetics: stability and plasticity during cellular differentiation. Trends Genet 2009;25(3):129–36.

[106] Yamanaka S, Blau HM. Nuclear reprogramming to a pluripotent state by three approaches. Nature 2010;465(7299):704–12.

[107] Bhutani N, Brady JJ, Damian M, Sacco A, Corbel SY, Blau HM. Reprogramming towards pluripotency requires AID-dependent DNA demethylation. Nature 2009;463(7284):1042–7.

[108] Skora AD, Spradling AC. Epigenetic stability increases extensively during Drosophila follicle stem cell differentiation. Proc Natl Acad Sci USA 2010;107(16):7389–94.

[109] Narva E, Autio R, Rahkonen N, Kong L, Harrison N, Kitsberg D, et al. High-resolution DNA analysis of human embryonic stem cell lines reveals culture-induced copy number changes and loss of heterozygosity. Nat Biotechnol 2010;28(4):371–7.

[110] Okita K, Ichisaka T, Yamanaka S. Generation of germline-competent induced pluripotent stem cells. Nature 2007;448(7151):313–7.

[111] Kim K, Doi A, Wen B, Ng K, Zhao R, Cahan P, et al. Epigenetic memory in induced pluripotent stem cells. Nature 2010;467(7313):285–90.

[112] Kim K, Zhao R, Doi A, Ng K, Unternaehrer J, Cahan P, et al. Donor cell type can influence the epigenome and differentiation potential of human induced pluripotent stem cells. Nat Biotechnol 2011;29(12):1117–9.

[113] Lister R, Pelizzola M, Kida YS, Hawkins RD, Nery JR, Hon G, et al. Hotspots of aberrant epigenomic reprogramming in human induced pluripotent stem cells. Nature 2011;471(7336):68–73.

[114] Cahan P, Daley GQ. Origins and implications of pluripotent stem cell variability and heterogeneity. Nat Rev Mol Cell Biol 2013;14(6):357–68.

[115] Zentner GE, Tesar PJ, Scacheri PC. Epigenetic signatures distinguish multiple classes of enhancers with distinct cellular functions. Genome Res 2011;21(8):1273–83.

[116] Creyghton MP, Cheng AW, Welstead GG, Kooistra T, Carey BW, Steine EJ, et al. Histone H3K27ac separates active from poised enhancers and predicts developmental state. Proc Natl Acad Sci USA 2010;107(50):21931–6.

[117] Zeng J, Konopka G, Hunt BG, Preuss TM, Geschwind D, Yi SV. Divergent whole-genome methylation maps of human and chimpanzee brains reveal epigenetic basis of human regulatory evolution. Am J Hum Genet 2012;91(3):455–65.

BIOINFORMATICS OF PERSONALIZED EPIGENETICS

CHAPTER

6

Computational Methods in Epigenetics

Vanessa Aguiar-Pulido[1,3,*], Victoria Suarez-Ulloa[2,*], Jose M. Eirin-Lopez[2], Javier Pereira[3], Giri Narasimhan[1]

[1]School of Computing & Information Sciences, Florida International University, Miami, FL, USA; [2]Department of Biological Sciences, Florida International University, North Miami, FL, USA; [3]Department of Information & Communication Technologies, University of A Coruña, A Coruña, Spain

OUTLINE

*The first two authors contributed equally to this chapter.

1. INTRODUCTION

Epigenetics is defined as the heritable changes in gene expression resulting from modifications in chromatin structure, without involving changes in the genetic information stored in DNA [1]. The understanding of epigenetics requires integrative analysis of disparate heterogeneous data to generate global interpretations and biological knowledge [2]. The epigenome arises from interactions among different epigenetic mechanisms, including discrete biomolecules (e.g., nucleic acid–protein interactions) as well as chemical modifications (i.e., DNA methylation and protein post-translational modifications, or PTMs), thus possessing an elaborate combinatorial complexity [1]. Peculiarities of each of the different epigenetic marks must be taken into account to understand the diverse nature of the data resulting from epigenomic studies. The analysis of each one of these marks involves specific techniques and work flows, resulting in different types of data (Figure 1).

The heterogeneity and scale of data of epigenomics studies pose serious challenges for computational analyses and information management, similar to those created by other complex systems (e.g., markets, social dynamics) [3]. Developing more efficient analyses to keep up with the pace of data production and expand the frontiers in epigenetics is the current task of bioinformatics. Fortunately, a number of efforts are aimed at standardizing and integrating heterogeneous data sources in such a way that they become suitable for meta-analysis and are openly available to the scientific community [4].

In this chapter, we describe the main characteristics of the various types of data generated during epigenetic studies, providing a description of the most common computational approaches used for their integrative analysis. Additionally, we cover substantial advances in biomedical research that are illustrated by the production of online resources and by the

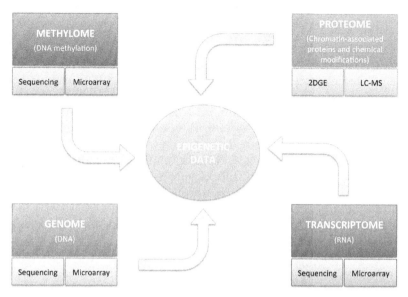

FIGURE 1 Heterogeneous types of data generated from various "omics" and specific techniques are included as epigenetic data. High-throughput techniques such as DNA sequencing and microarray-based analyses are currently dominant in functional genomics as well as in the study of the methylome, producing specific data formats that must be processed and standardized before being compared. Proteomics comprises all chromatin-related proteins of interest, most notably the histone family, including their possible chemical modifications and interactions. In proteomics, gel-based methods and mass spectrometry represent the main sources of high-throughput data. The combination of all these disparate data types constitutes the bulk of epigenetic information.

establishment of worldwide consortia for the standardization of molecular data. These contributions are critical to effectively linking genetic and epigenetic data with clinical information and pave the way toward a realistic future of personalized medicine.

2. EPIGENETIC PROFILING: HETEROGENEOUS DATA AND PARTICULAR CHALLENGES OF DIFFERENT EPIGENETIC FACTORS

Epigenetic profiling involves the coordinated study of diverse biological marks responsible for the transmission of epigenetic information, including but not limited to DNA methylation, histone variants and their PTMs, and chromatin remodeling complexes. Subsequently, some of the most relevant biological aspects of epigenetics research as well as the main characteristics of the data that they involve are addressed.

2.1 Methylation Patterns

Methylation is the best studied epigenetic mark, especially pertaining to CpG islands in the case of mammals [5]. A variable percentage of the CpG dinucleotides of the genome, ranging from 60% to 90%, is actually methylated. The remaining portion of CpGs are free of methylation, largely constituting the so-called CpG islands, which are usually associated with gene promoter regions [6]. The presence of methylation marks on DNA has been widely associated with repressive states of the chromatin. Moreover, it is hypothesized that its evolutionary origin was the neutralization of invading DNA by blocking its ability to be expressed [7]. The actual effect of the methylation transformation of DNA varies depending on a number of factors: the proximity of methylated CpG islands to the gene promoter, the density of those methylation marks, and the strength of the promoter itself [5].

Furthermore, the specific location of the methylation marks in relation to the promoter may display contradictory effects. This has been labeled as the "methylation paradox" [8]: although methylation of CpG islands in the promoter region is strongly associated with inhibition of transcription, it has been found that methylation of CpG islands downstream from the site of transcription initiation shows no inhibition effect. Furthermore, it has been reported that this may actually enhance the levels of transcription [9]. Although the specific molecular effects of methylation events may be difficult to predict, it is widely accepted that the patterns of methylation marks are nonrandomly associated with diseases (such as cancer), displaying a good level of specificity between tumor types in many cases [10].

The previous findings highlight the importance of an accurate mapping of methylated sites on the genome at a single-base resolution. Unbiased epigenetic mapping and annotation of genomes requires high-throughput sequencing techniques; however, more directed methods such as microarray platforms could also be used. The information of which nucleotides (cytosines, C's) are methylated on the genome is lost during PCR amplification and additional experimental steps are required, most notably the selective bisulfite transformation of unmethylated cytosine into thymine ($C \rightarrow T$ transformation). Each different procedure requires optimized bioinformatic data-processing techniques to produce standardized data summaries that allow comparisons across experiments; that is, the differentially methylated regions (DMRs) table. To produce these tables, raw data must be processed and controlled for quality and then mapped onto a reference genome, and last, an appropriate statistical analysis must be carried out to obtain a list of significantly differentially methylated genomic regions along with the absolute values of methylation levels.

The set of nucleic acid methylation modifications in an organism's genome or in a particular cell is referred to as the methylome. Methylome data processing must address common general problems that are posed by high-throughput sequencing or microarray-based techniques. Most notably, the bioinformatic analysis of bisulfite-treated DNA is challenging owing to the decrease in complexity of the sequences after $C \rightarrow T$ transformation. Genome-wide sequencing or enriched DNA libraries produce a collection of reads that must be aligned to a reference genome. To allow the alignment of bisulfite-transformed reads that account for C–T mismatches, different approaches can be used. One possible approach involves wildcard aligners, which change C's in the sequence to the degenerate IUPAC symbol "Y" (equivalent to both C and T). Another existing approach modifies the scoring matrix used by the alignment algorithm to prevent penalization of C–T mismatches. As a third possible approach, all C's can be converted into T's on both reads as well as on the reference genome. This way, the alignment is worked on a three-letter alphabet for both the template and the complementary strand sequences of the genome. All these methods present some bias owing to lower complexity in the sequence and consequently a lack of specificity in the alignment, causing useful good quality reads to be discarded.

To improve mapping efficiency, additional steps such as local realignment, analysis of sequence quality scores, and the application of statistical models of allele distribution can be carried out [11]. The visualization of this information can be performed with any available genome browser. Methylated positions are often represented using color codes and quantitative methylation data with bar charts are superposed. Once mapped, quantification of methylation at a single-base resolution is carried out and then associated with each genomic position. There are a number of well-established protocols for producing bisulfite reads that use high-throughput sequencing technologies such as methylC-seq [12] and reduced representation bisulfite sequencing [13]. Specific bioinformatics tools that focus in processing this type of data were developed [14–16]. Alternatively, protocols such as methylated DNA immunoprecipitation (Me-DIP) use a different approach for methylome analysis based on the use of specific antibodies for 5-methylcytosine. This protocol has been adapted for the use of both sequencing techniques (Me-DIP-seq [17]) and microarray technologies (Me-DIP-chip [18]). When dealing with microarray data, the critical steps to optimize the accuracy of the method include image processing and data normalization [19–21]. The final parameters calculated are the β-value and the M-value. The β-value represents a ratio calculated between the intensity of the methylated probe and the sum of the intensities of the methylated and unmethylated probes, while the M-value is defined as the $\log_2$ transform of the ratio between the intensities of the methylated and the unmethylated probes [22]. Many pipelines and specific software packages have been developed over time to carry out this initial processing of data. The R-Bioconductor repository is particularly useful since it offers up-to-date analysis packages written using the R statistical language [23].

Ultimately, the most common goal of methylome analysis is to find significant differences in methylation patterns when comparing several groups (i.e., diseased samples vs healthy ones, different tumor types, different developmental stages, etc.). Multiple hypothesis testing is required; methods such as t test or Wilcoxon rank sum test are commonly utilized as a basis. Afterward, an adjustment of the p-values obtained with these methods is carried out, always at the cost of reducing the statistical power of the analysis. The false discovery rate is currently the most used. Finally, a ranked list of DMRs is obtained for further analysis and interpretation.

2.2 Histone Proteins and Their Chemical Modifications

The study of histones has come a long way since they were simply considered as structural proteins. Although initially believed to be a simple physical support for the DNA within the cell nucleus, it is now clear that histones play critical functional roles [24]. Histone proteins are highly conserved throughout the different branches of the tree of life, being ubiquitous in eukaryotes and represented in some Archaea groups. In addition to canonical histone types (H1, H2A, H2B, H3, H4), several specialized histone variants have arisen during evolution. The recruitment of these variants into nucleosomes modulates the physicochemical properties of the chromatin structure, regulating the access of the transcription machinery to target genes [24].

In addition to the characterization of histone variants, posttranslational modifications in histones are considered fundamental epigenetic marks. Two main approaches can be used for the production of histone mark data: unbiased identification and quantitation of histone modifications using mass spectrometry methods, or genome-wide mapping of specific modified histones using chromatin immunoprecipitation (ChIP) techniques. Taken together, histones and their modifications offer an overwhelming range of possible combinatorial effects that remains to be fully understood. Several efforts have been made to tackle this complex regulatory mechanism, including the "histone code" hypothesis [25]. To unravel this hypothetical code, innovative analytical techniques in proteomics seek to find the patterns that would work as biomarkers of specific molecular processes.

In addition to the identification of histone genes and the genome-wide mapping of histones and PTMs, there are two main aspects of interest when considering histone proteomics: first, the quantification and comparison of expression levels and modification levels; second, the structural characterization and the simulation of the protein dynamics under specific conditions. In this section, these two issues are addressed, and an overview of the different types of data that can be produced in this context is provided.

2.2.1 *Quantitative Protein Analysis*

Current advances in proteomic techniques, using both gel-based and gel-free methods, allow high-throughput quantitative analysis of proteins. To date, the most widely used technique in proteomic analysis has been two-dimensional gel electrophoresis (2DGE). From a computational perspective, the analysis of 2DGE requires the development of image processing algorithms that allow the accurate reading and comparison of the gels, requiring the processing of thousands of spots that correspond to the various proteins separated by the technique [26]. The intensity of the spot correlates with the amount of protein present in the sample. Therefore, differential expression analysis can be carried out by aligning the spots obtained from a problem sample with those spots from a control sample with the aim of finding the correspondence and subsequently calculating the difference in their intensity levels [27].

Gel-based methods, however, are rapidly being displaced by "shotgun" methods, mainly involving liquid chromatography coupled to mass spectrometry (LC–MS) [28]. The nature of the data obtained from LC–MS analyses is substantially different from the data observed from the 2D gels. Mass spectrometry ionizes molecules and separates them according to their mass-to-charge ratio under an electrical field. Results are then recorded in the form of a mass spectrum. Mass spectra provide a graphical representation of the different masses of ions detected in the analysis. Different mass spectrum peaks correspond to different mass/charge ratios (usually the charge equals 1, thus the peak simply represents the mass of the ion), and the area under the peak corresponds to the quantity of ions detected. Automated analyses of mass spectra are possible by comparing the observed patterns of peaks with those stored in databases. Once the elements of interest are identified, absolute quantification analyses are possible using calibration curves, as well as relative quantification analyses.

LC–MS techniques do not allow just the identification of the proteins, but also the identification of their chemical modifications. Alternatively, ChIP-seq data on histone marks can also be considered quantitative, thus allowing differential analyses. Computational methods like that one described by Xu et al. compare the differences in read count between two sequencing libraries, using hidden Markov models (HMMs), with the objective of finding differential histone modification sites [29].

2.2.2 *Structural Modeling and Dynamic Simulation*

The traditional method for modeling the three-dimensional structure of proteins that have a well-defined crystal structure from X-ray diffraction or nuclear magnetic resonance analysis is comparative modeling (also known as homology modeling). In this method, an initial known structure of a homologous protein is used as a template to build the predicted structure of the target protein. This is possible because the protein structure is evolutionarily more conserved than the corresponding genetic coding sequence [30]. The structures of template proteins can be obtained in protein data bank (pdb) format files from the RCSB Protein Data Bank [31]. Three-dimensional modeling can be useful in the prediction of the dynamic effects caused by the structural and physicochemical variations. An application of these techniques in epigenetics would involve the prediction of changes in nucleosome and chromatin structure resulting from the replacement of canonical histones by histone variants. However, the study of histone variants poses specific challenges since the majority of the structural differences tend to accumulate in the most external and dynamic part of the protein chain [32], that is, the tails, which are especially difficult to study by X-ray diffraction methods and therefore difficult to model.

For dynamic studies, molecular dynamics simulations are frequently used [33,34]. In these methods, the known structure of the protein is translated into an array of coordinates for every atomic nucleus present. These are allowed to vibrate under a simulated force field while the instantaneous kinetic and potential energies for each atomic bond are calculated *ab initio*. These methods produce a very visual output by which the movement of the protein and the development of chemical processes can be observed, conveying a very important application for novel drug discovery and design [35,36]. These methods are applicable for the computational prediction of condensation levels of the DNA in the nucleosomes and, subsequently, to gauge the transcriptional accessibility of the genetic material, providing insights of great value for further experimental validation.

2.3 Nucleosome Positioning

Nucleosomes are the fundamental subunits of the chromatin, constituted by segments of DNA wrapped around a core structure of histone proteins. Nucleosome structure and positioning have direct implications for gene transcription since the majority of transcription factors cannot bind DNA packed by nucleosomes. Additionally, the positioning of nucleosomes can be actively modified by ATP-dependent remodeling factors that dynamically reorganize the structure of the chromatin [37]. Therefore, the accurate genome-wide mapping of nucleosomes and the analysis of their dynamics convey critical information about the epigenetic state of the cell.

The genome-wide analysis of specific chromatin components has been traditionally carried out using ChIP-based technologies (ChIP-seq or

ChIP-chip) [38]. ChIP-based techniques target specific proteins (usually histone variants including PTMs) using appropriate antibodies; that protein is then precipitated together with its associated DNA. More recently, the combination of a digestion step with DNases and high-throughput sequencing methods has been successfully used for this purpose, allowing a single-nucleotide resolution for the mapping of nucleosome positions [39]. With either one of these techniques, the bioinformatic analysis starts with the processing of high-throughput sequencing data followed by the mapping of those sequences on a reference genome to specify the protein-binding loci. The computational challenges of these methods are the common issues of short-read alignments [40,41].

When experimental data is lacking, however, it is still possible to perform computational predictions of nucleosome positioning based on motifs found on the genome sequence and thermodynamic properties of the chromatin [42,43].

2.4 Noncoding RNA

Part of what was considered years ago as "junk DNA" is nowadays an important focus of research for the scientific community. Although the inclusion of the ncRNA as an epigenetic mark at the same level of the methylome or the chromatin remains controversial, there is a general trend toward its acceptance [44–49], specifically regarding the role of long noncoding RNA in the epigenetic regulation of gene expression through interactions with chromatin-modifying proteins [50]. One of the most particular computational methods used in this field is the one that serves as the basis for comparative genomics. This approach is known as the "guilt by association" method and it implies functional inference through the observation of consistent coexpression events. In general, the construction of interaction networks in epigenetics is a problem that involves all different factors mentioned here as well as others that have been overlooked, such as chromatin-associated proteins (which may have a direct influence in the chromatin structure and dynamics) [51]. This problem involves difficulties well beyond the data processing related to the extraction of knowledge from heterogeneous data.

3. EPIGENETIC DATA INTEGRATION AND ANALYSIS

Data integration refers to the process by which a system combines information from different sources to make meaningful interpretations and produce relevant outcomes. Epigenetics is a field of research in which data integration is especially relevant, given the complex nature and interactions among the mechanisms responsible for various epigenetic marks. Data from various epigenomic studies, which may be obtained through different techniques, must be analyzed from a holistic perspective. Additional marks encompassing potential epigenetic relevance that can be considered include other chromatin-associated biomolecules such as histone-modifying enzymes and remodeling factors.

The number of existing bioinformatics resources created for this general purpose has steadily increased. These resources include tools, services, databases, standards, or even terminologies for each specific domain or area of expertise. Many publications highlight the importance and usefulness of properly integrating different types of biomedical data [52]. The large amount of information and diverse technology platforms raise multiple challenges, regarding not only data access, but also data processing [53]. More specifically, in the context of epigenetics, it is very likely that data-integrating approaches, with the aim of identifying functional genetic variability, represent a possible solution to the challenge of interpreting meaningfully the results of genome-wide profiling [54].

3.1 Quality and Format Standards for Data Integration

The most significant barrier for adopting a holistic perspective on data obtained from different sources is probably the standardization of methods and formats. Standardization in file formats allows the machine to find the specific pieces of information required for each step of the processing algorithm. This issue has been recognized and addressed through the establishment of consortia that specify necessary standards for the scientific community. The ultimate goal is to make all the produced data suitable for integrative analyses. The standards often refer to the metadata of the data sets (i.e., "data about the data") that are generated, information about the experiments that produced those data sets, and other general characteristics. For biomedical and molecular biology research, the MIBBI project is an important global endeavor with the objective of establishing data standards to aid collaboration and meta-analyses [4]. Consisting of many subprojects, MIBBI suggests protocols to standardize reporting of data. From the well-established MIAME project for gene expression data obtained with microarray technologies [55], to the more recent MIAPE for proteomics data reports [56], it also addresses data standards relevant for epigenetics. In each of these protocols for standards compliance, all the various possible technologies and preprocessing methods for the generation of data sets are considered. Specific data exchange formats are designed to allow automated interpretation of the various data sets, mainly using tag-based structures such as the widely known XML.

Data standardization is crucial in bioinformatics and there exist multiple emerging standards that go beyond the scope of this chapter. Chervitz et al. provide a detailed review of the most relevant standards that may be applied to epigenetics [57].

3.2 Data Resources

A database may be defined as a searchable collection of interrelated data. The amount of high-throughput biological data generated has increased exponentially [58]. Furthermore, these types of data (including sequences, microarray data sets, gel imaging files, etc.) are not published in a conventional manner anymore; they are stored in databases. In fact, many journals require researchers to upload their data to specific repositories prior to trying to publish their research work.

The need for storing and linking large-scale data sets has also consistently increased. Archiving, curating, analyzing, and interpreting all of these data sets represent a major challenge. Therefore, the development of methods that allow the proper storage, searching, and retrieval of information becomes critical. Databases represent the most efficient way of managing this glut of data. The construction of databases and tools that allow accessing the data will enable the scientific community to manage and share vast amounts of high-throughput biological information. Hence, support for large-scale analysis is essential. Access to data must be facilitated and data must be periodically updated. Specifically in epigenetics, browsers have been made available to the scientific community facilitating the access and integration of the data generated by some of the initiatives described in the next section. Among these, we can find the UCSC Genome Browser [59], the Roadmap Epigenomics Visualization Hub (VizHub), and the Human Epigenome Browser at Washington University [60] (Table 1). Finally, knowledge extracted from various fields involving, among others, different disciplines within epigenetics and general biology, as well as clinical medicine, must be linked.

Given all of the above, it is clear that databases have become vital for carrying out successful bioinformatics research. They make data available to researchers in a format that is understandable by a machine. Hence, analyses can be carried out automatically with computers, managing great amounts of data and providing user-friendly interfaces. Data will be stored in predefined formats making possible the automatic retrieval of information. Ultimately, valuable extra information could be extracted if data are properly linked to external resources. However, an important

TABLE 1 Epigenetic Browsers

Browser name	Web site
UCSC Genome Browser	http://www.epigenomebrowser.org
Human Epigenome Browser at Washington University	http://epigenomegateway.wustl.edu/
VizHub	http://vizhub.wustl.edu

challenge that must be taken into account is the proper anonymization of data, to protect the privacy of the subjects who participate in the studies from which the data are collected.

3.2.1 Large-Scale Projects and Consortia

There exist numerous initiatives that manage huge amounts of diverse biological data, such as the Encyclopedia of DNA Elements (ENCODE) [61] or the Human Epigenome Project (HEP) [62]. The first project involves researchers from all over the world and can be considered a continuation of the Human Genome Project. Its objective is to identify all functional elements in the human genome and is funded by the National Human Genome Research Institute. Under the ENCODE project, many computational approaches were developed to handle epigenomic data [63–66]. The second project, on the other hand, seeks the identification and classification of genome-wide DNA methylation patterns for all human genes, studied in different tissues, linking this information to diseases and environmental conditions. This project is an international endeavor of global interest, and it is funded by public funds as well as private investment via a consortium of genetic research organizations.

Other relevant initiatives include the NIH Roadmap Epigenomics Mapping Consortium [67], which was launched in 2008. The goal of this project was to develop publicly available resources (more specifically, reference epigenome maps from a variety of cell types) of human epigenomic data to foster basic biology and disease-oriented research. On this basis, two data repositories were made available: the National Center for Biotechnology Information (NCBI) Epigenome Gateway and the Epigenome Atlas (see Table 2 in the next section).

The NIH Roadmap Epigenomics Program is a member of the International Human Epigenome Consortium [68], a growing international effort to coordinate worldwide epigenome mapping and to disseminate experimental standards for epigenome characterization, officially presented in 2010.

Other U.S. initiatives include the Epigenetic Mechanisms in Cancer Think Tank [69], sponsored by the National Cancer Institute in 2004, and the American Association for Cancer Research Human Epigenome Task Force [70], which emerged from a series of workshops and included scientists from all over the world.

Epigenetic research has been funded by several entities outside the United States as well. The European Union has dedicated significant amount of resources (more than €50M) over the years. Numerous initiatives have been funded, such as the above-mentioned HEP, High-Throughput Epigenetic Regulatory Organization in Chromatin, and Epigenetic Treatment of Neoplastic Disease, to focus on general questions such as DNA methylation, chromatin profiling, and treatment of neoplastic disease,

TABLE 2   Epigenetic Resources

Resource name	Web site
Cancer Methylome System	http://cbbiweb.uthscsa.edu/KMethylomes/
DBCAT	http://dbcat.cgm.ntu.edu.tw/
CREMOFAC	http://www.jncasr.ac.in/cremofac/
EpimiR	http://bioinfo.hrbmu.edu.cn/EpimiR/
HEMD	http://mdl.shsmu.edu.cn/HEMD/
Histome	http://www.actrec.gov.in/histome/
HistoneHits	http://histonehits.org
Histone Database	http://research.nhgri.nih.gov/histones/
Human Epigenome Atlas	http://www.genboree.org/epigenomeatlas/
Human lincRNA Catalog	http://www.broadinstitute.org/genome_bio/human_lincrnas/
MeInfoText	http://bws.iis.sinica.edu.tw:8081/MeInfoText2/
MethBase	http://smithlabresearch.org/software/methbase/
MethDB	http://www.methdb.net/
MethyCancer	http://methycancer.psych.ac.cn/
MethyLogiX	http://www.methylogix.com/genetics/database.shtml.htm
NCBI Epigenomics Gateway	http://www.ncbi.nlm.nih.gov/epigenomics/
NGSMethDB	http://bioinfo2.ugr.es/NGSmethDB/
NONCODE	http://www.noncode.org/
PEpiD	http://wukong.tongji.edu.cn/pepid
PubMeth	http://www.pubmeth.org/

respectively. Furthermore, the European Commission created in 2004 the Epigenome Network of Excellence [71] with the objective of studying major epigenetic questions in the postgenomic era.

Asia has focused mainly on disease epigenomes through the organization of various international meetings, as well as the creation of the Japanese Society for Epigenetics. Australia has also contributed to the Human Epigenome Project by creating in 2008 the Australian Alliance for Epigenetics and holding several workshops. Finally, Canada tried to position itself at the forefront of international efforts by creating the Canadian Epigenetics, Environment and Health Research Consortium, which is funded by the Canadian Institutes of Health Research and multiple Canadian and international partners.

As a worldwide initiative, and with the aim of joining all possible efforts, the Alliance for the Human Epigenome and Disease [72] was created. The aim of this project was to provide high-resolution reference epigenome maps, which will be useful in basic and applied research, will have an impact on how many diseases are understood, and, ultimately, will lead to the discovery of new ways of controlling these diseases.

3.2.2 Data Models

3.2.2.1 Traditional Database Models

Traditionally, most databases have followed what is known as the "entity-relationship model." This model tries to describe the data using entities, which correspond to concepts or objects, and relationships that may exist between these. In general, this model leads to a relational database implementation.

Most databases are usually offered as part of a tool or a service, which in most cases is presented through a Web interface. Most of them provide free access via the Internet and/or allow researchers the visualization or downloading of data. There exist numerous resources of this type in the field of epigenetics [73] and, in many cases, they have been constructed as a result of text mining analyses (see Section 3.3.3). Some of these widely used resources are listed in Table 2.

The NCBI, the European Bioinformatics Institute (EMBL-EBI), and the DNA Data Bank of Japan represent the three most important and largest available resources regarding biomedical databases. Major databases included as part of the first resource are GenBank (for DNA sequences), Gene Expression Omnibus, and PubMed (bibliographic database of biomedical literature). The EMBL-EBI provides major bioinformatics resources such as Ensembl, UniProt, ArrayExpress, and Reactome, among others, as well as tools and services to browse and analyze these databases.

The existing epigenetic databases can be broadly classified into several categories according to the type of data they store. The first category includes DNA methylation databases. These databases are useful for studying the covalent modification of a cell's genetic material. Among these, we can find DBCAT, MethBase, MethDB, MethyLogiX, and NGSMethDB.

The second category contains all of those databases related to histone data. Histone databases are important for research in the compaction and accessibility of eukaryotic and probably Archaeal genomic DNA. Some examples of this type of database are Histome, HistoneHits, and Histone Database.

The third category comprises databases related to chromatin-associated factors. Although the molecules involved in these processes are not directly part of the chromatin, they do interact with it. Within this category, databases including chromatin remodeling factors or noncoding RNA data can be found: CREMOFAC, Human lincRNA Catalog, and NONCODE.

The fourth category includes epigenetic databases related to cancer. Most of them are cancer methylation databases, which are helpful for analyzing irregular methylation patterns correlated with cancer. Some of these databases are Cancer Methylome System, MeInfoText, MethyCancer, PEpiD, and PubMeth.

Finally, other more general databases including all types of epigenetic information can be found. Some examples are EpimiR, HEMD, and NCBI Epigenomics Gateway. All the resources mentioned are listed in Table 2.

3.2.2.2 Nontraditional Models

Recently, new models involving semistructured or nonstructured data have gained more attention in scientific fields. Big companies such as Google or Amazon have put a lot of effort into NoSQL resources. NoSQL, also known as Not Only SQL, provides a mechanism for storing and retrieving data that is modeled different from the tabular relations used in relational databases. In this regard, the biomedical field has taken advantage of the Semantic Web and, thus, has focused on developing resource description framework (RDF)-based solutions.

The Semantic Web can be seen as an extension of the World Wide Web. It allows people to share content by providing a standardized way of representing the relationships between Web pages. Thus, machines will be capable of understanding the meaning of hyperlinked information and the information will be given well-defined meaning.

In this context, the RDF model should be highlighted. The RDF, although it was originally designed as a metadata model, is being utilized as a general manner of describing concepts or modeling information (being widely implemented in Web resources) and represents the immediate future [74].

Based on the use of RDF, many approaches in computational bioinformatics have been developed. As of this writing, mashups are the most frequent ones. These Web pages or applications, taking advantage of the Semantic Web, use content from different data sources with the aim of creating one unique service that will be displayed by means of a single graphical interface. Therefore, the main objective is to make searches easier and data more useful.

To integrate and standardize different databases, there have been approaches, such as Bio2RDF [75], that try to help solve the problem of knowledge integration in bioinformatics by developing a mashup application. Other authors developed an ontology-driven mashup that integrates two resources of genomic information and three resources containing information of biological pathways [76], proving that Semantic Web technologies provide an effective framework for information integration in life sciences. Cheung et al. demonstrate the power of the Semantic Web by applying different tools to two different scenarios, concluding that these could be used by people without programming experience to accomplish useful data mashup over the Web [77].

3.3 Knowledge Discovery in Databases

The aim of knowledge discovery in databases (KDD) is to make sense out of data. Traditionally, this task was carried out manually. Nowadays, however, given the amount of data involved, various computational methods and techniques have become indispensable, taking advantage of the processing power offered by computers and turning a tedious process into a largely automatized procedure.

Trying to extract knowledge from data is a nontrivial process. KDD involves several stages, each with its own complexity: data acquisition and storage, data preprocessing and transformation, data mining, and data postprocessing (see Figure 2). Hence, this includes data preparation and selection, data cleaning, incorporating prior knowledge in data sets, and interpreting accurate solutions from the observed results. The use of data warehouses, understood as central repositories of information obtained from the integration of data from diverse data sources, can make some of these tasks easier to deal with. This type of data resource was designed to facilitate massive data analyses and, therefore, the reporting of results. However, the still increasing volume of data available introduces a new challenge, that is, algorithms must be scaled to support massive data analysis and management.

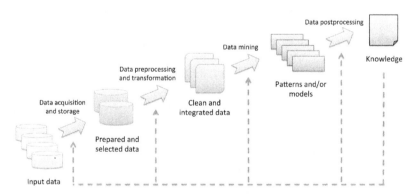

FIGURE 2   The knowledge discovery in databases process. The various stages of this process and the products obtained as a result of each stage are shown. Data are acquired and stored from different heterogeneous data sources. As part of this stage, it may be necessary to prepare and obtain a selection of the available input data. After that, the data goes through a preprocessing and transformation phase, in which they will be cleaned and integrated, obtaining a consistent data resource. Over this resource, data mining techniques can be applied, obtaining as a result different patterns and/or models. Finally, these results are postprocessed to extract the final product of the whole process, that is, knowledge.

3.3.1 Data Preprocessing

The data preprocessing stage involves not only selecting the data sources, but also handling missing data issues and altering the data if necessary. Data integration is not trivial, a convenient model must be chosen because this will have a direct impact on the performance of the bioinformatics pipeline.

3.3.1.1 Data Enrichment

Countless efforts have been made to integrate various resources and data existing over the Internet, especially in the biomedical field, and, for this purpose, standards and/or terminologies, such as ontologies, have been developed.

Ontologies can be defined as a set of concepts (terms) and the relationships among them as representing the consensual knowledge of a specific domain. Ontologies can be represented as graphs (with the nodes representing terms and the edges representing relationships) or as trees (with the nodes as terms and the branches representing hierarchical relationships).

Ontologies enable a clear and unified machine-readable vision of a domain that enables sharing, reusing (partially or totally), and extending knowledge. They are currently the most utilized form for representing biomedical knowledge. Furthermore, it is a field in which a lot of effort is being dedicated, with an increasing rate of usage. Within the Semantic Web (described in the previous section), they have become a key element since they make knowledge representation easier. They are also playing a major role regarding linked data, that is, a way of publishing structured data so that it can be interconnected and, hence, more useful. There currently exist more than 300 biomedical ontologies, and BioPortal [78] is the most important resource, providing access and tools for working with them.

Most studies involving epigenetics use Gene Ontology (GO) [79] to enrich data and draw conclusions from it. The GO project is a collaborative effort to address the need for consistent descriptions of gene products across databases. The GO project offers three structured, controlled vocabularies (ontologies) that describe gene products in terms of their associated (1) biological processes, (2) cellular components, and (3) molecular functions. Using GO terms across databases enables one to obtain more uniform queries. Apart from maintaining and developing these ontologies, other aims of the project are annotating genes and gene products, as well as assimilating and disseminating these data, or providing tools that facilitate accessing and utilizing the data and that allow functional interpretation of experimental data using GO, for example, via enrichment analysis.

GO-based analyses have been used to obtain clusters and to do functional analyses by looking for over- and underrepresentation of GO terms, possibly after combining genomic and transcriptomic data [80–83]. KEGG pathways have also been used in this context [84–86].

Finally, other existing approaches include Chromatin Regulation Ontology siRNA Screening, a new method that has been developed to identify writers and erasers of epigenetic marks [87]. In this work, the authors use this method to identify chromatin factors involved in histone H3 methylation and conclude that it facilitates the identification of drugs targeting epigenetic modifications.

3.3.1.2 Massive Data Set Analysis

Analysis of large data sets usually implies having to preprocess the data or represent it in certain ways. For instance, dimensionality reduction may be carried out to reduce the number of initial variables and obtain a more manageable data set [88]. Many techniques model data using graphs, networks, or matrices. These approaches are very powerful and allow hidden relationships to be found, as well as giving a new perspective of the data. Here, some examples of this are included.

Goh and Wong [89] present four scenarios in which building networks from proteomics data improves the results. They find that networks are convenient for identifying primary causes of cancer, given that they can reflect a structured hierarchy of molecular regulations. The typical network-based analysis framework for proteomics would include several stages. The first would involve data preprocessing, in which the data would be transformed into a network. The second would involve the usage of supervised or unsupervised methods such as those described above. Finally, the third stage would involve interpreting and evaluating the results obtained. Another example of this type of data representation is that proposed by Zheng et al. [90], in which they encode histone modification data as a Bayesian network for gene-regulatory network reconstruction.

Principal component analysis (PCA) has also been used to reduce data dimensionality. For example, Dyson et al. [91] and Figueroa et al. [92] use PCA on gene expression array data, while Volkmar et al. [93] use it on DNA methylation data. Cieślik and Bekiranov [63] take advantage of nonnegative matrix factorization to reduce the dimensionality of epigenetic data. Clustering techniques may also be used for reducing dimensionality [94].

3.3.2 Data Mining

The data mining stage, as part of the KDD process, involves choosing the most appropriate method or technique to be used for searching underlying patterns, as well as the creation of explicative and/or predictive models [95]. This stage comprises deciding which models and parameters might be appropriate for the overall KDD process. Also, searching for patterns of interest in a specific representational form (such as trees or rules) and applying tools such as regression or clustering are part of the data mining stage. Finally, the most important part is the interpretation of the relevant knowledge that can be drawn from the obtained results. This knowledge may then be used and/or incorporated into the bioinformatic analysis pipeline.

Two distinct approaches can be considered in data mining techniques: supervised and unsupervised methods. Supervised methods try to obtain relationships between a set of independent variables and a dependent variable. Therefore, the objective of these methods will be to infer a function from labeled training data. In this case, the computer's task will be the extraction of patterns from the input data to get the dependent or target variable. For this approach, two different types of problem can be identified. On one hand, the first type corresponds to those cases in which the dependent variable is categorical or nominal, usually referred to as a "class." On the other hand, the second type corresponds to those problems in which the dependent variable can take infinite numeric values, that is, a continuous variable. Thus, we call the first type classification problems and the second type regression problems.

Schäfer et al. present a Bayesian model for carrying out integrative analyses using "omics" data [96]. A Bayesian mixture model is utilized to compare and classify measurements of histone acetylation in order to identify DNA fragments obtained from ChIP analyses. Mo et al. [97] propose a framework for joint modeling of discrete and continuous variables obtained from integrated genomic, epigenomic, and transcriptomic profiling. Within this framework, the authors developed iCluster+. This method is capable of performing pattern discovery that integrates binary, categorical, and continuous data. It is based on different types of regression (linear and lasso regression), and it is used to extract novel biological information from integrated cancer genomic data for tumor classification and cancer gene identification. As a last example, Gonzalo et al. [98] apply logistic regression adjusted for different factors to colorectal cancer data, with the aim of evaluating the difference in DNA methylation data between two independent groups.

Unlike supervised techniques, when using unsupervised methods, no target variable is specified. In this case, instead of asking the computer to predict the value of a dependent variable out of a given data set (which corresponds to the independent variables), the question will be "which are the best four groups that can be made out of the data?" or "which variables are most likely to occur together?" On this basis, two types of problems can be identified. The first is clustering and it consists in grouping similar items together. The second is association analysis and it entails finding which features are most frequently found together.

Hierarchical clustering is by far the most popular method when trying to analyze epigenetic data. This technique groups data by creating a hierarchy of clusters, known as a cluster tree or dendrogram. Additionally, k-means is also frequently chosen to carry out this type of analyses. This technique partitions the observations into k clusters, in such a way that each observation will belong to the cluster that has the closest mean, known as a centroid. After that, centroids are updated as the mean value of the observations that belong to the cluster. This process is repeated iteratively until the centroids do not change.

DeltaGseg [99] is an R package that applies hierarchical clustering for preprocessing signals to perform estimations from multiple replicated series. Hence, molecular biologists/chemists will be able to gain physical insight into the molecular details that are not easily accessible by experimental techniques. Unsupervised hierarchical clustering was also utilized by Busche et al. [100] to cluster methylation levels and by Towle et al. [86] to analyze methylation patterns.

Zeller et al. [83] applied both hierarchical clustering and k-means to DNA methylation profiles to validate the results. Another study used both techniques to cluster transcription factors [101]. Clifford et al. [102] compared hierarchical clustering to other clustering techniques (k-means, k-medoids, and fuzzy clustering) to determine the most appropriate one for analyzing Illumina methylation data. Since no significant difference was found between the methods, a combination was proposed; the final output will be given by the method that achieves the best results in each case. McGaughey et al. applied k-means to methylation data and observed that genome-wide methylation signals can reliably distinguish tissues [103].

In contrast to these two widely used methods, Jung et al. developed a density-based PIWI-interacting RNA (piRNA) clustering algorithm named piClust [104]. This algorithm is provided as a Web service through a graphical interface. piClust works as follows: first of all, it determines the clustering parameters carrying out a k-dist analysis; then it clusters preprocessed and previously aligned reads; last, it scores and validates candidate piRNA clusters. Ucar et al. [105] present an unsupervised subspace-clustering algorithm, named "coherent and shifted bicluster identification," which was designed to identify combinatorial patterns of chromatin modification across a specific epigenome. It was believed that applying this tool to the epigenome would help in the understanding of the role of chromatin structure in gene expression regulation. Yu et al. [106] proposed an algorithm, named GATE, for clustering genomic sequences based on spatiotemporal epigenomic information. This algorithm is based on a probabilistic model that was developed to annotate the genome using temporal epigenomic data. Each cluster obtained, which was modeled as an HMM, represented a time series of related epigenomic states. Steiner et al. [107] used an artificial neural network, more specifically a self-organizing map, to perform clustering, multidimensional scaling, and visualization of epigenetic patterns.

Last, Bayesian methods are on the rise, becoming an interesting alternative for unsupervised classification. Zhang et al. [108] developed an adaptive clustering algorithm aimed at analyzing ovarian cancer genome-wide gene expression, DNA methylation, microRNA expression, and copy number alteration profiles following an integrative approach. The method proposed combines an adaptive algorithm based on the Bayesian information criterion with another deep clustering algorithm, which was published previously ("super k-means"). Finally, Wockner et al. [109] used a recursively partitioned mixture model to cluster DNA methylation data, with the aim of obtaining profiles that could be used as a future prognostic indicator of schizophrenia. This model combines a fuzzy clustering algorithm with a level-weighted version of the Bayesian information criterion.

3.3.3 Text Mining

Within the scope of data mining techniques, text mining or text data mining can be defined as the process of deriving high-quality information from text. This way, the information is obtained by observing patterns and/or trends through the application of various techniques, such as statistically based ones. Text mining usually entails structuring the input text (by parsing it, adding and/or removing linguistic features, and inserting the result into a database), deriving patterns within the preprocessed data, and, last, evaluating and interpreting the output. Therefore, it can be considered as a special case of data mining. Text mining is very useful in the process of building databases and allows automatizing literature search, which is usually done manually and is time-consuming [110–114].

Kolářik et al. [115] proposed an approach designed for the identification of histone modifications in biomedical literature with conditional random fields and for the resolution of known histone modification term variants by term standardization. As part of their work, these authors also developed a histone modification term hierarchy to be used in a semantic text retrieval system. They concluded that this approach significantly improves the retrieval of articles that describe histone modifications. Bin Raies et al. [116] presented an innovative text mining methodology based on the concept of position weight matrices for text representation and feature generation. This concept was applied in combination with the document-term matrix, with the purpose of accurately extracting associations between methylated genes and diseases from free text. This methodology is offered also as a Web tool called DEMGD.

Ongenaert and Dehaspe [117] proposed a tool for automatic literature retrieval and annotation of DNA methylation data named GoldMine. This tool, taking into account data introduced by the user (a list of genes, keywords, and highlighting terms), carries out a search over PubMed and then processes the results. Li and Liu [118] also perform text mining over PubMed, but in this case, to obtain a list of candidate biomarkers.

4. CONCLUSIONS AND FUTURE TRENDS

The need for processing and interpreting the huge volume of biological data being produced in the postgenomic era (especially that pertaining to the mechanisms underlying the epigenetic transmission and regulation of heritable information) is currently being addressed by the development of big international projects and standardization endeavors. With the fast growth of the field of bioinformatics, a new landscape of possibilities for massive generation of biological knowledge is in sight. The computational modeling of systems considering simultaneously all relevant factors in a time-resolved manner is indeed the ultimate frontier for epigenetic knowledge. The development of databases and specialized algorithms and software for dynamic simulations should enhance modeling and prediction in epigenetics. Given the amount of data involved and its predictable exponential growth, it is essential that researchers divide the work and that decentralized data storage is used. Multidisciplinary collaboration seems to be the most adequate way to cover all the possible research perspectives. Researchers usually have at their service high-performance computing systems, such as clusters, to carry out computationally expensive tasks. However, it will be essential to continue working on different types of data representations and making algorithms more efficient so that they are scalable for big data analysis. Not only will a decentralized approach be required to achieve this objective, but also the parallelization of the algorithms must be strongly considered. In this sense, technologies such as grid computing appear to be a very promising approach. This type of computing involves many networked, heterogeneous, and geographically dispersed computers, which will probably be loosely coupled, acting together to perform large tasks.

Although a considerable number of epigenetic resources are currently available, these repositories of information are not usually linked. This could be considered an example of what is known as "the functional silo syndrome." To avoid this and with the aim of integrating and linking as much information as possible to take the best out of it, the RDF represents an interesting solution. In combination with ontologies, and what is known as ontology-based data mining, this will very probably be involved in the future of computational epigenetics. Multiple initiatives have been created to move forward in this direction.

LIST OF ACRONYMS AND ABBREVIATIONS

2DGE	Two-dimensional gel electrophoresis
ChIP	Chromatin immunoprecipitation
DMR	Differentially methylated region
EMBL-EBI	European Bioinformatics Institute
GO	Gene Ontology
HEP	Human Epigenome Project
HMM	Hidden Markov model
KDD	Knowledge discovery in databases
LC–MS	Liquid chromatography coupled to mass spectrometry
NCBI	National Center for Biotechnology Information
ncRNA	Noncoding RNA
NIH	National Institutes of Health
PCA	Principal component analysis
piRNA	PIWI-interacting RNA
RDF	Resource description framework

References

[1] Allis CD, Jenuwein T, Reinberg D. Epigenetics. New York: Cold Spring Harbor Laboratory Press; 2007.

[2] Gomez-Cabrero D, Abugessaisa I, Maier D, Teschendorff A, Merkenschlager M, Gisel A, et al. Data integration in the era of omics: current and future challenges. BMC Syst Biol 2014;8(Suppl. 2):I1.

[3] Marx V. Biology: the big challenges of big data. Nature 2013;498(7453):255–60.

[4] Taylor CF, Field D, Sansone SA, Aerts J, Apweiler R, Ashburner M, et al. Promoting coherent minimum reporting guidelines for biological and biomedical investigations: the MIBBI project. Nat Biotechnol 2008;26(8):889–96.

[5] Bird A. The essentials of DNA methylation. Cell 1992;70(1):5–8.

[6] Ng HH, Bird A. DNA methylation and chromatin modification. Curr Opin Genet Dev 1999;9(2):158–63.

[7] Bestor TH. DNA methylation – evolution of a bacterial immune function into a regulator of gene-expression and genome structure in higher eukaryotes. Philo Trans R Soc Lond Ser B Biol Sci 1990;326(1235):179–87.

[8] Jones PA. The DNA methylation paradox. Trends Genet 1999;15(1):34–7.

[9] Liang G, Salem CE, Yu MC, Nguyen HD, Gonzales FA, Nguyen TT, et al. DNA methylation differences associated with tumor tissues identified by genome scanning analysis. Genomics 1998;53(3):260–8.

[10] Costello JF, Fruhwald MC, Smiraglia DJ, Rush LJ, Robertson GP, Gao X, et al. Aberrant CpG-island methylation has non-random and tumour-type-specific patterns. Nat Genet 2000;24(2):132–8.

[11] Bock C. Analysing and interpreting DNA methylation data. Nat Rev Genet 2012;13(10):705–19.
[12] Lister R, O'Malley RC, Tonti-Filippini J, Gregory BD, Berry CC, Millar AH, et al. Highly integrated single-base resolution maps of the epigenome in Arabidopsis. Cell 2008;133(3):523–36.
[13] Gu H, Smith ZD, Bock C, Boyle P, Gnirke A, Meissner A. Preparation of reduced representation bisulfite sequencing libraries for genome-scale DNA methylation profiling. Nat Protoc 2011;6(4):468–81.
[14] Krueger F, Andrews SR. Bismark: a flexible aligner and methylation caller for bisulfite-seq applications. Bioinformatics 2011;27(11):1571–2.
[15] Lutsik P, Feuerbach L, Arand J, Lengauer T, Walter J, Bock C. BiQ analyzer HT: locus-specific analysis of DNA methylation by high-throughput bisulfite sequencing. Nucleic Acids Res 2011;39(Suppl. 2):W551–6.
[16] Ryan DP, Ehninger D. Bison: bisulfite alignment on nodes of a cluster. BMC Bioinforma 2014;15:337.
[17] Zhao MT, Whyte JJ, Hopkins GM, Kirk MD, Prather RS. Methylated DNA immunoprecipitation and high-throughput sequencing (MeDIP-seq) using low amounts of genomic DNA. Cell Reprogr 2014;16(3):175–84.
[18] Hsu YW, Huang RL, Lai HC. MeDIP-on-Chip for methylation profiling. Methods Mol Biol 2015;1249:281–90.
[19] Do JH, Choi DK. Normalization of microarray data: single-labeled and dual-labeled arrays. Mol Cells 2006;22(3):254–61.
[20] Wang X, Ghosh S, Guo SW. Quantitative quality control in microarray image processing and data acquisition. Nucleic Acids Res 2001;29(15):E75–5.
[21] Yang YH, Dudoit S, Luu P, Lin DM, Peng V, Ngai J, et al. Normalization for cDNA microarray data: a robust composite method addressing single and multiple slide systematic variation. Nucleic Acids Res 2002;30(4):e15.
[22] Du P, Zhang X, Huang CC, Jafari N, Kibbe WA, Hou L, et al. Comparison of beta-value and M-value methods for quantifying methylation levels by microarray analysis. BMC Bioinforma 2010;11:587.
[23] Gentleman RC, Carey VJ, Bates DM, Bolstad B, Dettling M, Dudoit S, et al. Bioconductor: open software development for computational biology and bioinformatics. Genome Biol 2004;5(10):R80.
[24] Talbert PB, Henikoff S. Histone variants–ancient wrap artists of the epigenome. Nat Rev Mol Cell Biol 2010;11(4):264–75.
[25] Jenuwein T, Allis CD. Translating the histone code. Science 2001;293(5532):1074–80.
[26] Hoffmann F, Kriegel K, Wenk C. An applied point pattern matching problem: comparing 2D patterns of protein spots. Discrete Appl Math 1999;93(1):75–88.
[27] Dowsey AW, English JA, Lisacek F, Morris JS, Yang GZ, Dunn MJ. Image analysis tools and emerging algorithms for expression proteomics. Proteomics 2010;10(23):4226–57.
[28] Lambert JP, Ethier M, Smith JC, Figeys D. Proteomics: from gel based to gel free. Anal Chem 2005;77(12):3771–87.
[29] Xu H, Wei CL, Lin F, Sung WK. An HMM approach to genome-wide identification of differential histone modification sites from ChIP-seq data. Bioinformatics 2008;24(20):2344–9.
[30] Marti-Renom MA, Stuart AC, Fiser A, Sanchez R, Melo F, Sali A. Comparative protein structure modeling of genes and genomes. Annu Rev Biophys Biomol Struct 2000;29:291–325.
[31] Berman HM, Westbrook J, Feng Z, Gilliland G, Bhat TN, Weissig H, et al. The protein data bank. Nucleic Acids Res 2000;28(1):235–42.
[32] Eirín-López J, González-Romero R, Dryhurst D, Méndez J, Ausió J. Long-term evolution of histone families: old notions and new insights into their mechanisms of diversification across eukaryotes. In: Pontarotti P, editor. Evolutionary biology. Springer Berlin Heidelberg; 2009. p. 139–62.

[33] Biswas M, Voltz K, Smith JC, Langowski J. Role of histone tails in structural stability of the nucleosome. PloS Comput Biol 2011;7(12).

[34] Ettig R, Kepper N, Stehr R, Wedemann G, Rippe K. Dissecting DNA-histone interactions in the nucleosome by molecular dynamics simulations of DNA unwrapping. Biophys J 2011;101(8):1999–2008.

[35] Durrant JD, McCammon JA. Molecular dynamics simulations and drug discovery. BMC Biol 2011;9.

[36] Borhani DW, Shaw DE. The future of molecular dynamics simulations in drug discovery. J Comput Aided Mol Des 2012;26(1):15–26.

[37] Narlikar GJ, Sundaramoorthy R, Owen-Hughes T. Mechanisms and functions of ATP-dependent chromatin-remodeling enzymes. Cell 2013;154(3):490–503.

[38] Jiang C, Pugh BF. Nucleosome positioning and gene regulation: advances through genomics. Nat Rev Genet 2009;10(3):161–72.

[39] Henikoff JG, Belsky JA, Krassovsky K, MacAlpine DM, Henikoff S. Epigenome characterization at single base-pair resolution. Proc Natl Acad Sci USA 2011;108(45):18318–23.

[40] Langmead B, Trapnell C, Pop M, Salzberg SL. Ultrafast and memory-efficient alignment of short DNA sequences to the human genome. Genome Biol 2009;10(3):R25.

[41] Li H, Durbin R. Fast and accurate short read alignment with Burrows–Wheeler transform. Bioinformatics 2009;25(14):1754–60.

[42] Balasubramanian S, Xu F, Olson WK. DNA sequence-directed organization of chromatin: structure-based computational analysis of nucleosome-binding sequences. Biophys J 2009;96(6):2245–60.

[43] Yuan GC, Liu YJ, Dion MF, Slack MD, Wu LF, Altschuler SJ, et al. Genome-scale identification of nucleosome positions in S. cerevisiae. Science 2005;309:626–30.

[44] Brown JD, Mitchell SE, O'Neill RJ. Making a long story short: noncoding RNAs and chromosome change. Heredity 2012;108(1):42–9.

[45] Lee JT. Epigenetic regulation by long noncoding RNAs. Science 2012;338(6113):1435–9.

[46] Magistri M, Faghihi MA, St Laurent III G, Wahlestedt C. Regulation of chromatin structure by long noncoding RNAs: focus on natural antisense transcripts. Trends Genet 2012;28(8):389–96.

[47] Mercer TR, Mattick JS. Structure and function of long noncoding RNAs in epigenetic regulation. Nat Struct Mol Biol 2013;20(3):300–7.

[48] Morlando M, Ballarino M, Fatica A, Bozzoni I. The role of long noncoding RNAs in the epigenetic control of gene expression. ChemMedChem 2014;9(3):505–10.

[49] Whitehead J, Pandey GK, Kanduri C. Regulation of the mammalian epigenome by long noncoding RNAs. Biochimica Biophysica Acta General Subj 2009;1790(9):936–47.

[50] Backofen R, Vogel T. Biological and bioinformatical approaches to study crosstalk of long-non-coding RNAs and chromatin-modifying proteins. Cell Tissue Res 2014;356(3):507–26.

[51] Suarez-Ulloa V, Fernandez-Tajes J, Aguiar-Pulido V, Rivera-Casas C, Gonzalez-Romero R, Ausio J, et al. The CHROMEVALOA database: a resource for the evaluation of okadaic acid contamination in the Marine environment based on the chromatin-associated transcriptome of the mussel Mytilus galloprovincialis. Mar Drugs 2013;11(3):830–41.

[52] Goh WW, Wong L. Computational proteomics: designing a comprehensive analytical strategy. Drug Discov Today 2014;19(3):266–74.

[53] Hawkins RD, Hon GC, Ren B. Next-generation genomics: an integrative approach. Nat Rev Genet 2010;11(7):476–86.

[54] Heyn H. A symbiotic liaison between the genetic and epigenetic code. Front Genet 2014;5:113.

[55] Brazma A, Hingamp P, Quackenbush J, Sherlock G, Spellman P, Stoeckert C, et al. Minimum information about a microarray experiment (MIAME)-toward standards for microarray data. Nat Genet 2001;29(4):365–71.

[56] Martinez-Bartolome S, Binz PA, Albar JP. The minimal information about a proteomics Experiment (MIAPE) from the proteomics standards initiative. Methods Mol Biol 2014;1072:765–80.

III. BIOINFORMATICS OF PERSONALIZED EPIGENETICS

[57] Chervitz SA, Deutsch EW, Field D, Parkinson H, Quackenbush J, Rocca-Serra P, et al. Data standards for omics data: the basis of data sharing and reuse. Methods Mol Biol 2011;719:31–69.

[58] Shakya K, O'Connell MJ, Ruskin HJ. The landscape for epigenetic/epigenomic biomedical resources. Epigenetics Official J DNA Methylation Soc 2012;7(9):982–6.

[59] Kent WJ, Sugnet CW, Furey TS, Roskin KM, Pringle TH, Zahler AM, et al. The human genome browser at UCSC. Genome Res 2002;12(6):996–1006.

[60] Zhou X, Maricque B, Xie M, Li D, Sundaram V, Martin EA, et al. The human epigenome browser at Washington university. Nat Methods 2011;8(12):989–90.

[61] Consortium EP. The encode (ENCyclopedia of DNA Elements) project. Science 2004;306(5696):636–40.

[62] Rakyan VK, Hildmann T, Novik KL, Lewin J, Tost J, Cox AV, et al. DNA methylation profiling of the human major histocompatibility complex: a pilot study for the human epigenome project. PLoS Biol 2004;2(12):e405.

[63] Cieslik M, Bekiranov S. Combinatorial epigenetic patterns as quantitative predictors of chromatin biology. BMC Genomics 2014;15:76.

[64] Benveniste D, Sonntag HJ, Sanguinetti G, Sproul D. Transcription factor binding predicts histone modifications in human cell lines. Proc Natl Acad Sci USA 2014;111(37):13367–72.

[65] Rosenbloom KR, Dreszer TR, Long JC, Malladi VS, Sloan CA, Raney BJ, et al. ENCODE whole-genome data in the UCSC Genome Browser: update 2012. Nucleic Acids Res 2012;40(Database issue):D912–7.

[66] Podlaha O, De S, Gonen M, Michor F. Histone modifications are associated with transcript isoform diversity in normal and cancer cells. PLoS Comput Biol 2014;10(6):e1003611.

[67] Bernstein BE, Stamatoyannopoulos JA, Costello JF, Ren B, Milosavljevic A, Meissner A, et al. The NIH roadmap epigenomics mapping consortium. Nat Biotechnol 2010;28(10): 1045–8.

[68] Bae JB. Perspectives of international human epigenome consortium. Genomics Inform 2013;11(1):7–14.

[69] Sogn JA, Anton-Culver H, Singer DS. Meeting report: NCI think tanks in cancer biology. Cancer Res 2005;65(20):9117–20.

[70] Jones PA, Martienssen R. A blueprint for a human epigenome project: the AACR human epigenome workshop. Cancer Res 2005;65(24):11241–6.

[71] Akhtar A, Cavalli G. The epigenome network of excellence. PLoS Biol 2005;3(5):e177.

[72] American Association for Cancer Research Human Epigenome Task F, European Union NoESAB. Moving AHEAD with an international human epigenome project. Nature 2008;454(7205):711–5.

[73] Lim SJ, Tan TW, Tong JC. Computational epigenetics: the new scientific paradigm. Bioinformation 2010;4(7):331–7.

[74] Wang X, Gorlitsky R, Almeida JS. From XML to RDF: how semantic web technologies will change the design of 'omic' standards. Nat Biotechnol 2005;23(9):1099–103.

[75] Belleau F, Nolin MA, Tourigny N, Rigault P, Morissette J. Bio2RDF: towards a mashup to build bioinformatics knowledge systems. J Biomed Informatics 2008;41(5):706–16.

[76] Sahoo SS, Bodenreider O, Rutter JL, Skinner KJ, Sheth AP. An ontology-driven semantic mashup of gene and biological pathway information: application to the domain of nicotine dependence. J Biomed Informatics 2008;41(5):752–65.

[77] Cheung KH, Yip KY, Townsend JP, Scotch M. HCLS 2.0/3.0: health care and life sciences data mashup using Web 2.0/3.0. J Biomed Inform 2008;41(5):694–705.

[78] Whetzel PL, Noy NF, Shah NH, Alexander PR, Nyulas C, Tudorache T, et al. BioPortal: enhanced functionality via new Web services from the National Center for Biomedical Ontology to access and use ontologies in software applications. Nucleic Acids Res 2011;39(Web server issue):W541–5.

[79] Ashburner M, Ball CA, Blake JA, Botstein D, Butler H, Cherry JM, et al. Gene ontology: tool for the unification of biology. The Gene Ontology Consortium. Nat Genet 2000;25(1):25–9.

[80] Wippermann A, Klausing S, Rupp O, Albaum SP, Buntemeyer H, Noll T, et al. Establishment of a CpG island microarray for analyses of genome-wide DNA methylation in Chinese hamster ovary cells. Appl Microbiol Biotechnol 2014;98(2):579–89.

[81] Triff K, Konganti K, Gaddis S, Zhou B, Ivanov I, Chapkin RS. Genome-wide analysis of the rat colon reveals proximal-distal differences in histone modifications and proto-oncogene expression. Physiol Genomics 2013;45(24):1229–43.

[82] Kalari S, Jung M, Kernstine KH, Takahashi T, Pfeifer GP. The DNA methylation landscape of small cell lung cancer suggests a differentiation defect of neuroendocrine cells. Oncogene 2013;32(30):3559–68.

[83] Zeller C, Dai W, Curry E, Siddiq A, Walley A, Masrour N, et al. The DNA methylomes of serous borderline tumors reveal subgroups with malignant- or benign-like profiles. Am J Pathol 2013;182(3):668–77.

[84] Bajpai M, Kessel R, Bhagat T, Nischal S, Yu Y, Verma A, et al. High resolution integrative analysis reveals widespread genetic and epigenetic changes after chronic in-vitro acid and bile exposure in Barrett's epithelium cells. Genes Chromosomes Cancer 2013;52(12):1123–32.

[85] Akulenko R, Helms V. DNA co-methylation analysis suggests novel functional associations between gene pairs in breast cancer samples. Hum Mol Genet 2013; 22(15):3016–22.

[86] Towle R, Truong D, Hogg K, Robinson WP, Poh CF, Garnis C. Global analysis of DNA methylation changes during progression of oral cancer. Oral Oncol 2013;49(11):1033–42.

[87] Baas R, Lelieveld D, van Teeffelen H, Lijnzaad P, Castelijns B, van Schaik FM, et al. A novel microscopy-based high-throughput screening method to identify proteins that regulate global histone modification levels. J Biomol Screen 2014;19(2):287–96.

[88] Reutlinger M, Schneider G. Nonlinear dimensionality reduction and mapping of compound libraries for drug discovery. J Mol Graph Model 2012;34:108–17.

[89] Goh WW, Wong L. Networks in proteomics analysis of cancer. Curr Opin Biotechnol 2013;24(6):1122–8.

[90] Zheng J, Chaturvedi I, Rajapakse J. Integration of epigenetic data in Bayesian network modeling of gene regulatory network. In: Loog M, Wessels L, Reinders MT, de Ridder D, editors. Pattern recognition in bioinformatics. Berlin Heidelberg: Springer; 2011. p. 87–96.

[91] Dyson MT, Roqueiro D, Monsivais D, Ercan CM, Pavone ME, Brooks DC, et al. Genome-wide DNA methylation analysis predicts an epigenetic switch for GATA factor expression in endometriosis. PLoS Genet 2014;10(3):e1004158.

[92] Figueroa ME, Wouters BJ, Skrabanek L, Glass J, Li Y, Erpelinck-Verschueren CA, et al. Genome-wide epigenetic analysis delineates a biologically distinct immature acute leukemia with myeloid/T-lymphoid features. Blood 2009;113(12):2795–804.

[93] Volkmar M, Dedeurwaerder S, Cunha DA, Ndlovu MN, Defrance M, Deplus R, et al. DNA methylation profiling identifies epigenetic dysregulation in pancreatic islets from type 2 diabetic patients. EMBO J 2012;31(6):1405–26.

[94] Loss LA, Sadanandam A, Durinck S, Nautiyal S, Flaucher D, Carlton VE, et al. Prediction of epigenetically regulated genes in breast cancer cell lines. BMC Bioinforma 2010;11:305.

[95] Aguiar-Pulido V, Seoane JA, Gestal M, Dorado J. Exploring patterns of epigenetic information with data mining techniques. Curr Pharm Des 2013;19(4):779–89.

[96] Schafer M, Lkhagvasuren O, Klein HU, Elling C, Wustefeld T, Muller-Tidow C, et al. Integrative analyses for omics data: a Bayesian mixture model to assess the concordance of ChIP-chip and ChIP-seq measurements. J Toxicol Environ Health Part A 2012;75(8–10):461–70.

[97] Mo Q, Wang S, Seshan VE, Olshen AB, Schultz N, Sander C, et al. Pattern discovery and cancer gene identification in integrated cancer genomic data. Proc Natl Acad Sci USA 2013;110(11):4245–50.

[98] Gonzalo V, Lozano JJ, Alonso-Espinaco V, Olshen AB, Schultz N, Sander C, et al. Multiple sporadic colorectal cancers display a unique methylation phenotype. PloS One 2014;9(3):e91033.

[99] Low DH, Motakis E. deltaGseg: macrostate estimation via molecular dynamics simulations and multiscale time series analysis. Bioinformatics 2013;29(19):2501–2.

[100] Busche S, Ge B, Vidal R, Spinella JF, Saillour V, Richer C, et al. Integration of high-resolution methylome and transcriptome analyses to dissect epigenomic changes in childhood acute lymphoblastic leukemia. Cancer Res 2013;73(14):4323–36.

[101] Tian R, Feng J, Cai X, Zhang Y. Local chromatin dynamics of transcription factors imply cell-lineage specific functions during cellular differentiation. Epigenetics Official J DNA Methylation Soc 2012;7(1):55–62.

[102] Clifford H, Wessely F, Pendurthi S, Emes RD. Comparison of clustering methods for investigation of genome-wide methylation array data. Front Genet 2011;2:88.

[103] McGaughey DM, Abaan HO, Miller RM, Kropp PA, Brody LC. Genomics of CpG methylation in developing and developed zebrafish. G3 2014;4(5):861–9.

[104] Jung I, Park JC, Kim S. piClust: a density based piRNA clustering algorithm. Comput Biol Chem 2014;50:60–7.

[105] Ucar D, Hu Q, Tan K. Combinatorial chromatin modification patterns in the human genome revealed by subspace clustering. Nucleic acids research 2011;39(10):4063–75.

[106] Yu P, Xiao S, Xin X, Song CX, Huang W, McDee D, et al. Spatiotemporal clustering of the epigenome reveals rules of dynamic gene regulation. Genome Res 2013;23(2):352–64.

[107] Steiner L, Hopp L, Wirth H, Galle J, Binder H, Prohaska SJ, et al. A global genome segmentation method for exploration of epigenetic patterns. PloS One 2012;7(10):e46811.

[108] Zhang W, Liu Y, Sun N, Wang D, Boyd-Kirkup J, Dou X, et al. Integrating genomic, epigenomic, and transcriptomic features reveals modular signatures underlying poor prognosis in ovarian cancer. Cell Reports 2013;4(3):542–53.

[109] Wockner LF, Noble EP, Lawford BR, Young RM, Morris CP, Whitehall VL, et al. Genome-wide DNA methylation analysis of human brain tissue from schizophrenia patients. Transl Psychiatry 2014;4:e339.

[110] Ongenaert M, Van Neste L, De Meyer T, Menschaert G, Bekaert S, Van Criekinge W. PubMeth: a cancer methylation database combining text-mining and expert annotation. Nucleic Acids Res 2008;36(Database issue):D842–6.

[111] Fang YC, Huang HC, Juan HF. MeInfoText: associated gene methylation and cancer information from text mining. BMC Bioinforma 2008;9:22.

[112] Fang YC, Lai PT, Dai HJ, Hsu WL. MeInfoText 2.0: gene methylation and cancer relation extraction from biomedical literature. BMC Bioinforma 2011;12:471.

[113] Harmston N, Filsell W, Stumpf MP. What the papers say: text mining for genomics and systems biology. Hum Genomics 2010;5(1):17–29.

[114] Krallinger M, Leitner F, Valencia A. Analysis of biological processes and diseases using text mining approaches. Methods Mol Biol 2010;593:341–82.

[115] Kolarik C, Klinger R, Hofmann-Apitius M. Identification of histone modifications in biomedical text for supporting epigenomic research. BMC Bioinforma 2009;10 (Suppl. 1):S28.

[116] Bin Raies A, Mansour H, Incitti R, Bajic VB. Combining position weight matrices and document-term matrix for efficient extraction of associations of methylated genes and diseases from free text. PloS One 2013;8(10):e77848.

[117] Ongenaert M, Dehaspe L. Integrating automated literature searches and text mining in biomarker discovery. BMC Bioinforma 2010;11(Suppl. 5):O5.

[118] Li H, Liu C. Biomarker identification using text mining. Comput Math Methods Med 2012;2012:4.

SECTION IV

DIAGNOSTIC AND PROGNOSTIC EPIGENETIC APPROACHES TO PERSONALIZED MEDICINE

7

Epigenetic Biomarkers in Personalized Medicine

Fabio Coppedè[1], Angela Lopomo[1,2], Lucia Migliore[1]
[1]Department of Translational Research and New Technologies in Medicine and Surgery, University of Pisa, Pisa, Italy; [2]Doctoral School in Genetics, Oncology, and Clinical Medicine, University of Siena, Siena, Italy

OUTLINE

1. INTRODUCTION

Epigenetic marks, originally viewed as stable changes orchestrating differential gene expression among tissues and cell types in differentiated organisms, include DNA methylation, histone tail modifications, and chromatin remodeling processes. In addition, the regulation of gene expression mediated by noncoding RNA molecules, such as microRNAs (miRNAs) or long noncoding RNAs, is being increasingly recognized as an epigenetic mechanism [1].

It is now clear that epigenetic marks are less stable than originally believed and can be environmentally induced as a response to external stimuli in the physiological processes of learning and memory, but also occur during the life span in healthy tissues and play a pivotal role in the onset and progression of several complex human diseases, including cancer, autoimmune disorders, neurodegenerative diseases, obesity, psychiatric disorders, and many others [2]. More interestingly, as largely discussed in a previous section of this book, variability of epigenetic marks exists among individuals and among the different cell types of each individual. This is not surprising since epigenetic changes are largely viewed as the missing link between the environment and the genome, probably representing a cellular response to environmental exposures. As a consequence of this, the study of epigenetic biomarkers often requires isolation of the target tissue/organ and separation and analysis of different cell types [3]. Moreover, to fully understand the interindividual variability of epigenetic marks, large cohorts of individuals need to be investigated in different environmental settings or within the different stages leading from healthy conditions to early, intermediate, and advanced disease status. For this reason most of our knowledge of the human epigenetic variability derives from studies on surgically resected cancers, owing to the large availability of cancerous samples and surrounding healthy tissues that are routinely removed from hundreds of patients at different disease stages [4,5]. By contrast, our knowledge of the epigenetic basis of other diseases, such as neurodegenerative ones, is often limited by the availability of few well-preserved postmortem materials coupled with studies in tissues obtained from animal disease models [6,7].

Among epigenetic mechanisms, DNA methylation has gained particular interest, as DNA methylation profiles represent a more chemically and biologically stable source of molecular diagnostic information than RNA or most proteins, making this marker of particular interest for basic research studies as well as for either diagnostic or prognostic purposes. Moreover, in the case of cancer, circulating free DNA in the plasma or serum of the patient can be screened to search for easily detectable epigenetic biomarkers of the disease [8]. Similarly, tumor-derived noncoding RNAs can be screened for the same purpose [9].

Diagnostic tools, such as those based on the search for methylated septin 9 (*SEPT9*) in circulating DNA of colorectal cancer patients or for methylated vimentin (*VIM*) in stool DNA, are now available for certain cancers, and hundreds of methylated genes or aberrantly expressed noncoding RNAs are under investigation for their prognostic value [8,9]. Similarly, increasing evidence suggests that as stable changes in the DNA sequence, such as rare mutations and common polymorphisms, are of extreme relevance to the design of personalized therapeutic strategies, so also can epigenetically induced changes in gene expression be responsible for interindividual responses to several drugs, reducing their efficacy or enhancing their adverse effects, and should be taken into account in the design of personalized pharmacological interventions. In addition, most epigenetic marks are potentially reversible, opening the way to drugs that are able to mold the epigenome for the treatment of various human illnesses. As a consequence, pharmacoepigenetics represents a timely and very fascinating field of active research [10].

This chapter provides several examples of either already available epigenetic biomarkers or potential ones that could represent valuable tools for diagnostic (Table 1) and prognostic purposes (Table 2), as well as for the choice of the most proper and personalized therapeutic approach (Table 3).

TABLE 1   Examples of Proposed Epigenetic Diagnostic Biomarkers

Disease	Modification	Biomarker	Description	Reference
Hematological malignancies	DNA methylation	*TLX3, FOXE3*	Biomarkers of minimal residual disease in pediatric acute lymphoblastic leukemia	[35]
Colorectal cancer	DNA methylation	*SEPT9, ALX4, HLTF*	Biomarkers of diagnosis in blood	[45–49]
		THBD	Biomarkers of early diagnosis in blood	[50]
		APC, FBN1, MLH1, MGMT, SFRP1, SFRP2, VIM, WIF-1	Biomarkers of diagnosis in stool	[53–55]
	microRNA	miR-21, miR-92a, miR-135b	Potential diagnostic biomarkers in stool DNA	[59,60]

Continued

TABLE 1   Examples of Proposed Epigenetic Diagnostic Biomarkers—cont'd

Disease	Modification	Biomarker	Description	Reference
Gastric cancer	DNA methylation	CHRNA3, DOK1, GNMT, ALDH2, MTHFR	Potential biomarkers of disease status	[76]
Breast and ovarian cancer	DNA methylation	BRCA1, RASSF1, APC, CDKN2A (p14), CDKN2A (p16), DAPK	Biomarkers of early ovarian cancer	[86,87]
Prostate cancer	DNA methylation	HOXD3	Biomarker of diagnosis	[95]
		GSTP1	Biomarker in blood	[96]
Bladder cancer	microRNA	miR-152, miR-10a, miR-200b	Biomarkers in urinary DNA samples	[114]
Lung cancer	DNA methylation	CDKN2A (p16), PAX5, DAPK, RASSF1, GATA5, MGMT	Biomarkers in sputum DNA samples	[124]
		SEPT9	Biomarker of early diagnosis in plasma DNA	[44]
Thyroid cancer	DNA methylation	TSHR, NIS, E-CAD, ATM, AIT, TIMP3, DAPK, RARB2	Potential diagnostic biomarkers	[139]
Assisted reproduction	DNA methylation	MEST	Biomarker of oligozoospermia in semen DNA	[193]
		ANKRD30A, ELF5, PRAME, SPACA3, MAGEA1, MORC1	Potential biomarkers of male subfertility	[194]

TABLE 2 Examples of Epigenetic Prognostic Biomarkers

Disease	Modification	Biomarker	Description	Reference
Hematological malignancies	DNA methylation	ANGPT2	Biomarker of adverse prognosis	[36]
		CD34, RHOC, SCRN1, F2EL1, FAM92A1, MIR155HG, VWA8	Biomarkers of overall survival	[37]
Colorectal cancer	DNA methylation	CDKN2A (p14), RASSF1, APC,	Biomarkers of poor prognosis	[42]
		HOPX-β	Biomarkers of worse prognosis of stage II CRC	[42]
		IGFBP3, EVL, CD109, FLNC	Biomarkers of worse survival	[42]
		IGF2, LINE1	Biomarkers of worse prognosis	[42,62]
	Histone modification	H3K9me3	Biomarker of lymph node metastasis	[61]
Gastric cancer	DNA methylation	MGMT, CDKN2A (p16), RASSF2, FLNC	Biomarkers of poor survival	[77]
Breast and ovarian cancer	DNA methylation	BMP6	Biomarker of lymph node metastasis in breast cancer	[80]
		RASSF1	Biomarker of poor prognosis in breast cancer	[81]
		ESR1, CXCR4	Biomarker of aggressiveness in breast cancer	[82,83,84]
		BRCA1	Biomarker of advanced ovarian cancer	[88]
		ZIC1	Biomarker of poor progression-free survival in ovarian cancer	[89]

Continued

TABLE 2   Examples of Epigenetic Prognostic Biomarkers—cont'd

Disease	Modification	Biomarker	Description	Reference
Prostate cancer	DNA methylation	*APC, CDH1, CD44, EDNRB, GSTP1, MDR1, MT1G, PTGS2, RARB2, RASSF1, RUNX3*	Biomarkers of disease progression and prediction of cancer recurrence	[97–102]
	Histone modification	H3K18ac, H3K4me1, H3K4me2	Biomarkers of cancer recurrence	[104,105]
	microRNA	miR-141	Biomarker of metastasis	
		miR-31, miR-96, miR-205	Biomarkers of disease severity	[106]
		miR-125b, miR-205, miR-222	Biomarkers of tumor stage	[106]
Bladder cancer	DNA methylation	*APC, ARF, BCL2, CDH1, CDH13, CDKN2A (p16), EDNRB, OPCML, PMF1, RASSF1, TERT, TNFSR25*	Biomarkers of disease progression, poor prognosis, and prediction of cancer recurrence	[115–118]
	Histone modification	H3K4me1, H4K20me1, H4K20me2, H4K20me3	Biomarkers of advanced pathological stage	[119]
	microRNA	miR-129, miR-133b, miR-518c	Biomarkers of worse outcome	[120]
		miR-452	Biomarker of lymph node metastases and poor prognosis	[120]
Lung cancer	DNA methylation	*CDKN2A (p16), CDH13, APC, RASSF1*	Biomarkers of disease recurrence	[125]
		RASSF1, DAPK1	Biomarkers of poor prognosis	[126]

TABLE 2   Examples of Epigenetic Prognostic Biomarkers—cont'd

Disease	Modification	Biomarker	Description	Reference
	microRNA	miR-155, let-7a2	Biomarkers of poor survival	[122]
		miR-21	Biomarkers of poor prognosis	[122]
Thyroid cancer	DNA methylation	CDH1	Biomarker of metastasis	[139]
		TIMP3, SOX17, RAP1GAP	Biomarkers of tumor proliferation, invasion, and metastasis	[140–142]
		EI24, WT1	Biomarkers of disease recurrence	[143]
Esophageal cancer	DNA methylation	APC, CDH1, CDKN2A (p16), DAPK, ER, MGMT, TIMP-3	Biomarkers of reduced survival and tumor recurrence	[145]
		NELL1, TAC1, FHIT	Biomarkers of poor prognosis	[146]
		TLSC1	Biomarker of aggressiveness	[146]
		UCHL1	Biomarker of lymph node metastases	[146]
Kidney cancer	DNA methylation	APAF1, DAPK1, EPB41L3, JUP, PTEN	Biomarkers of poor outcome	[149]

2. CANCER

It is now clear that both mutations and epigenetic changes play fundamental roles in the onset and progression of cancer, in the tendency toward metastasis, and in the response to treatment [11]. More interestingly, it is believed that virtually all tumors harbor mutations in proteins that control the epigenome, and for every genetic mutation in a given patient's tumor the expression of hundreds of genes is epigenetically deregulated to drive tumorigenesis [12,13]. Global DNA hypomethylation and regional hypermethylation of CpG islands in or around the promoter region of hundreds of genes per tumor are commonly observed in cancer, the latter leading to decreased gene expression and working as an alternative mechanism to mutation by which silencing of tumor suppressor genes can occur [11].

TABLE 3　Examples of DNA Methylation Biomarkers of Response to Treatment

Disease	Biomarker	Description	Reference
Hematological malignancies	SMAD1	Biomarker of chemoresistance in diffuse large B cell lymphoma cells	[40]
	DPH1	Biomarker of immunotoxin resistance in acute lymphoblastic leukemia	[41]
Colorectal cancer	TFAP2E	Biomarker of nonresponsiveness to standard chemotherapy	[66]
	MGMT	Biomarker of response to alkylating agents	[67]
	SRBC	Candidate predictive biomarker for oxaliplatin resistance	[72]
Breast and ovarian cancer	RASSF1	Biomarker of docetaxel-based chemotherapy in breast cancer	[91]
	SFRP5, MLH1, ASS1, ESR2, MCJ, DAPK	Biomarkers of resistance to platinum-based therapy in ovarian cancer	[86]
	BRCA1	Biomarker of enhanced sensitivity to platinum-derived drugs in ovarian cancer	[93]
	MAL, ABCG2	Candidate biomarkers of chemoresistance in ovarian carcinoma cell lines	[86,94]
	GSTP1	Biomarker of response to and efficacy of doxorubicin treatment in breast cancer cells	[102]
Prostate cancer	GSTP1	Biomarker of DNMTi efficacy in human prostate cancer cells	[101]
Psychiatric and behavioral disorders	IL11	Biomarker of clinical response to antidepressants in depressive disorder	[171]
Obesity	TNF, ATP10A, WT1	Biomarkers of response to a balanced hypocaloric diet	[189,190]
	AQP9, DUSP22, HIPK3, TNNT1, TNNI3	Biomarkers of response to weight loss interventions in adolescents	[191]

Other frequently observed epigenetic changes in cancer are posttranslational modifications of the histone tails of nucleosomes, including global deacetylation and regional demethylation at lysine 4 of histone H3 at DNA-hypermethylated genes, leading to a condensed chromatin configuration [14], as well as altered expression of several noncoding RNAs occurring within disease progression, invasion, and metastasis [15]. Therefore, a complex interplay between DNA methylation, chromatin events, and other mechanisms that regulate gene expression levels occurs in cancer and is responsible for the inactivation of tumor suppressor genes, DNA repair genes, and many others to allow escape from immune response and apoptosis and resistance to chemotherapy [13].

2.1 Hematological Malignancies

Hematological malignancies were among the first pathological conditions to be identified as suitable candidates for pharmacological manipulation and therapeutic targeting by means of drugs acting on DNA methylation reactions [16]. DNA methylation is mediated by DNA methyltransferase enzymes (DNMTs) using S-adenosylmethionine as the methyl donor compound [17]. There are multiple families of DNMTs in mammals. Among them, DNMT1 is primarily involved in the maintenance of DNA methylation patterns during development and cell division, whereas DNMT3A and DNMT3B are the de novo methyltransferases and establish DNA methylation patterns during early development. The methylated DNA can be specifically recognized by a set of proteins called methyl-CpG binding proteins (MBPs), which contain a transcription repression domain to interact with other proteins and enhance DNA methylation-mediated transcriptional repression [18]. Enzymes of the TET (ten–eleven translocation) family promote DNA demethylation in mammalian cells, converting 5-methylcytosine to 5-hydroxymethylcytosine (5-hmC), a modified form of cytosine hydroxymethylated at the 5 position showing lower affinity to MBPs compared to 5-methylcytosine [19]. 5-hmC is also involved in regulating the pluripotency of stem cells and is connected to the processes of cellular development and carcinogenesis [20].

Studies performed in mice with conditional knockout of genes coding for either maintenance or de novo DNMTs have revealed a role for DNA methylation in mediating the self-renewal and differentiation of normal hematopoietic stem cells and leukemia stem cells [21]. Moreover, the large-scale sequencing of cancer genomes has allowed the identification of various mutations of *DNMT3A* and other regulators of DNA methylation in hematological malignancies [21–23]. Particularly, almost 44% of the patients with acute myeloid leukemia (AML) exhibit mutations in genes that regulate the methylation of genomic DNA and particularly in the genes encoding DNMT3A, isocitrate dehydrogenase 1 and 2 (IDH1 and IDH2), and TET oncogene family member 2 (TET2) [21–23].

Concerning the *DNMT3A* gene, recurrent mutations in AML include missense, frameshift, nonsense, and splice-site mutations and a partial deletion [24]. Among them, the most common type of missense mutation is predicted to affect the amino acid at R882 and is found in over half of the individuals with *DNMT3A* mutations [25]. A meta-analysis of the literature confirmed objectively the robustness of the association between mutant *DNMT3A* and worse prognosis of adults with de novo AML and suggested that, combined with other important genetic and epigenetic biomarkers, which are often missing in the available literature, *DNMT3A* mutations would contribute to a more precise clinical risk stratification and decision of treatment. Particularly, it was suggested that more studies are needed to clarify whether dose-intensified chemotherapy could improve the survival outcomes in patients with *DNMT3A* mutations [25].

Mutations in *IDH1* and *IDH2* have been identified in both glioblastoma multiforme and AML, are mutually exclusive, and were reviewed by Cairns and Max in 2013 [26]. For the *IDH1* gene, the R132H substitution in the active site of the protein constituted the vast majority of mutational events, while for *IDH2* the disease-associated mutations caused predominantly R172K and R140Q amino acid substitutions [26]. Concerning AML, approximately half of the identified *IDH* mutations occurred in *IDH2*, and both R172K and R140Q were common. The same mutations were observed in myeloproliferative neoplasms and myelodysplastic syndrome, both of which are conditions that can progress to AML. In addition, *IDH* mutations in AML are mutually exclusive with mutations in *TET2*, which are found in 10–20% of the cases [26]. Moreover, *TET2* mutations are common in angioimmunoblastic T cell lymphoma (AITL), one of the most common subtypes of peripheral T cell lymphoma, and *IDH2* mutations at R172 have been reported in 20–45% of AITL cases [27,28]. Concerning *IDH* mutations, they are linked to an increased production of d-2-hydroxyglutarate that inhibits several dioxygenases required for the correct methylation/demethylation of cytosine residues in DNA and of lysine residues in histone proteins [26]. TET dioxygenases catalyze the hydroxylation of 5-methylcytosine residues in DNA, and in vitro studies revealed that *IDH* mutations interfere with TET-driven demethylation of the DNA, resulting in elevated 5-methylcytosine levels. As a consequence, samples from cancer patients with *IDH* mutations or *TET2* mutations show similar patterns of global DNA hypermethylation [29]. The understanding of the biological implications of mutations in those enzymes is opening the way for clinical applications. For example, the *IDH* mutation status has a potential prognostic value as it is a strong predictor of progression and outcome in glioma [26], while d-2-hydroxyglutarate levels are evaluated as a potential biomarker of disease activity and therapeutic response in AML [30]. In addition, in 2012, the first small molecules that inhibit mutant IDH proteins were developed and are being tested in model systems of cancer [31,32].

Indeed, DNA methylation biomarkers capable of diagnosis and subtyping have been investigated in several hematopoietic tumors as well as prognostic ones [33]. We are now in the phase of validation of such biomarkers. For instance, in a genome-wide association study on pediatric acute lymphoblastic leukemia (ALL) 325 genes were found to be hypermethylated and downregulated and 45 genes hypomethylated and upregulated [34]. A validation study of some of these markers led to a reliable detection of DNA methylation of the *TLX3* and *FOXE3* genes, which allowed the monitoring of minimal residual disease in pediatric ALL patients [35]. In chronic lymphocytic leukemia low angiopoietin-2 (*ANGPT2*) methylation status was highly associated with adverse prognostic markers, shorter time to first treatment, and overall survival [36]. In AML, seven genes have been identified (*CD34, RHOC, SCRN1, F2EL1, FAM92A1, MIR155HG,* and *VWA8*) with promoter methylation and expression associated with overall survival [37].

The use of epigenetic modulating agents for cancer therapy is valuable for hematological malignancies. The DNA demethylating agents decitabine and 5-azacytidine received U.S. Food and Drug Administration (FDA) approval for myelodysplasia and AML, and subsequent studies revealed that while high doses of the drugs are toxic for the patients, low doses minimize toxicity while retaining the ability to bind to and inhibit DNMTs [16]. Another class of drugs working as epigenetic modulators are inhibitors of histone deacetylases (HDACs). As reviewed by Azad and coworkers [13], the HDAC inhibitors (HDACi's) vorinostat and romidepsin have both received FDA approval for hematological malignancies and particularly for the treatment of T cell cutaneous lymphoma, but the use of HDACi's alone had little success in clinical trials for solid tumors [13]. There is also indication that combinations of drugs working on DNA methylation and histone tail modifications could be beneficial in hematological malignancies. For example, a recent study in human malignant lymphoma cells revealed a combined effect of epigallocatechin-3-gallate and the HDACi trichostatin A in reactivating *CDKN2A* (*p16*) gene expression and decreasing cell proliferation [38].

An epigenetic mechanism of resistance to targeted therapy in T cell acute lymphoblastic leukemia (T-ALL) has been observed [39]. The response to γ-secretase inhibitors (GSIs), used to prevent NOTCH1 activation in T-ALL, is limited by a resistance mediated by an epigenetic mechanism. Indeed, relative to GSI-sensitive cells, persister cells activate distinct signaling and transcriptional programs and exhibit chromatin compaction. A knockdown screen identified chromatin regulators essential for persister viability, including bromodomain-containing protein 4, a chromatin reader protein that recognizes and binds acetylated histones and plays a key role in the transmission of epigenetic memory across cell divisions [39]. Studies in diffuse large B cell lymphoma (DLBCL) revealed that chemoresistance is associated with aberrant DNA methylation programming. Particularly, it was shown that prolonged exposure to low-dose DNMT inhibitors (DNMTi's) reprogrammed chemoresistant cells to become doxorubicin sensitive without major toxicity in vivo. Nine genes were recurrently hypermethylated in chemoresistant DLBCL [40]. Of these, the gene coding for the transcription factor SMAD1 was a critical contributor, and reactivation was required for chemosensitization [40]. A phase I clinical study with 5-azacytidine followed by standard chemoimmunotherapy confirmed that *SMAD1* demethylation is required for chemosensitization in DLBCL [40]. There is also an indication from in vitro studies that methylation of the promoter of *DPH1*, a gene coding for diphthamide biosynthesis protein 1, causes immunotoxin resistance in ALL [41]. These are some of several examples indicating that both immune- and chemoresistance in cancer cells are largely mediated by epigenetic events and suggesting that the optimization of dosing of agents exerting epigenetic properties and their combination with standard therapies could give these agents a promising place in cancer management.

2.2 Solid Tumors

Concerning solid tumors, the promise of using epigenetic drugs for cancer therapy has not yet been realized [13]. However, as reviewed by Azad and coworkers [13], laboratory and clinical studies revealed that these drugs do not work when doses that produce cytotoxic effects are used, suggesting that lower doses are likely to have a clinical effect in reprogramming malignant cells. In addition, especially when low doses are used, the epigenetic response might take longer to become apparent with respect to conventional chemotherapeutic agents, and combination of epigenetic compounds with standard chemotherapeutic agents might yield better results in clinical trials than the use of epigenetic drugs alone. Therefore, taking into account all the above considerations, epigenetic drugs could still have a promising future in the management of solid tumors [13]. Despite this, increasing evidence suggests that epigenetic biomarkers need to be taken into account for diagnosis, for prognosis, and in the design of the most proper personalized therapeutic approach in solid tumors [42], and several examples are provided below.

2.2.1 Colorectal Cancer

The search for noninvasive and cost-effective DNA methylation biomarkers of colorectal cancer (CRC) led researchers to develop blood-based and stool-based diagnostic tools [43]. One of the epigenetic changes discovered is methylation of the *SEPT9* gene. It is a suppressor gene whose expression defects are probably one of the causes of cancer development. Decrease in expression of this gene is most often connected with methylation of its promoter region, which induces cancer cell proliferation and migration [44]. This results in an acceleration of tumor growth and facilitates creation of distant metastases. Increased promoter methylation of *SEPT9* in circulating DNA appears to be particularly common in CRC patients [44]. Indeed, the presence of aberrantly methylated *SEPT9* in plasma is a valuable and minimally invasive blood-based PCR test, showing a sensitivity and a specificity of almost 90% in the detection of CRC [45–47], and represents a currently commercialized test, as it is able to detect CRC at all stages and locations [47]. To increase the sensitivity of this blood-based test, researchers have evaluated the possibility of including the methylation analysis of additional genes, such as *ALX4* and *HLTF* [48,49]. Others are searching for blood-based biomarkers other than *SEPT9*, and genome-scale approaches in blood DNA have revealed that the methylated thrombomodulin (*THBD*) gene detects 74% of stage I/II CRCs at a specificity of 80% [50] and that methylation of the syndecan 2 (*SDC2*) gene has a sensitivity of 92% for stage I CRC [51]. A stool-based test for the methylation analysis of the *VIM* gene is available in the United States and has a specificity and sensitivity of almost 80% [52]. Several other hypermethylated genes isolated from fecal DNA have been utilized as biomarkers for the detection of CRC or colorectal adenomas, including *APC, FBN1, MLH1, MGMT, SFRP1, SFRP2,* and *WIF1* [53–55]. Meta-analyses of the literature have revealed that the sensitivity for the detection of CRC or adenomas using those biomarkers ranged from 62% to 75% [53,56]. A new approach was performed by Holdenreider and coworkers [57], by using the enzyme-linked immunosorbent assay to investigate the relevance of 5-methylcytosine-modified DNA present in cell-free circulating nucleosomes. A reduced methylation of DNA in circulating nucleosomes was found in patients with CRC [57].

An interesting screening approach using stable miRNA molecules extracted from stool and blood samples has been proposed by Ahmed and collaborators [58]. The quantitative changes in the expression of a few cell-free circulatory mature miRNA molecules in plasma (miR-124, miR-127-3p, miR-138, miR-143, miR-146a, and miR-222), which exhibited reduced expression in plasma (and also in tissues) of patients with CRC, are associated with disease progression from early to later carcinoma stages and seem to be sensitive and specific markers for CRC [58]. Concerning stool samples, the levels of miR-135b were significantly higher in subjects with CRC or adenoma. Moreover, whereas patients with CRC had significantly higher stool miR-21 and miR-92a levels compared with normal controls, stool miR-92a, but not miR-21, was significantly higher in patients with polyps than in controls [59,60].

The diagnostic potential of histone tail modifications in CRC is limited if compared with DNA methylation and noncoding RNA biomarkers that can be easily detected in circulating blood or in fecal samples. An attempt to use them as potential CRC biomarkers in blood evaluated histone modifications in circulating nucleosomal DNA of CRC patients and revealed lower levels of trimethylation of histone 3 at lysine 9 (H3K9me3) [61].

We have reviewed prognostic CRC epigenetic biomarkers [42]. Among them, methylation of the *CDKN2A (p14)*, *RASSF1*, and *APC* genes defines a poor prognosis subset of CRC patients, and *HOPX-β* promoter methylation was associated with worse prognosis of stage II CRC patients and also with poor differentiation. Methylation of genes in the extracellular matrix remodeling pathway, such as *IGFBP3*, *EVL*, *CD109*, and *FLNC*, was associated with worse survival, and hypomethylation of the *IGF2* differentially methylated region in colorectal tumors was associated with poor prognosis [42]. A correlation between long interspersed nuclear element-1 (LINE1) extreme hypomethylation and earlier age of onset (<60 years) and poor prognosis was identified [62]. Methylation of miR-34b/c, miR-128, and miR-126 correlated with invasive tumors, lymphatic invasion, peritoneal dissemination, and angiogenesis in CRC [63–65]. In addition, H3K9me3 positively correlated with lymph node metastasis, while dimethylation of histone 3 at lysine 4 (H3K4me2) and acetylation of lysine 9 correlated with the tumor histological type [61].

Standard chemotherapies in CRC involve the use of oxaliplatin, irinotecan, and 5-fluorouracil, or cetuximab or panitumumab if *KRAS* is wild type. Hypermethylation of the gene encoding the transcription factor AP-2ε (*TFAP2E*) resulted in clinical nonresponsiveness to standard chemotherapy (5-fluorouracil, irinotecan, or oxaliplatin) [66]. *TFAP2E* chemoresistance is mediated through its downstream target gene *DKK4*, encoding a protein that antagonizes the canonical Wnt signaling pathway [66]. CRC patients nonresponsive to standard therapies are treated with dacarbazine, an alkylating agent that exerts its antitumor activity by inducing base pair mismatches. However, the clinical response to dacarbazine is confined to those tumors harboring epigenetic silencing of the *MGMT* gene [67]. *MGMT* codes for the DNA repair protein O^6-methylguanine-DNA methyltransferase, and silencing of this gene by promoter hypermethylation is reported in almost 40% of sporadic CRCs [68]. *MGMT* promoter methylation is also frequently observed in the normal colonic mucosa, suggesting that it might represent an early event in CRC that precedes and predisposes to the development of cancer [68,69]. The prognostic value of *MGMT* methylation is still uncertain, but *MGMT* hypermethylation is associated with a better response to chemotherapy in patients treated with alkylating agents such as dacarbazine [68].

The CpG island methylator phenotype (CIMP), characterized by high levels of DNA methylation, has been found in multiple precancerous and cancerous lesions, including colorectal adenomas, colorectal cancers, and duodenal adenocarcinomas [70]. A literature meta-analysis revealed that CIMP is independently associated with significantly worse prognosis in CRC patients. However, the CIMP's value as a predictive factor in assessing whether adjuvant 5-fluorouracil therapy will confer additional survival benefit to CRC patients is still uncertain and controversial [70]. However, a 2014 study revealed that patients with stage III, CIMP-positive, and mismatch repair-intact colon tumors have longer survival times than CIMP-negative patients when irinotecan is added to combination therapy with fluorouracil and leucovorin, an active metabolite of folic acid [71].

Oxaliplatin resistance in CRC cells was linked to the DNA methylation-associated inactivation of the BRCA1 interactor *SRBC* gene. Indeed *SRBC* overexpression or depletion in CRC cells gave rise to sensitivity or resistance to oxaliplatin, respectively, suggesting the need for future clinical studies to validate *SRBC* hypermethylation as a predictive marker for oxaliplatin resistance in CRC [72]. In addition, studies in CRC mice revealed that *Dnmt3a* is predominantly expressed in the stem/progenitor cell compartment of tumors and that deletion of *Dnmt3a* inhibits the earliest stages of intestinal tumor development [73], DNMTi's improved the effects of chemotherapeutic agents in CRC cells [74], and combinations of agents acting on both DNA methylation and histone tail modifications are effective in promoting apoptosis, inducing cell cycle arrest, and inducing DNA damage in CRC cells at low and physiologically achievable concentrations [75].

2.2.2 Gastric Cancer

The examples in solid tumors are not confined to CRC. Studies in gastric cancer revealed some genes as potential predictive biomarkers for early diagnosis or ongoing prognosis of the disease. By a genome-wide methylation screening some genes related to cell adhesion, ubiquitination, transcription, p53 regulation, and diverse signaling pathways have been identified as differentially methylated. In particular *CHRNA3*, *DOK1*, and *GNMT* were hypermethylated, whereas two genes (*ALDH2* and *MTHFR*) showed significant hypomethylation, in gastric tumors. Moreover, a global demethylation of CG sequences of cancer cells occurs, consistent with genome-wide hypomethylation usually found in tumors [76]. Methylation of *MGMT*, *CDKN2A* (*p16*), *RASSF2*, *MLH1*, *HAND1*, and *FLNC* was closely associated with poor survival in gastric cancer, particularly *MGMT*, *CDKN2A* (*p16*), *RASSF2* and *FLNC* [77]. In addition, promoter methylation of the somatostatin gene (*SST*), connected with a decrease in SST protein and RNA levels, is a common event associated with gastric carcinogens, and treatment with demethylating drugs, including 5-azacytidine, reverses the status of methylation and recovers *SST* mRNA expression [78].

2.2.3 Breast and Ovarian Cancer

It has also been shown that treatment with low doses of the DNA methyltransferase inhibitor 5-azacytidine in 63 cell lines from breast, ovarian, and colorectal cancer led to a significant enrichment for immunomodulatory pathways in all three cancers [79]. In addition, samples from selected patient biopsies showed upregulation of immunity-related genes after treatment with the drug, suggesting a broad immune-stimulatory role for DNA demethylating drugs in multiple cancers [79]. Epigenetic changes in several other genes play a role in breast and ovarian cancers. For example, hypermethylation of *BMP6* was higher in lymph node metastasis [80], and high *RASSF1* promoter methylation levels predict poor prognosis and are predictors of worse outcome in breast cancer patients [81]. Also *ESR1* methylation may play an important role in the pathogenesis of the more aggressive breast tumors [82,83]. A chemokine, CXCL12, interacting with its receptor, CXCR4, is able to promote cellular adhesion, survival, proliferation, and migration. It was observed that the loss of DNA methylation in the promoter region of *CXCR4* correlated with a more aggressive disease in terms of tumor stage, size, and grade [84]. Furthermore, epigenetic changes in *CXCR4* and its ligand correlated with shorter overall survival and disease-free survival in breast cancers [84]. By contrast, promoter hypermethylation of *FBXW7*, a F-box protein substrate of a ubiquitin ligase that is responsible for the proteasomal degradation of some oncoproteins (cyclin E, c-Myc, c-Jun, and Notch), is associated with a favorable prognosis in breast cancer [85].

At present, no single epigenetic biomarker can be considered able to accurately detect early ovarian cancer in either tissue or body fluids, although the analysis of the methylation status of multiple genes simultaneously in a blood-based assay could be regarded as a promising method for the molecular classification and prognosis of ovarian cancer. Among genes found with promoter hypermethylation in ovarian cancer specimens are *BRCA1*, *RASSF1*, *APC*, *CDKN2A* (*p14*), *CDKN2A* (*p16*), and *DAPK* [86]. Hypermethylation of *BRCA1* and *RASSF1* was, moreover, detected in serum, plasma, and peritoneal fluid from ovarian cancer patients [87]. It seems a common and relatively early event in ovarian tumorigenesis that can be detected in the serum DNA from patients with ovary-confined (stage IA or IB) tumors and in cytologically negative peritoneal fluid [87]. In addition, *BRCA1* methylation frequency was higher in advanced ovarian cancers (stages II and III) and lower in earlier ovarian cancers (stage I) and benign tumors, resulting in decreased gene expression with the progression of the disease [88]. A 2013 study analyzed the methylation status of cell-free serum DNA of seven candidate genes (*APC*, *RASSF1*, *CDH1*, *RUNX3*, *TFPI2*, *SFRP5*, and *OPCML*). A sensitivity and specificity of 85.3% and 90.5%, respectively, in stage I epithelial ovarian cancer was achieved [89]. Huang et al. conducted a methylome analysis of ovarian cancers and observed that many differential DNA methylation regions distinguished malignant from nonmalignant ovarian tissues and that hypermethylation and corresponding gene silencing of sonic hedgehog pathway members *ZIC1* and *ZIC4* correlated with increased proliferation, migration, and invasion; particularly *ZIC1* promoter hypermethylation correlated with poor progression-free survival [90].

Docetaxel is a microtubule-targeting agent and represents one of the most commonly used chemotherapeutic drugs in breast cancer. Promoter methylation of *RASSF1* is associated with response to docetaxel in breast cancer patients. Indeed, in vitro studies revealed that *RASSF1* has a cooperative activity in suppression of cancer cell growth and proliferation by enhancing docetaxel-induced cell cycle arrest, and breast cancer patients who did not respond to docetaxel-based chemotherapy had higher *RASSF1* promoter methylation levels than patients with partial or complete response [91]. Other epigenetic aspects of breast cancer therapy have been reviewed by Byler and coworkers [92]. Several studies showed that the combination of either demethylating agents or HDACi's with other cytotoxic drugs produces synergistic growth inhibition in breast cancer cells, both in cell cultures and in mouse models, suggesting that epigenetic drug treatment may sensitize drug-resistant breast cancer cells and breast cancer stem cells to other cytotoxic drugs [92]. Koukoura and coworkers [86] reviewed the epigenetic factors linked to chemoresistance in ovarian cancers. Chemotherapeutic strategies in ovarian cancer involve a combination of a platinum- and taxane-based therapy. Platinum resistance in ovarian cancer cells has been linked to DNA methylation-induced silencing of various genes, including *SFRP5*, which is a Wnt antagonist; the DNA repair *MLH1* gene; the arginine biosynthesis-related *ASS1* gene; *ESR2*, encoding the estrogen receptor β; the methylation-controlled DNAJ (*MCJ*) gene; and *DAPK*, which is involved in apoptosis [86]. Germ-line mutations in the *BRCA1* or *BRCA2* genes are associated with an increased risk of breast and ovarian cancer development, and a study performed in cancer cell lines and xenografted tumors revealed that *BRCA1* promoter hypermethylation-associated silencing predicts enhanced sensitivity to platinum-derived drugs to the same extent as *BRCA1* mutations. Particularly, *BRCA1* hypermethylation was a predictor of longer time to relapse and improved overall survival in ovarian cancer patients undergoing chemotherapy with cisplatin [93]. In parallel, hypomethylation and upregulation of several genes, including the myelin and lymphocyte protein (*MAL*) gene and the *ABCG2* multidrug transporter gene, were shown to occur during chemoresistance in ovarian carcinoma cell lines [86,94]. In vitro studies and clinical trials have provided evidence that DNA-demethylating agents such as azacytidine and decitabine are capable of reversing platinum resistance in ovarian cancer patients, and there are also encouraging results suggesting that combinations of inhibitors of DNMTs and HDACs, or combinations of epigenetic and standard chemotherapeutic drugs, can be more effective than each single agent alone [86].

2.2.4 *Prostate Cancer*

In a study on prostate cancer patients, of 80 biopsies with *HOXD3* promoter hypermethylation, 66 were confirmed to come from patients with cancer [95]. Moreover, *HOXD3* promoter hypermethylation is associated with clinicopathological features and is more frequent in older higher-risk patients [95]. As far as the possibility of detecting epigenetic biomarkers by minimally invasive methods in prostate cancer patients, hypermethylation of the glutathione *S*-transferase P1 (*GSTP1*) gene was detected in plasma samples from 27 of 31 (92.86%) patients with prostate cancer. The methylation status of this gene was also able to effectively discriminate between prostate cancer and benign prostatic hyperplasia patients [96]. Many studies in urological tumors have focused their attention on the identification of epigenetic prognostic biomarkers. In prostate cancer, elevated levels of promoter methylation of some genes (*APC, CDH1, EDNRB, GSTP1, MDR1, MT1G, PTGS2, RARB2, RASSF1, RUNX3*) have been associated with higher pathological tumor stage and Gleason score and with prediction of cancer recurrence after radical prostatectomy [97–100], such as *GSTP1* in serum [101] and *PTGS2* and *CD44* in tumor tissue [102]. A study observed that the use of the prostate cancer methylation assay of three genes (*GSTP1, RARB2,* and *APC*) could be useful to improve the accuracy of prostate cancer screening because it correlates with positive biopsy and Gleason score; the assay predictive accuracy was higher than that of age, digital rectal examination, family history, and PSA, so that it has the potential to add value to the biopsy decision to better identify men with prostate cancer [103]. High global levels of acetylation of histone 3 at lysine 18 (H3K18ac), and H3K4me1 and H3K4me2, correlate with increased risk of prostate cancer recurrence [104,105]. As for miRNAs, high serum levels of miR-141 were found in metastatic prostate cancer patients; miR-31, miR-96, and miR-205 correlated with Gleason score; and miR-125b, miR-205, and miR-222 correlated with tumor stage [106]. miR-34c levels were inversely correlated with tumor aggressiveness, grade, and metastasis formation and were able to discriminate high-risk patients from those with low risk of progression [107]. Moreover, a 2014 study provided clinical validation of an epigenetic assay, particularly a gene methylation assay of *GSTP1, APC,* and *RASSF1*, with significant negative predictive value of 88%; so this epigenetic assay might help, with other known risk factors, to decrease unnecessary repeat prostate biopsies [108]. As largely discussed in the previous section classic demethylating agents, such as azacytidine and decitabine, hold the potential for reprogramming somatic cancer cells demonstrating high therapeutic efficacy in hematological malignancies, but their use in the treatment of solid tumors often gave rise to undesired cytotoxic side effects. A preclinical study on prostate cancer revealed that decitabine was still able to induce a significant tumor mass reduction in prostate cancer models at a concentration that was 700 times lower than the therapeutic dose, thereby limiting the potential of inducing unexpected side effects and opening the way for translation to humans [109]. Similarly, individuals who develop metastatic castration-resistant prostate cancer are treated with therapies targeting androgen receptor (AR) signaling, but durable responses are limited, probably because of acquired resistance. However, a 2014 study demonstrated that targeting BET bromodomain-containing proteins was more efficacious than direct AR antagonism in mouse models of the disease, thus providing a novel epigenetic target for the disease [110]. Glutathione S-transferases (GSTs) are a family of enzymes that play an important role in detoxification by catalyzing the conjugation of many hydrophobic and electrophilic compounds with reduced glutathione. Among them GSTP1 plays an important role in protecting cells from cytotoxic and carcinogenic agents. The *GSTP1* gene is frequently methylated in prostate cancers, and in vitro studies have revealed that *GSTP1* methylation and expression status is indicative of DNMTi efficacy in human prostate cancer cells [111]. Similarly, *GSTP1* methylation was suggested to be predictive for the response to and efficacy of doxorubicin treatment in breast cancer cells [112].

2.2.5 Bladder Cancer

Also for bladder cancer (BlCa), a study identified three genes, *GDF15*, *TMEFF2*, and *VIM*, that are able to detect BlCa in both tissue and urine and discriminate it from prostate cancer and renal epithelial tumors with a sensitivity and a specificity of 94% and 100%, respectively; the methylation levels of these genes were significantly higher in BlCa tissues than in normal bladder mucosa [113]. Urinary miRNAs are under investigation as potential biomarkers for bladder cancer. In vitro studies in many bladder cancer cell lines and primary nonmalignant urothelial cells indicate that hypermethylation of miR-152, miR-10a, and miR-200b regulative DNA sequences could be used as epigenetic bladder cancer biomarkers [114]. Aberrant methylation of *APC, CDH1, CDH13, p16, EDNRB, OPCML, PMF1, RASSF1*, and *TNFSR25* was associated with disease progression and with parameters of poor prognosis [115–117]. Methylation levels of *ARF, BCL2, EDNRB, RASSF1*, and *TERT* were significantly associated with tumor stage and grade, and aberrant promoter methylation of *DAPK* was associated with tumor recurrence [118]. H3K4me1, H4K20me1, H4K20me2, and H4K20me3 immunoexpression levels were correlated with advanced pathological stage [119]. miR-129, miR-133b, and miR-518c were associated with worse outcome, and increased levels of miR-452 were found in bladder cancer patients with lymph node metastases and were associated with poor prognosis [120].

2.2.6 Lung Cancer

In 15–80% of lung cancers changes in DNA methylation of CpG islands are commonly found [121]. Among the genes frequently methylated are *FHIT, APC, CDKN2A (p16)*, and *RASSF1* [122]. For diagnostic purposes, especially in the case of lung tumors, sputum can be considered an excellent specimen derived by a noninvasive sampling. The analysis of methylation patterns in a panel of predetermined genes in sputum has been shown of predictive value for the eventual development of lung cancer among high-risk smokers [123]. The genes examined were involved in cell cycle regulation (*CDKN2A* and *PAX5*), apoptosis (*DAPK* and *RASSF1*), signal transduction (*GATA5*), and DNA repair (*MGMT*). Methylation of *CDKN2A* (*p16*) was one of the earliest epigenetic changes seen in the lungs of smokers [124]. In a following study, methylation patterns within sputum were found to approximate those seen in tumors much better than those detected in serum; sputum has thus been shown effectively to be a surrogate for tumor tissue to predict the methylation status of advanced lung cancer [124]. In addition to colorectal cancer patients, lung cancer patients also show methylation of the *SEPT9* promoter region in plasma DNA and this marker was assumed quite sensitive for early diagnosis of lung cancer [44].

In lung cancer, methylation of the promoter region of four genes (*CDKN2A* (*p16*), *CDH13*, *APC*, and *RASSF1*) in patients with stage I non small-cell lung cancer (NSCLC), surgically treated with curative intent, is a risk factor for early recurrence [125]. One of the first integrations of epigenetic biomarkers into clinical trials of patients with lung cancer gave interesting results. Tumor samples were analyzed for promoter methylation of *RASSF1* and *DAPK1* within the French Intergroup phase III trial of neoadjuvant treatment of stage I and II NSCLC and it was noticed that the methylation of both genes in stage I adenocarcinoma correlated with poor prognosis [126]. As for miRNA, it is emerging that selected miRNAs have the capacity to distinguish histological subtypes; it was observed that increased miR-155 and decreased let-7a2 in stage I correlated with poor survival, and miR-21 correlated with poor outcomes among early lung stage cancer [122].

Treatment strategies for lung cancer have changed from a "general and empiric" to a "personalized and evidence-based" approach according to the driver oncogenic mutation. However, some lung cancers do not respond to treatments targeting driver oncogenic mutations, while others show resistance to molecular-targeted drugs [127]. Also for lung cancers there is evidence that drug resistance could be partially mediated by epigenetic changes, as shown for cisplatin resistance in non small-cell lung cancer cell models [128]. Indeed, epigenetic regimens based on DNMTi or HDACi have been tested for non small-cell lung cancer, but none of them have been currently approved for the treatment of this disease, and other epigenetic strategies combining epigenetic drugs with standard chemotherapeutic agents are being investigated in clinical trials [129]. Similar combined strategies are under investigation for skin cancers, in which epigenetic drugs are being tested in conjunction with standard immuno-, chemo-, and radiotherapeutic strategies [130].

2.2.7 *Others*

Epigenetic drugs such as HDACi's and DNMTi's are being tested, either alone or combined with standard therapeutic strategies, in several other solid tumors including liver cancers [131], thyroid cancers [132], renal cancers [133], and others. Those studies are paralleled by an increasing amount of in vitro evidence in cell lines indicating that resistance and/or sensitivity to standard therapeutic drugs is largely mediated by epigenetic changes [134,135]. For example, it was shown that promoter methylation is responsible for loss of expression of the antioxidant enzyme glutathione peroxidase 3 (GPX3) in a wide spectrum of cancer cell lines (lung, ovarian, bladder, skin, breast, esophagus) as well as in primary tumors of the bladder and head and neck carcinomas, and *GPX3* promoter methylation correlated with resistance to cisplatin in head and neck cancer patients [136]. Overall, those studies have revealed that epigenetic changes are at the basis of chemoresistance, immunoresistance, and angiogenesis and highlighted the potential role of epigenetic drugs as novel molecules that, if properly administered in combination with standard therapies, could hold the promise of reversing several such effects also in solid tumors. Epigenetic biomarkers are of relevance also for diagnostic and prognostic purposes in those cancers. For example, papillary thyroid cancer (PTC), as well as a high number of endocrine-related tumors, shows tumor-specific methylation of many genes, among them *RASSF1* [137]. A quantitative evaluation of *RASSF1* methylation and its correlation with tumor characteristics was performed by Kunstman and collaborators [138]. *RASSF1* promoter hypermethylation was observed in nearly all PTC cases with respect to normal thyroid tissue. It was, moreover, associated with multifocality, and inversely correlated with extracapsular invasion [138]. Several other genes (*TSHR, NIS, CDH1, ATM, AIT, TIMP3, DAPK,* and *RARB2*) show an altered methylation profile in thyroid cancer. Dynamic changes in E-cadherin (*CDH1*) methylation status may occur during metastatic progression, causing the loss of E-cadherin expression that correlated with lymphocytic infiltration, extrathyroidal invasion, and the presence of metastases [139], and hypermethylation of *TIMP3, SOX17,* and *RAP1GAP* contributes to tumor proliferation and progression, extrathyroidal invasion, lymph node metastasis, and multifocality [140–142]. Elevated levels of methylation of two genes, *EI24* and *WT1*, known to participate in carcinogenesis, were associated with increased risk of recurrence of thyroid cancer [143], and increased gene expression levels of *EZH2* and *SMYD3*, two histone methyltransferases, have been detected in the more aggressive diseases, such as persistent disease, occurrence of metastases, and disease-related death [144].

In esophageal cancer, increased methylation in the genes *APC, CDH1, CDKN2A (p16), DAPK, ER, MGMT,* and *TIMP-3* was significantly associated with reduced survival and earlier tumor recurrence [145]. Aberrant methylation in *NELL1, TAC1,* and *FHIT* was associated with poor prognosis, and *TLSC1* methylation was associated with aggressive tumor behavior. Patients with high *UCHL1* methylation levels had poorer 5-year survival rates and also an increased incidence of lymph nodes metastases versus those with lower methylation values. In squamous cell carcinoma *APC* aberrant methylation has been associated with reduced survival of patients following treatment, *CDH1* methylation correlated with stage I, and methylated *ITGA4* with stage II [146]. *CDKN2A (p16)* is a promising candidate biomarker for predicting clinical outcome of oropharyngeal squamous cell carcinoma, especially for recurrence-free survival [147].

A noninvasive method of performing cervical cancer diagnosis could involve the DNA extraction from cells of the cervical epithelium, easily obtained by a cytobrush. With this method the methylation status of the *ZAR1* and *SFRP4* promoter regions as potential biomarkers for diagnosis of preneoplastic and neoplastic lesions of cervix was evaluated. The methylation frequency of both genes increased as the grade of lesion increased and the differences between normal and cervical cancer were statistically significant [148].

In kidney cancer, DNA methylation of *APAF1*, *DAPK1*, *EPB41L3*, *JUP*, and *PTEN* was indicated as a predictive marker of poor outcome [149], and methylation of several genes, including *APC*, *CDH1*, *RARB*, *TP53*, *FHIT*, *MUC1*, *PTEN*, *MGMT*, *KAI1*, and *SMAD4*, has been related to worse prognosis in hepatocellular carcinoma [150], for which, a 2013 study reported that the analysis of global hypomethylation of plasma DNA may be useful for its detection and for monitoring patients [151].

Several other examples are available in the literature, but could not be fully described here owing to limited space. For example, there is an increasing number of less common tumors in which epigenetic changes have been identified and for which studies have started on the use of gene-specific hypermethylation as a biomarker of diagnosis, prognosis, and response to treatment. For instance, hypermethylation of the *CDKN2A (p16)* gene promoter was found in half of patients with periocular sebaceous carcinoma and is associated with younger patient age [152]. Other pathological conditions than tumors have been widely investigated in search of epigenetic biomarkers. They are nearly all complex conditions ranging from infertility to behavioral diseases and neurodegeneration and others, and they are discussed in the next section.

3. NONCANCEROUS DISEASES

Aberrant epigenetic regulatory mechanisms are increasingly being recognized as a major element not only in cancer but also in the setting of noncancerous diseases, including neurodegenerative diseases, behavioral and psychiatric disorders, autoimmune diseases, obesity, infertility, and cardiovascular diseases, among others.

3.1 Neurodegenerative Diseases

Epigenetic mechanisms play a fundamental role in learning and memory processes, and some of the causative genes of Alzheimer disease (AD), Parkinson disease (PD), and Huntington disease (HD) code for products able to interfere with epigenetic regulatory proteins that could represent promising targets for pharmacological interventions [153]. Global changes in DNA methylation and histone tail modifications in AD are supported by postmortem analyses performed in human brains of affected individuals [154–157]. Mutations in the *SNCA* gene coding for α-synuclein cause autosomal dominant PD, *SNCA* polymorphisms are among the genetic risk factors for sporadic PD, and reduced *SNCA* methylation levels were observed in postmortem brains of sporadic PD patients [158,159]. Moreover, α-synuclein binds to DNMT1 and sequesters it in the cytoplasm [160], interacts with histones, and inhibits histone acetylation [161]. HD is caused by trinucleotide repeat expansion in the *HTT* gene coding for the huntingtin protein, and also mutant huntingtin directly interacts with histone acetyltransferases, leading to altered histone acetylation levels [162,163]. Studies performed in animal models of neurodegenerative diseases have highlighted the potential role of epigenetic drugs, mainly HDACi's, in ameliorating the cognitive symptoms and preventing or delaying the motor symptoms of the disease, thereby opening the way for a potential application in humans [153]. However, reasons for concern include the multitargeted and multicellular effects exerted by most of these agents and the possibility of unexpected side effects. Moreover, the scarce availability of postmortem human brain specimens and the difficulty in handling those samples to preserve epigenetic marks have limited our potential to discover diagnostic or prognostic epigenetic biomarkers for those conditions or the most proper targets of epigenetic drugs. For example, as we have reviewed [153], studies in postmortem brain regions of AD individuals searching for gene promoter methylation of candidate genes have been so far conflicting, and no single gene has unequivocally emerged as a solid biomarker of the disease. Several studies are, however, ongoing, searching for easily detectable epigenetic biomarkers of neurodegeneration in blood DNA, such as LINE1 methylation levels, but again data are still scarce and conflicting [164,165]. Further investigation is warranted to clarify the timing and the dosage regimen, as well as the most proper target molecules, in the context of the application of epigenetic drugs for each of several neurodegenerative diseases [153].

3.2 Psychiatric and Behavioral Disorders

Epigenetic modifications such as histone acetylation and deacetylation, as well as DNA methylation, can induce lasting and stable changes in gene expression and have therefore been implicated in promoting the adaptive behavioral and neuronal changes that accompany neurodevelopmental and psychiatric disorders such as autism spectrum disorders (ASDs), major depressive disorder, drug addiction, eating disorders, and schizophrenia, among others [166]. Many attempts are being made to search for peripheral epigenetic biomarkers in neurodegenerative or behavioral diseases. For instance, the expression of DNA-methylating/demethylating enzymes and schizophrenia candidate genes such as *BDNF* and *GCR* was found to be altered in the same direction in both brain and blood lymphocytes [167]. In adolescents affected by anorexia nervosa, whole-blood global DNA methylation was decreased with respect to controls, correlating with plasma leptin and steroid hormone levels [168]. In posttraumatic stress disorder patients, the impact of various early environments on disease-related genome-wide gene expression and DNA methylation in peripheral blood cells was assessed. Distinct genomic and epigenetic profiles were found in the presence or absence of exposure to childhood abuse [169].

The primary pharmacological treatments for schizophrenia are antipsychotics, which are, however, not effective enough to sufficiently treat the cognitive deficits observed in the patients. Therefore, the combination of epigenetic drugs with standard pharmacological interventions could represent a more effective strategy in ameliorating cognitive deficits [170]. Studies in patients with major depressive disorder revealed that DNA methylation in interleukin-11 (*IL11*) predicts clinical response to antidepressants, suggesting that it could represent a pharmacoepigenetic biomarker for the prediction of antidepressant response [171]. The role played by epigenetics in ASDs is still in its infancy, but accumulating data imply that ASDs may be "epigenopathies" [172]. Within this context epigenetic drugs targeting DNA methylation and histone deacetylation enzymes could be helpful in ASD therapy, but no clinical trial has yet been conducted [172]. Little is known still concerning the extent of epigenetic disruption established over long periods of malnourishment in women with anorexia nervosa, suggesting that large genome-wide studies of epigenetic modifications, encompassing both DNA methylation and other epigenetic marks, are required to determine the degree to which anorexia nervosa is associated with specific epigenetic changes, potentially modifiable through appropriate treatments that improve nutrition [173]. Concerning drug addiction, studies in animals have revealed that chronic drug exposure alters gene expression in the brain and produces long-term changes in neural networks that underlie compulsive drug taking and seeking, and there is increasing evidence for a role of epigenetic changes in mediating the addictive potential of various drugs of abuse, including cocaine, amphetamine, and alcohol. For example, there is evidence for epigenetic mechanisms that regulate brain-derived neurotrophic factor (*BDNF*) gene expression following chronic cocaine exposure [174,175].

Identifying epigenetic signatures that define psychostimulant addiction may lead to novel, efficacious treatments for drug craving and relapse.

3.3 Autoimmune Diseases

Early studies performed at the end of the twentieth century showed that CD4$^+$ T cells treated with 5-azacytidine induced a lupus-like syndrome if injected into mice [176], revealing that epigenetic changes are involved in autoimmunity. It is now clear that epigenetic changes in hundreds of genes are observed in systemic lupus erythematosus, rheumatoid arthritis, Sjögren syndrome, psoriasis, multiple sclerosis, systemic sclerosis, autoimmune thyroid diseases, and many other autoimmune disorders [177]. For example, global hypomethylation, reduced DNMT1, and increased MBD2 levels are common findings in CD4$^+$ T cells of patients with systemic lupus erythematosus, followed by histone tail modifications, a deregulated miRNA profile, and aberrant expression of hundreds of genes [178–180]. Global DNA hypomethylation and reduced DNMT1 levels were also observed in synovial fibroblasts of patients with rheumatoid arthritis, followed by changes in histone tail modifications and miRNA expression, overall accounting for a deregulated expression of over 200 genes [181]. Over 1100 CpG sites showed differential methylation levels between skin samples of patients with psoriasis and healthy controls, and deregulation of almost 100 miRNAs occurred in psoriatic skin cells, and several similar examples are available for other autoimmune disorders [182]. Selective inhibitors of certain HDAC proteins showed anti-inflammatory and antirheumatic effects in in vitro studies and in animal models of those disorders [183]. For example, there is evidence that small molecules that modulate the function of the HDAC sirtuin 1 could represent potential therapeutic molecules for autoimmune inflammatory diseases such as rheumatoid arthritis [184].

3.4 Cardiovascular Diseases

Cardiovascular disease (CVD) is considered to be the leading cause of death worldwide, and there is increasing evidence of a role of epigenetics in CVD [185]. Particularly, histone tail modifications have been reported in animal models of cardiac hypertrophy, heart failure, and arrhythmias [185]. Studies performed in vitro and with animal disease models revealed that HDACi's have a protective role against hypertrophy [186,187] and reversed atrial fibrosis and atrial arrhythmia vulnerability in mice [188]. There is also increasing evidence for a role of aberrant DNA methylation in atherosclerosis and artery disease [185]. However, as of this writing there are no significant reports of epigenetic contribution in clinical practice or therapeutics in CVD [185].

3.5 Obesity

Epigenetic markers are increasingly being recognized as a new tool for understanding the influence of lifestyle factors on obesity phenotypes. Indeed, it was shown that promoter methylation of the human tumor necrosis factor-α (TNF) gene could be involved in the predisposition to lose body weight after following a balanced hypocaloric diet [189]. Other potential epigenetic biomarkers of weight loss after an energy-restriction intervention were the ATP10A and WT1 genes [190]. Moreover, changes in DNA methylation that could be associated with a better weight loss response after a multidisciplinary intervention program have been investigated in Spanish obese or overweight adolescents, and the study revealed that five regions located in or near the AQP9, DUSP22, HIPK3, TNNT1, and TNNI3 genes showed differential methylation levels between high and low responders to a multidisciplinary weight loss intervention [191]. Collectively those studies suggest that hypocaloric-diet-induced weight loss in humans could alter the DNA methylation status of specific genes and that those changes may be used as epigenetic markers to predict the weight loss response in obese humans.

3.6 Male Subfertility and Assisted Reproduction

Aberrant sperm DNA methylation patterns, mainly in imprinted genes, have been associated with male subfertility and oligospermia. Functional analysis of sperms from men undergoing assisted reproduction showed a wide increase in methylation level in spermatogenesis-related genes and hypomethylation in inflammation- and immune response-related genes [192]. Studies have shown associations of aberrant gene-specific DNA methylation in spermatozoa of men with idiopathic infertility. The DNA methylation of the maternally imprinted gene MEST was significantly associated with oligozoospermia, decreased bitesticular volume, and increased follicle-stimulating hormone levels [193]. Because in human semen there is a high concentration of cell-free seminal DNA, the detection of altered methylation levels can be achieved also in the case of defects in testicular sperm production or epididymal sperm maturation. Indeed testis- and epididymis-specific genes have been identified with altered methylation. Among genes with promoter hypomethylation are MORC1 (related to spermatogenesis), ELF5 (involved in cell proliferation), PRAME (associated with the regulation of apoptosis), and SPACA3 (correlated with the sperm–egg recognition); the testis-specific hypomethylated genes include MAGEA1 and ANKRD30A [194]. The proposal that epigenetic analysis can supplement traditional semen parameters and has the potential to provide new insights into the etiology of male subfertility is indeed intriguing [192].

4. CONCLUDING REMARKS

The increasing burden of evidence linking epigenetic changes to almost all complex multifactorial diseases has been paralleled, in recent years, by a growing number of studies suggesting the possible translation of epigenetic biomarkers into clinical practice. Particularly, epigenetic biomarkers are increasingly suggested as potential diagnostic and prognostic tools, as well as valuable markers for the choice of the most proper therapeutic approach. In this chapter we have described several examples of either validated or suggested epigenetic diagnostic, prognostic, and pharmacoepigenetic markers mainly deriving from studies in human cancers, even though examples are increasingly available also for noncancerous diseases. The detection of the epigenetic signature offers not only the possibility of diagnosing a pathological condition but also of providing a better classification of cancer subtypes and monitoring over time. In fact some of these epigenetic changes occur early in the process of carcinogenesis, others may interfere later, having a role in tumor growth, invasion, and metastasis, and therefore can provide both diagnostic and prognostic tools as well as having a predictive potential for drug treatment efficiencies. In addition, most of the epigenetic signatures are reversible, and reversing the abnormal cancer epigenome with drugs acting on DNA methylation and histone tail modifications could therefore offer the possibility of affecting multiple altered signaling networks in cancer, including those involved in cancer initiation and progression as well as those responsible for the resistance to conventional chemotherapeutic approaches. Increasing evidence suggests the potential of epigenetic drugs in noncancerous conditions, such as neurodegenerative, behavioral, and autoimmune disorders, and examples of pharmacoepigenetic biomarkers of response to treatment are available for psychiatric conditions and suggested for other noncancerous diseases.

We are, however, at the beginning of a very promising and fascinating journey. Indeed, many advances have been made in the implementation of epigenetic biomarkers in laboratory diagnostics, but in many cases further prospective validation studies are required before clinical application can take place. Among urgent needs are the search for detection techniques combining high accuracy with low cost, minimal invasiveness, and standardization of protocols, as well as the establishment of reliable reference values. To validate completely these biomarkers, moreover, prospective studies are needed on large cohorts of patients and trials on the reproducibility of the methods in different laboratories. Constant research in this area is highly desirable over the long term because the application of epigenetic biomarkers definitely has the potential to significantly improve patient care.

References

[1] Boland MJ, Nazor KL, Loring JF. Epigenetic regulation of pluripotency and differentiation. Circ Res 2014;115(2):311–24.

[2] Brunet A, Berger SL. Epigenetics of aging and aging-related disease. J Gerontol A Biol Sci Med Sci 2014;69(Suppl. 1):S17–20.

[3] Jacoby M, Gohrbandt S, Clausse V, Brons NH, Muller CP. Interindividual variability and co-regulation of DNA methylation differ among blood cell populations. Epigenetics 2012;7(12):1421–34.

[4] Coppedè F. The role of epigenetics in colorectal cancer. Expert Rev Gastroenterol Hepatol 2014;8(8):935–48.

[5] Kaz AM, Wong CJ, Dzieciatkowski S, Luo Y, Schoen RE, Grady WM. Patterns of DNA methylation in the normal colon vary by anatomical location, gender, and age. Epigenetics 2014;9(4):492–502.

[6] Condliffe D, Wong A, Troakes C, Proitsi P, Patel Y, Chouliaras L, et al. Cross-region reduction in 5-hydroxymethylcytosine in Alzheimer's disease brain. Neurobiol Aging 2014;35(8):1850–4.

[7] Coppieters N, Dieriks BV, Lill C, Faull RL, Curtis MA, Dragunow M. Global changes in DNA methylation and hydroxymethylation in Alzheimer's disease human brain. Neurobiol Aging 2014;35(6):1334–44.

[8] Potter NT, Hurban P, White MN, Whitlock KD, Lofton-Day CE, Tetzner R, et al. Validation of a real-time PCR-based qualitative assay for the detection of methylated SEPT9 DNA in human plasma. Clin Chem 2014;60(9):1183–91.

[9] Shivapurkar N, Weiner LM, Marshall JL, Madhavan S, Deslattes Mays A, Juhl H, et al. Recurrence of early stage colon cancer predicted by expression pattern of circulating microRNAs. PLoS One 2014;9(1):e84686.

[10] Ivanov M, Barragan I, Ingelman-Sundberg M. Epigenetic mechanisms of importance for drug treatment. Trends Pharmacol Sci 2014;35(8):384–96.

[11] Baylin SB, Jones PA. A decade of exploring the cancer epigenome–biological and translational implications. Nat Rev Cancer 2011;11(10):726–34.

[12] Herman JG, Baylin SB. Gene silencing in cancer in association with promoter hypermethylation. N Engl J Med 2003;349(21):2042–54.

[13] Azad N, Zahnow CA, Rudin CM, Baylin SB. The future of epigenetic therapy in solid tumours-lessons from the past. Nat Rev Clin Oncol 2013;10(5):256–66.

[14] Hashimoto H, Vertino PM, Cheng X. Molecular coupling of DNA methylation and histone methylation. Epigenomics 2010;2(5):657–69.

[15] Hayes J, Peruzzi PP, Lawler S. MicroRNAs in cancer: biomarkers, functions and therapy. Trends Mol Med 2014;20(8):460–9.

[16] Tsai HC, Li H, Van Neste L, Cai Y, Robert C, Rassool FV, et al. Transient low doses of DNA-demethylating agents exert durable antitumor effects on hematological and epithelial tumor cells. Cancer Cell 2012;21(3):430–46.

[17] Jones PA. Functions of DNA methylation: islands, start sites, gene bodies and beyond. Nat Rev Genet 2012;13(7):484–92.

[18] Fournier A, Sasai N, Nakao M, Defossez PA. The role of methyl-binding proteins in chromatin organization and epigenome maintenance. Brief Funct Genomics 2012;11(3):251–64.

[19] Guo JU, Su Y, Zhong C, Ming GL, Song H. Hydroxylation of 5-methylcytosine by TET1 promotes active DNA demethylation in the adult brain. Cell 2011;145(3):423–34.

[20] Münzel M, Globisch D, Carell T. 5-Hydroxymethylcytosine, the sixth base of the genome. Angew Chem Int Ed Engl 2011;50(29):6460–8.

[21] Li KK, Luo LF, Shen Y, Xu J, Chen Z, Chen SJ. DNA methyltransferases in hematologic malignancies. Semin Hematol 2013;50(1):48–60.

[22] Wakita S, Yamaguchi H, Omori I, Terada K, Ueda T, Manabe E, et al. Mutations of the epigenetics-modifying gene (DNMT3a, TET2, IDH1/2) at diagnosis may induce FLT3-ITD at relapse in de novo acute myeloid leukemia. Leukemia 2013;27(5):1044–52.

[23] Im AP, Sehgal AR, Carroll MP, Smith BD, Tefferi A, Johnson DE, et al. DNMT3A and IDH mutations in acute myeloid leukemia and other myeloid malignancies: associations with prognosis and potential treatment strategies. Leukemia 2014;28(9):1774–83.

[24] Ley TJ, Ding L, Walter MJ, McLellan MD, Lamprecht T, Larson DE, et al. DNMT3A mutations in acute myeloid leukemia. N Engl J Med December 2010;363(25):2424–33.

[25] Tie R, Zhang T, Fu H, Wang L, Wang Y, He Y, et al. Association between DNMT3A mutations and prognosis of adults with de novo acute myeloid leukemia: a systematic review and meta-analysis. PLoS One 2014;9(6):e93353.

[26] Cairns RA, Mak TW. Oncogenic isocitrate dehydrogenase mutations: mechanisms, models, and clinical opportunities. Cancer Discov 2013;3(7):730–41.

[27] Cairns RA, Iqbal J, Lemonnier F, Kucuk C, de Leval L, Jais JP, et al. IDH2 mutations are frequent in angioimmunoblastic T-cell lymphoma. Blood 2012;119(8):1901–3.

[28] Lemonnier F, Couronné L, Parrens M, Jaïs JP, Travert M, Lamant L, et al. Recurrent TET2 mutations in peripheral T-cell lymphomas correlate with TFH-like features and adverse clinical parameters. Blood 2012;120(7):1466–9.

[29] Figueroa ME, Abdel-Wahab O, Lu C, Ward PS, Patel J, Shih A, et al. Leukemic IDH1 and IDH2 mutations result in a hypermethylation phenotype, disrupt TET2 function, and impair hematopoietic differentiation. Cancer Cell 2010;18(6):553–67.

[30] Fathi AT, Sadrzadeh H, Borger DR, Ballen KK, Amrein PC, Attar EC, et al. Prospective serial evaluation of 2-hydroxyglutarate, during treatment of newly diagnosed acute myeloid leukemia, to assess disease activity and therapeutic response. Blood 2012;120(23):4649–52.

[31] Popovici-Muller J, Saunders JO, Salituro FG, Travins JM, Yan S, Zhao F, et al. Discovery of the first potent inhibitors of mutant IDH1 that lower tumor 2-HG in vivo. ACS Med Chem Lett 2012;3(10):850–5.

[32] Rohle D, Popovici-Muller J, Palaskas N, Turcan S, Grommes C, Campos C, et al. An inhibitor of mutant IDH1 delays growth and promotes differentiation of glioma cells. Science 2013;340(6132):626–30.

[33] Liersch R, Müller-Tidow C, Berdel WE, Krug U. Prognostic factors for acute myeloid leukaemia in adults–biological significance and clinical use. Br J Haematol 2014;165(1):17–38.

[34] Chatterton Z, Morenos L, Mechinaud F, Ashley DM, Craig JM, Sexton-Oates A, et al. Epigenetic deregulation in pediatric acute lymphoblastic leukemia. Epigenetics 2014;9(3):459–67.

[35] Chatterton Z, Burke D, Emslie KR, Craig JM, Ng J, Ashley DM, et al. Validation of DNA methylation biomarkers for diagnosis of acute lymphoblastic leukemia. Clin Chem 2014;60(7):995–1003.

[36] Martinelli S, Kanduri M, Maffei R, Fiorcari S, Bulgarelli J, Marasca R, et al. ANGPT2 promoter methylation is strongly associated with gene expression and prognosis in chronic lymphocytic leukemia. Epigenetics 2013;8(7):720–9.

[37] Marcucci G, Yan P, Maharry K, Frankhouser D, Nicolet D, Metzeler KH, et al. Epigenetics meets genetics in acute myeloid leukemia: clinical impact of a novel seven-gene score. J Clin Oncol 2014;32(6):548–56.

[38] Wu DS, Shen JZ, Yu AF, Fu HY, Zhou HR, Shen SF. Epigallocatechin-3-gallate and trichostatin A synergistically inhibit human lymphoma cell proliferation through epigenetic modification of p16INK4a. Oncol Rep 2013;30(6):2969–75.

[39] Knoechel B, Roderick JE, Williamson KE, Zhu J, Lohr JG, Cotton MJ, et al. An epigenetic mechanism of resistance to targeted therapy in T cell acute lymphoblastic leukemia. Nat Genet 2014;46(4):364–70.

[40] Clozel T, Yang S, Elstrom RL, Tam W, Martin P, Kormaksson M, et al. Mechanism-based epigenetic chemosensitization therapy of diffuse large B-cell lymphoma. Cancer Discov 2013;3(9):1002–19.
[41] Hu X, Wei H, Xiang L, Chertov O, Wayne AS, Bera TK, et al. Methylation of the DPH1 promoter causes immunotoxin resistance in acute lymphoblastic leukemia cell line KOPN-8. Leuk Res 2013;37(11):1551–6.
[42] Coppedè F, Lopomo A, Spisni R, Migliore L. Genetic and epigenetic biomarkers for diagnosis, prognosis and treatment of colorectal cancer. World J Gastroenterol 2014;20(4):943–56.
[43] Tänzer M, Balluff B, Distler J, Hale K, Leodolter A, Röcken C, et al. Performance of epigenetic markers SEPT9 and ALX4 in plasma for detection of colorectal precancerous lesions. PLoS One 2010;5(2):e9061.
[44] Powrózek T, Krawczyk P, Kucharczyk T, Milanowski J. Septin 9 promoter region methylation in free circulating DNA-potential role in noninvasive diagnosis of lung cancer: preliminary report. Med Oncol 2014;31(4):917.
[45] Grützmann R, Molnar B, Pilarsky C, Habermann JK, Schlag PM, Saeger HD, et al. Sensitive detection of colorectal cancer in peripheral blood by septin 9 DNA methylation assay. PLoS One 2008;3(11):e3759.
[46] Kostin PA, Zakharzhevskaia NB, Generozov EV, Govorun VM, Chernyshov SV, Shchelygin IuA. Hypermethylation of the CDH1, SEPT9, HLTF and ALX4 genes and their diagnostic significance in colorectal cancer. Vopr Onkol 2010;56(2):162–8.
[47] Tang D, Liu J, Wang DR, Yu HF, Li YK, Zhang JQ. Diagnostic and prognostic value of the methylation status of secreted frizzled-related protein 2 in colorectal cancer. Clin Invest Med 2011;34(2):E88–95.
[48] Lange CP, Campan M, Hinoue T, Schmitz RF, van der Meulen-de Jong AE, Slingerland H, et al. Genomescale discovery of DNA-methylation biomarkers for bloodbased detection of colorectal cancer. PLoS One 2012;7(11):e50266.
[49] Oh T, Kim N, Moon Y, Kim MS, Hoehn BD, Park CH, et al. Genome-wide identification and validation of a novel methylation biomarker, SDC2, for blood-based detection of colorectal cancer. J Mol Diagn 2013;15(4):498–507.
[50] Yang H, Xia BQ, Jiang B, Wang G, Yang YP, Chen H, et al. Diagnostic value of stool DNA testing for multiple markers of colorectal cancer and advanced adenoma: a meta-analysis. Can J Gastroenterol 2013;27(8):467–75.
[51] Guo Q, Song Y, Zhang H, Wu X, Xia P, Dang C. Detection of hypermethylated fibrillin-1 in the stool samples of colorectal cancer patients. Med Oncol 2013;30(4):695.
[52] Itzkowitz S, Brand R, Jandorf L, Durkee K, Millholland J, Rabeneck L, et al. A simplified, noninvasive stool DNA test for colorectal cancer detection. Am J Gastroenterol 2008;103(11):2862–70.
[53] Antelo M, Balaguer F, Shia J, Shen Y, Hur K, Moreira L, et al. A high degree of LINE-1 hypomethylation is a unique feature of early-onset colorectal cancer. PLoS One 2012;7(9):e45357.
[54] Zhang H, Zhu YQ, Wu YQ, Zhang P, Qi J. Detection of promoter hypermethylation of Wnt antagonist genes in fecal samples for diagnosis of early colorectal cancer. World J Gastroenterol 2014;20(20):6329–35.
[55] Rhee YY, Kim MJ, Bae JM, Koh JM, Cho NY, Juhnn YS, et al. Clinical outcomes of patients with microsatellite-unstable colorectal carcinomas depend on L1 methylation level. Ann Surg Oncol 2012;19(11):3441–8.
[56] Hur K, Cejas P, Feliu J, Moreno-Rubio J, Burgos E, Boland CR, et al. Hypomethylation of long interspersed nuclear element-1 (LINE-1) leads to activation of proto-oncogenes in human colorectal cancer metastasis. Gut 2014;63(4):635–46.
[57] Holdenrieder S, Dharuman Y, Standop J, Trimpop N, Herzog M, Hettwer K, et al. Novel serum nucleosomics biomarkers for the detection of colorectal cancer. Anticancer Res 2014;34(5):2357–62.

[58] Ahmed FE, Ahmed NC, Vos PW, Bonnerup C, Atkins JN, Casey M, et al. Diagnostic microRNA markers to screen for sporadic human colon cancer in stool: I. Proof of principle. Cancer Genomics Proteomics 2013;10(3):93–113.

[59] Wu CW, Ng SS, Dong YJ, Ng SC, Leung WW, Lee CW, et al. Detection of miR-92a and miR-21 in stool samples as potential screening biomarkers for colorectal cancer and polyps. Gut 2012;61(5):739–45.

[60] Wu CW, Ng SC, Dong Y, Tian L, Ng SS, Leung WW, et al. Identification of microRNA-135b in stool as a potential noninvasive biomarker for colorectal cancer and adenoma. Clin Cancer Res 2014;20(11):2994–3002.

[61] Gezer U, Holdenrieder S. Post-translational histone modifications in circulating nucleosomes as new biomarkers in colorectal cancer. In Vivo 2014;28(3):287–92.

[62] Ahn JB, Chung WB, Maeda O, Shin SJ, Kim HS, Chung HC, et al. DNA methylation predicts recurrence from resected stage III proximal colon cancer. Cancer 2011;117(9):1847–54.

[63] Kamimae S, Yamamoto E, Yamano HO, Nojima M, Suzuki H, Ashida M, et al. Epigenetic alteration of DNA in mucosal wash fluid predicts invasiveness of colorectal tumors. Cancer Prev Res (Phila) 2011;4(5):674–83.

[64] Takahashi Y, Iwaya T, Sawada G, Kurashige J, Matsumura T, Uchi R, et al. Up-regulation of NEK2 by MicroRNA-128 methylation is associated with poor prognosis in colorectal Cancer. Ann Surg Oncol 2014;21(1):205–12.

[65] Zhang Y, Wang X, Xu B, Wang B, Wang Z, Liang Y, et al. Epigenetic silencing of miR-126 contributes to tumor invasion and angiogenesis in colorectal cancer. Oncol Rep 2013;30:1976–84.

[66] Ebert MP, Tänzer M, Balluff B, Burgermeister E, Kretzschmar AK, Hughes DJ, et al. TFAP2E-DKK4 and chemoresistance in colorectal cancer. N Engl J Med 2012;366(1):44–53.

[67] Amatu A, Sartore-Bianchi A, Moutinho C, Belotti A, Bencardino K, Chirico G, et al. Promoter CpG island hypermethylation of the DNA repair enzyme MGMT predicts clinical response to dacarbazine in a phase II study for metastatic colorectal cancer. Clin Cancer Res 2013;19(8):2265–72.

[68] Minoo P. Toward a molecular classification of colorectal Cancer: the role of MGMT. Front Oncol 2013;3:266.

[69] Coppedè F, Migheli F, Lopomo A, Failli A, Legitimo A, Consolini R, et al. Gene promoter methylation in colorectal cancer and healthy adjacent mucosa specimens: correlation with physiological and pathological characteristics, and with biomarkers of one-carbon metabolism. Epigenetics 2014;9(4):621–33.

[70] Juo YY, Johnston FM, Zhang DY, Juo HH, Wang H, Pappou EP, et al. Prognostic value of CpG island methylator phenotype among colorectal cancer patients: a systematic review and meta-analysis. Ann Oncol 2014;25(12):2314–27.

[71] Shiovitz S, Bertagnolli MM, Renfro LA, Nam E, Foster NR, Dzieciatkowski S, et al. CpG island methylator phenotype is associated with response to adjuvant irinotecan-based therapy for stage 3 Colon Cancer. Gastroenterology 2014;147(3):637–45.

[72] Moutinho C, Martinez-Cardús A, Santos C, Navarro-Pérez V, Martínez-Balibrea E, Musulen E, et al. Epigenetic inactivation of the BRCA1 interactor SRBC and resistance to oxaliplatin in colorectal cancer. J Natl Cancer Inst 2014;106(1):djt322.

[73] Weis B, Schmidt J, Maamar H, Raj A, Lin H, Tóth C, et al. Inhibition of intestinal tumor formation by deletion of the DNA methyltransferase 3a. Oncogene 2014;0.

[74] Flis S, Gnyszka A, Flis K. DNA methyltransferase inhibitors improve the effect of chemotherapeutic agents in SW48 and HT-29 colorectal cancer cells. PLoS One 2014;9(3):e92305.

[75] Saldanha SN, Kala R, Tollefsbol TO. Molecular mechanisms for inhibition of colon cancer cells by combined epigenetic-modulating epigallocatechin gallate and sodium butyrate. Exp Cell Res 2014;324(1):40–53.

[76] Balassiano K, Lima S, Jenab M, Overvad K, Tjonneland A, Boutron-Ruault MC, et al. Aberrant DNA methylation of cancer-associated genes in gastric cancer in the European Prospective Investigation into Cancer and Nutrition (EPIC-EURGAST). Cancer Lett 2011;311(1):85–95.

[77] Shi J, Zhang G, Yao D, Liu W, Wang N, Ji M, et al. Prognostic significance of aberrant gene methylation in gastric cancer. Am J Cancer Res 2012;2(1):116–29.

[78] Shi X, Li X, Chen L, Wang C. Analysis of somatostatin receptors and somatostatin promoter methylation in human gastric cancer. Oncol Lett 2013;6(6):1794–8.

[79] Li H, Chiappinelli KB, Guzzetta AA, Easwaran H, Yen RW, Vatapalli R, et al. Immune regulation by low doses of the DNA methyltransferase inhibitor 5-azacitidine in common human epithelial cancers. Oncotarget 2014;5(3):587–98.

[80] Barekati Z, Radpour R, Lu Q, Bitzer J, Zheng H, Toniolo P, et al. Methylation signature of lymph node metastases in breast cancer patients. BMC Cancer 2012;12:244.

[81] Martins AT, Monteiro P, Ramalho-Carvalho J, Costa VL, Dinis-Ribeiro M, Leal C, et al. High RASSF1A promoter methylation levels are predictive of poor prognosis in fine-needle aspirate washings of breast cancer lesions. Breast Cancer Res Treat 2011;129(1):1–9.

[82] Izadi P, Mehrdad N, Foruzandeh F, Reza NM. Association of poor prognosis subtypes of breast cancer with estrogen receptor alpha methylation in Iranian women. Asian Pac J Cancer Prev 2012;13(8):4113–7.

[83] Ramezani F, Salami S, Omrani MD, Maleki D. CpG island methylation profile of estrogen receptor alpha in Iranian females with triple negative or non-triple negative breast cancer: new marker of poor prognosis. Asian Pac J Cancer Prev 2012;13(2):451–7.

[84] Ramos EA, Grochoski M, Braun-Prado K, Seniski GG, Cavalli IJ, Ribeiro EM, et al. Epigenetic changes of CXCR4 and its ligand CXCL12 as prognostic factors for sporadic breast cancer. PLoS One 2011;6(12):e29461.

[85] Akhoondi S, Lindström L, Widschwendter M, Corcoran M, Bergh J, Spruck C, et al. Inactivation of FBXW7/hCDC4-β expression by promoter hypermethylation is associated with favorable prognosis in primary breast cancer. Breast Cancer Res 2010;12(6):R105.

[86] Koukoura O, Spandidos DA, Daponte A, Sifakis S. DNA methylation profiles in ovarian cancer: implication in diagnosis and therapy (Review). Mol Med Rep 2014;10(1):3–9.

[87] Ibanez de Caceres I, Battagli C, Esteller M, Herman JG, Dulaimi E, Edelson MI, et al. Tumor cell-specific BRCA1 and RASSF1A hypermethylation in serum, plasma, and peritoneal fluid from ovarian cancer patients. Cancer Res 2004;64(18):6476–81.

[88] Wang YQ, Yan Q, Zhang JR, Li SD, Yang YX, Wan XP. Epigenetic inactivation of BRCA1 through promoter hypermethylation in ovarian cancer progression. J Obstet Gynaecol Res 2013;39(2):549–54.

[89] Zhang Q, Hu G, Yang Q, Dong R, Xie X, Ma D, et al. A multiplex methylation-specific PCR assay for the detection of early-stage ovarian cancer using cell-free serum DNA. Gynecol Oncol 2013;130(1):132–9.

[90] Huang RL, Gu F, Kirma NB, Ruan J, Chen CL, Wang HC, et al. Comprehensive methylome analysis of ovarian tumors reveals hedgehog signaling pathway regulators as prognostic DNA methylation biomarkers. Epigenetics 2013;8(6):624–34.

[91] Gil EY, Jo UH, Jeong H, Whang YM, Woo OH, Cho KR, et al. Promoter methylation of RASSF1A modulates the effect of the microtubule-targeting agent docetaxel in breast cancer. Int J Oncol 2012;41(2):611–20.

[92] Byler S, Goldgar S, Heerboth S, Leary M, Housman G, Moulton K, et al. Genetic and epigenetic aspects of breast cancer progression and therapy. Anticancer Res 2014;34(3):1071–7.

[93] Stefansson OA, Villanueva A, Vidal A, Martí L, Esteller M. BRCA1 epigenetic inactivation predicts sensitivity to platinum-based chemotherapy in breast and ovarian cancer. Epigenetics 2012;7(11):1225–9.

[94] Lee PS, Teaberry VS, Bland AE, Huang Z, Whitaker RS, Baba T, et al. Elevated MAL expression is accompanied by promoter hypomethylation and platinum resistance in epithelial ovarian cancer. Int J Cancer 2010;126(6):1378–89.

[95] Chen LN, Rubin RS, Othepa E, Cer C, Yun E, Agarwal RP, et al. Correlation of HOXD3 promoter hypermethylation with clinical and pathologic features in screening prostate biopsies. Prostate 2014;74(7):714–21.

[96] Dumache R, Puiu M, Motoc M, Vernic C, Dumitrascu V. Prostate cancer molecular detection in plasma samples by glutathione S-transferase P1 (GSTP1) methylation analysis. Clin Lab 2014;60(5):847–52.

[97] Ellinger J, Bastian PJ, Jurgan T, Biermann K, Kahl P, Heukamp LC, et al. CpG island hypermethylation at multiple gene sites in diagnosis and prognosis of prostate cancer. Urology 2008;71(1):161–7.

[98] Jerónimo C, Henrique R, Hoque MO, Mambo E, Ribeiro FR, Varzim G, et al. A quantitative promoter methylation profile of prostate cancer. Clin Cancer Res 2004;10(24):8472–8.

[99] Jerónimo C, Henrique R, Hoque MO, Ribeiro FR, Oliveira J, Fonseca D, et al. Quantitative RARbeta2 hypermethylation: a promising prostate cancer marker. Clin Cancer Res 2004;10(12 Pt 1):4010–4.

[100] Enokida H, Shiina H, Urakami S, Igawa M, Ogishima T, Li LC, et al. Multigene methylation analysis for detection and staging of prostate cancer. Clin Cancer Res 2005;11(18):6582–8.

[101] Bastian PJ, Palapattu GS, Lin X, Yegnasubramanian S, Mangold LA, Trock B, et al. Preoperative serum DNA GSTP1 CpG island hypermethylation and the risk of early prostate-specific antigen recurrence following radical prostatectomy. Clin Cancer Res 2005;11(11):4037–43.

[102] Woodson K, O'Reilly KJ, Ward DE, Walter J, Hanson J, Walk EL, et al. CD44 and PTGS2 methylation are independent prognostic markers for biochemical recurrence among prostate cancer patients with clinically localized disease. Epigenetics 2006;1(4):183–6.

[103] Baden J, Adams S, Astacio T, Jones J, Markiewicz J, Painter J, et al. Predicting prostate biopsy result in men with prostate specific antigen 2.0 to 10.0 ng/ml using an investigational prostate cancer methylation assay. J Urol 2011;186(5):2101–6.

[104] Ellinger J, Kahl P, von der Gathen J, Rogenhofer S, Heukamp LC, Gütgemann I, et al. Global levels of histone modifications predict prostate cancer recurrence. Prostate 2010;70(1):61–9.

[105] Bianco-Miotto T, Chiam K, Buchanan G, Jindal S, Day TK, Thomas M, et al. Global levels of specific histone modifications and an epigenetic gene signature predict prostate cancer progression and development. Cancer Epidemiol Biomarkers Prev 2010;19(10):2611–22.

[106] Schaefer A, Jung M, Mollenkopf HJ, Wagner I, Stephan C, Jentzmik F, et al. Diagnostic and prognostic implications of microRNA profiling in prostate carcinoma. Int J Cancer 2010;126(5):1166–76.

[107] Hagman Z, Larne O, Edsjö A, Bjartell A, Ehrnström RA, Ulmert D, et al. miR-34c is downregulated in prostate cancer and exerts tumor suppressive functions. Int J Cancer 2010;127(12):2768–76.

[108] Partin AW, Van Neste L, Klein EA, Marks LS, Gee JR, Troyer DA, et al. Clinical validation of an epigenetic assay to predict negative histopathological results in repeat prostate biopsies. J Urol 2014;192(4):1081–7.

[109] Naldi I, Taranta M, Gherardini L, Pelosi G, Viglione F, Grimaldi S, et al. Novel epigenetic target therapy for prostate cancer: a preclinical study. PLoS One 2014;9(5):e98101.

[110] Asangani IA, Dommeti VL, Wang X, Malik R, Cieslik M, Yang R, et al. Therapeutic targeting of BET bromodomain proteins in castration-resistant prostate cancer. Nature 2014;510(7504):278–82.

[111] Chiam K, Centenera MM, Butler LM, Tilley WD, Bianco-Miotto T. GSTP1 DNA methylation and expression status is indicative of 5-aza-2'-deoxycytidine efficacy in human prostate cancer cells. PLoS One 2011;6(9):e25634.

[112] Dejeux E, Rønneberg JA, Solvang H, Bukholm I, Geisler S, Aas T, et al. DNA methylation profiling in doxorubicin treated primary locally advanced breast tumours identifies novel genes associated with survival and treatment response. Mol Cancer 2010;9:68.

[113] Costa VL, Henrique R, Danielsen SA, Duarte-Pereira S, Eknaes M, Skotheim RI, et al. Three epigenetic biomarkers, GDF15, TMEFF2, and VIM, accurately predict bladder cancer from DNA-based analyses of urine samples. Clin Cancer Res 2010;16(23):5842–51.

[114] Köhler CU, Bryk O, Meier S, Lang K, Rozynek P, Brüning T, et al. Analyses in human urothelial cells identify methylation of miR-152, miR-200b and miR-10a genes as candidate bladder cancer biomarkers. Biochem Biophys Res Commun 2013;438(1):48–53.

[115] Maruyama R, Toyooka S, Toyooka KO, Harada K, Virmani AK, Zöchbauer-Müller S, et al. Aberrant promoter methylation profile of bladder cancer and its relationship to clinicopathological features. Cancer Res December 2001;61(24):8659–63.

[116] Yates DR, Rehman I, Abbod MF, Meuth M, Cross SS, Linkens DA, et al. Promoter hypermethylation identifies progression risk in bladder cancer. Clin Cancer Res 2007;13(7):2046–53.

[117] Duarte-Pereira S, Paiva F, Costa VL, Ramalho-Carvalho J, Savva-Bordalo J, Rodrigues A, et al. Prognostic value of opioid binding protein/cell adhesion molecule-like promoter methylation in bladder carcinoma. Eur J Cancer 2011;47(7):1106–14.

[118] Friedrich MG, Weisenberger DJ, Cheng JC, Chandrasoma S, Siegmund KD, Gonzalgo ML, et al. Detection of methylated apoptosis-associated genes in urine sediments of bladder cancer patients. Clin Cancer Res 2004;10(22):7457–65.

[119] Schneider AC, Heukamp LC, Rogenhofer S, Fechner G, Bastian PJ, von Ruecker A, et al. Global histone H4K20 trimethylation predicts cancer-specific survival in patients with muscle-invasive bladder cancer. BJU Int 2011;108(8 Pt 2):E290–6.

[120] Veerla S, Lindgren D, Kvist A, Frigyesi A, Staaf J, Persson H, et al. MiRNA expression in urothelial carcinomas: important roles of miR-10a, miR-222, miR-125b, miR-7 and miR-452 for tumor stage and metastasis, and frequent homozygous losses of miR-31. Int J Cancer 2009;124(9):2236–42.

[121] Tessema M, Belinsky SA. Mining the epigenome for methylated genes in lung cancer. Proc Am Thorac Soc 2008;5(8):806–10.

[122] Nana-Sinkam SP, Powell CA. Molecular biology of lung cancer: diagnosis and management of lung cancer, 3rd ed: American College of Chest Physicians evidence-based clinical practice guidelines. Chest 2013;143(5 Suppl.):e30S–9S.

[123] Belinsky SA, Liechty KC, Gentry FD, Wolf HJ, Rogers J, Vu K, et al. Promoter hypermethylation of multiple genes in sputum precedes lung cancer incidence in a high-risk cohort. Cancer Res 2006;66(6):3338–44.

[124] Belinsky SA, Grimes MJ, Casas E, Stidley CA, Franklin WA, Bocklage TJ, et al. Predicting gene promoter methylation in non-small-cell lung cancer by evaluating sputum and serum. Br J Cancer 2007;96(8):1278–83.

[125] Brock MV, Hooker CM, Ota-Machida E, Han Y, Guo M, Ames S, et al. DNA methylation markers and early recurrence in stage I lung cancer. N Engl J Med 2008;358(11):1118–28.

[126] de Fraipont F, Levallet G, Creveuil C, Bergot E, Beau-Faller M, Mounawar M, et al. An apoptosis methylation prognostic signature for early lung cancer in the IFCT-0002 trial. Clin Cancer Res 2012;18(10):2976–86.

[127] Suda K, Mitsudomi T. Successes and limitations of targeted cancer therapy in lung cancer. Prog Tumor Res 2014;41:62–77.

[128] Zhang YW, Zheng Y, Wang JZ, Lu XX, Wang Z, Chen LB, et al. Integrated analysis of DNA methylation and mRNA expression profiling reveals candidate genes associated with cisplatin resistance in non-small cell lung cancer. Epigenetics 2014;9(6).

[129] Langevin SM, Kratzke RA, Kelsey KT. Epigenetics of lung cancer. Transl Res 2015; 165(1):74–90.

[130] Lee JJ, Murphy GF, Lian CG. Melanoma epigenetics: novel mechanisms, markers, and medicines. Lab Invest 2014;94(8):822–38.

[131] Lee YH, Seo D, Choi KJ, Andersen JB, Won MA, Kitade M, et al. Antitumor effects in hepatocarcinoma of isoform-selective inhibition of HDAC2. Cancer Res 2014;74(17):4752–61.

[132] Bernet V, Smallridge R. New therapeutic options for advanced forms of thyroid cancer. Expert Opin Emerg Drugs 2014;19(2):225–41.

[133] Ramakrishnan S, Pili R. Histone deacetylase inhibitors and epigenetic modifications as a novel strategy in renal cell carcinoma. Cancer J 2013;19(4):333–40.

[134] Peters I, Dubrowinskaja N, Abbas M, Seidel C, Kogosov M, Scherer R, et al. DNA methylation biomarkers predict progression-free and overall survival of metastatic renal cell cancer (mRCC) treated with antiangiogenic therapies. PLoS One 2014;9(3):e91440.

[135] Zhu J, Wang Y, Duan J, Bai H, Wang Z, Wei L, et al. DNA Methylation status of Wnt antagonist SFRP5 can predict the response to the EGFR-tyrosine kinase inhibitor therapy in non-small cell lung cancer. J Exp Clin Cancer Res 2012;31:80.

[136] Chen B, Rao X, House MG, Nephew KP, Cullen KJ, Guo Z. GPx3 promoter hypermethylation is a frequent event in human cancer and is associated with tumorigenesis and chemotherapy response. Cancer Lett 2011;309(1):37–45.

[137] Rodríguez-Rodero S, Delgado-Álvarez E, Fernández AF, Fernández-Morera JL, Menéndez-Torre E, Fraga MF. Epigenetic alterations in endocrine-related cancer. Endocr Relat Cancer 2014;21(4):R319–30.

[138] Kunstman JW, Korah R, Healy JM, Prasad M, Carling T. Quantitative assessment of RASSF1A methylation as a putative molecular marker in papillary thyroid carcinoma. Surgery 2013;154(6);1255–61; discussion 1261–2.

[139] Jensen K, Patel A, Hoperia V, Larin A, Bauer A, Vasko V. Dynamic changes in E-cadherin gene promoter methylation during metastatic progression in papillary thyroid cancer. Exp Ther Med 2010;1(3):457–62.

[140] Feng H, Cheung AN, Xue WC, Wang Y, Wang X, Fu S, et al. Down-regulation and promoter methylation of tissue inhibitor of metalloproteinase 3 in choriocarcinoma. Gynecol Oncol 2004;94(2):375–82.

[141] Zuo H, Gandhi M, Edreira MM, Hochbaum D, Nimgaonkar VL, Zhang P, et al. Down-regulation of Rap1GAP through epigenetic silencing and loss of heterozygosity promotes invasion and progression of thyroid tumors. Cancer Res 2010;70(4):1389–97.

[142] Li JY, Han C, Zheng LL, Guo MZ. Epigenetic regulation of Wnt signaling pathway gene SRY-related HMG-box 17 in papillary thyroid carcinoma. Chin Med J Engl 2012;125(19):3526–31.

[143] Mancikova V, Buj R, Castelblanco E, Inglada-Pérez L, Diez A, de Cubas AA, et al. DNA methylation profiling of well-differentiated thyroid cancer uncovers markers of recurrence free survival. Int J Cancer 2014;135(3):598–610.

[144] Sponziello M, Durante C, Boichard A, Dima M, Puppin C, Verrienti A, et al. Epigenetic-related gene expression profile in medullary thyroid cancer revealed the overexpression of the histone methyltransferases EZH2 and SMYD3 in aggressive tumours. Mol Cell Endocrinol 2014;392(1–2):8–13.

[145] Brock MV, Gou M, Akiyama Y, Muller A, Wu TT, Montgomery E, et al. Prognostic importance of promoter hypermethylation of multiple genes in esophageal adenocarcinoma. Clin Cancer Res 2003;9(8):2912–9.

[146] Kaz AM, Grady WM. Epigenetic biomarkers in esophageal cancer. Cancer Lett 2014;342(2):193–9.

[147] Al-Kaabi A, van Bockel LW, Pothen AJ, Willems SM. p16INK4A and p14ARF gene promoter hypermethylation as prognostic biomarker in oral and oropharyngeal squamous cell carcinoma: a review. Dis Markers 2014;2014:260549.

[148] Brebi P, Hoffstetter R, Andana A, Ili CG, Saavedra K, Viscarra T, et al. Evaluation of ZAR1 and SFRP4 methylation status as potentials biomarkers for diagnosis in cervical cancer: exploratory study phase I. Biomarkers 2014;19(3):181–8.

[149] Jerónimo C, Henrique R. Epigenetic biomarkers in urological tumors: a systematic review. Cancer Lett 2014;342(2):264–74.

[150] Tänzer M, Liebl M, Quante M. Molecular biomarkers in esophageal, gastric, and colorectal adenocarcinoma. Pharmacol Ther 2013;140(2):133–47.

[151] Chan KC, Jiang P, Chan CW, Sun K, Wong J, Hui EP, et al. Noninvasive detection of cancer-associated genome-wide hypomethylation and copy number aberrations by plasma DNA bisulfite sequencing. Proc Natl Acad Sci USA 2013;110(47):18761–8.

[152] Liau JY, Liao SL, Hsiao CH, Lin MC, Chang HC, Kuo KT. Hypermethylation of the CDKN2A gene promoter is a frequent epigenetic change in periocular sebaceous carcinoma and is associated with younger patient age. Hum Pathol 2014;45(3):533–9.

[153] Coppedè F. The potential of epigenetic therapies in neurodegenerative diseases. Front Genet 2014;5:220.

[154] Ding H, Dolan PJ, Johnson G. Histone deacetylase 6 interacts with the microtubule-associated protein tau. J Neurochem 2008;106(5):2119–30.

[155] Mastroeni D, Grover A, Delvaux E, Whiteside C, Coleman PD, Rogers J. Epigenetic changes in Alzheimer's disease: decrements in DNA methylation. Neurobiol Aging 2010;31(12):2025–37.

[156] Gräff J, Rei D, Guan JS, Wang WY, Seo J, Hennig KM, et al. An epigenetic blockade of cognitive functions in the neurodegenerating brain. Nature 2012;483(7388):222–6.

[157] Chouliaras L, Mastroeni D, Delvaux E, Grover A, Kenis G, Hof PR, et al. Consistent decrease in global DNA methylation and hydroxymethylation in the hippocampus of Alzheimer's disease patients. Neurobiol Aging 2013;34(9):2091–9.

[158] Jowaed A, Schmitt I, Kaut O, Wüllner U. Methylation regulates alpha-synuclein expression and is decreased in Parkinson's disease patients' brains. J Neurosci 2010;30(18):6355–9.

[159] Matsumoto L, Takuma H, Tamaoka A, Kurisaki H, Date H, Tsuji S, et al. CpG demethylation enhances alpha-synuclein expression and affects the pathogenesis of Parkinson's disease. PLoS One 2010;5(11):e15522.

[160] Desplats P, Spencer B, Coffee E, Patel P, Michael S, Patrick C, et al. Alpha-synuclein sequesters Dnmt1 from the nucleus: a novel mechanism for epigenetic alterations in Lewy body diseases. J Biol Chem 2011;286(11):9031–7.

[161] Harrison IF, Dexter DT. Epigenetic targeting of histone deacetylase: therapeutic potential in Parkinson's disease? Pharmacol Ther 2013;140(1):34–52.

[162] Steffan JS, Kazantsev A, Spasic-Boskovic O, Greenwald M, Zhu YZ, Gohler H, et al. The Huntington's disease protein interacts with p53 and CREB-binding protein and represses transcription. Proc Natl Acad Sci USA 2000;97(12):6763–8.

[163] Jiang H, Poirier MA, Liang Y, Pei Z, Weiskittel CE, Smith WW, et al. Depletion of CBP is directly linked with cellular toxicity caused by mutant huntingtin. Neurobiol Dis 2006;23(3):543–51.

[164] Bollati V, Galimberti D, Pergoli L, Dalla Valle E, Barretta F, Cortini F, et al. DNA methylation in repetitive elements and Alzheimer disease. Brain Behav Immun 2011;25(6):1078–83.

[165] Hernández HG, Mahecha MF, Mejía A, Arboleda H, Forero DA. Global long interspersed nuclear element 1 DNA methylation in a colombian sample of patients with late-onset Alzheimer's disease. Am J Alzheimers Dis Other Demen 2014;29(1):50–3.

[166] Mahgoub M, Monteggia LM. Epigenetics and psychiatry. Neurotherapeutics 2013;10(4):734–41.

[167] Guidotti A, Auta J, Davis JM, Dong E, Gavin DP, Grayson DR, et al. Toward the identification of peripheral epigenetic biomarkers of schizophrenia. J Neurogenet 2014;28(1–2):41–52.

[168] Tremolizzo L, Conti E, Bomba M, Uccellini O, Rossi MS, Marfone M, et al. Decreased whole-blood global DNA methylation is related to serum hormones in anorexia nervosa adolescents. World J Biol Psychiatry 2014;15(4):327–33.

[169] Mehta D, Klengel T, Conneely KN, Smith AK, Altmann A, Pace TW, et al. Childhood maltreatment is associated with distinct genomic and epigenetic profiles in posttraumatic stress disorder. Proc Natl Acad Sci USA 2013;110(20):8302–7.

[170] Cha DS, Kudlow PA, Baskaran A, Mansur RB, McIntyre RS. Implications of epigenetic modulation for novel treatment approaches in patients with schizophrenia. Neuropharmacology 2014;77:481–6.

[171] Powell TR, Smith RG, Hackinger S, Schalkwyk LC, Uher R, McGuffin P, et al. DNA methylation in interleukin-11 predicts clinical response to antidepressants in GENDEP. Transl Psychiatry 2013;3:e300.

[172] Siniscalco D, Cirillo A, Bradstreet JJ, Antonucci N. Epigenetic findings in autism: new perspectives for therapy. Int J Environ Res Public Health 2013;10(9):4261–73.

[173] Saffrey R, Novakovic B, Wade TD. Assessing global and gene specific DNA methylation in anorexia nervosa: a pilot study. Int J Eat Disord 2014;47(2):206–10.

[174] Wong CC, Mill J, Fernandes C. Drugs and addiction: an introduction to epigenetics. Addiction 2011;106(3):480–9.

[175] Schmidt HD, McGinty JF, West AE, Sadri-Vakili G. Epigenetics and psychostimulant addiction. Cold Spring Harb Perspect Med 2013;3(3):a012047.

[176] Quddus J, Johnson KJ, Gavalchin J, Amento EP, Chrisp CE, Yung RL, et al. Treating activated CD4+ T cells with either of two distinct DNA methyltransferase inhibitors, 5-azacytidine or procainamide, is sufficient to cause a lupus-like disease in syngeneic mice. J Clin Invest 1993;92(1):38–53.

[177] Coppedè F, Migliore L. Epigenetics of autoimmune diseases. Molecular mechanisms and physiology of disease: implications for epigenetics and health. Springer Science + Business Media New York; 2014. p. 151–73.

[178] Balada E, Ordi-Ros J, Serrano-Acedo S, Martinez-Lostao L, Vilardell-Tarrés M. Transcript overexpression of the MBD2 and MBD4 genes in CD4+ T cells from systemic lupus erythematosus patients. J Leukoc Biol 2007;81(6):1609–16.

[179] Zhu X, Liang J, Li F, Yang Y, Xiang L, Xu J. Analysis of associations between the patterns of global DNA hypomethylation and expression of DNA methyltransferase in patients with systemic lupus erythematosus. Int J Dermatol 2011;50(6):697–704.

[180] Qin HH, Zhu XH, Liang J, Yang YS, Wang SS, Shi WM, et al. Associations between aberrant DNA methylation and transcript levels of DNMT1 and MBD2 in CD4+T cells from patients with systemic lupus erythematosu. Australas J Dermatol 2013;54(2):90–5.

[181] Karouzakis E, Gay RE, Michel BA, Gay S, Neidhart M. DNA hypomethylation in rheumatoid arthritis synovial fibroblasts. Arthritis Rheum 2009;60(12):3613–22.

[182] Roberson ED, Liu Y, Ryan C, Joyce CE, Duan S, Cao L, et al. A subset of methylated CpG sites differentiate psoriatic from normal skin. J Invest Dermatol 2012;132(3 Pt 1):583–92.

[183] Akimova T, Beier UH, Liu Y, Wang L, Hancock WW. Histone/protein deacetylases and T-cell immune responses. Blood 2012;119(11):2443–51.

[184] Kong S, Yeung P, Fang D. The class III histone deacetylase sirtuin 1 in immune suppression and its therapeutic potential in rheumatoid arthritis. J Genet Genomics 2013;40(7):347–54.

[185] Abi Khalil C. The emerging role of epigenetics in cardiovascular disease. Ther Adv Chronic Dis 2014;5(4):178–87.

[186] Antos CL, McKinsey TA, Dreitz M, Hollingsworth LM, Zhang CL, Schreiber K, et al. Dose-dependent blockade to cardiomyocyte hypertrophy by histone deacetylase inhibitors. J Biol Chem 2003;278(31):28930–7.

[187] Kee HJ, Sohn IS, Nam KI, Park JE, Qian YR, Yin Z, et al. Inhibition of histone deacety-lation blocks cardiac hypertrophy induced by angiotensin II infusion and aortic band-ing. Circulation 2006;113(1):51–9.

[188] Liu F, Levin MD, Petrenko NB, Lu MM, Wang T, Yuan LJ, et al. Histone-deacetylase inhibition reverses atrial arrhythmia inducibility and fibrosis in cardiac hypertrophy independent of angiotensin. J Mol Cell Cardiol 2008;45(6):715–23.

[189] Campión J, Milagro FI, Goyenechea E, Martínez JA. TNF-alpha promoter meth-ylation as a predictive biomarker for weight-loss response. Obes (Silver Spring) 2009;17(6):1293–7.

[190] Milagro FI, Campión J, Cordero P, Goyenechea E, Gómez-Uriz AM, Abete I, et al. A dual epigenomic approach for the search of obesity biomarkers: DNA methylation in relation to diet-induced weight loss. FASEB J 2011;25(4):1378–89.

[191] Moleres A, Campión J, Milagro FI, Marcos A, Campoy C, Garagorri JM, et al. Differential DNA methylation patterns between high and low responders to a weight loss intervention in overweight or obese adolescents: the EVASYON study. FASEB J 2013;27(6):2504–12.

[192] Schütte B, El Hajj N, Kuhtz J, Nanda I, Gromoll J, Hahn T, et al. Broad DNA methyla-tion changes of spermatogenesis, inflammation and immune response-related genes in a subgroup of sperm samples for assisted reproduction. Andrology 2013;1(6):822–9.

[193] Kläver R, Tüttelmann F, Bleiziffer A, Haaf T, Kliesch S, Gromoll J. DNA methylation in spermatozoa as a prospective marker in andrology. Andrology 2013;1(5):731–40.

[194] Wu C, Ding X, Li H, Zhu C, Xiong C. Genome-wide promoter methylation profile of human testis and epididymis: identified from cell-free seminal DNA. BMC Genomics 2013;14:288.

8

Epigenetic Fingerprint

Leda Kovatsi[1], Athina Vidaki[2], Domniki Fragou[1],
D. Syndercombe Court[2]

[1]Laboratory of Forensic Medicine and Toxicology, School of Medicine, Aristotle University of Thessaloniki, Thessaloniki, Greece; [2]Faculty of Biological Sciences and Medicine, King's College London, London, UK

OUTLINE

OUTLINE

The aim of this chapter is to introduce advanced undergraduate and postgraduate students of forensic science, forensic practitioners, and university researchers to new ideas in the advancement of genetics to aid criminal justice. It introduces several areas where epigenetics could be employed to aid an investigator in circumstances where the suspect cannot readily be identified.

1. INTRODUCTION

On September 10, 1994, Alec Jeffreys understood that he was looking at what had initially seemed to be just a complex set of bands on an X-ray image, but which had revealed a pattern that explained the inheritance of genetic material from a mother and father to their child. He had been researching highly variable "mini-satellites"—short series of around 10 to 60 nucleobases, repeated in tandem, that are found in many locations throughout the genome. Here, the "genetic fingerprint" was born [1]. This resolution of a problematic paternity test was rapidly followed by the identification of Colin Pitchfork as the person who had most likely raped and murdered two young girls—charges for which he was subsequently convicted.

The complexity of banding patterns resulted in difficulties in their interpretation within forensic casework, leading to many arguments when this technology was brought to court. Within a few years, the fingerprint was replaced by the "genetic profile." The genetic profile used microsatellites consisting of repeated short nucleobase sequences from two to six in number, known as *short tandem repeats* (STRs). STRs from different loci were characterized by scientists in the United States and the United Kingdom [2,3]; these and others have since been commercialized within large multiplexes to provide high discrimination between individuals. STR loci have been used not only because of their polymorphic diversity but also because of their location within noncoding DNA, ensuring that personal private information is not revealed.

STR profiling continues to be the main tool in human identification today and has formed the basis of forensic DNA databases worldwide. Despite their power, knowledge of the DNA profile alone leaves many forensic questions unanswered, for example, what is the tissue source of the DNA from which the profile was obtained? Chemical tests have been utilised for many years to suggest the presence of body fluids of particular forensic interest: blood, saliva, and semen. These tests are predominantly presumptive and destructive, such that a DNA profile cannot be obtained from the same material [4].

More recent work has focused on resolving this through methods that coextract both RNA and DNA [5], the former being selected from

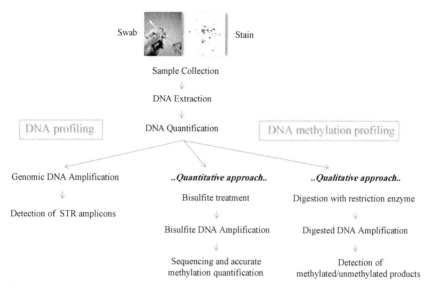

FIGURE 1 Potential forensic workflow for co-analysis of DNA and DNA methylation profiling.

candidate gene targets to provide tissue-specific mRNA assays [6,7], with the latter providing the necessary and essential DNA profile and allowing collateral analysis (Figure 1).

The finished human genome was first published in 2004, allowing researchers to examine areas of the DNA that influence gene expression. The factors are known as epigenetic (outside genetics) and there are two main types employed within forensic analysis:

- RNA-mediated—tissue-specific microRNA (miRNA) sequences that are involved in the regulation of gene expression by interacting with mRNA and have been employed in the forensic arena [8].
- DNA methylation—in which cytosine (normally) bases within CpG islands become methylated during development and through a variety of environmental factors. The methylated sites are chromosome specific and characteristic methylation patterns associated with particular tissues, diseases, or phenotypic traits, can be identified [9].

1.1 Biology of DNA Methylation

DNA methylation is one of the most important epigenetic modifications, with an important role in regulating gene expression. Because this does not change the DNA sequence, normal development is ensured. It involves the addition of a methyl group ($-CH_3$) at the 5' position of cytosine that is followed by guanine (CpG). On a genome-wide scale, most CpG sites are methylated (60–90%), while the unmethylated ones are often grouped in genomic areas known as "CpG islands" (300–3000 bp long with >55% GC content) [9]. These CpG islands are usually located near the regulatory regions of a lot of human genes. It is generally believed that DNA methylation is linked with condensed heterochromatin and silencing of gene expression; however, the opposite is also possible [10].

It is DNA methylation that we wish to concentrate on throughout the rest of this chapter and will illustrate the potential utility of epigenetics in forensic examination. It cannot, however, be over-emphasised, that the utility of any technologies to be used within a forensic examination and subsequently presented in court must be proven through a rigorous validation process.

1.2 Scenario

Mrs A, a 72-year-old widow, was admitted to hospital in a confused state where she was found to be suffering from severe dehydration and multiorgan failure, from which she subsequently died a few days later. Her daughter had called on the house because her mother had not responded to her telephone calls and called an ambulance when she found her mother in a collapsed state. All she would say was "Don't let him near me," again and again. Her mother had been complaining of abdominal pains recently and her daughter was trying to get her to go to the GP, but Mrs A was reluctant as she didn't want to bother her doctor over something that would probably get better.

A postmortem was requested and the pathologist noticed a characteristic patterning in Mrs A's fingernails, known as Mee's lines, suggestive of arsenic poisoning and the coroner and the police were informed.

Mrs A lived in a large detached house where she and her husband had lived for more than 40 years until her husband's death 5 years ago. Mrs A had not wanted to leave her home and her daughter visited her every other Sunday. Her husband drove her there and then went to the pub while she and her mother had lunch together.

The police undertook an examination of the home, looking for a possible source of arsenic. They found nothing related to the latter but Mrs A's desk seemed to have had its lock forced and there was a red-brown stain near the lock which tested positive for blood using a chemical test and an asthma inhaler on the table nearby. Mrs A did not suffer from asthma. Neighbors talked about a young man who sometimes came round and took Mrs A out in his car but they hadn't seen him for some time. Mrs A's daughter did not know who this was but said that her mother sometimes talked about "that lovely young man." Mrs A's solicitor told police that Mrs A had made an appointment to discuss changing her will.

A DNA examination of the asthma pump revealed a partial DNA profile from a male. A swab of the stain near the lock produced a full profile, consistent with it being from the same person who had used the inhaler. A search of the DNA database revealed no matches—the police had no leads to go on and had no idea who the young male visitor was. Could an epigenetic investigation assist in the investigation and narrow the list of potential suspects? Standard DNA profiling would, of course, be required to confirm a match with the crime stain.

2. IDENTIFYING A STAIN'S TISSUE SOURCE

The stain near the desk lock appears to be blood and it would be useful to be able to attribute the DNA profile to the blood, to provide any credence to the hypothesis that the owner of the pump also forced the lock on the desk, potentially injuring himself while doing this.

Recovering a biological stain at the crime scene and linking it with a suspect through DNA typing is in many cases crucial for the outcome of a trial. Locating potential biological material can usually be achieved through visual examination; however, for successful recovery, a test that indicates the presence of a specific body fluid is also employed in situ [11]. These so-called "presumptive tests" include biochemical or immunological reagents that react to specific elements in each fluid, for example the Kastle–Meyer test is a color test based on the peroxidase-like activity of the heme group and can successfully give a positive reaction for blood [11]. Although these tests are cheap and easy to perform, it is important to understand that they are not entirely specific and require a large amount of biological material.

In more complex cases, or when reconstructing the events that have taken place at a crime scene, it would be beneficial to be able to identify the cellular origin of a stain through a confirmatory test. The presence of specific body fluids could indicate particular types of crime. For instance the presence of semen or vaginal fluid could be linked to a sexual assault and, if it can be directly linked to a DNA profile, this information can be used to either exclude or convict a suspect. The development of a genetic body fluid/tissue identification system that is also compatible with current DNA profiling technologies would be very advantageous. Such tests have to be very sensitive, specific, and maintain the integrity of DNA evidence.

In the last decade, forensic scientists have focused their efforts on the use of messenger RNA (mRNA) assays; they have been shown to be valuable since some mRNA molecules are expressed in a cell-specific manner [12–14]. mRNA profiling using tissue-specific markers is generally considered as a very good option as these markers are very sensitive (give a positive reaction with as low as $0.05\,\mu l$ stain [12]) and can be multiplexed, allowing for simultaneous identification of all forensically relevant tissues. However, there are issues regarding tissue-to-tissue specificity and when applying these tests in extensively degraded samples where RNA is more difficult to obtain. A method that would not consume additional sample and at the same time can exploit the stability of the DNA molecule would be preferred. Such constraints are particularly important when re-examining "cold cases," where only DNA has been retained and current methods cannot be used. DNA-based tests could potentially overcome the limitations of existing methods and provide a direct link between the recovered DNA and its source. It is known that DNA methylation is one of the mechanisms responsible for cell differentiation and differential gene expression [15,16] and tissue-specific methylation patterns have been reported [17,18]. Therefore, DNA methylation profiling could be very helpful in confirming the presence of a specific body fluid/tissue.

In the forensic field, Frumkin et al. were the first to explore the possibility of DNA methylation-based forensic tissue identification [19]. One nanogram of blood, saliva, semen, skin, urine, menstrual blood, and vaginal DNA were treated with methylation-sensitive restriction enzymes and analyzed for 205 individual CpG islands that could potentially show differential methylation levels. A total of 38 genomic loci demonstrated differential amplification patterns [19]. Since highly methylated loci were protected from enzymatic digest and amplified with much higher efficiency than loci with lower methylation levels, the authors considered only ratios of methylation levels in their analysis. Employing only seven tissue-specific loci in pairs, each tissue gaves a distinct methylation profile; nevertheless natural interindividual variation also needs to be taken into account. Interestingly, Gomes et al. tried to reproduce their results on identification of skin using the two proposed loci but failed as the methylation profile of the skin and saliva they tested was the same [20]. Consequently, the suggested loci require further investigation.

Employing a similar approach, Wasserstrom et al. developed a DNA methylation-based semen test (Nucleix DSI-Semen kit), which could successfully distinguish between semen and non-semen samples and could potentially replace the time-consuming microscopic examination of casework samples [21]. A panel of five genomic loci that were believed to demonstrate substantial DNA methylation differences between semen and all other body fluids tested were selected and tested on 135 DNA samples. The accuracy of the kit was high and the accompanying software also developed by the authors made the subsequent data analysis more reliable. LaRue et al. performed a more comprehensive validation study on the kit's performance and illustrated that the required starting DNA material can be as low as 62 pg [22].

Nevertheless, a test that could simultaneously identify all body fluids would be preferable. Lee et al. tested previously reported tissue-specific differentially methylated regions (tDMRs) identified through the literature and proposed a different methodological approach that included bisulfite sequencing [23]. After testing pooled DNA from blood, semen, saliva, menstrual blood, and vaginal fluid, the authors identified two testis-specific DMRs (*DACT1* and *USP49*) that could potentially be applied for semen identification. They also tested another three tDMRs (*HOXA4*, *PFN3*, and *PRMT2*) which displayed varying degrees of methylation. As the authors suggested, the presence of an unmethylated clone in the *HOXA4* tDMR could possibly be used to exclude the presence of blood and the *PFN3* tDMR could potentially be used for the identification of vaginal secretions [23]. Although the results were promising, sex differences and interindividual variations were once again observed. To further validate these tDMRs, as well as developing a more sensitive multiplex assay system, the authors employed a methylation-specific restriction enzyme PCR (MSRE-PCR) technique using four out of the five previously proposed markers (*DACT1*, *USP49*, *PFN3*, and *PRMT2*) [24]. Their results using blood, semen, and saliva samples were similar, showing semen-specific hypomethylation only, both in young and elderly men. The discrimination power regarding the other body fluids was increased, however, when they integrated the proposed DNA methylation markers with selected body fluid-specific microbial DNA markers that could allow for identification of saliva and vaginal fluid [25]. The previously reported multiplex MSRE-PCR assay was also modified by the addition of a new semen-specific marker (*L81528*).

Most methods mentioned above use a restriction enzyme-based qualitative approach rather than provide accurate quantitative methylation results. Following a similar protocol, developed by Paliwal et al., which allowed for quantitative detection of DNA methylation states in minute amounts of DNA from body fluids [26], Madi and her colleagues [27] proposed a sensitive method that could be applied in traces levels of forensic samples. Utilising bisulfite pyrosequencing, they could accurately evaluate the relative quantity of methylated cytosines of various adjacent CpG sites at four genomic loci (*C20orf117*, *BCAS4*, *ZC3H12D*, and *FGF7*). *C20orf117* was found to be blood-specific, *ZC3H12D* and *FGF7* showed sperm-specific methylation, whereas *BCAS4* seemed to be saliva-specific although it had initially been proposed as a semen-specific marker [27]. It is, however, notable that the methylation differences between tissues were not always high (~40%) which could complicate the analysis of mixed body fluid stains.

Moreover, Ma et al. screened six potentially blood-specific tDMRs using methylation sensitive difference analysis and Sequenom Massarray (MS-RDA) technology [28]. The authors identified two fragments showing blood-specific hypomethylation and four fragments showing blood-specific hypermethylation; however no menstrual blood samples were tested so it is unclear if these tDMRs are only venous blood-specific. Finally, a recent study explored the potential of next generation (massive parallel) sequencing (NGS) for tissue source identification [29]. The authors modified a previously published CE-based method and achieved a concordance rate of 15/16 using only 1 ng of DNA from four different tissues: semen, saliva, skin epidermis, and blood. The method included digestion with a methylation sensitive restriction endonuclease (HhaI),

PCR amplification of 10 loci in total, NGS multiplex library preparation and sequencing using an Illumina MiSeq instrument.

As expected, the identification of biologically complex mixtures of body fluids in diverse ratios, with the exception of semen, is challenging using epigenetic markers; however the above studies show promising results, with several working at pg levels of DNA, which is necessary for forensic applications [21,24,25].

Using all potential blood-specific methylation markers published in the literature, such as the HOXA4 tDMR [23], C20orf117 [27], and the six blood-specific tDMRs reported by Ma et al. [28], investigators were able to identify the tissue source of the stain near the desk lock as being blood. For example, the DNA from the stain was found to be methylated for the HOXA4 locus indicating the possibility of blood present, whereas most of the CpG sites of the C20orf117 locus were also found to be highly methylated (>60%) matching the reported blood-specific pattern. The positive link between the stain and the DNA profile in the scenario has led to police viewing this particular donor with more suspicion than before and now want to find out more about him.

3. ESTIMATING BIOLOGICAL AGE

Could the person who has left the stain be the young man who visited Mrs A on occasions? The investigators ask whether it would be possible to determine the age of the donor.

The ability to accurately estimate a person's chronological age would be a great advantage in police investigations as it could provide significant investigative leads and potentially narrow down the number of suspects. Especially in those cases where an eye witness is not available, such a test could work as a "DNA witness" providing more information about the individual in question. Developing an age prediction model is a major challenge for forensic scientists as they would need to be able to apply and validate it using minute or degraded samples consisting of a range of tissues and body fluids.

Aging is a very complex process influenced by various genetic, lifestyle, and environmental factors. It causes a variety of modifications and adjustments in tissues and organs that accumulate over an individual's lifetime. These alterations can be examined at a molecular level either as changes in the composition of metabolites [30] or through gene expression [31]. In a medical setting, these age-related factors provide useful information for "personalised" medicine as they are used as risk factors in the development of many diseases, such as cancer [32,33]. In a forensic scenario, and in cases where a human skeleton is recovered, a rough estimation of the age-at-death can be achieved through various morphological age-associated changes in the skeleton or in dentition [34]; however most of these methodologies are relative, producing an estimate with a large age range.

Meissner and Ritz-Timme summarised four main mechanisms relating to aging which provide promise in their application within a forensic scenario [35]. These include: the accumulation of DNA damage as reflected through the amount of deletions seen in mitochondrial DNA; the shortening of telomeres that occurs along with cell division; and two essential protein alterations related to the aging process—aspartic acid racemisation and advanced glycation endproducts [35]. A further mechanism that has been proposed to reliably estimate biological age is the use of T-cell DNA rearrangements [36,37]. Intervening DNA sequences in the T-cell receptors (TCR) genes are deleted; these form episomal DNA molecules, called signal joint TCR excision circles (sjTRECs). It is believed that the number of sjTRECs declines as age increases. Zubakov et al. developed a robust and sensitive real-time quantitative PCR protocol; however, the standard error of the estimate remains high at ±8.9 years [36]. Despite the efforts of scientists, all proposed methods have limitations and, in most cases, exhibit low accuracy. They are more likely to suggest an age group (generation) than accurate age. In addition, many methods currently employed are destructive of the tissue.

Epigenetic analysis could serve as an alternative or supplementary method since DNA methylation is known to be one of the mechanisms responsible for cell differentiation and the cellular response to aging [38–40]. It is generally suggested that there is an increase in global epigenetic drift with age [41]. Changes in DNA methylation patterns due to aging are quickly observed during the first months of an individual's life and throughout childhood [42,43]. Monozygotic (MZ) twins can serve as an ideal model to study these dynamic epigenetic marks as they share the same DNA sequence and start life with almost identical methylation patterns [44,45]. Moreover, in order to identify specific age-associated differentially methylated CpG sites, scientists have chosen to perform genome-wide studies using microarrays that enable analysis of more than 450,000 CpGs at the same time [46]. However epigenetic changes can also be tissue-specific, so it is preferable to study DNA methylation patterns in each forensic-related tissue separately.

Blood is one of the most commonly found body fluids at crime scenes and it would be helpful in an investigation to be able to predict the donor's Blood is one of the most commonly age from small blood stains. Hannum et al. performed one of the largest genome-wide methylation studies to date using the whole blood of 656 individuals aged 19–101 years old [46]. The authors built a quantitative model using 71 highly age-predictive markers with a correlation between true and predicted age of 0.96 and an error of 3.9 years. As might be expected, nearly all of the markers applied in the model lay within or near genes with a known link to age-related conditions including DNA damage, cancer, and Alzheimer's disease. Interestingly they noticed that the methylome of men appeared to age approximately 4% faster than that of women (mirroring the increased longevity of females) and there were always a few individuals that appeared to be aging faster or slower than what the model would predict.

Crime scene samples are often of limited quantity and quality and will not provide sufficient DNA for epigenome-wide approaches. In an attempt to narrow down the number of age-associated markers needed for accurate prediction, Weidner et al. performed a comprehensive analysis of methylation profiles and found that the methylation levels of only three CpGs—located in the genes *ITGA2B*, *ASPA*, and *PDE4C*—were enough to create an epigenetic-aging-signature [47]. They developed a bisulfite pyrosequencing protocol that allowed age prediction with a mean absolute deviation from chronological age of less than 5 years after testing 151 whole blood samples. Similarly, applying a very different methodological approach (methylation-sensitive representational difference analysis (MS-RDA)), Yi et al. also identified eight gene fragments in which the degree of cytosine methylation was significantly correlated with age [48]. They tested 105 donors altogether and by using two CpG sites from each fragment (16 in total) they built a regression model that explained 95% of the variance in age.

Apart from blood, other tissues, such as buccal epithelium and brain, have also been used for age-associated methylation analysis [44,49]. Bocklandt et al. performed a genome-wide methylation analysis using saliva samples of 34 pairs of male identical twins (21–55 years old) using Illumina's HumanMethylation27 microarray [44]. The authors identified a total of 88 novel loci that are significantly correlated with age. Using only three CpG sites—located in the genes *EDARADD*, *TOM1L1*, and *NPTX2*—they built a regression model that was linear with age over a range of five decades which explained 73% of the variance (average accuracy of 5.2 years).

As previously mentioned, each tissue or body fluid shows a different age-associated DNA methylation pattern. It would be very useful to develop an age prediction test that could be applied in all human tissues and Kock and Wagner have gone some way toward this. They analyzed several publicly available DNA methylation data sets that examined more than 27,000 CpG sites in 13 different cell types [50]. Initially they identified 431 age-associated hypermethylated, and 25 age-associated hypomethylated CpG sites. Next, they chose a subset of 5 markers—located in the genes *TRIM58*, *KCNQ1DN*, *NPTX2*, *BIRC4BP*, and *GRIA2*—to be integrated into their epigenetic-aging-signature test. One of these CpG sites (*NPTX2*), interestingly, is also included in the previous study by Bocklandt et al. [44]. Based on these five CpG sites, their predictions had an average precision of ±9.3 years [50].

Predicting age across a broad spectrum of human tissues and cell types appears to be a very challenging task; using a much larger data set could, however, potentially overcome the difficulties and provide a more accurate age prediction. Horvath developed a multitissue predictor of age by employing 8000 samples from publicly available Illumina DNA methylation array data sets (both 27K and 450K) comprised of 51 healthy tissues and cell types [51]. He used a regression analysis that avoids over-fitting of the model (penalised regression with elastic net analysis) and selected 353 CpGs: he named their collection the "epigenetic clock." Across all test data, the age correlation was 0.96 with an error of 3.6 years, which appears very promising. Individual models according to tissue type were also built to improve accuracy [50]. The age predictor worked well even in heterogeneous tissues such as whole blood, buccal epithelium, uterine cervix, and saliva, but in sperm the methylated-DNA age was significantly lower than the chronological age of the donor.

As shown above, accurate molecular age estimation from biological materials is not a simple task as aging is biologically complex. However, age-associated DNA methylation profiling is very promising and future research has the potential to influence our understanding of aging and allow for more accurate predictions.

In order to estimate the age of the blood stain donor, the investigators analyzed the methylation levels of the three age-associated CpG sites reported by Weidner et al. [47]. To the investigating officer's surprise, the age estimation they got back from the blood stain was for a man aged in his fifties (53 ± 5 years), which did not fit with the young man they were expecting.

4. EXPOSURE TO THE HEAVY METALS, ARSENIC AND LEAD

The suspicion that Mrs A might have been exposed to arsenic led the investigating team to perform a toxicological analysis both on the clinical blood and urine samples collected during her hospitalization, as well as on the postmortem samples (hair, nails, and various organs) collected upon autopsy. Although environmental exposure to arsenic is almost unknown in the west, a common cause is drinking contaminated well water, or occupational exposure. Furthermore, they decided to examine the male DNA for epigenetic phenomena that are induced following heavy metal exposure. Although environmental exposure seems unlikely the team wishes to rule this out and any potential link with the unknown male, if this was the case.

4.1 Arsenic

Arsenic (As) is either ingested by humans through contaminated food or water or can be inhaled in the form of airborne particles. Clinical symptoms of As toxicity include several forms of cancer (liver, lung, bladder, and skin), neurological and cardiovascular diseases, and diabetes. Inorganic arsenic is more toxic than organic and exists in two forms, trivalent As (III) and oxidized, pentavalent As (V). Arsenic exposure can have genetic and epigenetic effects and can induce cell proliferation, differentiation, and apoptosis [52].

Prolonged exposure to high doses of As was found to cause hypomethylation of the *p53* and *p16* promoter sites, in contrast to acute exposure which induced hypermethylation of the same promoter sites. The individuals under study had been exposed to As via drinking water and the DNA was extracted from whole blood [53]. Hypomethylation of the *p53* gene has also been shown to play a role in lung cancer, secondary to arsenic exposure [54]. In a study conducted on people with arseniasis (chronic arsenic poisoning), extensive methylation of the *p16* promoter was shown in DNA extracted from peripheral blood mononuclear leukocytes. Furthermore, the level of methylation correlated with the severity of clinical symptoms. However, no correlation was found with the duration of exposure. The authors suggested that this particular epigenetic change could be used as a biomarker for As exposure [55]. Apart from heavy-metal exposure, alterations in the methylation status of *p53* and *p16* has been extensively revealed in many different forms of cancer such as leukemia and gastric cancer [56,57].

A similar exposure to As was shown to cause genomic hypermethylation. Mononuclear cells from peripheral blood were again used as the source of DNA [58]. In a more recent study, hypomethylation of various promoters was reported in people exposed to drinking water containing As. The authors concluded that if a direct relationship was proven between As exposure and DNA methylation (as an underlying mechanism for the adverse effects), then a personal DNA methylation profile could serve as a biomarker [59]. When human prostate epithelial cells were chronically exposed to arsenic, hypermethylation of the *MLH1* promoter was observed, leading to decreased expression of the gene and thus, increased cell growth and transformation [60].

The aforementioned studies were conducted on a large number of cancer patients with a medical history of prolonged As exposure. It has been shown in the past that cancer is associated with global DNA hypomethylation, whereas specific transcription sites in the genome are hypermethylated. This epigenetic mechanism can explain the link between As exposure and cancer [52,61].

Arsenic may also influence the epigenome in terms of histone modifications. In arsenic-exposed individuals, acetylation of histone3K9 decreased globally in blood mononuclear cells while its methylation increased [62]. This finding supported the conclusions of an earlier study performed on human UROtsa cells, according to which As causes a decrease in the acetylation of histones 3 and 4 [63].

4.2 Lead

Humans are exposed to lead (Pb) through contaminated food, water, soil, and air. Its adverse effects mainly target the central nervous system but also the hematopoietic system, the kidneys, and the bones. In children

Pb can cause severe mental disorders and learning disabilities while in some extreme cases, even coma and death. Lead is thought to exert its toxic effects through disruption of cell signaling and neurotransmission pathways as well as changes in the DNA methylation status [64].

In a recent study it was shown that hypermethylation of the *p16* promoter in the blood of lead-exposed individuals correlated with the concentration of Pb in their blood [65]. Three years later, a study on Pb toxicity was conducted both in vitro and in vivo. When cell lines were used, significant hypomethylation of LINE-1 was observed ($p = 0.009$). Exposed individuals exhibited major differences in the methylation status when compared to controls ($p < 0.001$) and lead concentration was inversely proportional to the level of LINE-1 methylation ($p < 0.001$). The authors concluded that LINE-1 promoter methylation might be used as a biomarker of lead toxicity, especially in cases of occupational exposure to Pb [66]. This finding was reinforced by a study carried out in 2009 on lead-exposed pregnant women, where hypomethylation of both LINE-1 and Alu promoters was observed in the umbilical cord blood. Furthermore, this decrease in methylation correlated inversely with Pb levels in maternal bones. LINE-1 and Alu exist in large copy numbers and occupy 25% of the human genome and 40% of the methylated CpGs; therefore they have been extensively used for methylation studies. In this particular study, it was suggested that Pb exposure, even at the beginning of life, can induce epigenetic effects that may appear later in life [67]. Furthermore, Pb could cause changes in the methylation status associated with oxidative damage, by modulating DNA methyltransferase activity [68]. In particular, it can increase homocysteine levels, inhibit DNA methyltransferases and therefore, reduce DNA methylation levels [69].

Similar effects of Pb exposure have been reported in studies conducted on Alzheimer's disease (AD). When DNA methyltransferase activity is affected during the early stages of brain development, then hypomethylation of gene promoter sites associated with AD, such as the β-amyloid precursor protein (APP), is observed. This imprint can be triggered later in life to mark the onset of AD through an increase in the levels of APP and β-amyloid causing oxidative damage of DNA [70]. A 20% inhibition of DNA methyltransferase was observed in the brain of primates exposed to Pb. This inhibition could affect the methylation status of various, rich in CpGs promoter sites, such as *APP* and *BACE1* [71,72]. Neurological damage, as a result of lead exposure, can also be caused by impairment of DNA methylation through inhibition of insulin-like growth factor one stimulated activity of methionine synthase. Growth retardation and impairment in cognitive development have been reported as adverse effects of Pb toxicity [73].

The above strongly suggests that lead exposure causes hypomethylation of promoter sites (and therefore changes in the gene transcription status) by inhibiting DNA methyltransferases. Moreover, early life exposure can create an epigenetic mark (change) which might exert its effect later in life, as in the case of AD. Epigenetic alterations caused by lead seem to be involved mainly in impaired cognitive development and various neurological disorders [74]. Hypomethylation of various promoter sites can be indicative of lead exposure when studied in stains from crime scenes or in the organs and blood of victims, giving valuable insight on environmental factors that have affected the life and death of individuals.

The suspicion of arsenic poisoning of Mrs A was confirmed. ICP-MS analysis of the clinical and post-mortem biological samples of Mrs A revealed extremely high, lethal, concentrations of arsenic. Interestingly, epigenetic analysis of the DNA isolated from the blood stain, from the unknown middle-aged man, revealed hypermethylation of the p16 promoter and hypomethylation of the LINE-1 and Alu promoters. The police were now looking for someone who fitted with this profile—perhaps a plumber or construction worker—and who also was in receipt of medication for asthma. In this case, conventional toxicological analysis, combined with advanced epigenetic screening, provided useful information leading the investigation one step closer to the truth and the perpetrator.

It is important for the reader to recognise the epigenetic tools discussed here are part of an immature science and still under development for forensic use.

5. MZ TWINS

The investigative focus now led police to study the family in more detail. The damage to the desk where Mrs A kept her personal papers, along with no evidence of a break in, and the information that Mrs A was apparently intending to visit her solicitor to alter her will stimulated this. The son-in-law, Mr Z, was brought in for questioning. He was in the right age group at 58 and was also an asthma sufferer. Although he was out of work he had worked in the print industry on the compositor's floor from the age of 16 until he was made redundant a few years ago. A DNA profile obtained from Mr Z was shown to match the blood stain in the house, and that found on the asthma pump. Mr Z denied being involved and said that he never went into the house, despite taking his wife to visit her mother regularly. He had fallen-out with his mother-in-law some time ago and had not visited for some years. When challenged by the strong DNA evidence Mr Z revealed that he had an identical twin Mr Y, who lived locally and who had also been employed in the print. Mr Z thought that it was possible that Mr Y could have been invited in to the house by his mother-in-law as she would not be able to recognise that the visitor was not Mr Z. How could the police discriminate between them?

The study of epigenetic phenomena in MZ twins can lead to a better understanding of the discordances observed between the twins in relation to various disease states and can help forensic scientists discriminate between the otherwise genetically identical pair.

The genetic approach toward discriminating between males, either related or unrelated, with a high power of discrimination, is based on mutations. One such approach is the rapidly mutating Y chromosome STRs [75]. Another approach is to locate the rare mutations that occur shortly after the division of the human blastocyst into two, during the generation of MZ twins. However, this approach utilizes ultradeep next generation sequencing and is very expensive and laborious, but can discriminate between MZ twins utilizing somatic differences especially in semen samples [76,77].

The answer to this major forensic issue of MZ twin discrimination could be provided by epigenetic studies. A very recent study suggested the possibility of differentiating between MZ twins by genome-wide methylation analysis of their peripheral blood by BeadChip technology. The researchers identified 92 "best" CpG sites as candidates for demonstrating such differences in the methylation status. Studying these CpG sites could enhance our ability to distinguish between MZ twins based on the epigenome, since they both share the exact same DNA sequence. The possible forensic implications of these findings are of extreme importance since crime solving in cases where MZ twins are involved could be made simpler if the blood sample from a crime scene could be uniquely matched to only one individual from a pair of MZ twins [45]. Further, Li et al. (2011) used similar methodology to identify 377 CpG sites which showed discordance in the methylation profile between MZ twins [78]. In the same year it was reported that in 56.2% of 16 MZ twin pairs, major differences in DNA methylation were observed between twins and the authors concluded that their findings could be used for individual identification of MZ twins [79]. DNA methylation differences within MZ twins, but this time at birth, were also reported in a study carried out on 22 twin pairs by whole genome DNA methylation analysis. However, methylation differences were small in CpG-dense regions and were more significant the further away from CpG islands [80].

Methylation studies, carried out by BeadChip analysis of DNA extracted from peripheral blood, revealed that Mr Z and his brother could be distinguished. A targeted approach was then used to analyze the blood stain for the identified differences and the methylation pattern revealed was shown to match the Mr Z and not his brother and he was charged with the murder of Mrs A.

A court in Boston in 2014 is poised to consider the results of a test on an accused rapist and his identical twin brother, although the judge has yet to agree to its admissibility in the case.

6. CLINICAL PROFILING

Although Mr Z's motive could have been financial, trying to hurry the demise of his mother-in-law to benefit his wife and their family, it is also possible that an individual who is prepared to take such an action may also be suffering from some form of medical condition and epigenetic analysis can be utilized to provide other clues about the perpetrator of a crime.

Unraveling the genome-wide and site-specific methylation profile of stains found in crime scenes could help forensic scientists build up or give a hint of the clinical profile of the donor of the stain, whether it is the perpetrator's or belongs to another, unidentified person. Such a profile could include various psychiatric disorders or autoimmune diseases that have been studied in discordant MZ twins or even, as mentioned in the section on heavy metals, various disease states related to environmental exposure.

Some studies have focused on the differences between MZ twins in cases where one twin has developed a psychiatric disorder, such as schizophrenia or bipolar disorder, and the other twin does not share the same trait. Moreover, MZ twin studies have been used to identify possible biomarkers in various disease states such as cancer, diabetes, thyroidism, and drug abuse, since it is likely that many of the observable differences between the otherwise identical pair can be shown to be associated with disease state. Genome-wide DNA methylation of peripheral blood samples have revealed hypomethylation in a particular promoter, *ST6GALNAC1*, in affected psychotic individuals [81]. This specific locus is involved in protein glycosylation, as a member of the sialyltransferease family of molecules, with a key role in mediating cell–cell interactions. The mean DNA methylation difference was 6%, with some families showing a 20% reduction. These differences were supported by postmortem analysis of brain tissue samples which revealed even more extensive hypomethylation (>25%) in a subset of psychosis patients which were compared to healthy individuals. The authors concluded that hypomethylation of the specific locus leads to increased gene expression which is in agreement with findings from another study where duplication of this gene was reported in a case of schizophrenia [81,82]. Therefore, epigenetic alterations of the CpG sites of that specific gene promoter can lead to different phenotypes of psychiatric illnesses between MZ twins. Similar findings, although in a different genomic region, were reported in 2003 in a study performed on two pairs of schizophrenia discordant MZ twins. Peripheral blood samples were analyzed by direct sequencing of bisulfite treated DNA for the 5-end regulatory region of the dopamine D2 receptor gene (*DRD2*). The scientists concluded that the individual with the developed trait was epigenetically more similar to the individual from the other twin pair with the same trait, than to his unaffected discordant co-twin [83]. Global DNA hypomethylation was also reported in patients suffering from schizophrenia, when compared to controls [84]. However, this study failed to show any significant difference between pairs of MZ or dizygotic twins, in terms of whole genome methylation status. On the other hand, the findings suggested hypomethylation in the *SOX10* gene promoter between discordant twins which could be involved in the pathogenesis of schizophrenia and related psychiatric disorders. In 2006, a group of scientists identified a region, peptidylprolyl isomerase E-like (*PPIEL*), with reduced DNA methylation patterns, which correlated with increased expression levels in discordant MZ twins which suffered from bipolar II disorder [85]. The method used was methylation-sensitive representational difference analysis carried out in lymphocytes separated from peripheral blood. The suggested role of PPIEL is mainly neuronal, perhaps in neuroendocrine systems or dopamine neurotransmission, although the etiology remains unclear. Another psychiatric disorder, depression, has also been studied epigenetically. Depression has been associated with the serotonin transporter gene, *SLC6A4*. When the methylation status of this gene promoter was studied between discordant twins, it was found that a 10% hypermethylation correlated with a 4.4-fold increase in the Beck Depressive Inventory II score [86]. The study was carried out in peripheral blood leukocytes by pyrosequencing. Once again, a difference in the methylation status of a promoter region associated with a psychiatric disorder, was shown between MZ twins.

Differences in the methylation status (associated with different phenotypes) in otherwise genetically identical individuals, was also shown in relation to autism spectrum disorder. Moreover, the degree of methylation was correlated with autistic traits when these were quantitatively rated. Differentially methylated regions were found in genes previously associated with the disorder, as well as in unrelated genes. Given that epigenetic phenomena, such as DNA methylation, can alter gene expression, these findings could explain the predisposition to the disease for some individuals and the discordance observed in some cases between MZ twins [87]. It should of course be clarified that autism does not suggest criminal intent or behavior, but is just another example of how epigenetics could help identify a clinical condition.

Differences in DNA methylation have also been observed between MZ twins presenting with different phenotypes for psoriasis [88]. CD4+ and CD8+ cells were isolated from blood lymphocytes from MZ twins and cultured. Subsequent DNA methylation analysis, combined with gene expression analysis, revealed differences between affected and unaffected twins in genes involved in the pathogenesis of the disease such as *IL13, TNFSF11, PTHLH, ALOX5AP*. Another autoimmune disease, scleroderma, was investigated in terms of genome-wide DNA methylation in MZ twins by methylation-specific immunoprecipitation in peripheral blood DNA [89]. It was concluded that the affected twins showed either hypermethylation or hypomethylation in various gene-specific sites on the X chromosome, explaining the female predominance of the disease. The authors were able to link these epigenetic phenomena with the mode of onset and action of the disease and in particular cell proliferation, apoptosis, inflammation and oxidative stress. Furthermore, they suggested that genome-wide methylation studies could be used as biomarkers for scleroderma since affected individuals exhibited different findings compared to unaffected individuals.

In order to study the role of epigenetics in MZ twins discordant for asthma, regulatory and effector T cell (Treg and Teff) subsets were isolated from peripheral blood samples from 21 twin pairs [90]. Increased methylation was observed (by bisulfite treatment of extracted DNA and sequencing) in two loci, namely Forkhead box P3 (*FOXP3*) and interferon gamma-c (*IFNc*) in the affected individuals. The observed hypermethylation correlated with decreased protein expression and therefore impaired function of the Treg and Teff respectively, explaining the clinical manifestation of asthma. Different methylation profiles were also reported in 60 gene regions in MZ twins discordant for primary biliary cirrhosis [91]. Fourteen of these genes were also differentially expressed. Once again, a different methylation profile between MZ twins was reported and possible biomarkers for the disease were identified.

7. CONCLUSIONS

This hypothetical forensic case has illustrated the potential use of an epigenetic approach to assist an investigator within the criminal justice system. Only a few of the potential applications have been mentioned and it is likely that it will be some time before such an approach could become reality. Most are in their infancy, with respect to forensic analysis and, to our knowledge, none have been presented in court. There are several limitations to the use of these methods within criminal justice in the future:

- Availability of suitable material
- Ethical and legal considerations
- Validation
- Court acceptance

Many molecular biology techniques in current use require availability of significant amounts of genomic DNA in a pure state. Forensic stains will often be limited in size, potentially containing DNA in the low picogram amounts, often degraded, or recovered from substrates that limit analysis, and often involving mixtures from two or more individuals. Further, any investigation of this poor quality material must prioritise the ability to get a DNA profile, sufficient to identify the individual, as it is the latter that will finally provide sufficient evidence that the accused is the responsible person in a crime, rather than someone who just fits the profile. The acceptance of use of an approach that reveals private information about an individual, such as their medical history, will be a significant ethical challenge. Many scientists researching the use of epigenetic approaches in forensic analysis focus their proposed tools on externally visible characteristics (EVCs) to avoid this privacy issue. Legislation would need to be changed—some jurisdictions deny scientists analyzing areas of DNA from within a gene and, of course, even a prediction of an EVC will be looking within a gene. Even if these hurdles are overcome, any technique to be used within the criminal justice system must undergo significant validation to ensure that it is robust, sensitive, and accurate. The forensic capabilities promised by epigenetics are probably some way off.

References

[1] Jeffreys AJ, Wilson V, Thein SL. 'Hypervariable/minisatellite' regions in human DNA. Nature 1985;314:67–73.

[2] Edwards A, Civitello A, Hammond HA, Caskey CT. DNA typing and genetic mapping with trimeric and tetrameric tandem repeats. Am J Hum Genet 1991;49:746–56.

[3] Kimpton CP, Gill P, Walton A, Urquhart A, Millican ES, Adams M. Automated DNA profiling employing multiplex amplification of short tandem repeat loci. PCR Methods Appl 1993;3:13–22.

[4] Raymond MA, Lecompte MH, Gunn PR. Forensic biology. In: Freckleton L, Selby H, editors. Expert evidence. Melbourne: Thomson Reuters; 2011. p. 81.

[5] Alvarez M, Juusola J, Ballantyne J. An mRNA and DNA co-isolation method for forensic casework samples. Anal Biochem 2004;336:289–98.

[6] Haas C, Hanson E, Bar W, Banemann R, Bento AM, Berti A, et al. mRNA profiling for the identification of blood – results of a collaborative EDNAP exercise. Foren Sci Int Genet 2011;5:21–6.

[7] Hanson EL, Ballantyne J. Highly specific mRNA biomarkers for the identification of vaginal secretions in sexual assault investigations. Sci Justice 2013;53:14–22.

[8] Zubakov D, Boersma AWM, Choi Y, Kuijk PF, Wiemer EAC, Kayser M. MicroRNA markers for forensic body fluid identification obtained from microarray screening and quantitative RT-PCR confirmation. Int J Leg Med 2010;124:217–26.

[9] Espada, Esteller M. DNA methylation and the functional organization of the nuclear compartment. Stem Cell Devel Biol 2010;21:238–46.

[10] Newell-Price, Clark A, King P. DNA methylation and silencing of gene expression. Trends Endo Metab 2000;11:142–8.

[11] Virkler K, Ledney IK. Analysis of body fluids for forensic purposes: from laboratory testing to non-destructive rapid confirmatory identification at a crime scene. Foren Sci Int 2009;188:1.

[12] Lindenbergh A, de Pagter M, Ramdayal G, Visser M, Zubakov D, Kayser M, et al. A multiplex (m)RNA-profiling system for the forensic identification of body fluids and contact traces. Foren Sci Int Genet 2012;6:565–77.

[13] Haas C, Hanson E, Anjos MJ, Ballantyne KN, Banemann R, Bhoelai B, et al. RNA/DNA co-analysis from human menstrual blood and vaginal secretion stains: results of a fourth and fifth collaborative EDNAP exercise. Foren Sci Int Genet 2014;8:203–12.

[14] Fleming RI, Harbison S. The development of a mRNA multiplex RT-PCR assay for the definitive identification of body fluids. Foren Sci Int Genet 2010;4:244–56.

[15] Plachot C, Lelievre SA. DNA methylation control of tissue polarity and cellular differentiation in the mammary epithelium. Exp Cell Res 2004;298:122–32.

[16] Song F, Mahmood S, Ghosh S, Liang P, Smiraglia DJ, Nagase H, et al. Tissue specific differentially methylated regions (TDMR): changes in DNA methylation during development. Genomics 2009;93:130–9.

[17] llingworth R, Kerr A, DeSousa D, Jorgensen H, Ellis P, Stalker J, et al. A novel CpG island set identifies tissue-specific methylation at developmental gene loci. PloS Biol 2008;6:37–51.

[18] De Bustos C, Ramos E, Young JM, Tran RK, Menzel U, Langford CF, et al. Tissue-specific variation in DNA methylation levels along human chromosome 1. Epigenetics Chromatin 2009;2:7.

[19] Frumkin D, Wasserstrom A, Budowle B, Davidson A. DNA methylation-based forensic tissue identification. Foren Sci Int Genet 2011;5:517–24.

[20] Gomes I, Kohlmeier F, Schneider PM. Genetic markers for body fluid and tissue identification in forensics. Foren Sci Int Genet Suppl Ser 2011;3:e469–70.

[21] Wasserstrom A, Frumkin D, Davidson A, Shpitzen M, Herman Y, Gafny R. Demonstration of DSI-semen–A novel DNA methylation-based forensic semen identification assay. Foren Sci Int Genet 2013;7:136–42.

[22] LaRue BL, King JL, Budowle B. A validation study of the Nucleix DSI-Semen kit–a methylation-based assay for semen identification. Int J Leg Med 2013;127:299–308.

[23] Lee HY, Park MJ, Choi A, An JH, Yang WI, Shin KJ. Potential forensic application of DNA methylation profiling to body fluid identification. Int J Leg Med 2012;126:55–62.

[24] An JH, Choi A, Shin KJ, Yang WI, Lee HY. DNA methylation-specific multiplex assays for body fluid identification. Int J Leg Med 2013;127:35–43.

[25] Choi A, Shin KJ, Yang WI, Lee HY. Body fluid identification by integrated analysis of DNA methylation and body fluid-specific microbial DNA. Int J Leg Med 2013;128:33–41.

[26] Paliwal A, Vaissiere T, Herceg Z. Quantitative detection of DNA methylation states in minute amounts of DNA from body fluids. Methods 2010;52:242–7.

[27] Madi T, Balamurugan K, Bombardi R, Duncan G, McCord B. The determination of tissue-specific DNA methylation patterns in forensic biofluids using bisulfite modification and pyrosequencing. Electrophoresis 2012;33:1736–45.

[28] Ma LL, Yi SH, Huang DX, Mei K, Yang RZ. Screening and identification of tissue-specific methylation for body fluid identification. Foren Sci Int Genet - Suppl Ser 2013;4:e37–8.

[29] Bartling CM, Hester ME, Bartz J, Heizer Jr E, Faith SAA. Next-generation sequencing approach to epigenetic-based tissue source attribution. Electrophoresis 2014;35:3096–101.

[30] Menni C, Kastenmuller G, Petersen AK, Bell JT, Psatha M, Tsai PC, et al. Metabolomic markers reveal novel pathways of ageing and early development in human populations. Int J Epidem 2013;42:1111–9.

[31] Glass D, Vinuela A, Davies MN, Ramasamy A, Parts L, Knowles D, et al. Gene expression changes with age in skin, adipose tissue, blood and brain. Genome Biol 2013;14:R75.

[32] Teschendorff AE, Menon U, Gentry-Maharaj A, Ramus SJ, Weisenberger DJ, Shen H, et al. Age-dependent DNA methylation of genes that are suppressed in stem cells is a hallmark of cancer. Genome Res 2010;20:440–6.

[33] Tsai HC, Baylin SB. Cancer epigenetics: linking basic biology to clinical medicine. Cell Res 2011;21:502–17.

[34] Lynnerup N, Kjeldsen H, Zweihoff R, Heegaard S, Jacobsen C, Heinemeier J. Ascertaining year of birth/age at death in forensic cases: a review of conventional methods and methods allowing for absolute chronology. Foren Sci Int 2010;201:74–8.

[35] Meissner C, Ritz-Timme S. Molecular pathology and age estimation. Foren Sci Int 2010;203:34–43.

[36] Zubakov D, Liu F, van Zelm MC, Vermeulen J, Oostra BA, van Duijin CM, et al. Estimating human age from T-cell DNA rearrangements. Curr Biol 2010;20:1–2.

[37] Ou X, Zhao H, Sun H, Yang Z, Xie B, Shi Y, et al. Detection and quantification of the age-related sjTREC decline in human peripheral blood. Int J Leg Med 2011;125:603–8.

[38] Gentilini D, Mari D, Castaldi D, Remondini D, Ogliari G, Ostan R, et al. Role of epigenetics in human aging and longevity: genome-wide DNA methylation profile in centenarians and centenarians' offspring. Age (Dordr) 2013;35:1961–73.

[39] Bell JT, Tsai PC, Yang TP, Pidsley R, Nisbet J, Glass D, et al. Epigenome-wide scans identify differentially methylated regions for age and age-related phenotypes in a healthy ageing population. PLoS Genet 2012;8:e1002629.

[40] Day K, Waite LL, Thalacker-Mercer A, West A, Bamman MM, Brooks JD, et al. Differential DNA methylation with age displays both common and dynamic features across human tissues that are influenced by CpG landscape. Genome Biol 2013;14:R102.

[41] Teschendorff AE, West J, Beck S. Age-associated epigenetic drift: implications, and a case of epigenetic thrift? Hum Mol Genet 2013;22:R7–15.

[42] Martino D, Loke YJ, Gordon L, Ollikainen M, Cruickshank MN, Saffery R, et al. Longitudinal, genome-scale analysis of DNA methylation in twins from birth to 18 months of age reveals rapid epigenetic change in early life and pair-specific effects of discordance. Genome Biol 2013;14:R42.

[43] Alisch RS, Barwick BG, Chopra P, Myrick LK, Satten GA, Conneely KN, et al. Age-associated DNA methylation in pediatric populations. Genome Res 2012;22:623–32.

[44] Bocklandt S, Lin W, Sehl ME, Sanchez FJ, Sinsheimer JS, Horvath S, et al. Epigenetic predictor of age. PLoS ONE 2011;6:1–6.

[45] Li C, Zhao S, Zhang N, Zhang S, Hou Y. Differences of DNA methylation profiles between monozygotic twins' blood samples. Mol Biol Rep 2013;40:5275–80.

[46] Hannum G, Guinney J, Zhao L, Zhang L, Hughes G, Sadda S, et al. Genome-wide methylation profiles reveal quantitative views of human aging rates. Mol Cell 2013;49:359–67.

[47] Weidner CI, Lin Q, Koch CM, Eisele L, Beier F, Ziegler P, et al. Aging of blood can be tracked by DNA methylation changes at just three CpG sites. Genome Biol 2014;15:1–11.

[48] Yi SH, Xu LC, Mei K, Yang RZ, Huang DX. Isolation and identification of age-related DNA methylation markers for forensic age-prediction. Foren Sci Int Genet 2014;11:117–25.

[49] Hernandez DG, Nalls MA, Gibbs JR, Arepalli S, van der Brug M, Chong S, et al. Distinct DNA methylation changes highly correlated with chronological age in the human brain. Hum Mol Genet 2011;20:1164–72.

[50] Koch CM, Wagner W. Epigenetic-aging-signature to determine age in different tissues. Aging 2011;3:1–10.

[51] Horvath S. DNA methylation age of human tissues and cell types. Genome Biol 2013;14:1.

[52] Salnikow K, Zhitkovich A. Genetic and epigenetic mechanisms in metal carcinogenesis and cocarcinogenesis: nickel, arsenic, and chromium. Chem Res Toxicol 2008;21:28–44.

[53] Chanda S, Dasgupta UB, Guhamazumder D, Gupta M, Chaudhuri U, Labiri S, et al. DNA hypermethylation of promoter of gene p53 and p16 in arsenic-exposed people with and without malignancy. Toxicol Sci 2006;89:431–7.

[54] van Breda SG, Claessen SM, Lo K, van Herwijnen M, Brauers KJ, Lisanti S, et al. Epigenetic mechanisms underlying arsenic-associated lung carcinogenesis. Arch Toxicol September 9, 2014 [Epub ahead of print].

[55] Zhang AH, Bin HH, Pan XL, Xi XG. Analysis of p16 gene mutation, deletion and methylation in patients with arseniasis produced by indoor unventilated-stove coal usage in Guizhou, China. J Toxicol Environ Health 2007;70:970–5.

[56] Bodoor K, Haddad Y, Alkhateeb A, Al-Abbadi A, Dowairi M, Magableh A, et al. DNA hypermethylation of cell cycle (p15 and p16) and apoptotic (p14, p53, DAPK and TMS1) genes in peripheral blood of leukemia patients. Asian Pac J Cancer Prev 2014;15:75–84.

[57] Qu Y, Dang S, Hou P. Gene methylation in gastric cancer. Clin Chim Acta 2013;424:53–65.

[58] Majumdar S, Chanda S, Ganguli B, Mazumder DN, Lahiri S, Dasgupta UB. Arsenic exposure induces genomic hypermethylation. Environ Toxicol 2010;25:315–8.

[59] Bailey KA, Wu MC, Ward WO, Smeester L, Rager JE, Garcia-Vargas G, et al. Arsenic and the epigenome: interindividual differences in arsenic metabolism related to distinct patterns of DNA methylation. J Biochem Mol Toxicol 2013;27:106–15.

[60] Treas J, Tyagi T, Singh KP. Chronic exposure to arsenic, estrogen, and their combination causes increased growth and transformation in human prostate epithelial cells potentially by hypermethylation-mediated silencing of MLH1. Prostate 2013;73:1660–72.

[61] Baylin SB, Herman JG. DNA hypermethylation in tumorigenesis: epigenetics joins genetics. Trends Genet 2000;16:168–74.

[62] Brocato J, Costa M. 10th NTES Conference: nickel and arsenic compounds alter the epigenome of peripheral blood mononuclear cells. J Trace Elem Med Biol 2014;S0946–672X(14):00057–61.

[63] Chu F, Ren X, Chasse A, Hickman T, Zhang L, Yuh J, et al. Quantitative mass spectrometry reveals the epigenome as a target of arsenic. Chem Biol Interact 2011;192:113–7.

[64] Verstraeten SV, Aimo L, Oteiza PI. Aluminium and lead: molecular mechanisms of brain toxicity. Arch Toxicol 2008;82:789–802.

[65] Kovatsi L, Georgiou E, Ioannou A, Haitoglou C, Tzimagiorgis G, Tsoukali H, et al. p16 promoter methylation in Pb2+ -exposed individuals. Clin Toxicol 2010;48:123–8.

[66] Li C, Yang X, Xu M, Zhang J, Sun N. Epigenetic marker (LINE-1 promoter) methylation level was associated with occupational lead exposure. Clin Toxicol 2013;51:225–9.

[67] Pilsner JR, Hu H, Ettinger A, Sanchez BN, Wright RO, Cantonwine D, et al. Influence of prenatal lead exposure on genomic methylation of cord blood DNA. Environ Health Perspect 2009;117:1466–71.

[68] Valinluck V, Tsai HH, Rogstad DK, Burdzy A, Bird A, Sowers LC. Oxidative damage to methyl-CpG sequences inhibits the binding of the methyl-CpG binding domain (MBD) of methyl-CpG binding protein 2 (MeCP2). Nucleic Acids Res 2004;32:4100–8.

[69] Yi P, Melnyk S, Pogribna M, Pogribny IP, Hine RJ, James SJ. Increase in plasma homocysteine associated with parallel increases in plasma S-adenosylhomocysteine and lymphocyte DNA hypomethylation. J Biol Chem 2000;175:29318–23.

[70] Zawia NH, Lahiri DK, Cardozo-Pelaez F. Epigenetics, oxidative stress, and Alzheimer disease. Free Radic Biol Med 2009;46:1241–9.

[71] Wu J, Basha MR, Brock B, Cox DP, Cardozo-Pelaez F, McPherson CA, et al. Alzheimer's disease (AD)-like pathology in aged monkeys after infantile exposure to environmental metal lead (Pb): evidence for a developmental origin and environmental link for AD. J Neurosci 2008;28:3–9.

[72] Wu J, Basha MR, Zawia NH. The environment, epigenetics and amyloidogenesis. J Mol Neurosci 2008;34:1–7.

[73] Waly M, Olteanu H, Banerjee R, Choi SW, Mason JB, Parker BS, et al. Activation of methionine synthase by insulin-like growth factor-1 and dopamine: a target for neurodevelopmental toxins and thimerosal. Molec Psych 2004;9:358–70.

[74] Fragou D, Fragou A, Kouidou S, Njau S, Kovatsi L. Epigenetic mechanisms in metal toxicity. Toxicol Mech Meth 2011;21:343–52.

[75] Ballantyne KN, Ralf A, Aboukhalid R, Achakzai NM, Anjos MJ, Aybu Q, et al. Towards male individualization with rapidly mutating Y-chromosomal STRs. Hum Mutat 2014;35:1021–32.

[76] Budowle B. Molecular genetic investigative leads to differentiate monozygotic twins. Investig Genet 2014;5:11.

[77] Weber-Lehmann J, Schilling E, Gradl G, Richter DC, Wiehler J, Rolf B. Finding the needle in the haystack: differentiating "identical" twins in paternity testing and forensics by ultra-deep next generation sequencing. Foren Sci Int Genet 2014;9:42–6.

[78] Li C, Zhang S, Que T, Li L, Zhao S. Identical but not the same: the value of DNA methylation profiling in forensic discrimination within monozygotic twins. Foren Sci Int Genet Suppl Ser 2011;3:e337–8.

[79] Sahin K, Yilmaz S, Temel A, Gozukirmizi N. DNA methylation analyses of monozygotic twins. Curr Opin Biotech 2011;22:S105.

[80] Gordon L, Joo JE, Powell JE, Ollikainen M, Novakovic B, Li X, et al. Neonatal DNA methylation profile in human twins is specified by a complex interplay between intrauterine environmental and genetic factors, subject to tissue-specific influence. Genome Res 2012;22:1395–406.

[81] Dempster EL, Pidsley R, Schalkwyk LC, Owens S, Georgiades A, Kane F, et al. Disease-associated epigenetic changes in monozygotic twins discordant for schizophrenia and bipolar disorder. Hum Mol Genet 2011;20:4786–96.

[82] Xu B, Roos JL, Levy S, van Rensburg EJ, Gogos JA, Karayiorgou M. Strong association of de novo copy number mutations with sporadic schizophrenia. Nat Gen 2008;40:880–5.

[83] Petronis A, Gottesman II, Kan P, Kennedy JL, Basile VS, Paterson AD, et al. Monozygotic twins exhibit numerous epigenetic differences: clues to twin discordance? Schizophr Bull 2003;29:169–78.

[84] Bonsch D, Wunschel M, Lenz B, Janssen G, Weisbrod M, Sauer H. Methylation matters? Decreased methylation status of genomic DNA in the blood of schizophrenic twins. Psych Res 2012;198:533–7.

[85] Kuratomi G, Iwamoto K, Bundo M, Kusumi I, Kato N, Iwata N, et al. Aberrant DNA methylation associated with bipolar disorder identified from discordant monozygotic twins. Mol Psych 2008;13:429–41.

[86] Zhao J, Goldberg J, Bremner JD, Vaccarino V. Association between promoter methylation of serotonin transporter gene and depressive symptoms: a monozygotic twin study. Psychosom Med 2013;75:523–9.

[87] Wong CC, Meaburn EL, Ronald A, Price TS, Jeffries AR, Schalkwyk LC, et al. Methylomic analysis of monozygotic twins discordant for autism spectrum disorder and related behavioural traits. Mol Psych 2014;19:495–503.

[88] Gervin K, Vigeland MD, Mattingsdal M, Hammere M, Nygård H, Olsen AO, et al. DNA methylation and gene expression changes in monozygotic twins discordant for psoriasis: identification of epigenetically dysregulated genes. PLoS Genet 2012;8:e1002454.

[89] Selmi C, Feghali-Bostwick CA, Lleo A, Lombardi SA, De Santis M, Cavaciocchi F, et al. X chromosome gene methylation in peripheral lymphocytes from monozygotic twins discordant for scleroderma. Clin Exp Immun 2012;169:253–62.

[90] Runyon RS, Cachola LM, Rajeshuni N, Hunter T, Garcia M, Ahn R, et al. Asthma discordance in twins is linked to epigenetic modifications of T cells. PLoS One 2012;7:e48796.

[91] Selmi C, Cavaciocchi F, Lleo A, Cheroni C, De Francesco R, Lombardi SA, et al. Genome-wide analysis of DNA methylation, copy number variation, and gene expression in monozygotic twins discordant for primary biliary cirrhosis. Front Immun 2014;5:128.

9

Epigenetics of Personalized Toxicology

Alexandre F. Aissa, Lusânia M.G. Antunes

Department of Clinical Analyses, Toxicology and Food Sciences, School of
Pharmaceutical Sciences of Ribeirão Preto, University of São Paulo (USP),
Ribeirão Preto, São Paulo, Brazil

OUTLINE

OUTLINE

This chapter focuses on the epigenetics of personalized toxicology, which has led to the development of proposals for new approaches in the treatment of toxic effects based on genetic variation. What is known about personalized treatments for human intoxications? How may basic knowledge of epigenetics assist in understanding the effects of exposure to toxicants? The ultimate goals of the epigenetics of personalized toxicology are to understand and translate science into practical personalized health promotion.

1. INTRODUCTION

Genetic mechanisms involve irreversible changes in the DNA sequence, including deletions and chromosomal and gene mutations, and are associated with the initiation and progression of many abnormalities [1]. These phenomena may affect the sequence of nucleotides, thereby altering the expression of genes and facilitating the onset of illnesses. However, genetic variation can explain the risk of the development of only a few of a large number of nontransmissible disorders [2]. Thus, it is known that gene regulation by epigenetic mechanisms has equal importance in the development of diseases [3]. The term *epigenetics* is defined as the study of inheritable changes in gene expression and chromatin configuration that do not alter the nucleotide sequence [4].

According to Nebert and colleagues [5], toxicogenetics is the study of how genetic variability can influence the individual response to a toxicant, while toxicogenomics emphasizes the effects of toxicants on gene expression and the downstream effects of this alteration [5].

Scientists have long been investigating genotoxicity and its effects on the sequence of nucleotides as major factors that are related to the effects of environmental chemicals on human health. However, major advances have been made in the several decades since knowledge beyond the genome—the epigenome—was discovered. Thus, epigenomics is the study of the set of epigenetic modifications in the genome. Therefore, two new concepts must be considered in the study of the toxicity of compounds that are related to epigenetic mechanisms: toxicoepigenetics and toxicoepigenomics.

The division of terms aims to facilitate the handling and classification of information from studies that investigate each of these concepts but that address one theme: how toxicology and epigenetic mechanisms are related. In this regard, toxicoepigenetics is the study of how epigenetic variability influences the individual response to the toxicant and toxicoepigenomics is related to the effects of toxicants on epigenetic marks and, consequently, to the alteration of gene expression.

Similar to genetic variation, in which polymorphisms of genes cause different individual responses after the same dose and period of exposure to a toxicant, epigenetic changes can also provide a mechanism for response variation. These mechanisms lead to a need for studies that associate epigenetic changes with toxicological responses.

There are no studies describing the personalized epigenetic profile in humans in regions of DNA near genes that are related to the absorption, metabolism, transport, and excretion of toxicants. However, Feinberg and colleagues [6] demonstrated that these signatures are present in gene regions that are correlated with body mass index. The authors examined global DNA methylation in samples from 74 participants who provided two DNA samples at time points that were 11 years apart. These authors identified 227 regions in which methylation patterns are highly variable among individuals. Moreover, these authors were able to identify two types of variably methylated regions (VMRs): VMRs that were dynamic over the intervening 11 years and VMRs that remained stable over time.

The study of Feinberg and colleagues [6] gives an example of using epigenetic signatures to identify genes that may have changes in their expression due to epigenetic variation. Thus, individuals with the same allele may exhibit significant differences in toxicological response not due to genetic variability but to its personalized epigenetic signature. According to the two results that were related to VMRs, some changes are dynamic, varying throughout life, while others are relatively stable [6]. Therefore, we can assume that epigenetic variations after exposure may be stable throughout the life of an individual, making it part of his or her personalized epigenetics or acquired epigenetic variability. Some epigenetic modifications that are induced by exposure may therefore change how the individual responds after new environmental exposure. This information makes the relationship between toxicoepigenetics and toxicoepigenomics even closer, i.e., they are dependent on one another.

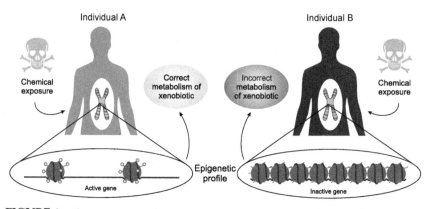

FIGURE 1   The epigenetic profile (e.g., chromatin configuration) regulates gene expression, compromising the correct metabolism of xenobiotics. Thus, individuals with two different epigenetic patterns exposed to the same chemical may have different toxicological responses.

Although studies are scarce that identify distinct epigenetic signatures in genes related to the absorption, metabolism, transport, and excretion of toxicants, several studies with candidate genes have demonstrated that this variation is also present in this group of genes. It has been demonstrated that these genes show a high degree of interindividual variation that is attributable to regulation by diverse genetic, nongenetic, and epigenetic mechanisms [7].

Thereby, individual epigenetic variation is an increasingly recognized mechanism for altered toxicant response. Therefore, personalized toxicology could be defined as an approach based on the knowledge about the epigenetic signature from an individual that could permit a better treatment decision that is tailored to the individual patient after toxic exposure [2]. This chapter focuses on the epigenetics of personalized toxicology, which has led to the development of proposals for new approaches in the treatment of toxic effects based on genetic variation and epigenetic variation (Figure 1). What is known about personalized treatments for human intoxications? How may basic knowledge in epigenetics assist in understanding the effects of exposure to toxicants? The ultimate goals of the epigenetics of personalized toxicology are to understand and translate science into practical personalized health promotion.

2. BASIC EPIGENETICS

The cells in a multicellular organism carry the same genetic information encoded in their DNA sequence but demonstrate high morphological and functional diversity. This variety of functions is influenced by the control of the differentiation of gene expression in each cell type by epigenetic

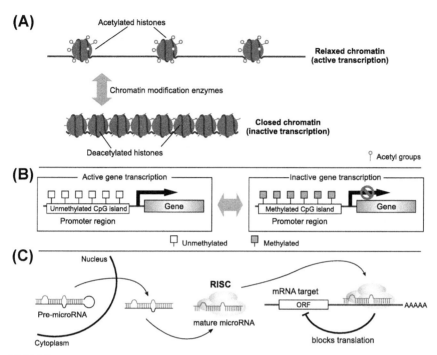

FIGURE 2 Schematic representation of epigenetic modifications. (A) Removal of acetyl groups from histone tails, inducing the histones to wrap more tightly around the DNA and interfering with the transcription of genes by blocking access by transcription factors. (B) Addition of a methyl group to CpG dinucleotides, blocking the promoters at which activating transcription factors should bind, repressing the transcription of the gene. (C) MicroRNA binding to complementary sequences on the target mRNA results in the repression of target gene expression through a block in protein translation or altered mRNA stability. ORF, open reading frame; RISC, RNA-induced silencing complex.

mechanisms [8]. Accordingly, epigenetics has elucidated several pathways in the study of genetic programming as related to cellular events, such as differentiation, cell response to stress, and abnormal cell growth. These studies have resulted in significant advances in the understanding of the development of human diseases [9].

The main epigenetic events involve DNA methylation, noncoding RNAs, and histone modifications (Figure 2). These processes affect the stability of gene transcription, the folding of the DNA strand, the positioning of nucleosomes in chromatin organization, and ultimately cellular organization. The configuration of chromatin plays an important role in gene expression. Regions of less-compacted DNA are more accessible to transcription. Thus, the same gene sequence can often be expressed or silenced according to chromatin compaction [10]. Despite all of the epigenetic pathways modulating the transcription of chromatin, the molecular events by which epigenetic information is transmitted through cell division are not yet fully understood.

In multicellular organisms, the ability of epigenetic marks to persist during development to be potentially transmitted to future generations may cause the wide variety of different phenotypes that arise from the same ancestral genotype [11]. Furthermore, epigenetic changes occurring during embryonic development may persist throughout adult life. A long period between environmental stimulus and the onset of disease is widely recognized in the etiology of certain cancers and should be considered in other approaches, for example, in the toxicology response.

A great deal of evidence, including epidemiological data and data from extensive clinical and experimental studies, has indicated that events occurring early in life play a role in an organism's susceptibility to chronic diseases. Studies using animal models have shown that the induction and stability of the change in the phenotype of the offspring due to epigenetic changes involve the DNA methylation and covalent modification of histones, indicating that these epigenetic changes are highly gene-specific [12].

During embryonic development, every cell, tissue, and organ acquires different prototypes of gene expression that are mediated by epigenetic factors, including DNA methylation. Indeed, the mammalian genome undergoes a profound reprogramming of methylation patterns in germ cells during the early stages of the preimplantation embryo [13]. At the time of fertilization, the methylation pattern that is obtained from the parents is deleted by a process called demethylation, which affects the entire genome; during deployment, standards are set again through de novo methylation by the DNMT3B enzyme [14]. Thus, embryonic and fetal exposure to chemicals through the maternal organism can affect these epigenetic phenomena, influencing fetal development and even the individual's health in adulthood [15].

DNA methylation is catalyzed by the family of enzymes called DNA methyltransferases (i.e., DNMT1, DNMT3A, DNMT3B, and DNMT3L), resulting in changes in chromatin structure and changes in gene expression [16]. Furthermore, acetyltransferase, deacetylase, and methyltransferase enzymes can modify the N-terminal tails of histones, which can in turn alter the accessibility of transcription factors, thereby preventing or permitting gene expression. Changes in the expression of genes encoding these enzymes could interfere with their activity in DNA methylation, thereby creating a chain of events whereby the gene expression of several genes is compromised [17,18]. Among eukaryotic species, methylation occurs predominantly in the cytosines at the 5' end of guanines; these cytosines are known as CpG dinucleotides, and their distribution in the mammalian genome sequence is not random [19].

It has been suggested that more than half of the CpG islands participate in transcriptional regulation [19]. There is a common inverse relationship between the degree of methylation of a CpG island and the regulation of gene transcription [20]. Two mechanisms have been proposed to explain the possible influence of cytosine methylation on gene expression [21]. First, the addition of the methyl group to cytosine causes a change in the DNA molecule, preventing the binding of transcription factors with recognition sites in CpG dinucleotides. A class of proteins known as "methyl-binding proteins" then bind to methylated CpGs and prevent the access of transcription factors to their regulatory elements. Both of these mechanisms will thus prevent the gene from being expressed.

In addition to gene-specific CpG island methylation, the genome-wide DNA methylation of repetitive elements is a major contributor of global DNA methylation patterns and has been investigated in relation to a variety of human diseases. Approximately 45% of the human genome is composed of repetitive elements that include 1 million Alu sequences occupying approximately 10% of the genome and long interspersed nuclear element-1 (LINE1) elements that represent a huge genomic percentage [22]. In addition, tandem repeats, including DNA satellites (i.e., SAT-α), as well these Alu and LINE1 interspersed repeated DNA sequences, are generally located in centromeres or centromere-adjacent heterochromatin and contain many CpG dinucleotides. The methylation signature of these sequences is a major contributor to global DNA methylation patterns [23].

Changes in epigenetic mechanisms promote the development of an abnormal phenotype and the development of genetic events, such as DNA breaks, chromosomal instability, and mutations, that contribute to the development of diseases such as cancer [4]. Improper DNA methylation such as hypermethylation or hypomethylation that results in chromosomal instability is the most common epigenetic modification in human cancers. Studies have demonstrated that aberrant DNA methylation is present in many steps of carcinogenesis, even in the late stages of chronic inflammation [24]. In fact, global DNA methylation patterns have been investigated in relation to a variety of human diseases [25].

Centromeric regions of chromosomes are heterochromatic and are within repeated sequences in tandem. These regions are epigenetically silenced by the methylation of histones and by the hypermethylation of global DNA, thereby permitting a low frequency of recombination and a repression of transcription [26,27]. However, the hypomethylation of repetitive DNA sequences in the centromeric and pericentromeric regions of chromosomes is highly related to chromosomal instability [22]. The hypomethylation of DNA can affect the binding of kinetochore spindle fibers, preventing correct chromosome segregation and inducing chromosomal instability [28].

Posttranslational modifications of histones that regulate chromatin configuration are closely related to gene expression [29]. The modification of histones is an important mechanism for regulation processes, such as gene expression, replication, DNA repair, chromatin condensation, segregation, and apoptosis. Histones are no longer considered simple proteins that package DNA but rather are currently recognized as regulators of chromatin dynamics. These proteins are subject to the methylation and acetylation of lysine, serine and threonine phosphorylation, the ubiquitination of lysine, and glycosylation, carboxylation, and sumoylation, which are carried out through chromatin alterations [30,31].

Moreover, modifications to euchromatin, such as through the acetylation of histones H3 and H4 or the di- or trimethylation of lysine 4 on histone H3, are associated with active transcription. In addition, the di- or trimethylation of lysine 9 on histone H3 and the trimethylation of lysine 27 on histone H3 are involved in gene repression. Histone acetyltransferases and histone methyltransferases add acetyl and methyl groups, respectively, to the tails of histones, while histone deacetylase and demethylase enzymes remove such groups [32,33].

MicroRNAs (miRNAs) can be used to support the emerging idea that noncoding RNAs are as important for gene regulation as are proteins. Since the first discovery of miRNAs as regulators of the development of the nematode *Caenorhabditis elegans*, thousands of genes that are related to miRNAs have been identified in animal and plant genomes [34]. It is believed that more than half of the human transcriptome is regulated by miRNAs, indicating that this mechanism of posttranscriptional control is incorporated at almost all levels of the gene cascade [35,36]. Changes in the activity of miRNA contribute to the development of many diseases, including cancer, cardiovascular disease, and neurological disorders. Furthermore, owing to the regulatory power of this mechanism, miRNAs have been tested as targets for the treatment of certain diseases and infections [37].

MiRNAs are not translated into proteins; however, they can modulate protein synthesis by binding to the 3′ untranslated region (UTR) of the transcript encoding the protein that contains the sequence that is complementary to the miRNA sequence region, called the "seed," which contains two to seven nucleotides. The link between miRNA and mRNA obeys the law of Watson–Crick complementarity but does not always require complete homology binding. Indeed, complete homology between the miRNA and its target mRNA initiates mRNA degradation [38].

This delicate control over gene expression by miRNA suggests a broad application of miRNA–mRNA interactions with possible divergent (one miRNA with many targets) or convergent (many miRNAs with a single target) mechanisms as well as a complex regulation of protein synthesis [39]. In addition, as partial pairing between an miRNA and a target is often sufficient, a huge list of genes as potential targets with partially complementary sequences can be predicted, the production of which is not a simple task [40].

It is evident in animals that most miRNAs form only partial connections with their targets; most target mRNAs that have been studied to date are regulated through 3′ UTR interactions. Most cases involve multiple links that contain nucleotide mismatches and bulges [40]. The central region of the miRNA may also bind to a target. The apparent flexibility in the rules of connection between the miRNA and its target suggests that other factors in addition to capacity pairing can mediate interactions between miRNAs and their targets [40].

3. INDIVIDUAL EPIGENETIC VARIATION IN RESPONSE TO TOXICANTS

Epigenetic variability can be acquired before birth when the mother's body is exposed to environmental factors that are capable of inducing epigenetic modifications in both the maternal organism and the embryo. Instead of carrying a change in the sequence of nucleotides, such as a mutation, the individual can acquire an aberrant epigenetic pattern, such as an altered pattern of DNA methylation or histone modification. Considering that epigenetic changes are transmitted to the daughter cells after division, this modification may persist throughout the adult life of an individual that was exposed only through the maternal environment during initial development.

This interindividual epigenetic variation, affecting multiple cell types, involves what are called "metastable epialleles" [41]. The environment, such as maternal nutrition during early pregnancy, induces persistent and systemic epigenetic changes in metastable epialleles in animals [42–44]. It was demonstrated by Dominguez-Salas and colleagues [45] that this effect can also be observed in humans. The authors showed that variations in methyl-donor nutrient intake during early pregnancy influenced 13 relevant plasma biomarkers; several of these maternal biomarkers predict the postnatal pattern of methylation in metastable epialleles in lymphocytes and hair follicles in infants [45]. According to the authors, these results were the first to show that in humans, individual epigenetic variation may be induced during intrauterine development. That is, changes in the methylation pattern of genes, as well as the modification of the histones that are localized in these genes, induce a mechanism resulting in individual variability.

When two subjects are exposed to the same dose of any chemical, different responses may occur as a result of interindividual variation in the subjects' genetic susceptibility. However, we now know that genetic variability is not the only factor that is responsible for this difference. Epigenetic variability also provides information about the response. In addition, direct exposure to xenobiotics can also induce epigenetic changes.

Humans are exposed to various toxic substances on a daily basis, including pollutants, occupational chemicals, drugs, heavy metals, and irradiation. Exposure to chemicals in the workplace and in the living environment may have a toxic effect, resulting in changes in the health status of people who work and live in such environments. Mercury, lead, toluene, and trichloroethylene are industrial toxic agents that fall within the category of chemicals that can cause deterministic toxic effects (dose–response), are widely diffused, and/or have very high intrinsic toxicity. Among the chemical agents that are capable of causing mutagenesis, carcinogenesis, and teratogenesis are substances of great importance, such as occupational benzene, arsenic, hexavalent chromium, and antineoplastics.

For example, methemoglobinizing agents are capable of inducing the oxidation of the iron atoms of hemoglobin from the ferrous (Fe^{2+}) to the ferric (Fe^{3+}) state, resulting in a pigment called methemoglobin. Methemoglobin cannot bind to oxygen, carbon dioxide, or carbon monoxide because of the positive charge of iron. Mutations in this gene can induce a condition called methemoglobinemia, which is characterized by high levels of methemoglobin in red blood cells. Furthermore, individuals with a mutation in the *CYB5R3* gene have an increased risk of developing this condition when exposed to methemoglobinizing agents, such as nitrates and chlorates [46]. More importantly, *CYB5R3* gene expression can be influenced by changes in the pattern of DNA methylation and histone modification, as well as the dimethylation of lysine 4 of histone H3 (H3K4me2) and the trimethylation of lysine 27 of histone H3 (H3K27me3) in the livers of mice [47]. The changes in DNA methylation and histone modifications are associated with the age-related expression pattern of the *CYB5R3* gene in liver cells during development [47], indicating that in addition to the genetic variability that is induced, for example, by mutation in *CYB5R3*, epigenetic changes that are associated with this gene may also indicate a mechanism of variation when the individual with the abnormal epigenetic modification also presents susceptibility to methemoglobinemia upon exposure to methemoglobinemia inducers.

The cytochrome P450 (CYP) superfamily of drug-metabolizing genes represents a major example of variation in individual responses after chemical exposure. Because this family of genes contains CpG islands, DNA methylation can affect the regulation of expression of these genes. The *CYP3A4* gene, which encodes the major drug-metabolizing enzyme in humans, exhibits a high interindividual variation in liver expression that can result in interindividual differences in drug metabolism and associated adverse drug effects. Kacevska and colleagues [48] investigated the influence of DNA methylation on interindividual *CYP3A4* expression in 72 adult and 7 fetal human livers using bisulfite sequencing. The authors identified highly variable CpG methylation sites in adult livers that correspond to important *CYP3A4* transcription factor binding sites. In addition, greater CpG hypermethylation was observed within these regulatory regions in fetal livers compared to adult livers. The findings of Kacevska and colleagues [48] provide novel insight into *CYP3A4* regulation with possible implications for understanding interindividual differences in the drug response, as dynamic DNA methylation elements are likely to be associated with key regulatory *CYP3A4* promoter regions and may potentially contribute to the commonly observed interindividual expression.

The study of Dejeux and colleagues [49] proposed molecular explanations for differential responses to chemotherapy in patients with breast cancer and suggested that it is valuable to consider the methylation profile of genes related to drug metabolism for patient management and treatment decisions. In this study, the DNA methylation profiles of genes that are associated with survival and drug response were investigated in 75 samples from advanced breast cancer, and the results were associated with the clinical and molecular parameters in patients with locally advanced breast cancer. Progressive disease during doxorubicin treatment was correlated with a lack of methylation at the *ABCB1* promoter during treatment with the antitumoral drug doxorubicin. In addition, the DNA methylation status at the promoters of *GSTP1*, *FOXC1*, and *ABCB1* correlated with the survival of patients. Moreover, the *GSTP1* and *FOXC1* methylation profiles exhibited independent prognostic markers that were associated with survival.

The DNA methylation profile in the regulatory regions of the *PITX2* gene, which is associated with cell proliferation, was analyzed in 241 node-positive, estrogen receptor-positive breast cancer samples. In these cases, patients who have a poor outcome may be recommended for currently available anthracycline-free alternative treatment options, whereas patients with a good outcome using anthracycline are adequately treated. The promoter methylation of *PITX2* was correlated with clinical outcome among different breast cancer patient populations for anthracycline-based chemotherapy, indicating that a well-defined panel of DNA methylation markers enables outcome prediction in each case [50].

Tumor cells with a defect in the DNA-damage response gene *BRCA1* are thought to be more sensitive to the DNA-damaging agents that are used in chemotherapy. Stefansson and colleagues [51] showed that in addition to mutation in *BRCA1*, DNA methylation is involved in this differential response. In xenografted tumors and cancer cell lines, the hypermethylation of the promoter region of *BRCA1* is associated with gene silencing and predicts enhanced sensitivity to platinum-derived drugs. Interestingly, *BRCA1* hypermethylation was found to be a predictor of a longer time to relapse and improved the overall survival in ovarian cancer patients undergoing chemotherapy with cisplatin.

A good example of a personalized toxicology response that is influenced by epigenetic changes is the case of the drug temozolomide, which is used in the treatment of patients with glioblastoma. The standard treatment for glioblastoma usually comprises a daily administration of temozolomide with radiation therapy for 6 weeks. Temozolomide is considered a safe drug but, in some cases, can cause cytotoxicity by acting at the O^6 position of guanine, causing DNA damage that is associated with its myelosuppressive effect [52,53]. Lombardi and colleagues [54] studied the genetic and epigenetic factors that are associated with the myelosuppressive effect of temozolomide by analyzing the promoter methylation of the *MGMT* gene, which is responsible for repairing guanine lesions. These authors reported that the promoter methylation of *MGMT* in the hematopoietic cell system was a predictor of severe myelotoxicity [54].

The gene *GPX3* plays a pivotal role in the detoxification of hydrogen peroxide. It was demonstrated by Chen and colleagues [55] that methylation of the promoter of this gene is associated with chemoresistance, as observed in head and neck cancer; therefore, the methylation status of *GPX3* could be used as a potential prognostic indicator for head and neck cancer patients who are treated with the antitumoral drug cisplatin.

The toxicological response related to epigenetic variability has been interestingly explored *in vivo* by verifying changes in the epigenetic patterns in different strains of rodents by exposing the animals to the same dose and time of exposure. This type of approach is interesting in the study of toxicological mechanisms because under natural conditions, it is difficult to measure the dose or the period of occupational exposure in humans. 1,3-Butadiene is a common carcinogenic environmental contaminant and is considered an epigenotoxic agent (i.e., it affects DNA and histone methylation). Koturbash and colleagues [56] used a panel of genetically diverse inbred mice (NOD/LtJ, CAST/EiJ, A/J, WSB/EiJ, PWK/PhJ, C57BL/6J, and 129S1/SvImJ) to assess whether 1,3-butadiene-induced epigenotoxic events may be subject to interstrain differences. The authors observed that epigenetic effects of 1,3-butadiene were most prominent in C57BL/6J mice, with a loss of global DNA methylation and a loss of trimethylation of histone H3 lysine 9, histone H3 lysine 27, and histone H4 lysine 20, accompanied by the dysregulation of liver gene expression, indicating hepatotoxicity. These authors also observed an increase in histone methylation in the absence of changes in gene expression and DNA methylation in the CAST/EiJ strain. The findings of Koturbash and colleagues [56] demonstrate that interstrain susceptibility to genotoxicity from a well-known environmental carcinogen may be due to strain-specific epigenetic events in response to exposure.

In addition to the individual epigenetic variability that is inherent to an individual, thereby providing differences in the toxicological response as described above, the toxicity of a compound may be attributed to its ability to induce epigenetic alterations. Considering that these epigenetic changes may be permanent after exposure, it becomes important to investigate what changes are usually related to epigenetic variability that is acquired after environmental exposure [57]. For example, Chen and colleagues [58] demonstrated in mice that the transient neonatal activation of the gene *Nr1i3 (Car)*, which is a central regulator of the drug/xenobiotic metabolism, results in epigenetic memory and a permanent change in liver drug metabolism. The exposure of mice to the *Nr1i3*-specific ligand 1,4-bis[2-(3,5-dichloropyridyloxy)] benzene during the neonatal period persistently induced the expression of the *Nr1i3* target genes *Cyp2B10* and *Cyp2C37* throughout the life of the exposed mice. As a result, during adulthood, these mice showed a permanent increase in histone H3 lysine 4 mono-, di-, and trimethylation and a decrease in H3K9 trimethylation within the *Cyp2B10* locus and manifested a permanent reduction in the sensitivity to zoxazolamine treatment. Thus, it is also important to investigate the epigenetic changes that are induced by toxic exposure.

4. EPIGENETIC CHANGES THAT ARE INDUCED BY TOXIC EXPOSURE

The absorption, metabolism, distribution, and excretion of xenobiotics are mediated by enzymes that are encoded by drug-processing genes, namely uptake transporters, phase I and phase II enzymes, and efflux transporters [59]. These genes are regulated by epigenetic mechanisms; thus, epigenetic variability could influence the individual response after exposure to a toxicant (Figure 1).

Intracellular toxicants, as well as a wide variety of drugs, may be transported back into the extracellular space by P-glycoprotein (an ATP-dependent xenobiotic transporter), which is present in hepatocytes, renal proximal tubular cells, and intestinal epithelium and is encoded by the *ABCB1* gene [60]. The main focus of research involving the *ABCB1* gene has been the role of this gene in drug resistance. Allelic variants that provide a higher expression of this gene are responsible for multidrug resistance, mainly involving chemotherapy drugs [61]. The reduced methylation of the 5' region of the *ABCB1* gene in a human T cell leukemia cell line is associated with increased gene expression and thus the multidrug resistance phenotype [62].

Alcohol consumption induces changes in brain gene expression in humans and animal models that could contribute to alcohol dependence [63]. Moreover, decreased DNA methylation and increased histone H3K4 trimethylation have been demonstrated in alcoholics, suggesting that epigenetic mechanisms play an important role in alcohol addiction [64]. This information indicates that epigenetic intervention could be used to alter gene expression that is abnormally induced in the brain through alcohol abuse. Moreover, epigenetics is progressing toward the clarification of the primary mechanisms of toxicity and addiction.

A predetermined aberrant epigenetic profile, such as that induced by gestational exposure to alcohol, can be a determining factor for the development of alcohol addiction. Alcohol dependence is a multifactorial polygenic disease; it involves interactions between multiple genes and between genes and the environment. The SNCA gene codes α-synuclein, which is involved in the regulation of dopamine release and transport [65]. The reuptake of dopamine, which is caused by changes in α-synuclein expression, may alter dopamine-mediated neuronal signaling; it has been suggested that this signaling is the main mechanism of craving, withdrawal, and the underlying pathways of alcohol addiction [66]. Moreover, it has been proposed that the decreased expression of α-synuclein predisposes individuals to susceptibility to psychiatric diseases or drug abuse [67]. A hypermethylated promoter is generally associated with reduced gene expression, and the promoter region of α-synuclein is hypermethylated in patients with alcoholism [68]. This possible epigenetic effect that is related to the consumption of alcohol is reinforced by a study demonstrating a genetic variation—a polymorphism in the SNCA gene—related to blood oxygenation level-dependent responses during functional magnetic resonance imaging for alcohol dependence [69].

DNA methylation can also be altered by the improper function of enzymes called DNA methyltransferases (DNMTs), which methylate DNA. During the reproductive period, several environmental factors can alter the intrauterine environment, thereby influencing the health and the risk of disease later in life. It has been demonstrated in rats that perinatal exposure to alcohol increases the DNMT activity in the hippocampus of the offspring [70].

In contrast to efflux transporters, such as P-glycoprotein, the hepatic uptake of various endo- and xenobiotics occurs through the family of organic anion-transporting polypeptides (OATPs) [71]. OATPs are sodium-independent uptake transporters that are capable of transporting organic anions, such as bile acids, bilirubin, steroid hormone conjugates, thyroid hormones, prostaglandins, clinically used drugs, and toxins [72]. Polymorphisms in these OATP transporters have significant functional consequences for drug-related toxicity [73]. In addition, epigenetic modifications may influence the expression of genes of the OATP family, influencing transport processes and triggering an imbalance in OATP substrates, resulting in increased drug toxicity and adverse drug reactions [72,74].

The study of DNA methylation and histone acetylation profiles of some genes from the mouse and human OATP families has demonstrated a clear association between tissue-specific expression and epigenetic profiles for Oatp1a1, Oatp1a4, Oatp1a6, Oatp1b2, and Oatp1c1 in mice and OATP1B1 and OATP1B3 in humans, demonstrating the epigenetic regulation of this gene family [71]. Moreover, the gene expression of OATP1B3 is regulated by DNA methylation in cancer cell lines [75].

In addition to the importance of genetic and epigenetic variants that are related to xenobiotic transporters, the most important changes are related

259

TABLE 1   CYP Genes and Their Associated Effects

Genes	Epigenetic effect	References
CYP1A1	Promoter region hypomethylated in placenta from smokers	[76]
	Different methylation patterns between male and female mice	[80]
CYP1A2	DNA methylation in interindividual variation in human liver	[83]
	DNA methylation controls the mRNA level in human livers	[84]
	Regulated by histone modification in mouse hepatocytes	[85]
CYP1B1	Induction is dependent on methylation of promoter region of the gene	[89]
	Gene expression regulation by miR-27b	[90]
CYP2E1	Different methylation patterns between male and female mice	[80]
	Methylation of the gene is related to its lack of transcription in fetal liver	[86]
CYP7B1	Different methylation patterns between male and female mice	[80]

to the family of metabolizing genes, such as CYP (Table 1). For example, in humans, tobacco exposure significantly decreases the methylation of the promoter region of *CYP1A1*, and this effect is associated with increased gene expression in the placenta [76]. This effect, which is observed in placenta, suggests that smoking during the reproductive period could affect the fetus if the tobacco metabolites are not excreted, thereby compromising the epigenetic pattern of the fetus. This altered epigenetic profile could persist after birth and throughout the individual's life span [77].

In fact, the field of the developmental origins of health and disease focuses on how events in early life could favor health or disease in later life [78]. Moreover, epigenetic events early in life could also be related to epigenetic variability. The cells of monozygotic twins at birth can present a differential pattern of gene expression in response to the external environment, probably induced by epigenetic mechanisms. This variation in gene expression is accentuated in dichorionic twins compared to monochorionic twins, suggesting that even small differences in the intrauterine environment can disturb epigenetic marks and consequently alter the gene expression profile [79].

Male and female cells that are obtained from embryos prior to sexual differentiation show different patterns of gene expression and, as a result, different responses when exposed to toxins *in vitro*. This differential response to toxins has been partially attributed to variations in the methylation pattern of CYP genes, such as *CYP1A1, CYP2E1,* and *CYP7B1,* between male and female cells [80].

The enzyme cytochrome P450 1A2 is encoded by the gene *CYP1A2* and is involved in an NADPH-dependent electron transport pathway that metabolizes a large number of drugs. Moreover, this enzyme metabolically activates numerous promutagens and procarcinogens. The expression of this gene shows substantial interindividual differences, and this variability has significant consequences for drug efficacy and cancer susceptibility. In addition to the importance of genetic variation in the variable expression of *CYP1A2*, epigenetic events also play a pivotal role [81,82]. The reduced expression of *CYP1A2* is associated with the hypermethylation of the gene, and this epigenetic state of the transporter can vary among individuals. This reduced expression of *CYP1A2* affects drug efficacy and risk in the development of cancer [83,84]. In addition to DNA methylation, the expression of *CYP1A2* can also be controlled by histone acetylation [85]. Moreover, the methylation of the *CYP2E1* gene is related to its lack of transcription in human fetal liver [86].

Environmental compounds have a carcinogenic effect when bioactivated by the cytochrome P450 monooxygenase system, which promotes the metabolic initiation of the compound to its electrophilic intermediates [87]. The *CYP1B1* gene is involved in the metabolization of a large amount of procarcinogens to carcinogens, such as polycyclic aromatic hydrocarbons. Moreover, *CYP1B1* is activated by dioxin, one of the most potent known carcinogens [88]; the *CYP1B1* gene is silenced by the hypermethylation of its promoter in HepG2 cells, which results in changes in some of the mechanisms of induction of *CYP1B1* by dioxin. The demethylation of the promoter by 5-azacytidine, a demethylation agent, results in the induction of *CYP1B1* [89]. *CYP1B1* is posttranscriptionally regulated by miR-27b. In humans, there is an inverse association between the expression levels of miR-27b and the CYP1B1 enzyme, resulting in the decreased expression of miR-27b in cancerous tissues accompanied by a high level of the CYP1B1 protein [90].

5. EPIGENETIC EVALUATION OF TOXICITY

Exposure to environmental factors, especially during sensitive periods, such as development, alters the phenotype through epigenetic changes in the affected tissues. Since 1995, the recognition of the importance of epigenetics has advanced, with intense research and practice relating to toxicology and safety assessment.

Many studies regarding epigenetics, including sequencing projects involving international collaborations, have a main objective of identifying the healthy profile (i.e., normal profile) of the epigenome in various tissues and cell types, as well as epigenetic variability resulting from exposure to a drug or toxicant. Thus, although evaluating the genotoxicity and mutagenicity of compounds is of fundamental importance in drug screening studies, the epigenetic information is equally important as an end point with regard to evaluating the toxicity of chemicals.

Currently, the difficulties that are involved in the development of wider epigenetic research are being overcome through advances in high-throughput and genome-wide profiling technologies, which have allowed an extremely comprehensive identification of methylation patterns, histone modification, and miRNA expression, thereby increasing the total knowledge of the epigenome. These approaches have facilitated the generation of knowledge of individual epigenetic profiles, thereby establishing a profile of susceptibility that is attributed to epigenetic variation.

In fact, the field of epigenetics has become a subject of great interest in toxicology. In addition to the challenge of gathering knowledge regarding genetic variability, epigenetic variation represents an even greater challenge for toxicologists [91]. In addition, scientific effort is required to elucidate how epigenetic variability arises in response to drug and toxin exposure with the hope of differentiating adverse and nonadverse epigenetic alterations; this type of endeavor is attributed to toxicology.

The main end points that must be considered in a drug screening study should be those that demonstrate a direct epigenetic effect, such as changes in DNA methylation, histone modification, and expression of miRNAs. Furthermore, the overhead epigenetic effects should also be considered, including changes in the expression of methyltransferase enzymes as well as the enzymes that are responsible for maintaining the configuration of chromatin via covalent histone modification.

One of the most frequently used approaches for DNA methylation analysis, which distinguishes between methylated and unmethylated DNA, involves restriction endonucleases. After drug exposure, cells are incubated with the isoschizomer enzymes HpaII–MspI and SmaI–XmaI, which recognize CCGG and CCCGGG, respectively; HpaII and SmaI lack activity when a methyl group is present in their recognition site [20]. McrBc digestion can also be used, as this enzyme cuts only methylated DNA, including CpG islands.

Global CpG DNA methylation can be evaluated *in vitro* or *in vivo* after treatment with chemicals/drugs. For this purpose, the pyrosequencing of LINE1 can be used. LINE1 is a retrotransposon element that can be moved within DNA and that amplifies itself through an RNA intermediate, with an estimated 500,000 copies per genome [92]. LINE1 represents a reliable parameter for measuring DNA methylation to evaluate the epigenetic effects of toxins/drugs and the risk assessment related to drugs, food, and environmentally relevant pollutants [93].

Bisulfite sequencing has become the "gold standard" for analyzing altered DNA methylation profiles [94]. This method was used in a toxicity model in which zebrafish liver cells were studied to identify the

DNA methylation alteration that was induced by the nonmutagenic model substance 5-azacytidine, as well as a selection of environmental pollutants, such as sodium arsenite, 2,3,7,8-tetrachlorodibenzo-*p*-dioxin, 17α-ethinylestradiol, and diethylstilbestrol. Moreover, methylation-sensitive high-resolution melting technology was used in this study in combination with the drug exposure of cells and was found to be an excellent technology for early contaminant screening methods that permit DNA methylation alterations to be quantified with a high-throughput capacity [95].

It is also possible to use epigenome-wide technologies, such as the Illumina HumanMethylation BeadChip®, which covers 99% of the RefSeq genes, with an average of 17 CpG sites per gene region distributed across the promoter, 5' UTR, first exon, gene body, and 3' UTR. In addition, this chip covers 96% of the CpG islands, with additional coverage in the island shores and their flanking regions.

Another useful method is methylated DNA immunoprecipitation [96]. In this technique, the use of an antibody against the methylated cytosine (5-methylcytosine) permits methylated DNA fragments to be isolated. The captured DNA can be quantified using microarray technology, sequencing, or combined polymerase chain reaction (PCR) and a luciferase-fused zinc finger protein [97–99].

The differential nuclear distribution patterns of methylcytosine and genomic DNA (gDNA) can also be assessed by a high-content imaging-based approach called "3D quantitative DNA methylation imaging." This technology permits the detection of drug-induced DNA demethylation and concurrent heterochromatin decondensation/reorganization in cells. This technology is based on the differential in situ analysis of the relevant nuclear structures that are represented by methylated CG dinucleotides and gDNA and combines three steps: (1) the visualization of cellular targets by nondestructive assays, such as immunofluorescence and fluorescence in situ hybridization; (2) high-resolution scanning microscopy realized through different imaging modalities; and (3) computerized three-dimensional (3D) image analysis utilizing advanced nuclear segmentation, signal extraction, and image cytometry [100,101]. The use of 3D quantitative DNA methylation imaging is a powerful precursor to a series of fully automatable assays that use chromatin structure and higher organization as novel pharmacodynamic biomarkers for various epigenetic drug actions. There has been great interest in the search for drugs that act on enzymes that are involved not only in DNA methylation but also in histone modification. This search is evidenced by several studies that have investigated the mechanism of action and toxicity of compounds or drugs in the modification of histones [102–106].

The combination of technical analysis and transcriptomics of epigenetic information is of invaluable relevance to the field of toxicology

[107]. The methylation and acetylation of histones has been used to study the effects of the toxicants valproate and trichostatin A on the epigenetic switch between reversible and irreversible drug effects *in vitro* [108]. Specifically, Western blot and chromatin immunoprecipitation have been used to quantify the acetylation and methylation of histones, respectively. In addition, the mRNAs of some important genes have been quantified using approaches such as quantitative real-time PCR and microarray analysis. These analyses demonstrated that altered phenotypes are reliably reflected by transcriptome data from a disturbed/stressed differentiation model. Moreover, histone modification in the promoter region of key genes was identified as a possible persistence detector that acts in the transient stage before short-term cellular adaptation to established toxicity [108].

In addition to DNA methylation and histone modification, miRNAs have been a subject of considerable interest in studies of toxicity assessment. The main reason for this interest is that miRNAs could be interesting biomarkers of exposure and of the effects after exposure to drugs and/or toxicants. Because they are somewhat easily detectable in the plasma of exposed individuals or in individuals who already have the disease, miRNAs have become one of the most studied epigenetic mechanisms. Not only can the mere presence of miRNAs provide important information regarding the toxicity of exposure, but a change in the expression of these small RNAs may also indicate a consequent change in the expression of target genes. For example, the role of miRNAs as posttranscriptional regulators of cytochrome P450 genes and nuclear receptors has been demonstrated, providing evidence for the importance of miRNAs in drug toxicity [109]. The reliability of the analysis of miRNAs as a strategy to identify the effectiveness of drugs has been demonstrated by a growing number of studies using *in vivo* rather than *in vitro* methods for drug screening, which is the recommended first step [110,111].

Currently, using the microarray technique, one can obtain the complete profile of miRNA expression in a tissue or cell type; this is known as the miRNAome. This technique implies that in a study of drug screening, an almost complete list of possible epigenetic alterations, as demonstrated by changes in the expression of miRNAs, may provide important toxicological information regarding the mechanism of drug action. For example, after treatment with phenobarbital, the rat liver showed modified expression of miR-200a/200b/429 and miR-96/182 clusters. Moreover, an association was reported between perturbations of miR-29b and global DNA methylation induced by phenobarbital. These data indicate that miRNAs play a significant role in the adult rodent liver's response to phenobarbital treatment [112].

In addition to plasma, other bodily fluids have been used to quantify miRNAs after drug exposure. Urinary miRNA profiling was proposed for the identification of biomarkers after cisplatin-induced kidney injury in rats [113]. The miRNA profile in urine can also be used to identify patients with acute T-cell-mediated rejection in renal allograft recipients to predict long-term kidney function [114].

6. EPIGENETIC TOXICITY OF ENVIRONMENTAL CHEMICALS

6.1 Metals

6.1.1 Aluminum

Aluminum (Al) is the third most abundant element in the environment (8%) and the most abundant metal. Because of its high reactivity, Al is always combined with other elements and is never found in its natural metallic state. In the environment, Al is found in the form of oxides, hydroxides, silicates existing in clays and micas, and water–soluble complex forms, such as sulfates, nitrates, and chlorides, in the presence of dissolved organic matter [115].

Aluminum is considered a genotoxic metal that is responsible for causing both DNA alterations and epigenetic effects and for playing a key role in breast cancer [116]. Aluminum sulfate treatment significantly upregulates nuclear factor-κB (NF-κB)-sensitive miR-146a and downregulates the *CFH* gene in human neural cells [117]. This information indicates that in the brain of patients with Alzheimer disease, the NF-κB-sensitive miRNA-146a-mediated modulation of *CFH* gene expression may contribute to inflammatory responses in aluminum-stressed human neural cells.

The epigenetic effect of aluminum in primary human neural cells has also been demonstrated in relation to the increased expression of neuronal miR-9, miR-125b, and miR-128 with the exposure of the cells to aluminum sulfate. It is important to note that these miRNAs also exhibit increased expression in the brain in cases of Alzheimer disease, supporting the theory that aluminum sulfate induces the reactive oxygen species-mediated upregulation of specific regulatory elements and pathogenic genes that cause genotoxicity; these mechanisms bring about the progressive dysfunction and apoptotic cell death of brain cells [118].

The suggestion of an aluminum-driven interaction of NF-κB and miRNAs was reinforced by a study that demonstrated that the incubation of microglial cells with aluminum sulfate results in the upregulation of NF-κB-sensitive miR-34a, thereby downregulating the expression of the gene *TREM2*. *TREM2* is essential for the sensing, recognition, phagocytosis, and clearance of noxious cellular debris from brain cells, including neurotoxic Aβ42 peptides [119].

6.1.2 Arsenic

There is increasing evidence to the effect that arsenic (As) can deregulate epigenetic mechanisms that are related to toxicity and carcinogenesis [120]. In its trivalent and pentavalent forms, arsenic is widely distributed in nature and is found in organic and inorganic compounds. Significant differences in the toxicokinetics and toxicity of various arsenic compounds have been identified. The daily intake of arsenic varies widely, with intervals ranging from micrograms to milligrams; in addition, arsenic is influenced by the consumption of seafood and naturally occurs in water with high concentrations of the metal [115].

It is important to note that arsenic is a human carcinogen that has a weak mutagenic effect. Therefore, it has been proposed that the carcinogenic effect of arsenic is related to its activity in epigenetic mechanisms, whereby it may induce genomic instability, thereby promoting the development of cancer. Evidence suggests that prenatal arsenic exposure has serious, long-term consequences through epigenetic mechanisms. A positive association between DNA methylation in LINE1 repeated elements and high concentrations of arsenic were reported in both maternal and fetal leukocytes. Furthermore, although to a lesser extent, an association between the levels of arsenic and methylation of CpG sites was also found in the promoter region of the *CDKN2A* (*P16*) gene.

Research using the umbilical cord blood of a newborn demonstrated differential patterns of expression of 12 miRNAs that are involved in adaptive immune response signaling due to maternal exposure to inorganic arsenic. This effect was associated with high levels of inorganic arsenic in maternal urine [121]. Cells that are exposed to arsenic have inhibited DNA methyltransferase activity, leading to DNA hypomethylation, which has been proposed as the mechanism of carcinogenesis due to arsenic exposure [122].

The potential epigenetic component of the action of arsenic was evaluated in the promoters of 13,000 human genes by measuring the state of histone acetylation. Looking at arsenical-induced malignant transformation, it was demonstrated that changes in histone H3 acetylation were associated with the expression of the associated gene. Another association was identified between the DNA hypermethylated and hypoacetylated promoters. The authors proposed that these promoters were targeted specifically by arsenicals and probably occur in important regions of arsenical-induced malignant transformation. These data suggest that arsenicals may participate in tumorigenesis by altering the chromatin configuration in the gene region [123].

Exposure to arsenite increases H3K9 dimethylation (H3K9me2) and decreases H3K27me3 in human lung carcinoma A549 cells. These events are related to the silencing of gene expression. Meanwhile, H3K4 trimethylation (H3K4me3), which is associated with active gene expression, was increased by arsenite. Moreover, histone modification was induced by very low dose (0.1 µM) arsenite. These data suggest that the alteration of specific histone methylation, which is usually associated with both gene silencing and activating marks, represents a mechanism of action by which arsenic induces carcinogenesis [124].

6.1.3 Lead

Lead (Pb) is a heavy metal that has been used by humans for more than 4000 years. Over time, it was realized that the contamination of food that was produced using lead in industrialized regions had devastating effects. In addition, organisms that live in aquatic environments capture and accumulate the lead existing in water and sediment. The epigenetic toxicity of lead is associated with DNA methylation and histone modification, with no definitive results so far in terms of the altered expression of miRNAs.

Exposure to Pb during childhood has been associated with behavioral problems and learning disabilities. Moreover, animal studies have suggested that these effects are related to problems in neuronal differentiation in the developing brain. In fact, it was demonstrated in human embryonic stem cells that lead partially compromises neuronal differentiation by inducing changes in the DNA methylation of genes that are related to brain development [125]. Lead exposure resulted in increased locomotor activity in rats, which was most likely induced by histone modification, as there was an increase in the levels of histone acetylation in the hippocampus. Meanwhile, no changes were observed in the expression of other proteins that are associated with neurological diseases [126].

6.1.4 Cadmium

Cadmium (Cd) is a heavy metal that is widely used in industry and has raised considerable concern because it is classified as a human carcinogenic compound that is involved in environmental and occupational exposure. The toxic effects of Cd are associated with accumulation over time in a variety of tissues, including the liver, bladder, stomach, and renal and hematopoietic systems. Dietary cadmium exposure has been associated with DNA hypomethylation in peripheral blood, although this association is modified by *DNMT1* genotypes, indicating that the genetic variant could interfere with the epigenetic effects of cadmium. The analysis of gene-specific promoter methylation and global DNA methylation was measured in the peripheral blood of nonsmoking Argentinean women through pyrosequencing and the measurement of associated cadmium concentrations in the blood. When genotyping for *DNMT1*, it was demonstrated that urinary cadmium was inversely associated with global DNA methylation but not with gene-specific methylation. *DNMT1* polymorphisms modified associations between urinary cadmium and global DNA methylation, showing stronger hypomethylation with increasing urinary cadmium concentrations [127].

Cadmium can partially cross the placental barrier and therefore is associated with harmful effects in newborns. A methylated CpG island recovery assay that permits the assessment of more than 4.6 million sites spanning 16,421 CpG islands was used to study the relationship between methylation in leukocytes and the levels of cadmium during pregnancy in mother–newborn pairs. In both fetal and maternal DNA, the gene methylation levels in the promoter region were associated with levels of cadmium, providing evidence for distinct DNA methylation profiles that are associated with exposure to cadmium [128].

6.1.5 Nickel

Exposure to nickel (Ni) may occur through the inhalation of air, the ingestion of food and water, or skin contact. Dermal exposure can cause contact dermatitis, and nickel often touches the skin because it is commonly used in jewelry and clothing ornaments containing metal. Another important route of exposure to nickel is through tobacco. The primary route of occupational exposure is respiratory inhalation: the metal is mainly found in the form of insoluble compounds of dust, aerosols that are formed from solutions of soluble compounds, and nickel carbonyl vapors.

Nickel has potent carcinogenic activity, as there is an increased risk of lung cancer in individuals occupationally exposed to nickel [129,130]. The epigenetic contribution to the tumorigenesis that is induced by nickel has been shown by studies demonstrating gene-specific methylation alteration and histone modification. The nickel-induced epithelial–mesenchymal transition, which is considered the main mechanism that is involved in the pathogenesis of lung fibrosis and tumor metastasis, was accompanied by the downregulation of the E-cadherin gene, which is induced by the hypermethylation of the gene's promoter region [131].

The consistent potential of nickel to silence genes through aberrant DNA-induced methylation was also demonstrated by another study that reported the downregulation of *RAR-β2*, *RASSF1A*, and *CDKN2A*. This downregulation was induced via the hypermethylation of the 5′ region of these genes in rats that were exposed to nickel [132].

Histone modification due to nickel exposure has also been demonstrated. H3K4me3, as evaluated in peripheral blood mononuclear cells, was elevated in Chinese subjects with occupational exposure to nickel, while H3K9me2 was decreased. Moreover, urinary concentrations of nickel were positively associated with H3K4me3 [133]. The treatment of human lung adenocarcinoma cells with nickel increased H3K4me3 in both the promoters and the coding regions of several genes, increasing gene expression. This effect was accompanied by the cross-linking of chromatin in the coding regions immediately downstream of the transcription start sites of some nickel-induced genes, demonstrating the effect of nickel in the modification of chromatin [134].

6.2 Pesticides

Pesticides are toxic chemicals that are designed to kill or repel insects, molds, unwanted animals, or unwanted plants. In addition to being harmful to pests, pesticides can affect humans, causing nausea, headache, dizziness, and skin allergy. Many pesticides are also linked to chronic diseases and conditions such as cancer, birth defects, neurological and reproductive imbalances, and the development of sensitivity to chemicals. Chemically sensitive individuals, such as the elderly, pregnant women, newborns, and children, are especially vulnerable to the toxic effects of pesticides.

The toxicological response after exposure to pesticides involves several mechanisms, including epigenetic events. In addition, individual epigenetic variability should also be considered after pesticide exposure, as changes in methylation or histone modification could affect gene expression, thereby impairing the metabolism of xenobiotics and enhancing their toxic effect.

Direct experimental evidence of these epigenetic effects was put forward by a study that evaluated the genome-wide DNA methylation of DNA from cells that were exposed to diazinon, an organophosphate that has been associated with cancers. DNA methylation was demonstrated in 1069 CpG sites in 984 genes, mainly with tumor suppressor activity [135]. A cross-sectional study demonstrated that the serum concentration of organochlorines, the major constituent of some pesticides, was inversely correlated with global methylation levels in leukocyte DNA in women [136].

It was demonstrated in rats that epigenetic effects after pesticide exposure affect not only the exposed individual but also its future generations. The incidence of disease was evaluated in the F1 and F3 generations from rat mothers that were exposed to a "pesticide mixture" (permethrin pesticide and *N,N*-diethyl-*meta*-toluamide insect repellent). It was demonstrated that the descendants were affected by the pesticide, with a more potent impact in generation F3, showing pubertal abnormalities, testis disease, and ovarian disease. The sperm epigenome from F3 showed 363 differential DNA methylation regions, termed epimutations [137].

Exposure to the endocrine-disrupting pesticide methoxychlor downregulates the expression of some genes through hypermethylation in the ovaries of exposed rats, and this effect has been associated with female infertility [138]. The detrimental epigenetic effect of methoxychlor has also been demonstrated according to exposure during fetal and neonatal periods, which caused the hypermethylation of 10 genes, including the ovarian gene estrogen receptor-β in rats. This effect was accompanied by an increase in the expression of the *DNMT3B* gene, whose enzyme is responsible for the methylation of DNA [139].

6.3 Endocrine Disruptors

Endocrine disruptors are a group of chemicals present in the environment that are capable of interfering with the endocrine systems of humans and other animals, affecting health, growth, and reproduction. Exposure to endocrine disruptors during critical periods, such as during the reproductive stage when methylation patterns are resettling, may induce significant changes in the exposed organism that can persist throughout its life span. Endocrine disruption modifies the behavior of the endocrine system by damaging or changing the function of an endocrine organ or altering endocrine metabolism. Among the many disruptor chemicals with known epigenetic toxicity are bisphenol A (BPA) and phthalates.

6.3.1 Bisphenol A

BPA is a synthetic xenoestrogen that is used to cross-link chemicals in the production of polycarbonate plastics. When plastic food containers that are made with BPA are heated, the BPA can leach into the food product. In 2005, a U.S. study found BPA in 95% of the 394 adult study participants' urine samples [140]. Mounting evidence shows the influence of BPA on epigenetic events, with the induction of hypo- or hypermethylation of DNA, histone modification, and mechanisms involving the action of miRNAs in humans and rodents [141].

By measuring the CpG methylation in 2500 loci of the fetal mouse forebrain, the study detected that the expression of two functionally related genes changed with exposure to BPA and that this alteration was correlated with the methylation status [142]. The development of neoplastic and preneoplastic lesions in adulthood can be associated with the epigenetic effects of prenatal exposure to BPA. In the mammary glands of rat pups, maternal exposure to BPA induced 7412 differentially methylated global DNA segments, which were mainly observed at postnatal day 21, with profound effects in gene expression that were observed mainly at postnatal day 50 [143].

Transgenerational inheritance of obesity, reproductive disease, and sperm epimutations also trigger epigenetic effects that are induced by exposure to a mixture of endocrine disruptors, such as BPA, di(2-ethylhexyl) phthalate (DEHP), and dibutyl phthalate. The incidence of adult-onset disease was evaluated in F1- and F3-generation rats whose mother was exposed to the chemicals. A differential DNA methylation in 197 gene promoters whose functions correlate with pathologies was identified [144].

6.3.2 Phthalates

Phthalates are plasticizers that are used in the manufacture of polyvinyl chloride plastics to render them softer and more flexible. These chemicals are easily released into the environment because they are loosely bound to the plastic. Human exposure occurs through ingestion, inhalation, and dermal contact. The most toxic effects are related to skin absorption through the application of makeup and ingestion through contaminated foods.

The epigenetic effect of phthalates was demonstrated through maternal exposure to DEHP in mice. DEHP significantly increased DNA methylation levels and the expression of DNA methyltransferases [145]. DEHP and butyl benzyl phthalate can also interfere with the epigenetic regulation of the immune system. Exposing plasmacytoid dendritic cells to phthalates *in vitro* suppressed the CpG-induced expression of genes that are related to immunity against infection [146].

7. EPIGENETIC TOXICOLOGY OF DRUG INTOXICATION

Ingested chemicals are metabolized in the liver through the conversion of these substances to their metabolites to avoid any toxic effects. In some cases of high exposure or ingestion, the organ is unable to metabolize all of the chemicals, causing an imbalance in the metabolism that then leads to intoxication. The effects of this overdose are reflected throughout the organism, and in some cases, toxicity induces death. Similar effects are observed when a drug or pharmaceutical results in a powerful adverse reaction or side effect.

The main hepatic-metabolizing mechanism is cytochrome P450, which is composed of a massive set of encoding genes and a wide range of metabolizing enzymes. An important point concerning the metabolism of chemical substances such as medicines is the ability of the gene to produce an enzyme that is completely effective in the metabolism of the substance, which is the case for the majority of the population. However, sometimes a gene (allele) may not encode an enzyme that is fully efficient. This genetic variation provides a different response in the metabolism of a drug and, more importantly, in relation to an overdose. These effects can also be associated with potentiated side effects.

Individual genotyping could provide information about the allele, thereby offering information about the individual's vulnerability to intoxication [147]. In practice, the use of this approach is still in its infancy [148]. However, at present, a period of immeasurable technological growth is occurring that could allow the extrapolation of this type of data from several studies, demonstrating the effectiveness of this methodology. This technology will permit the planning of future applications of these methods with the implementation of the proper cautionary measures. Even more interesting, several studies have reported that in addition to the significance of genetic components, the epigenetic pattern has a similar importance. With individual epigenetic information in hand, adverse drug reactions from overdosing can be avoided; moreover, drug use by individuals with an epigenetic modification that compromises the correct metabolism of that drug should also be avoided, as this could induce adverse drug reactions. In this case, the epigenetic modification of genes that are responsible for the transport and metabolism of drugs is the target.

Several genes that are involved in drug transport and metabolism, such as *CYP2C9*, *CYP2C19*, *CYP2D6*, and *OATP1B1*, have polymorphisms that alter the metabolism of drugs, thereby contributing to drug resistance or to the induction of adverse drug reactions [149,150]. More importantly, variations in drug metabolism by genes such as *CYP1A2* and *CYP3A4* cannot be clarified only according to genetic polymorphisms, suggesting that epigenetic changes have similar or even more importance than genetic variation in the drug metabolism response.

8. INTERACTION BETWEEN GENETIC AND EPIGENETIC VARIATION IN TOXICOLOGY

In some cases, either the genetic variation that is conferred by genetic polymorphisms or the epigenetic variation that is brought about by changes in the pattern of methylation, histone modification, or action of miRNAs can explain different (genetic or epigenetic) responses that are affected by individual variability. Another factor that has been studied is the interaction between genetic and epigenetic variants. It has been demonstrated that some genetic variants are associated with DNA methylation at CpG sites, suggesting a mechanism by which methylation profiles can be transmitted across generations. So-called methylation quantitative trait loci provide evidence for the interaction between genetic and epigenetic events [151]. Thus, individual genetic variation (e.g., polymorphisms) may also confer a specific pattern of methylation that is derived from individual genetic variation; epigenetic variation may be linked to genetic variation, indicating that in a personalized toxicology approach based on individual variability, genetic or epigenetic variation should not be considered alone but should be considered together.

Because encompassing both genetics and epigenetics is a new approach, no evidence of methylation quantitative trait loci associated with toxicology-related genes has been reported so far. However, some studies have reported interesting results. Using genome-wide genotype data from 268 African Americans and 143 European Americans, one study demonstrated that DNA methylation—which can be influenced by sequence variants—partially explains the association between genetic variation and alcohol dependence [152].

It has also been proposed that the risk for human cancer may be related to the DNA methylation of quantitative trait loci. The genome-scale DNA methylation profiles of 3649 primary human tumors were compared with patient cancer risk genotype data, comprising a comprehensive methylation quantitative trait loci profile that was associated with 21% of the cancer risk polymorphisms in the examined data set [153].

One of the approaches that has been used in studies investigating the genetic variation that is associated with the risk of developing diseases is genome-wide association studies (GWASs). These studies search the genome for polymorphisms that occur more regularly in people with a specific disease than in healthy people. Hundreds or thousands of polymorphisms can be investigated at the same time. These data help researchers to identify genes that may contribute to a person's risk of developing a certain disease. Despite the successful use of GWASs in linking loci variation with common diseases, a considerable amount of the causality remains unexplained. The search for these gaps has now been exploited by epigenome-wide association studies (EWASs). The goal of EWASs is to identify changes in the epigenome at particular loci that are correlated with a phenotype of interest. Again, this type of approach is still in its infancy, and there are no data regarding the epigenetic-related variability in interindividual differences in drug response, but some results clearly suggest an underlying epigenetic predisposition to common human diseases [154,155]. These types of data demonstrate that in addition to the influence on gene expression, epigenetic events such as DNA methylation should also be considered to clarify individual susceptibility to disease, drug adverse reactions, and toxicant exposure due to variation in the genetic sequence.

9. CONCLUSION

Knowledge of the genetic variability of the individual has led to the development of proposals for new approaches to the treatment of toxic effects based on genetic variation. In addition, studies have demonstrated that genetic and epigenetic variation are critical effective approaches. Epigenetic changes in the patterns of DNA methylation and histone modification and the altered actions of miRNAs as induced by direct environmental exposure alter the individual's epigenetic profile. When these changes occur during the fetal period, they may persist throughout the life span of the individual, creating an epigenetic profile that cannot be ignored by those in the field of toxicology. Personalized toxicological approaches based on genetics and epigenetic profiles must be developed. The information that has been obtained so far suggests that personalized toxicological treatment is not only an important tool but also an essential one for the future.

Glossary

Epigenome The complete collection of epigenetic marks, such as DNA methylation and histone modifications, and other molecules that can transmit epigenetic information, such as noncoding RNAs, that exist in a cell at any given point in time.

Epimutation Abnormal transcriptional profile of genes caused by changes in the epigenetic marks (e.g., DNA methylation).

Genome the complete set of genetic information in a cell. In humans, the genome consists of 23 pairs of chromosomes. Each set of 23 chromosomes contains approximately 3.1 billion bases of DNA sequence.

Hypermethylation An increase in normal methylation levels.

Hypomethylation A decrease in normal methylation levels.

Luciferase An enzyme that oxidizes luciferin to emit light in bioluminescent animals. The "luciferase-fused zinc finger protein" is a method that fuses zinc finger protein with firefly luciferase to detect DNA sequences.

Metastable epialleles An epiallele at which the epigenetic state can switch and establishment is a probabilistic event. Epialleles with an identical sequence of nucleotides may encode distinct phenotypes as a consequence of their epigenetic differences.

Methylation Adding a methyl group to DNA. Occurs predominantly in the cytosines located 5′ to guanines known as CpG dinucleotides.

Microarray analysis Technology used to study the expression of thousands of genes at once.

MicroRNAs Small noncoding RNA molecules, generally 21 to 24 nucleotides in length, that are produced by cleavage of double-stranded RNA arising from small hairpins within RNA that is mostly single stranded. The microRNAs bind imperfectly to mRNA molecules and inhibit their translation.

Nucleosome Fundamental structural subunit of chromatin. Consists of about 150 base pairs of DNA sequence wrapped around a core of eight histone proteins.

P-glycoprotein A plasma membrane protein that acts as a localized drug transport mechanism, actively exporting drugs out of the cell.

RefSeq Reference sequence database that provides a comprehensive, integrated, nonredundant, well-annotated set of sequences, including genomic DNA, transcripts, and proteins.

Western blot Laboratory technique used to detect a specific protein in a sample.

LIST OF ACRONYMS AND ABBREVIATIONS

ABCB1	ATP-binding cassette, subfamily B (MDR/TAP), member 1
BPA	Bisphenol A
BRCA1	Breast cancer 1, early onset
CDKN2A/P16	Cyclin-dependent kinase inhibitor 2A
CFH	Complement factor H
CYB5R3	Cytochrome b_5 reductase 3
CYP	Cytochrome P450
CYP1A1	Cytochrome P450, family 1, subfamily A, polypeptide 1
CYP1A2	Cytochrome P450, family 1, subfamily A, polypeptide 2
CYP1B1	Cytochrome P450, family 1, subfamily B, polypeptide 1
CYP2B10	Cytochrome P450, family 2, subfamily B, polypeptide 10
CYP2C19	Cytochrome P450, family 2, subfamily C, polypeptide 19
CYP2C37	Cytochrome P450, family 2, subfamily C, polypeptide 37
CYP2C9	Cytochrome P450, family 2, subfamily C, polypeptide 9
CYP2D6	Cytochrome P450, family 2, subfamily D, polypeptide 6

CYP2E1	Cytochrome P450, family 2, subfamily E, polypeptide 1
CYP3A4	Cytochrome P450, family 3, subfamily A, polypeptide 4
CYP7B1	Cytochrome P450, family 7, subfamily B, polypeptide 1
DEHP	Di(2-ethylhexyl) phthalate
DNMT1	DNA methyltransferase 1
DNMT3A	DNA methyltransferase 3α
DNMT3B	DNA methyltransferase 3β
DNMT3L	DNA methyltransferase 3-like
DNMTs	DNA methyltransferases
EWASs	Epigenome-wide association studies
FOXC1	Forkhead box C1
gDNA	Genomic DNA
GPX3	Glutathione peroxidase 3
GSTP1	Glutathione *S*-transferase Pi 1
GWASs	Genome-wide association studies
H3K27me3	Trimethylation of lysine 27 of histone H3
H3K4me2	Dimethylation of lysine 4 of histone H3
H3K4me3	Trimethylation of lysine 4 of histone H3
H3K9me2	Dimethylation of lysine 9 of histone H3
LINE1	Long interspersed nuclear element 1
MGMT	O^6-methylguanine-DNA methyltransferase
NF-κB	Nuclear factor-κB
NR1I3	Nuclear receptor subfamily 1, group I, member 3
OATP1A1/SLCO1A1	Solute carrier organic anion transporter family, member 1A1
t*OATP1A4/SLCO1A4*	Solute carrier organic anion transporter family, member 1A4
OATP1A6/SLCO1A6	Solute carrier organic anion transporter family, member 1A6
OATP1B1/SLCO1B1	Solute carrier organic anion transporter family, member 1B1
OATP1B2/SLCO1B2	Solute carrier organic anion transporter family, member 1B2
OATP1B3/SLCO1B3	Solute carrier organic anion transporter family, member 1B3
OATP1C1/SLCO1C1	Solute carrier organic anion transporter family, member 1C1
OATPs	Organic anion-transporting polypeptides
PCR	Polymerase chain reaction
PITX2	Paired-like homeodomain 2

PVC	Polyvinyl chloride
qPCR	Quantitative real-time PCR
RAR-β2	Retinoic acid receptor β2
RASSF1A	Ras association (RalGDS/AF-6) domain family member 1
ROS	Reactive oxygen species
SNCA	Synuclein α
TREM2	Triggering receptor expressed on myeloid cells 2
TSSs	Transcription start sites
UTR	Untranslated region

References

[1] Haluskova J. Epigenetic studies in human diseases. Folia Biol-Prague 2010;56(3):83–96.
[2] Maher B. Personal genomes: the case of the missing heritability. Nature 2008;456 (7218):18–21.
[3] Luco RF, Allo M, Schor IE, Kornblihtt AR, Misteli T. Epigenetics in alternative pre-mRNA splicing. Cell 2011;144(1):16–26.
[4] Sawan C, Vaissiere T, Murr R, Herceg Z. Epigenetic drivers and genetic passengers on the road to cancer. Mutat Res-Fund Mol M 2008;642(1–2):1–13.
[5] Nebert DW, Zhang G, Vesell ES. Genetic risk prediction: individualized variability in susceptibility to toxicants. Annu Rev Pharmacol Toxicol 2013;53:355–75.
[6] Feinberg AP, Irizarry RA, Fradin D, Aryee MJ, Murakami P, Aspelund T, et al. Personalized epigenomic signatures that are stable over time and covary with body mass index. Sci Transl Med 2010;2(49):49–67.
[7] Rieger JK, Klein K, Winter S, Zanger UM. Expression variability of absorption, distribution, metabolism, excretion-related microRNAs in human liver: influence of nongenetic factors and association with gene expression. Drug Metabo Dispos Biol Fate Chem 2013;41(10):1752–62.
[8] Riddihough G, Zahn LM. What is epigenetics? Science 2010;330(6004):611.
[9] Portela A, Esteller M. Epigenetic modifications and human disease. Nat Biotechnol 2010;28(10):1057–68.
[10] Ptashne M. On the use of the word 'epigenetic'. Curr Biol CB 2007;17(7):R233–6.
[11] Kaminsky ZA, Tang T, Wang S-C, Ptak C, Oh GHT, Wong AHC, et al. DNA methylation profiles in monozygotic and dizygotic twins. Nat Genet 2009;41(2):240–5.
[12] Gluckman PD, Hanson MA, Cooper C, Thornburg KL. Effect of in utero and early-life conditions on adult health and disease. N Engl J Med 2008;359(1):61–73.
[13] Reik W, Dean W, Walter J. Epigenetic reprogramming in mammalian development. Science 2001;293(5532):1089–93.
[14] Tang WY, Ho SM. Epigenetic reprogramming and imprinting in origins of disease. Rev Endocr Metab Dis 2007;8(2):173–82.
[15] Dolinoy DC, Das R, Weidman JR, Jirtle RL. Metastable epialleles, imprinting, and the fetal origins of adult diseases. Pediatr Res 2007;61(5 Pt 2):30R–7R.
[16] Song SH, Han SW, Bang YJ. Epigenetic-based therapies in cancer: progress to date. Drugs 2011;71(18):2391–403.
[17] Chen T, Ueda Y, Dodge JE, Wang Z, Li E. Establishment and maintenance of genomic methylation patterns in mouse embryonic stem cells by Dnmt3a and Dnmt3b. Mol Cell Biol 2003;23(16):5594–605.

[18] Goll MG, Bestor TH. Eukaryotic cytosine methyltransferases. Annu Rev Biochem 2005;74:481–514.

[19] Murphy SK, Jirtle RL. Imprinted genes as potential genetic and epigenetic toxicologic targets. Environ Health Perspect 2000;108:5–11.

[20] Costello JF, Fruhwald MC, Smiraglia DJ, Rush LJ, Robertson GP, Gao X, et al. Aberrant CpG-island methylation has non-random and tumour-type-specific patterns. Nat Genet 2000;24(2):132–8.

[21] Singal R, Ginder GD. DNA methylation. Blood 1999;93(12):4059–70.

[22] Ehrlich M. DNA methylation in cancer: too much, but also too little. Oncogene 2002;21(35):5400–13.

[23] Yang AS, Estecio MR, Doshi K, Kondo Y, Tajara EH, Issa JP. A simple method for estimating global DNA methylation using bisulfite PCR of repetitive DNA elements. Nucleic Acids Res 2004;32(3):e38.

[24] Kanai Y. Genome-wide DNA methylation profiles in precancerous conditions and cancers. Cancer Sci 2010;101(1):36–45.

[25] Pogribny IP, Beland FA. DNA hypomethylation in the origin and pathogenesis of human diseases. Cell Mol Life Sci CMLS 2009;66(14):2249–61.

[26] Peters AHFM, Kubicek S, Mechtler K, O'Sullivan RJ, Derijck AAHA, Perez-Burgos L, et al. Partitioning and plasticity of repressive histone methylation states in mammalian chromatin. Mol Cell 2003;12(6):1577–89.

[27] Grunau C, Buard J, Brun ME, De Sario A. Mapping of the juxtacentromeric heterochromatin-euchromatin frontier of human chromosome 21. Genome Res 2006; 16(10):1198–207.

[28] Luzhna L, Kathiria P, Kovalchuk O. Micronuclei in genotoxicity assessment: from genetics to epigenetics and beyond. Front Genet 2013;4:131.

[29] Lee JH, Khor TO, Shu LM, Su ZY, Fuentes F, Kong ANT. Dietary phytochemicals and cancer prevention: Nrf2 signaling, epigenetics, and cell death mechanisms in blocking cancer initiation and progression. Pharmacol Ther 2013;137(2):153–71.

[30] Zhang Y, Moriguchi H. Chromatin remodeling system, cancer stem-like attractors, and cellular reprogramming. Cell Mol Life Sci 2011;68(21):3557–71.

[31] Nemeth A, Langst G. Chromatin higher order structure: opening up chromatin for transcription. Brief Funct Genomics Proteomics 2004;2(4):334–43.

[32] Shi Y. Histone lysine demethylases: emerging roles in development, physiology and disease. Nat Rev Genet 2007;8(11):829–33.

[33] Barski A, Cuddapah S, Cui KR, Roh TY, Schones DE, Wang ZB, et al. High-resolution profiling of histone methylations in the human genome. Cell 2007;129(4):823–37.

[34] Kozomara A, Griffiths-Jones S. miRBase: integrating microRNA annotation and deep-sequencing data. Nucleic Acids Res 2011;39:D152–7.

[35] Bartel DP. MicroRNAs: target recognition and regulatory functions. Cell 2009;136(2): 215–33.

[36] Rigoutsos I. New tricks for animal MicroRNAs: targeting of amino acid coding regions at conserved and nonconserved sites. Cancer Res 2009;69(8):3245–8.

[37] Jackson A, Linsley PS. The therapeutic potential of microRNA modulation. Discov Med 2010;9(47):311–8.

[38] Small EM, Olson EN. Pervasive roles of microRNAs in cardiovascular biology. Nature 2011;469(7330):336–42.

[39] Guo HL, Ingolia NT, Weissman JS, Bartel DP. Mammalian microRNAs predominantly act to decrease target mRNA levels. Nature 2010;466(7308):835–U66.

[40] Pasquinelli AE. NON-CODING RNA MicroRNAs and their targets: recognition, regulation and an emerging reciprocal relationship. Nat Rev Genet 2012;13(4):271–82.

[41] Rakyan VK, Blewitt ME, Druker R, Preis JI, Whitelaw E. Metastable epialleles in mammals. Trends Genet 2002;18(7):348–51.

[42] Waterland RA, Jirtle RL. Transposable elements: targets for early nutritional effects on epigenetic gene regulation. Mol Cell Biol 2003;23(15):5293–300.

[43] Waterland RA, Dolinoy DC, Lin JR, Smith CA, Shi X, Tahiliani KG. Maternal methyl supplements increase offspring DNA methylation at Axin Fused. Genesis 2006;44(9):401–6.

[44] Dolinoy DC, Huang D, Jirtle RL. Maternal nutrient supplementation counteracts bisphenol A-induced DNA hypomethylation in early development. Proc Natl Acad Sci USA 2007;104(32):13056–61.

[45] Dominguez-Salas P, Moore SE, Baker MS, Bergen AW, Cox SE, Dyer RA, et al. Maternal nutrition at conception modulates DNA methylation of human metastable epialleles. Nat Commun 2014;5:3746.

[46] Prchal JT, Gregg XT. Red cell enzymes. ASH Educ Program Book 2005;2005(1):19–23.

[47] Li Y, Zhong X-b. Epigenetic regulation of developmental expression of Cyp2d genes in mouse liver. Acta Pharm Sin B 2012;2(2):146–58.

[48] Kacevska M, Ivanov M, Wyss A, Kasela S, Milani L, Rane A, et al. DNA methylation dynamics in the hepatic CYP3A4 gene promoter. Biochimie 2012;94(11):2338–44.

[49] Dejeux E, Ronneberg J, Solvang H, Bukholm I, Geisler S, Aas T, et al. DNA methylation profiling in doxorubicin treated primary locally advanced breast tumours identifies novel genes associated with survival and treatment response. Mol Cancer 2010;9(1):68.

[50] Hartmann O, Spyratos F, Harbeck N, Dietrich D, Fassbender A, Schmitt M, et al. DNA methylation markers predict outcome in node-positive, estrogen receptor-positive breast cancer with adjuvant anthracycline-based chemotherapy. Clin Cancer Res Off J Am Assoc Cancer Res 2009;15(1):315–23.

[51] Stefansson OA, Villanueva A, Vidal A, Martí L, Esteller M. BRCA1 epigenetic inactivation predicts sensitivity to platinum-based chemotherapy in breast and ovarian cancer. Epigenetics Off J DNA Methylation Soc 2012;7(11):1225–9.

[52] Armstrong TS, Cao YM, Scheurer ME, Vera-Bolanos E, Manning R, Okcu MF, et al. Risk analysis of severe myelotoxicity with temozolomide: the effects of clinical and genetic factors. Neuro Oncol 2009;11(6):825–32.

[53] Sabharwal A, Waters R, Danson S, Clamp A, Lorigan P, Thatcher N, et al. Predicting the myelotoxicity of chemotherapy: the use of pretreatment O-6-methylguanine-DNA methyltransferase determination in peripheral blood mononuclear cells. Melanoma Res 2011;21(6):502–8.

[54] Lombardi G, Rumiato E, Bertorelle R, Saggioro D, Farina P, Della Puppa A, et al. Clinical and genetic factors associated with severe hematological toxicity in glioblastoma patients during radiation plus temozolomide treatment: a prospective study. Am J Clin Oncol 2013, Published ahead of print. http://dx.doi.org/10.1097/COC.0b013e3182a790ea.

[55] Chen B, Rao X, House MG, Nephew KP, Cullen KJ, Guo Z. GPx3 promoter hypermethylation is a frequent event in human cancer and is associated with tumorigenesis and chemotherapy response. Cancer Lett 2011;309(1):37–45.

[56] Koturbash I, Scherhag A, Sorrentino J, Sexton K, Bodnar W, Swenberg JA, et al. Epigenetic mechanisms of mouse interstrain variability in genotoxicity of the environmental toxicant 1,3-butadiene. Toxicol Sci Off J Soc Toxicol 2011;122(2):448–56.

[57] Ivanov M, Barragan I, Ingelman-Sundberg M. Epigenetic mechanisms of importance for drug treatment. Trends Pharmacol Sci 2014;35(8):384–96.

[58] Chen WD, Fu X, Dong B, Wang YD, Shiah S, Moore DD, et al. Neonatal activation of the nuclear receptor CAR results in epigenetic memory and permanent change of drug metabolism in mouse liver. Hepatology (Baltimore, Md) 2012;56(4):1499–509.

[59] Klaassen CD, Lu H, Cui JY. Epigenetic regulation of drug processing genes. Toxicol Mech Method 2011;21(4):312–24.

[60] Schinkel AH. P-Glycoprotein, a gatekeeper in the blood–brain barrier. Adv Drug Deliv Rev 1999;36(2–3):179–94.

[61] Zu B, Li Y, Wang X, He D, Huang Z, Feng W. MDR1 gene polymorphisms and imatinib response in chronic myeloid leukemia: a meta-analysis. Pharmacogenomics 2014;15(5):667–77.

[62] Kantharidis P, El-Osta A, deSilva M, Wall DM, Hu XF, Slater A, et al. Altered methylation of the human MDR1 promoter is associated with acquired multidrug resistance. Clin Cancer Res Off J Am Assoc Cancer Res 1997;3(11):2025–32.

[63] Mulligan MK, Ponomarev I, Hitzemann RJ, Belknap JK, Tabakoff B, Harris RA, et al. Toward understanding the genetics of alcohol drinking through transcriptome meta-analysis. Proc Natl Acad Sci USA 2006;103(16):6368–73.

[64] Ponomarev I, Wang S, Zhang L, Harris RA, Mayfield RD. Gene coexpression networks in human brain identify epigenetic modifications in alcohol dependence. J Neurosci Off J Soc Neurosci 2012;32(5):1884–97.

[65] Janeczek P, Lewohl JM. The role of α-synuclein in the pathophysiology of alcoholism. Neurochem Int 2013;63(3):154–62.

[66] Self DW, Nestler EJ. Relapse to drug-seeking: neural and molecular mechanisms. Drug Alcohol Depend 1998;51(1–2):49–60.

[67] Oksman M, Tanila H, Yavich L. Brain reward in the absence of alpha-synuclein. Neuroreport 2006;17(11):1191–4.

[68] Bonsch D, Lenz B, Kornhuber J, Bleich S. DNA hypermethylation of the alpha synuclein promoter in patients with alcoholism. Neuroreport 2005;16(2):167–70.

[69] Wilcox CE, Claus ED, Blaine SK, Morgan M, Hutchison KE. Genetic variation in the alpha synuclein gene (SNCA) is associated with BOLD response to alcohol cues. J Stud Alcohol Drugs 2013;74(2):233–44.

[70] Perkins A, Lehmann C, Lawrence RC, Kelly SJ. Alcohol exposure during development: impact on the epigenome. Int J Dev Neurosci 2013;31(6):391–7.

[71] Imai S, Kikuchi R, Kusuhara H, Sugiyama Y. DNA methylation and histone modification profiles of mouse organic anion transporting polypeptides. Drug Metab Dispos Biol Fate Chem 2013;41(1):72–8.

[72] Svoboda M, Riha J, Wlcek K, Jaeger W, Thalhammer T. Organic anion transporting polypeptides (OATPs): regulation of expression and function. Curr Drug Metab 2011;12(2):139–53.

[73] Gong IY, Kim RB. Impact of genetic variation in OATP transporters to drug disposition and response. Drug Metab Pharmacokinet 2013;28(1):4–18.

[74] Clarke JD, Cherrington NJ. Genetics or environment in drug transport: the case of organic anion transporting polypeptides and adverse drug reactions. Expert Opin Drug Metab Toxicol 2012;8(3):349–60.

[75] Imai S, Kikuchi R, Tsuruya Y, Naoi S, Kusuhara H, Sugiyama Y, et al. Epigenetic regulation of organic anion transporting polypeptide 1B3 in cancer cell lines. Pharm Res 2013;30(11):2880–90.

[76] Suter M, Abramovici A, Showalter L, Hu M, Shope CD, Varner M, et al. In utero tobacco exposure epigenetically modifies placental CYP1A1 expression. Metab Clin Exp 2010;59(10):1481–90.

[77] Waterland RA, Jirtle RL. Early nutrition, epigenetic changes at transposons and imprinted genes, and enhanced susceptibility to adult chronic diseases. Nutrition 2004;20(1):63–8.

[78] Calkins K, Devaskar SU. Fetal origins of adult disease. Curr Problems Pediatr Adolesc Health Care 2011;41(6):158–76.

[79] Gordon L, Joo JH, Andronikos R, Ollikainen M, Wallace EM, Umstad MP, et al. Expression discordance of monozygotic twins at birth: effect of intrauterine environment and a possible mechanism for fetal programming. Epigenetics Off J DNA Methylation Soc 2011;6(5):579–92.

[80] Penaloza CG, Estevez B, Han DM, Norouzi M, Lockshin RA, Zakeri Z. Sex-dependent regulation of cytochrome P450 family members Cyp1a1, Cyp2e1, and Cyp7b1 by methylation of DNA. FASEB J Off Publ Fed Am Soc Exp Biol 2014;28(2):966–77.

[81] Gunes A, Dahl ML. Variation in CYP1A2 activity and its clinical implications: influence of environmental factors and genetic polymorphisms. Pharmacogenomics 2008;9(5):625–37.

[82] Zhou S-F, Wang B, Yang L-P, Liu J-P. Structure, function, regulation and polymorphism and the clinical significance of human cytochrome P450 1A2. Drug Metab Rev 2010;42(2):268–354.

[83] Hammons GJ, Yan-Sanders Y, Jin B, Blann E, Kadlubar FF, Lyn-Cook BD. Specific site methylation in the 5'-flanking region of CYP1A2-Interindividual differences in human livers. Life Sci 2001;69(7):839–45.

[84] Ghotbi R, Gomez A, Milani L, Tybring G, Syvanen A-C, Bertilsson L, et al. Allele-specific expression and gene methylation in the control of CYP1A2 mRNA level in human livers. Pharmacogenomics J 2009;9(3):208–17.

[85] Jin B, Ryu DY. Regulation of CYP1A2 by histone deacetylase inhibitors in mouse hepatocytes. J Biochem Mol Toxicol 2004;18(3):131–2.

[86] Jones SM, Boobis AR, Moore GE, Stanier PM. Expression of CYP2E1 during human fetal development: methylation of the CYP2E1 gene in human fetal and adult liver samples. Biochem Pharmacol 1992;43(8):1876–9.

[87] Guengerich FP. Metabolic activation of carcinogens. Pharmacol Ther 1992;54(1):17–61.

[88] Mandal PK. Dioxin: a review of its environmental effects and its aryl hydrocarbon receptor biology. J Comp Physiol B Biochem Syst Environ Physiol 2005;175(4):221–30.

[89] Beedanagari SR, Taylor RT, Bui P, Wang F, Nickerson DW, Hankinson O. Role of epigenetic mechanisms in differential regulation of the dioxin-inducible human CYP1A1 and CYP1B1 genes. Mol Pharmacol 2010;78(4):608–16.

[90] Tsuchiya Y, Nakajima M, Takagi S, Taniya T, Yokoi T. MicroRNA regulates the expression of human cytochrome P450 1B1. Cancer Res 2006;66(18):9090–8.

[91] Gant TW. Skills and training for the 21st century chemical toxicologist. Chem Res Toxicol 2011;24(7):985–7.

[92] Rodic N, Burns KH. Long interspersed element-1 (LINE-1): passenger or driver in human neoplasms? PLoS Genet 2013;9(3):e1003402.

[93] Florea AM. DNA methylation pyrosequencing assay is applicable for the assessment of epigenetic active environmental or clinical relevant chemicals. Biomed Res Int 2013;2013:10. Article ID 486072. http://dx.doi.org/10.1155/2013/486072.

[94] Reed K, Poulin ML, Yan L, Parissenti AM. Comparison of bisulfite sequencing PCR with pyrosequencing for measuring differences in DNA methylation. Anal Biochem 2010;397(1):96–106.

[95] Farmen E, Hultman MT, Angles d'Auriac M, Tollefsen KE. Development of a screening system for the detection of chemically induced DNA methylation alterations in a zebrafish liver cell line. J Toxicol Environ Health Part A 2014;77(9–11):587–99.

[96] Duenas-Gonzalez A, Alatorre B, Gonzalez-Fierro A. The impact of DNA methylation technologies on drug toxicology. Expert Opin Drug Metab Toxicol 2014;10(5):637–46.

[97] Seifert M, Cortijo S, Colome-Tatche M, Johannes F, Roudier F, Colot V. MeDIP-HMM: genome-wide identification of distinct DNA methylation states from high-density tiling arrays. Bioinformatics 2012;28(22):2930–9.

[98] Jacinto FV, Ballestar E, Esteller M. Methyl-DNA immunoprecipitation (MeDIP): hunting down the DNA methylome. BioTechniques 2008;44(1):35, 7, 9 passim.

[99] Hiraoka D, Yoshida W, Abe K, Wakeda H, Hata K, Ikebukuro K. Development of a method to measure DNA methylation levels by using methyl CpG-binding protein and luciferase-fused zinc finger protein. Anal Chem 2012;84(19):8259–64.

[100] Gertych A, Oh JH, Wawrowsky KA, Weisenberger DJ, Tajbakhsh J. 3-D DNA methylation phenotypes correlate with cytotoxicity levels in prostate and liver cancer cell models. BMC Pharmacol Toxicol 2013;14:11.

[101] Tajbakhsh J. DNA methylation topology: potential of a chromatin landmark for epigenetic drug toxicology. Epigenomics 2011;3(6):761–70.

[102] Binda C, Valente S, Romanenghi M, Pilotto S, Cirilli R, Karytinos A, et al. Biochemical, structural, and biological evaluation of tranylcypromine derivatives as inhibitors of histone demethylases LSD1 and LSD2. J Am Chem Soc 2010;132(19):6827–33.
[103] Chang Y, Zhang X, Horton JR, Upadhyay AK, Spannhoff A, Liu J, et al. Structural basis for G9a-like protein lysine methyltransferase inhibition by BIX-01294. Nat Struct Mol Biol 2009;16(3):312–7.
[104] Kubicek S, O'Sullivan RJ, August EM, Hickey ER, Zhang Q, Teodoro ML, et al. Reversal of H3K9me2 by a small-molecule inhibitor for the G9a histone methyltransferase. Mol Cell 2007;25(3):473–81.
[105] Greiner D, Bonaldi T, Eskeland R, Roemer E, Imhof A. Identification of a specific inhibitor of the histone methyltransferase SU(VAR)3-9. Nat Chem Biol 2005;1(3):143–5.
[106] Wissmann M, Yin N, Muller JM, Greschik H, Fodor BD, Jenuwein T, et al. Cooperative demethylation by JMJD2C and LSD1 promotes androgen receptor-dependent gene expression. Nat Cell Biol 2007;9(3):347–53.
[107] Eglen RM, Reisine T. Screening for compounds that modulate epigenetic regulation of the transcriptome: an overview. J Biomol Screen 2011;16(10):1137–52.
[108] Balmer NV, Klima S, Rempel E, Ivanova VN, Kolde R, Weng MK, et al. From transient transcriptome responses to disturbed neurodevelopment: role of histone acetylation and methylation as epigenetic switch between reversible and irreversible drug effects. Archives Toxicol 2014;88(7):1451–68.
[109] Yokoi T, Nakajima M. microRNAs as mediators of drug toxicity. Annu Rev Pharmacol 2013;53:377–400.
[110] Juhasz K, Gombos K, Szirmai M, Gocze K, Wolher V, Revesz P, et al. Very early effect of DMBA and MNU on MicroRNA expression. Vivo 2013;27(1):113–7.
[111] Hegde VL, Tomar S, Jackson A, Rao R, Yang X, Singh UP, et al. Distinct microRNA expression profile and targeted biological pathways in functional myeloid-derived suppressor cells induced by Delta9-tetrahydrocannabinol in vivo: regulation of CCAAT/enhancer-binding protein alpha by microRNA-690. J Biol Chem 2013;288(52):36810–26.
[112] Koufaris C, Wright J, Osborne M, Currie RA, Gooderham NJ. Time and dose-dependent effects of phenobarbital on the rat liver miRNAome. Toxicology 2013;314(2–3):247–53.
[113] Pavkovic M, Riefke B, Ellinger-Ziegelbauer H. Urinary microRNA profiling for identification of biomarkers after cisplatin-induced kidney injury. Toxicology October 3, 2014;324:147–57. http://dx.doi.org/10.1016/j.tox.2014.05.005.
[114] Lorenzen JM, Volkmann I, Fiedler J, Schmidt M, Scheffner I, Haller H, et al. Urinary miR-210 as a mediator of acute T-cell mediated rejection in renal allograft recipients. American journal of transplantation. Off J Am Soc Transplant Am Soc Transpl Surg 2011;11(10):2221–7.
[115] Salgado PET. Metais em Alimentos. In: Oga S, Camargo MMA, Batistuzzo JAO, editors. Fundamentos de toxicologia. 3rd ed. São Paulo: Atheneu; 2008.
[116] Darbre PD. Aluminium, antiperspirants and breast cancer. J Inorg Biochem 2005;99(9):1912–9.
[117] Pogue AI, Li YY, Cui JG, Zhao Y, Kruck TP, Percy ME, et al. Characterization of an NF-kappaB-regulated, miRNA-146a-mediated down-regulation of complement factor H (CFH) in metal-sulfate-stressed human brain cells. J Inorg Biochem 2009;103(11):1591–5.
[118] Lukiw WJ, Pogue AI. Induction of specific micro RNA (miRNA) species by ROS-generating metal sulfates in primary human brain cells. J Inorg Biochem 2007;101(9):1265–9.
[119] Alexandrov PN, Zhao Y, Jones BM, Bhattacharjee S, Lukiw WJ. Expression of the phagocytosis-essential protein TREM2 is down-regulated by an aluminum-induced miRNA-34a in a murine microglial cell line. J Inorg Biochem 2013;128:267–9.
[120] Ren X, McHale CM, Skibola CF, Smith AH, Smith MT, Zhang L. An emerging role for epigenetic dysregulation in arsenic toxicity and carcinogenesis. Environ Health Perspect 2011;119(1):11–9.

[121] Rager JE, Bailey KA, Smeester L, Miller SK, Parker JS, Laine JE, et al. Prenatal arsenic exposure and the epigenome: altered microRNAs associated with innate and adaptive immune signaling in newborn cord blood. Environ Mol Mutagen 2014;55(3):196–208.

[122] Zhao CQ, Young MR, Diwan BA, Coogan TP, Waalkes MP. Association of arsenic-induced malignant transformation with DNA hypomethylation and aberrant gene expression. Proc Natl Acad Sci USA 1997;94(20):10907–12.

[123] Jensen TJ, Novak P, Eblin KE, Gandolfi AJ, Futscher BW. Epigenetic remodeling during arsenical-induced malignant transformation. Carcinogenesis 2008;29(8):1500–8.

[124] Zhou X, Sun H, Ellen TP, Chen H, Costa M. Arsenite alters global histone H3 methylation. Carcinogenesis 2008;29(9):1831–6.

[125] Senut MC, Sen A, Cingolani P, Shaik A, Land SJ, Ruden DM. Lead exposure disrupts global DNA methylation in human embryonic stem cells and alters their neuronal differentiation. Toxicological sciences. Off J Soc Toxicol 2014;139(1):142–61.

[126] Luo M, Xu Y, Cai R, Tang Y, Ge MM, Liu ZH, et al. Epigenetic histone modification regulates developmental lead exposure induced hyperactivity in rats. Toxicol Lett 2014;225(1):78–85.

[127] Hossain MB, Vahter M, Concha G, Broberg K. Low-level environmental cadmium exposure is associated with DNA hypomethylation in Argentinean women. Environ Health Perspect 2012;120(6):879–84.

[128] Sanders AP, Smeester L, Rojas D, DeBussycher T, Wu MC, Wright FA, et al. Cadmium exposure and the epigenome: exposure-associated patterns of DNA methylation in leukocytes from mother-baby pairs. Epigenetics Off J DNA Methylation Soc 2014;9(2):212–21.

[129] Report of the international committee on nickel carcinogenesis in man. Scand J Work Environ Health 1990;16(1):1–82.

[130] Beveridge R, Pintos J, Parent ME, Asselin J, Siemiatycki J. Lung cancer risk associated with occupational exposure to nickel, chromium VI, and cadmium in two population-based case-control studies in Montreal. Am J Ind Med 2010;53(5):476–85.

[131] Wu CH, Tang SC, Wang PH, Lee H, Ko JL. Nickel-induced epithelial-mesenchymal transition by reactive oxygen species generation and E-cadherin promoter hypermethylation. J Biol Chem 2012;287(30):25292–302.

[132] Zhang J, Zhang J, Li M, Wu Y, Fan Y, Zhou Y, et al. Methylation of RAR-beta2, RASSF1A, and CDKN2A genes induced by nickel subsulfide and nickel-carcinogenesis in rats. Biomed Environ Sci BES 2011;24(2):163–71.

[133] Arita A, Niu J, Qu Q, Zhao N, Ruan Y, Nadas A, et al. Global levels of histone modifications in peripheral blood mononuclear cells of subjects with exposure to nickel. Environ Health Perspect 2012;120(2):198–203.

[134] Tchou-Wong KM, Kiok K, Tang Z, Kluz T, Arita A, Smith PR, et al. Effects of nickel treatment on H3K4 trimethylation and gene expression. PLoS One 2011;6(3):e17728.

[135] Zhang X, Wallace AD, Du P, Lin S, Baccarelli AA, Jiang H, et al. Genome-wide study of DNA methylation alterations in response to diazinon exposure in vitro. Environ Toxicol Pharmacol 2012;34(3):959–68.

[136] Itoh H, Iwasaki M, Kasuga Y, Yokoyama S, Onuma H, Nishimura H, et al. Association between serum organochlorines and global methylation level of leukocyte DNA among Japanese women: a cross-sectional study. Sci Total Environ 2014;490:603–9.

[137] Manikkam M, Tracey R, Guerrero-Bosagna C, Skinner MK. Pesticide and insect repellent mixture (permethrin and DEET) induces epigenetic transgenerational inheritance of disease and sperm epimutations. Reprod Toxicol (Elmsford, NY) 2012;34(4):708–19.

[138] Zama AM, Uzumcu M. Targeted genome-wide methylation and gene expression analyses reveal signaling pathways involved in ovarian dysfunction after developmental EDC exposure in rats. Biol Reprod 2013;88(2):52.

[139] Zama AM, Uzumcu M. Fetal and neonatal exposure to the endocrine disruptor methoxychlor causes epigenetic alterations in adult ovarian genes. Endocrinology 2009;150(10):4681–91.

[140] Calafat AM, Kuklenyik Z, Reidy JA, Caudill SP, Ekong J, Needham LL. Urinary concentrations of bisphenol A and 4-nonylphenol in a human reference population. Environ Health Perspect 2005;113(4):391–5.

[141] Singh S, Li SS. Epigenetic effects of environmental chemicals bisphenol a and phthalates. Int J Mol Sci 2012;13(8):10143–53.

[142] Yaoi T, Itoh K, Nakamura K, Ogi H, Fujiwara Y, Fushiki S. Genome-wide analysis of epigenomic alterations in fetal mouse forebrain after exposure to low doses of bisphenol A. Biochem Biophys Res Commun 2008;376(3):563–7.

[143] Dhimolea E, Wadia PR, Murray TJ, Settles ML, Treitman JD, Sonnenschein C, et al. Prenatal exposure to BPA alters the epigenome of the rat mammary gland and increases the propensity to neoplastic development. PLoS One 2014;9(7):e99800.

[144] Manikkam M, Tracey R, Guerrero-Bosagna C, Skinner MK. Plastics derived endocrine disruptors (BPA, DEHP and DBP) induce epigenetic transgenerational inheritance of obesity, reproductive disease and sperm epimutations. PLoS One 2013;8(1):e55387.

[145] Wu S, Zhu J, Li Y, Lin T, Gan L, Yuan X, et al. Dynamic effect of di-2-(ethylhexyl) phthalate on testicular toxicity: epigenetic changes and their impact on gene expression. Int J Toxicol 2010;29(2):193–200.

[146] Kuo CH, Hsieh CC, Kuo HF, Huang MY, Yang SN, Chen LC, et al. Phthalates suppress type I interferon in human plasmacytoid dendritic cells via epigenetic regulation. Allergy 2013;68(7):870–9.

[147] Manini AF, Jacobs MM, Vlahov D, Hurd YL. Opioid receptor polymorphism A118G associated with clinical severity in a drug overdose population. J Med Toxicol Off J Am Coll Med Toxicol 2013;9(2):148–54.

[148] Peoc'h K, Megarbane B. Can mu-opioid receptor A118G gene polymorphism be predictive of acute poisoning severity in the emergency department? J Med Toxicol Off J Am Coll Med Toxicol 2013;9(3):292–3.

[149] Ingelman-Sundberg M, Sim SC, Gomez A, Rodriguez-Antona C. Influence of cytochrome P450 polymorphisms on drug therapies: pharmacogenetic, pharmacoepigenetic and clinical aspects. Pharmacol Ther 2007;116(3):496–526.

[150] Kacevska M, Ivanov M, Ingelman-Sundberg M. Perspectives on epigenetics and its relevance to adverse drug reactions. Clin Pharmacol Ther 2011;89(6):902–7.

[151] Heyn H. A symbiotic liaison between the genetic and epigenetic code. Front Genet 2014;5:113.

[152] Zhang H, Wang F, Kranzler HR, Yang C, Xu H, Wang Z, et al. Identification of methylation quantitative trait loci (mQTLs) influencing promoter DNA methylation of alcohol dependence risk genes. Hum Genet 2014;133(9):1093–104.

[153] Heyn H, Sayols S, Moutinho C, Vidal E, Sanchez-Mut JV, Stefansson OA, et al. Linkage of DNA methylation quantitative trait loci to human cancer risk. Cell Rep 2014;7(2):331–8.

[154] Verma M. Epigenome-wide association studies (EWAS) in Cancer. Curr Genomics 2012;13(4):308–13.

[155] Rakyan VK, Down TA, Balding DJ, Beck S. Epigenome-wide association studies for common human diseases. Nat Rev Genet 2011;12(8):529–41.

ENVIRONMENTAL PERSONALIZED EPIGENETICS

10

Environmental Contaminants and Their Relationship to the Epigenome

Andrew E. Yosim[1], Monica D. Nye[1], Rebecca C. Fry[1,2]

[1]Department of Environmental Sciences and Engineering, University of North Carolina, Chapel Hill, NC, USA; [2]Curriculum in Toxicology, School of Medicine, University of North Carolina, Chapel Hill, NC, USA

OUTLINE

1. INTRODUCTION

Exposure to toxic environmental contaminants can increase the risk for both cancer and noncancer endpoints and is estimated to contribute over 80 million disability-adjusted life years (DALYs) to the global burden of disease (reviewed in [1]). For many environmental contaminants, a precise mechanism of action of disease causation is unknown. Moreover, while some environmental agents directly damage DNA resulting in potentially mutagenic lesions, others do not, thus complicating an understanding of their modes of action in disease causation particularly in relationship to cancerous endpoints. The field of epigenetics, which translates to "above or on top of the genome," focuses on unraveling the mechanistic basis of disease not explained by direct action at a base pair/ sequence level. Epigenetic alterations do not directly modify DNA bases, yet can impact gene expression by modulating transcriptional activity and the levels of subsequently encoded proteins. Three well-studied epigenetic mechanisms include DNA methylation (i.e., 5-methyl cytosine), histone modification, and altered microRNAs (miRNAs) expression. Research supports the hypothesis that environmental contaminants/ factors can influence all three of these mechanisms (reviewed in [2]) and thus alter the epigenome.

An increasing number of environmental contaminants are implicated in disease, and many of these are suspected to have underlying epigenetic modes of action that contribute in part to their toxicities (reviewed in [2]). For the purposes of this chapter, contaminants and environmental factors were prioritized based on known detrimental impacts on human health including their association with either cancer or noncancer outcomes. Specifically, here we highlight known associations between 11 environmental contaminants/factors and the epigenome including: (1) air pollutants; (2) alcohol; (3) toxic metals/metalloids including (a) inorganic arsenic, (b) cadmium, (c) lead, (d) chromium, (e) mercury, and (f) nickel; (4) antibiotics; (5) antidepressants; and (6) cigarette smoke. Particular emphasis was made to highlight research relating prenatal and early life exposures to contaminants and their associated epigenetic modifications. However, for recent comprehensive reviews of the selected contaminants/factors and their broader association with modifications to the epigenome, readers can refer to the following publications by type of contaminant/factor: inorganic arsenic [3], cadmium [4], nickel [5], chromium, mercury, and lead [4], air pollution [6], alcohol [7], antibiotics, and antidepressants [8].

2. OVERVIEW OF EPIGENETIC MECHANISMS AND METHODS FOR ASSESSMENT

The three commonly studied epigenetic events are DNA methylation, histone modifications, and miRNA expression, and are detailed briefly in the following section. DNA methylation is the addition of a methyl group to the 5′ position of cytosine to form 5-methyl cytosine (5mC). While the majority of DNA methylation events occur at CpG sites or cytosines proximal to guanines, other nucleotide combinations may also be methylated such as CpA, CpC, and CpT, though these are estimated to occur far less frequently [9]. The addition of methyl groups occurs through the action of enzymes known as DNA methyltransferases (DNMTs), which transfer the methyl group from the donor S-adenosylmethionine (SAM). Traditionally, DNMT3a and DNMT3b are thought to be responsible for de novo methylation, whereas DNMT1 maintains DNA methylation marks. Studies assessing DNA methylation marks traditionally focus on either gene-specific methylation marks (often in the promoter region of genes) or more comprehensive measures across the genome. Genome measurements can be either "global," which assesses the DNA methylation status of highly conserved elements such as Line-1 or Alu, or "genome-wide," which measures the methylation status of individual CpG sites across thousands of genes.

While the mechanisms of DNA *de*methylation are still largely unknown, it has been proposed that the TET family of proteins may be actively involved in both passive and active demethylation. The proteins TET1 and TET3 have been shown to oxidize 5mC into 5-hydroxymethylcytosine (5hmC), which is further oxidized into 5-formylcytosine (5fC) and 5-carboxylcytosine (5caC) [10]. The enzyme thymine DNA glycosylase (TDG) can convert both 5fC and 5caC into an abasic site that is further processed to an unmethylated cytosine via base excision repair.

DNA hypermethylation that occurs within the promoter region of genes is traditionally thought of as a gene-silencing mechanism (reviewed in [11]). Interestingly, DNA methylation does not always confer gene silencing. For example, in cases where DNA methylation occurs within gene bodies, such marks have been associated with gene activation (reviewed in [12]), thus highlighting the complexity of these relationships and their apparent dependence on genomic positioning. The promoter regions of several tumor suppressors have been observed to be frequently hypermethylated in many cancers, and these "marks" have been used for both diagnostic and prognostic purposes (reviewed in [13]). Furthermore, toxicological studies have shown associations between environmental contaminants and aberrant patterns of DNA methylation with these modifications implicated in both cancer and noncancer endpoints (reviewed in [2,4]).

Clinically, therapeutics are being designed to specifically target aberrant methylation marks including nucleoside and nonnucleoside analog DNMT inhibitors (DNMTi) (reviewed in [14]). The most common DNMTi utilize cytosine analogs that are modified at the 5′ position and incorporated into the DNA, inhibiting the addition of methyl groups by DNMT enzymes (reviewed in [15]). Additionally, the DNA methylation status of target genes is currently being tested to predict the efficacy of several chemotherapeutics. For example, the mutation status of the oncogene Kirsten rat sarcoma viral oncogene homolog (*KRAS*) is used to predict whether individuals will respond to treatment that inhibits epidermal growth factor receptor (*EGFR*)-driven tumor growth in colorectal cancer (reviewed in [16]). Studies have shown that the methylation status of several genes, including O^6-methylguanine-DNA methyltransferase (*MGMT*) hypermethylation, is associated with mutations to *KRAS* [17]. *MGMT* plays a critical role in DNA damage repair and is responsible for direct removal of the DNA adduct O^6-methylguanine, and consequently aberrant methylation of *MGMT* is suspected to directly impact the potential for *KRAS* mutations. Furthermore, the methylation status of other oncogenes or tumor suppressors such as cyclin-dependent kinase inhibitor 2A (*CDKN2A*), breast cancer 1, early onset (*BRCA1*), and Ras association (RalGDS/AF-6) domain family member 1 (*RASSF1A*) have directly been used as measures of cancer diagnosis and prognosis (reviewed in [18]). Consequently, the methylation status of specific genes may be of clinical relevance in predicting cancer outcomes, as well as efficacy of chemotherapeutics.

In addition to DNA methylation as a modification that can impact gene expression, posttranslational modifications to the N-terminal tail of histone proteins impact the structure of chromatin resulting in remodeling and impacting access of transcriptional machinery to DNA. Three of the most well-studied histone modifications include histone methylation, acetylation, and phosphorylation, although several other modification classes exist including isomerization, sumoylation, ubiquitylation, deamination, and ribosylation of a variety of amino acids (reviewed in [19]). Histone modifications often regulate transcription, where in general lysine acetylation is associated with the loosely packed and open euchromatin, while lysine methylation is associated with the tightly packed chromatin, heterochromatin. However, such posttranslational mechanisms have also been shown to recruit other nonhistone proteins, such as transcription factors that subsequently influence messenger RNA (mRNA) abundance.

Patterns of histone modifications are maintained by a variety of enzymes including histone acetyltransferases (HAT), methyltransferases (HMT), deacetylases (HDAC), and demethyltransferases (HDMT).

Histone modifications may alter transcriptional activity of genes and subsequently the expression of proteins. Evidence suggests a variety of environmental contaminants including but not limited to toxic metals, cigarette smoke, and alcohol are associated with altered histone modifications, and such modifications have been implicated as mechanisms linking environmental exposures to disease (reviewed in [2,20]). Clinically, decreased levels of global histone modifications have been associated with increased cancer severity, and specific histone modifications have been used as prognostic indicators for a variety of cancers including the lung, breast, prostate, and kidney (reviewed in [21]). Research into therapeutics with the ability to inhibit the activity of various histone-modifying enzymes including HAT, HMT, HDAC, and HDMT is ongoing. Additionally, global levels of histone modifications such as dimethylation of histone H3 lysine 4 (H3K4me2), histone H3 lysine 9 (H3K9me2), or acetylation of histone H3 lysine 18 (H3K18ac) may be used to predict recurrence and clinical outcomes in a variety of carcinomas including lung, prostate, and kidney cancers (reviewed in [21]).

In addition to DNA methylation and histone modifications as forms of epigenetic modifications, miRNAs are noncoding RNAs that can posttranscriptionally regulate the genome by targeting mRNAs for degradation. These small RNAs are approximately 22 nucleotides long and serve generally as negative regulators of transcriptional targets, although miRNAs may also play a role in activating transcription or translation (reviewed in [22]). Transcriptional targets are degraded via complementary base pairing of the 5′ region of miRNA with the 3′ untranslated region (UTR) of mRNA, where a highly conserved "seed" approximately seven nucleotides in length is used for recognition and binding to the mRNA. A single miRNA may regulate many different mRNAs, and one mRNA is often regulated by a network of different miRNAs. Noncoding RNAs are implicated in organismal development, apoptotic pathways, and carcinogenesis (reviewed in [23]). A number of tumors displaying specific miRNA signatures including cancers of the lung, kidney, breast, and colon have been identified (reviewed in [24]). In fact, given the frequency with which cancers are associated with changes in miRNA abundance, tumors have been classified by their miRNA profiles with a high degree of specificity (reviewed in [25]). The role of miRNA dysregulation in cancer is of particular interest, as a large number of environmental toxicants have been associated with changes in miRNA expression (reviewed in [26]). Clinically, changes in miRNA abundance may help to inform biological mechanisms linking exposure to disease and serve as diagnostic and prognostic biomarkers. For example, miR-26 levels are useful for predicting both the course of hepatocellular carcinoma and an individual's response to interferon-α treatment, while miR-21 levels have been used to predict the efficacy of chemotherapy for the treatment of adenocarcinomas (reviewed in [24]). Given the links between miRNA dysregulation and disease, the use of miRNA profiling may serve as a valuable tool in determining inter-individual responsivity to clinical treatments.

The ability to study changes to the epigenome has been vastly improved by the advent of technologies that enable genome-wide study of chromatin modifications, histone modifications, DNA methylation, and miRNA levels. For example, bisulfite sequencing of chromatin-immunoprecipitated (BisChIP-seq) DNA allows for study of the DNA methylation status of histone-modified DNA [27]. Sequential chromatin immunoprecipitation followed by bisulfite sequencing (ChIP-BS-seq) enables an understanding of the interactions between DNA methylation and chromatin. The oxidative bisulfite sequencing (oxBS-Seq) technique allows for distinguishability between cytosine (C), methylcytosine (5mC), and hydroxymethylcytosine (5hmC) (reviewed in [27]). In addition, recent advances in high-throughput technologies are enabling study of the epigenome at a systems level (reviewed in [28]). Specifically, DNA microarrays are useful to interrogate the entire human DNA methylome, and advances in array technology are increasing the number of noncoding RNA, histone modifications, chromatin modifications, and proteins that can be assayed to many thousands of molecules in a single experiment.

3. AIR POLLUTION–ASSOCIATED CHEMICALS AND THE EPIGENOME

The World Health Organization (WHO) has estimated that annually, approximately seven million individuals die prematurely as a result of exposure to air pollutants [29]. This estimate approximates almost one out of eight deaths worldwide, resulting in air pollution's ranking as the largest single health risk due to an environmental contaminant [29]. In addition to increased mortality from heart disease, stroke, respiratory infection, and chronic obstructive pulmonary disease, both indoor and outdoor air pollutants are associated with increased incidence of cardiovascular, respiratory, and cerebrovascular disease (reviewed in [30,31]). Exposure to air pollutants is associated with reductions in lung function, increased medication usage, and acute cardiovascular incidents (reviewed in [32]). While many substances are considered air-based pollutants, for the purposes of this chapter, we have prioritized studies detailing the effects of particulate matter (PM), ozone, black carbon (BC), sulfate, and nitrogen dioxide. PM refers to particles or liquid constituents and can encompass airborne metals, dust, acids, or organic matter. PM is often categorized based on particulate diameter of interest, as the size of the pollutant is often correlated with the extent a particulate travels through the respiratory system, thus influencing the PM's associated health effects (reviewed in [33]). Two of the most well-studied categories of PM are coarse particles (PM_{10}) and fine particles ($PM_{2.5}$). $PM_{2.5}$ is often a byproduct of combustion, and is thought to be a more serious threat to human health than PM_{10}. Evidence suggests that PMs may act through the activation of pulmonary inflammation or immune response pathways, blood coagulation, direct translocation of PM into the circulatory system, production of reactive oxygen species, and systemic inflammation, in addition to epigenetic mechanisms (reviewed in [30,34]). To date, a limited number of studies have investigated the associations between ambient air pollution and epigenetic modifications and whether such modifications are predictive of increased disease susceptibility. Research has shown that $PM_{2.5}$ exposure in adults is associated with promoter hypomethylation of inducible nitric oxide synthase (*iNOS*) in peripheral blood lymphocytes (PBLs) (reviewed in [35]). Similarly, *iNOS* hypomethylation has been observed in PBLs collected from children exposed to $PM_{2.5}$ [36]. This hypomethylation had functional impact where children with *iNOS* hypermethylation had higher levels of fractional exhaled nitric oxide (FeNO), a measure of exhaled nitric oxide, while children with lower *iNOS* methylation levels showed no association between FeNO and $PM_{2.5}$ exposure. FeNO is often used as a measure of airway inflammation, and such results may demonstrate associations between gene-specific methylation following $PM_{2.5}$ exposure and disease states related to cardiovascular or respiratory health (reviewed in [37]).

In addition to $PM_{2.5}$, a number of other air pollutants have been studied for their association with epigenetic alterations. Together with our colleagues, we have shown that humans exposed to 0.4 ppm O_3 displayed changes in the abundance of 10 miRNAs in sputum: miR-132, miR-143, miR-145, miR-62 199a*, miR-199b-5p, miR-222, miR-223, miR-25, miR-424, and miR-582-5p [38]. These miRNAs are associated with inflammation and immune cell response, and of particular interest to personalized epigenetics, baseline measures of the miRNAs were associated with individual responses to the exposure and therefore may be utilized as predictors of interindividual response and susceptibility to O_3. In a separate cross-sectional study, 181 children from a community with high levels of air pollution in Fresno, CA were compared with a community with lower levels in Palo Alto, CA, and air pollution was found to be associated with hypermethylation of forkhead box transcription factor 3 (*FOXP3*) and interferon-gamma (*IFNG*) in peripheral blood T cells [39]. Both *FOXP3* and *IFNG* play roles in asthma and other respiratory conditions given their influence on regulatory T cells and several inflammatory pathways. Childhood asthma has been associated with gene-specific DNA methylation patterning following exposure to both

BC and sulfate (reviewed in [35]). As a consequence, it is predicted that epigenetic dysregulation may help to inform the link between childhood exposure to air pollution and a range of respiratory conditions and subsequent health outcomes.

In addition to studies assessing epigenetic alterations associated with childhood or adult air pollution exposure, researchers have also assessed the epigenetic consequences of prenatal exposure. For example, placental tissue from 240 newborns was assessed, and exposure to $PM_{2.5}$ during pregnancy was associated with global hypomethylation in the placenta, with the largest association seen during the first trimester [40]. A similar trend was noted among a cohort of 164 newborns, where exposure to polycyclic aromatic hydrocarbons was associated with global hypomethylation in cord blood lymphocytes [41]. Alterations to global DNA methylation levels as a consequence of prenatal exposure may be particularly detrimental to the health of the newborn as the growing fetus progresses through critical windows of development and de novo methylation marks are established (reviewed in [12]).

Clinically, air pollution–induced changes to the epigenome may ultimately be useful to inform disease susceptibility. As a specific example, in a study assessing childhood exposure to nitrogen dioxide, methylation of the gene adrenoceptor beta 2, surface (*ADRB2*) in peripheral blood samples was associated with asthma status [42]. Furthermore, children with hypermethylation of the 5′ UTR of *ADRB2* had higher risk of severe asthma following SO_2 exposure, while exposure was not correlated with disease among children with hypomethylation of *ADRB2*. As a result, the methylation status of *ADRB2* may serve as a useful predictive biomarker for an asthmatic individual's response to air pollution. In a separate study, individuals with *TLR-2* promoter hypermethylation or global hypermethylation were more susceptible (as measured via intercellular adhesion molecule 1 (ICAM-1) and vascular cell adhesion molecule 1 (VCAM-1)) to a range of air pollutants including NO_2, CO, and BC [43]. These results are promising, as they suggest that DNA methylation patterning may be predictive of disease response following exposure to specific air pollutants. While future research is needed to replicate and substantiate these findings, gene-specific epigenetic alterations have the possibility of predicting whether vulnerable populations, including those with respiratory conditions, may have increased susceptibility to different air pollutants.

4. ALCOHOL AND THE EPIGENOME

Classical genetic models of alcohol dependence and alcohol-associated carcinogenesis have traditionally focused on specific genes tied to the metabolism of alcohol, such as alcohol dehydrogenase and aldehyde dehydrogenase, which play an active role in the breakdown of alcohol (reviewed in [44]). Recently, several genome-wide association studies have been conducted, identifying dozens of possible genes with polymorphisms associated with alcohol dependence (reviewed in [45]). In addition to the study of genetic variants, research into alcohol-associated effects on the epigenome has focused on genes that may underlie addiction, modulate cravings, or explain interindividual differences in alcohol-associated diseases such as liver or colon cancer. Presently, a consensus has not been reached related to alcohol's global effects on DNA methylation, as both hyper- and hypomethylation have been reported in PBLs (reviewed in [7]). Gene-specific promoter methylation levels in PBLs have been assessed in alcohol-dependent patients and several genes have been shown to be differentially methylated, including vasopressin (*AVP*), atrial natriuretic peptide (*ANP*), Solute Carrier Family 6 (Neurotransmitter Transporter), member 3 (SLC6A3), dopamine receptor D4 (*DRD4*), homocysteine-induced ER protein (*HERP*), neural cell adhesion molecule 1 (*NCAM1*), nerve growth factor, prodynorphin (*PDYN*), pro-opiomelanocortin (*POMC*), alpha synuclein (*SNCA*), and thioredoxin reductase 1 (*TXNRD1*) [7].

Current evidence suggests that epigenetic marks may inform therapies or predict individuals at heightened susceptibility to alcohol dependence. As a specific example of this, classical models of the heritability of alcohol dependence have focused on a polygenic model that relies on single nucleotide polymorphisms (SNPs) associated with alcohol dependence. Recent work has shown DNA methylation of these SNPs in the brains of deceased human alcoholics to be associated with changes in their predicted alcohol dependence, as no- or low-risk alleles could be reversed via methylating marks [46]. The DNA hypermethylation status of the SLC6A3 in PBLs has been observed in individuals seeking treatment for alcohol dependence, compared with control subjects [47]. In addition, DNA methylation levels of SLC6A3 were inversely associated with alcohol cravings, suggesting interindividual differences in susceptibility to alcohol dependence. In addition to using epigenetic marks/alterations to inform susceptibility to alcohol dependence, these data may also be useful in the identification or prognosis of alcohol-related cancers. For example, promoter regions of the genes *RASSF1*, *MGMT*, and retinoblastoma 1 (*RB1*) are hypermethylated in hepatocellular carcinomas likely associated with alcohol intake (reviewed in [48]). Consumption of alcohol has been correlated with colon cancer and inversely related to global DNA methylation (reviewed in [48]). The association of alcohol with aberrant DNA methylation patterning may be partly explained by its role in the reduction of folate and a subsequent reduction in the levels of the methyl donor SAM (reviewed in [49]). Reductions in SAM are associated with the inhibition of DNA methyltransferases (reviewed in [7]), and may help to explain alcohol's associated DNA methylation patterns. Epidemiologic evidence supports these associations, as individuals with high consumption of alcohol have lower expression of *DNMT3a/DNMT3b*, and folate insufficiency has been associated with increased risk for a number of cancers (reviewed in [50]). In addition, individuals with low folate and high alcohol consumption had a greater frequency of hypermethylation in colorectal tumors of a number of genes involved in carcinogenesis including *CDKN2A*, mutL homolog 1 (*MLH1*), *MGMT*, and *RASSF1A* (reviewed in [48]). As research continues into the epigenetic mechanisms underlying alcohol's physiological effects, epigenetic marks and modifications may ultimately be used to predict individual susceptibility to alcohol dependence or alcohol-associated disease. Ultimately, this information may be useful in the identification of targeted therapeutics to treat alcohol-associated conditions.

5. METALS AND THE EPIGENOME

5.1 Inorganic Arsenic

More than 200 million individuals worldwide are chronically exposed to inorganic arsenic at levels that exceed the WHO's limit of 10 ppb inorganic arsenic in water (reviewed in [51]). Chronic exposure to this metalloid occurs primarily through water contaminated by naturally occurring sources of arsenic within the earth's crust. Additional exposure comes from current and previous anthropogenic activities where individuals are exposed through soil, air, food, and/or occupational exposures (reviewed in [51]). Chronic exposure to inorganic arsenic is associated with both cancer and noncancer endpoints including diabetes, cardiovascular disease, neuropathy, and cancers of the skin, lung, kidney, bladder, and liver (reviewed in [51]). There are numerous mechanisms of action suspected to underlie arsenic's toxicity with some of the most well studied being the generation of oxidative stress, dysregulation of DNA repair, and epigenetic alterations, including changes to DNA methylation, histones, and miRNAs (reviewed in [52]). Following exposure to inorganic arsenic, the body metabolizes the toxicant via a series of oxidation and reduction reactions, resulting in several intermediate metabolites. The methylation process requires SAM as one of the primary methyl donors for the specific arsenic methyltransferase (AS3MT) (reviewed in [52]). As mentioned previously, SAM is also a methyl donor required for DNMTs. Exposure to inorganic arsenic is known to influence global DNA methylation levels (reviewed in [4]). In addition to changes in the patterns of global DNA methylation, inorganic arsenic has been associated with changes in gene-specific DNA methylation, histone modifications, and changes in miRNA abundance [3,4].

Epidemiologic studies have assessed inorganic arsenic-associated epigenetic alterations and the link between these mechanisms and disease.

Of particular interest, prenatal inorganic arsenic exposure has been associated with global and gene-specific changes in DNA methylation (reviewed in [3]). Gene-specific DNA methylation analysis has uncovered a number of genes that are differentially methylated in peripheral blood samples associated with inorganic arsenic including: cyclin D1 (*CCND1*), *CDKN2A/TP16*, death-associated protein kinase 1 (*DAPK*), *KRAS*, *RASSF1A*, and tumor protein 53 (*TP53*) (reviewed in [4]). While the genes *TP53*, *CDKN2A*, *DAPK*, and *RASSF1A* are tumor suppressors, the genes *CCND1* and *KRAS* are oncogenes. In relationship to disease, epigenetic modifications to *CDKN2A* have been associated with colon, lung, lymphoma, bladder, brain, and esophageal cancers, while hypermethylation of *DAPK* has been implicated in lung, lymphoma, gastric, cervical, and bladder cancers (reviewed in [13,53]). Furthermore, promoter hypermethylation of *RASSF1A* has been associated with kidney, lung, liver, thyroid, and ovarian cancers (reviewed in [11,53]). In addition to DNA methylation, our lab has shown prenatal exposure to arsenic is associated with increased expression of 12 miRNAs in cord blood samples: let-7a, miR-107, miR-126, miR-16, miR-17, miR-195, miR-20a, miR-20b, miR-26b, miR-454, miR-96, and miR-98 [54]. Canonical pathway analysis of the altered miRNAs identified that many play roles in cancer, inflammatory response, and diabetes, suggesting that such arsenic-associated miRNA changes may help to inform arsenic-associated disease states.

Currently, the use of epigenetic data in the clinical setting builds upon global and gene-specific epigenetic alterations and their associations with cancer or other noncancerous health outcomes. A number of the previously mentioned arsenic-associated genes are currently used for diagnostic purposes, independent of arsenic-associated disease. These include the use of sputum, urine, and nipple aspirate to measure the DNA methylation of *CDKN2A* in lung, prostate, and breast cancers, while sputum and nipple aspirate samples are used to measure *DAPK* and *RASSF1A* methylation in lung and breast cancer (reviewed in [18]). The methylation status of the genes *DAPK*, *CDKN2A*, and *RASSF1A* are also currently used for prognostic purposes (reviewed in [18]), as hypermethylation of these tumor suppressors is associated with higher odds ratios for cancer-related mortality in lung, brain, and colorectal cancers (reviewed in [55]).

5.2 Cadmium

Cadmium is a naturally occurring toxic metal present in the earth's crust as well as a well-studied carcinogen and toxicant. Exposure to cadmium occurs primarily through dietary intake or cigarette smoke, and many individuals worldwide may also be at increased risk due to occupational exposures (reviewed in [56]). Moreover, in the last several decades, anthropogenic utilization of cadmium has led to increased human exposure from a number of sources including waste/emissions from mining, electroplating, and battery production, and such exposures are particularly frequent in developing countries (reviewed in [57]). Cadmium is classified by the International Agency for Research on Cancer (IARC) as a Group I carcinogen and associated with cancers of the liver, bone, kidney, and pancreas, as well as being a risk factor for several noncancer endpoints including cardiovascular and kidney disease (reviewed in [56,58]). Several mechanisms of action for cadmium's toxicity have been proposed including the inhibition of DNA repair, generation of reactive oxygen species, alterations of cell cycle progression, and epigenetic alterations (reviewed in [59]).

Epidemiologic studies have assessed cadmium's associated epigenetic alterations and the link between these mechanisms and disease. Of particular interest, prenatal cadmium exposure has been associated with global and gene-specific changes in DNA methylation. Among infants exposed in utero, cadmium exposure was associated with altered promoter DNA methylation in PBLs among a diverse set of 61 genes [60]. In a separate prenatal cohort, researchers identified 54 genes differentially methylated in cord blood in response to cadmium [61]. Many of the 54 dysregulated genes are involved in cellular development, cell cycle regulation, and apoptosis, implying that exposure-associated aberrant DNA methylation, if associated with functional changes, may impact an individual's health. This is especially important given the accumulation of cadmium in the placenta and susceptibility of developing fetuses to developmental toxicants.

Several in vitro studies have assessed gene-specific changes in DNA methylation associated with cadmium exposure. Among transformed human prostate epithelial cells, cadmium exposure resulted in hypermethylation of the tumor suppressors CDKN2A and RASSF1A [62]. In a separate study assessing peripheral lung epithelial cells, chronic subcytotoxic cadmium exposure was also associated with decreased expression of CDKN2A [63], suggesting the tumor suppressor may be a key gene of interest for cadmium-induced carcinogenesis. Cadmium exposure has been shown to result in promoter hypermethylation within the DNA repair genes MutS homolog 2 (MSH2), excision repair cross-complementation group 1 (ERCC1), X-ray repair complementing defective repair 1 (XRCC1), and 8-oxoguanine DNA glycosylase (OGG1) [64].

The aberrant methylation of various tumor suppressors and DNA repair genes may be of considerable interest in the clinical setting, as such alterations are potentially reversible and may serve as therapeutic targets. Hypermethylation of CDKN2A and RASSF1A is associated with a number of cancers including prostate, lung, and breast, and the methylation status of these tumor suppressors has been used in predicting disease outcome (reviewed in [18]). Additionally, the epigenetic dysregulation of DNA repair genes is a common element in cancer. As a result, such methylation marks may ultimately serve to inform an individual's risk of cancer and as targets for possible therapeutic interventions.

5.3 Lead

Lead is a ubiquitous nondegradable metal that occurs in the environment through natural and anthropogenic sources. Worldwide, common sources of lead exposure include contaminated water, lead-containing gasoline/pigments, solder, as well as its use in piping, toys, and ceramics (reviewed in [65]). In response to growing evidence that chronic lead exposure was associated with intellectual impairment, the US government mandated the removal of lead from paints and gasoline over four decades ago. As a result, blood lead levels in children fell dramatically. Nevertheless, lead exposure remains a threat to human health—in the United States, almost half a million children exceed the Center for Disease Control and Prevention's level of concern for lead of $5\,\mu g/dL$; worldwide, it is estimated that over 240 million individuals exceed this level of concern (reviewed in [66]). This is particularly important, as epidemiological evidence has shown that blood lead levels below $5\,\mu g/dL$ were associated with detrimental effects on the developing brain (reviewed in [67]), suggesting the lack of any truly "safe" blood lead level.

In addition to the health effects listed above, lead is known to impact human health more broadly and is associated with acute and chronic effects to the nervous, immune, cardiovascular, and renal systems (reviewed in [68]), as well as cancers of the lung, stomach, kidney, and brain [65]. Childhood lead exposure is related to behavioral problems, delayed neurodevelopment, cognitive impairment, and is even predictive of later life violence [67,69]. Similar health outcomes have been noted following prenatal exposure, suggesting that maternal exposure may predispose children to intellectual impairment or future states of disease. The mechanisms by which lead exposure is associated with developmental neurotoxicity and disease are well understood and numerous, with some proposed mechanisms including substitution for zinc or calcium, disruptions to the glutamatergic and dopaminergic systems, and disturbance of the blood–brain barrier (reviewed in [70]). Additionally, recent evidence suggests epigenetic mechanisms that further inform the relationships between lead exposure and disease. Presently, studies of humans exposed to lead and associated epigenetic modifications are limited (reviewed in [4]). In a study based in Mexico, maternal tibia lead levels were associated with hypomethylation in newborn cord blood [71]. While such global hypomethylation of the developing fetus may be attributable to maternal exposure during pregnancy, it is important to note that lead is reabsorbed from the bone into the blood during pregnancy [72]. Because bone contains almost 95% of the cumulative lead stored in the body, pregnancy may represent a particularly critical period during which lead exposure can occur, as the mother's lead burden is mobilized into the placenta. As a result, it is possible that the fetus may experience substantial exposure to the toxic metal irrespective of maternal exposure during development.

In addition to changes in global DNA methylation, gene-specific DNA methylation analyses have identified gene targets of interest. Lead-exposed embryonic stem cells and neural progenitor cells were associated with the hypermethylation of a number of genes including: ephrin-A2 (*EFNA2*), glutamate receptor, ionotropic, kainate 4 (*GRIK4*), and lim homebox 3 (*LXH3*). These genes play roles in neurodevelopment, and aberrant methylation may help to inform the link between lead exposure and developmental neurotoxicity. In another study, researchers found that individuals with high blood lead concentrations had *CDKN2A* hypermethylation in peripheral blood, compared with individuals who had low blood lead levels. In addition to the associations between hypermethylation of *CDKN2A* and cancers of lung, prostate, and breast (reviewed in [18]), CDKN2A is often overexpressed during neurodegeneration [73]. While additional research is warranted, the methylation status of these or other genes may ultimately inform additional molecular mechanisms underlying lead's cancerous and neurodegenerative effects. Given the prevalence of exposure worldwide, lead-associated epigenetic profiles may aid in predicting individuals at heightened risk for disease from chronic or occupational exposure.

5.4 Chromium, Mercury, and Nickel

Chromium, mercury, and nickel represent three ubiquitous toxic metals in the environment associated with a range of adverse health effects including both cancer and noncancer endpoints (reviewed in [74,75]). While much research has focused on the investigation of the health effects associated with these three metals, the mechanisms underlying such health effects are still under active study. For example, hexavalent chromium can induce mutagenic DNA adducts as well as other nonmutagenic cellular impacts, including dysregulation of DNA repair, activation of signaling pathways, and epigenetic mechanisms (reviewed in [75,76]).

Epidemiologic studies have assessed the epigenetic alterations associated with the toxic metals chromium and mercury. Among human cohorts, elevated mercury levels have been associated with promoter hypomethylation of selenoprotein P plasma 1 (*SEPP1*) in samples of hair and urine, which is known to play a role in oxidative stress pathways

[77]. Hexavalent chromium has been associated with global DNA hypomethylation in peripheral blood, and lung tumor samples from workers showed chromium exposure to be associated with promoter hypermethylation of the tumor suppressor genes *CDKN2A*, *MLH1*, *MGMT*, and adenomatous polyposis coli (*APC*) (reviewed in [4]).

To our knowledge, only a few human studies have assessed epigenetic alterations associated with nickel exposure. In two separate cohorts of occupationally exposed workers, nickel exposure was associated with hypermethylation of the cell growth regulator, cyclin-dependent kinase inhibitor 2B (*CDKN2B*) [78], as well as increased levels of histone H3 lysine 4 trimethylation (H3K4me3) and decreased histone H3 lysine 9 dimethylation (H3K9me2) [79]. While in vivo studies of nickel's epigenetic alterations may be sparse, these results are demonstrative of nickel's epigenetic mechanisms of action, as in vitro models of the metal have been associated with alterations in global histone modification levels, repression of histone-modifying enzymes, induction of aberrant DNA methylation patterning, and dysregulation of the family of DNA-demethylating TET proteins [80]. For example, treatment of human bronchial epithelial cells was associated with promoter hypermethylation of cadherin 1, Type 1, E-Cadherin (*CDH1*), while exposure was associated with promoter hypermethylation of *MGMT* in a transformed human bronchial epithelial cell line (reviewed in [5]). In a more recent study, human immortalized bronchial epithelial cells (BEAS-2B) exposed to nickel in vitro displayed increased H3K9me2 [81]. This same study noted that H3K9me2, a repressive mark, was associated with the reorganization of the H3K9me2 domain, which resulted in the spread of the H3K9me2 domain into actively transcribed regions resulting in gene silencing. As a result, the reorganization of H3K9me2 marks may serve as yet another viable epigenetic mechanism by which nickel exposure may be tied to aberrant gene silencing and associated with negative health outcomes.

While further research is still needed to establish the epigenetic alterations associated with exposure to chromium, mercury, and nickel, the current data suggest that epigenetic modifications may inform mechanisms of carcinogenicity. Exposure to both chromium and nickel is associated with hypermethylation of *MGMT* (reviewed in [5]) in tumor samples. Promoter hypermethylation of *MGMT* has been associated with lung, colorectal, and gastric cancers, among others, and has been shown to be a useful tool in their prognosis (reviewed in [14]). Additionally, the methylation status of *MGMT* has been useful in predicting the efficacy of alkylating chemotherapeutics. In addition to the methylation of *MGMT*, the tumor suppressor *MLH1* is another promising gene with implications for clinical practice. As discussed above, exposure to hexavalent chromium is associated with promoter hypermethylation of *MLH1* in lung cancer tumors (reviewed in [4]). In addition, chromium treatment among an immortalized human lung carcinoma line was directly related to dimethylation of H3K9 and inversely related to trimethylation of H3K4 in the promoter region of *MLH1* (reviewed in [4]). Dysregulation of *MLH1* has been associated with hereditary colorectal cancer, suggesting these methylation marks and histone modifications may be useful for diagnostic or prognostic purposes [55]. In addition to *MLH1*, both chromium and nickel exposures have been associated with DNA methylation among a range of other tumor suppressors including *CDKN2A*, *MGMT*, and *APC* (reviewed in [4,5]). If DNA methylation is associated with functional changes in the transcriptome or proteome, these marks may serve as future targets for therapeutics including the treatment or diagnosis of metals-induced cancers.

6. PHARMACOLOGICAL AGENTS AND THE EPIGENOME

6.1 Antibiotics

Antibiotic use during pregnancy is common and has been associated with both lowered birth weight and abnormal weight later in childhood [82]. Decreased birth weight is associated with later health outcomes, as abnormal birth weight is a risk factor for childhood and adult obesity, type 2 diabetes, cardiovascular disease, and various cancers [83]. While the mechanism linking antibiotic use and lowered birth weight is unknown, epigenetic mechanisms underlie this relationship.

Imprinted genes involved in growth and development, which are epigenetically regulated via DNA methylation or histone modifications, have been studied to examine whether differentially methylated regions mediate the association of antibiotic use during pregnancy with resulting birth weight. Almost 200 genes, or 1% of the genome, are predicted to be "imprinted" (reviewed in [84]). Imprinted genes, unlike their nonimprinted counterparts, are expressed in a parent-of-origin-dependent manner. As a result, one parental allele is silenced and becomes nonfunctional while the other copy is expressed (reviewed in [85]). Consequently, imprinted genes are particularly susceptible to a range of exogenous and endogenous chemicals or stressors that may impact the genome or epigenome (reviewed in [85]). Genomic imprinting is controlled through DNA methylation and histone modifications, and as a result, imprinted genes may be particularly susceptible to environmental toxicants that may modify the enzymes controlling these two epigenetic mechanisms (reviewed in [84]). Moreover, given the relationship between environmental contaminants and their role in disease, imprinted genes are being investigated as potential biomarkers or therapeutic targets.

The DNA methylation profiles of five imprinted genes—H19 gene (*H19*), insulin-like growth factor 2 (*IGF2*), maternally expressed gene 3 (*MEG3*), paternally expressed gene 3 (*PEG3*), and pleomorphic adenoma gene-like 1 (*PLAGL1*)—are altered in infants whose mothers used antibiotics during pregnancy [83]. In addition to maternal antibiotic use, altered DNA methylation in *PLAGL1* was correlated with decreased birth weight. *PLAGL1* is a paternally expressed imprinted gene that may act to regulate the expression of many imprinted and nonimprinted genes [86]. Both *PLAGL1* and many of the genes it regulates are involved in growth and development, which may link epigenetic dysregulation of *PLAGL1* with altered fetal growth. As a result, antibiotic use associated with aberrant methylation of *PLAGL1* has the potential to impact a much larger set of genes, and may support the association between antibiotic exposure and intrauterine growth effects (reviewed in [87]).

6.2 Antidepressants

Over the last several decades, the use of antidepressant medication in the United States has increased dramatically. Given their ability to produce stable physiological changes, epigenetic mechanisms have been proposed. While epigenetic alterations associated with antidepressant use are well documented in live rodent models, only a few epidemiologic studies have investigated the epigenetic effects. Among 25 depressed subjects, the antidepressant citalopram was associated with decreased trimethylation of histone 3, lysine 27 (H3K27me3) in the promoter region of brain-derived neurotrophic factor (*BDNF*) in peripheral blood samples [88]. H3K27me3 is often correlated with decreased transcription, and consequently, decreased H3K27me3 (associated with functional changes to BDNF expression) may have long-term implications, as dysregulation of *BDNF* is related to a number of neurological conditions/mood disorders and may be used as a possible target for clinical treatment (reviewed in [89]). In addition, antidepressant use during pregnancy was associated with increased cord blood hypermethylation of *H19* DMRs in a race-dependent manner [90]. While the later life health effects associated with *H19* and *IGF2* hypermethylation are under study, as previously discussed, aberrant methylation of the clustered genes *H19* and *IGF2* is associated with a range of cancer and noncancer endpoints (reviewed in [85]).

In addition to cohort-based studies, research using in vitro model systems is increasing the amount of information on relationships between antidepressant use and epigenetic alterations. In addition to its antiepileptic and mood-stabilizing properties, valproic acid is thought to inhibit HDAC, as demonstrated among human embryonal kidney cells where it was associated with acetylation of H3 histone proteins, as well as active demethylation of transfected DNA [91]. Addition studies by the same group queried gene-specific targets and identified both partial promoter demethylation and promoter hyperacetylation of H3 and H4 histone proteins in melanoma antigen family B, 2 (*MAGEB2*) and matrix metallopeptidase 2 (gelatinase A, 72 kDa gelatinase, 72 kDa type IV collagenase) (*MMP2*) [92]. Importantly, both genes are associated with tumor formation and metastasis, and may suggest that antidepressants that are active demethylating agents or that act to inhibit HDAC should be studied for their roles in cancer susceptibility and progression. Such relationships may extend beyond antidepressants, as the antiepileptic drugs topiramate and the primary metabolite of levetiracetam were associated with HDAC inhibition as measured by increased H4 histone acetylation in an in vitro cervical cancer model (reviewed in [93]). Taken together, these data suggest that epigenetic alterations associated with antidepressant use may aid researchers in determining interindividual differences in treatment efficacy. While there is some indication of tissue conservation for the methylation of some genes [94], further research is still needed in order to determine whether these convenient samples (e.g., peripheral blood) are reflective of the epigenetic marks found in the target tissue, particularly the brain.

7. CIGARETTE SMOKE AND THE EPIGENOME

Cigarette smoke is predicted to become the largest potentially preventable cause of morbidity and mortality in the United States. While the use of tobacco products by adults is a risk factor for many cancer and noncancer endpoints, in utero exposure is particularly detrimental and associated with changes in fetal development and adult-onset diseases (reviewed in [95]). Despite numerous nationwide public health campaigns to communicate the adverse birth outcomes associated with prenatal exposure to cigarette smoke, a significant percentage of women smoke throughout their pregnancies [96]. In utero cigarette smoke exposure is associated with a range of fetal and childhood health effects including: delayed fetal development, lowered birth weight, childhood neurodevelopmental impairment, and increased risk for asthma (reviewed in [95]). In addition to early life health outcomes, prenatal cigarette smoke exposure has been linked with adult-onset diseases such as coronary heart disease, type 2 diabetes, obesity, and cancers of the lung, esophagus, bladder, and stomach (reviewed in [97]). Numerous studies have proposed potential mechanisms linking tobacco smoke to disease including the formation of reactive oxygen species, dysregulation of DNA repair pathways, chromosome instability, and changes in cell cycle progression [98]. In addition, recent evidence suggests epigenetic mechanisms may inform the relationship between many of the health effects associated with in utero cigarette smoke exposure.

Prenatal cigarette smoke exposure may be associated with global hypomethylation [98]. As previously noted, global hypomethylation is related to a number of cancers, and such dysregulation of homeostatic methylation and gene expression may be predictive of future health outcomes. A genome-wide study identified 623 genes differentially methylated in response to maternal smoking, and many of these genes were associated with functional changes to their gene expression [99].

Gene-specific studies have noted a range of genes differentially methylated in response to in utero cigarette smoke exposure. Among smoking mothers, researchers noted promoter hypomethylation of cytochrome P450, family 1, subfamily A, polypeptide 1 (CYP1A1) in placental tissue that was correlated with increased gene expression [100]. In addition to metabolizing a range of carcinogenic compounds found in cigarette smoke, CYP1A1 polymorphisms have been used to predict cancer susceptibility, and dysregulation of its gene product is found in many different smoking-related tumors (reviewed in [101]). This result was replicated by another epigenome-wide study that found a number of genes were differentially methylated including CYP1A1, aryl-hydrocarbon receptor repressor (AHRR), major histocompatibility complex, class II, DP beta 2 (HLA-DPB2), and growth factor independent 1 transcription repressor (GFI1) [102]. AHRR, like CYP1A1, is involved in the detoxification of compounds found in cigarette smoke, and dysregulation of the gene has been implicated in colon, lung, and breast cancers [103].

Furthermore, in a case-control study of children whose mothers smoked during pregnancy, the gene AXL receptor tyrosine kinase (AXL) was hypermethylated using buccal swab samples [104]. Hypermethylation of AXL has been reported in several malignancies including cancers of the lung and breast (reviewed in [105]). Recent investigations suggest that AXL may have promising potential as a therapeutic target for cancer therapies [105]. Researchers have also noted changes in gene-specific methylation at imprinted gene sites. Infants with in utero cigarette smoke exposure displayed hypermethylation of IGF2 and imprinted maternally expressed transcript (H19) DMRs in cord blood compared with control infants [106]. The smoking-associated hypermethylation was most pronounced in males, where 20% of male newborns had low birth weight likely mediated by smoking-associated DNA methylation of IGF2. Aberrant methylation of IGF2 may be associated with additional health effects, as IGF2 expression is associated with overgrowth disorders, obesity, and cancer (reviewed in [107]).

In addition to changes in DNA methylation, exposure to cigarette smoke during pregnancy results in decreased expression of miR-16, miR-21, and miR-146a in human placental tissue (reviewed in [4]). Importantly, the abundance of miR-16 and miR-21 was decreased among infants with low birth weight, which supports the finding that miR-16 expression can be used to predict increased risk for newborns born small for gestational age (reviewed in [108]). In addition, miR-21 is overexpressed in a number of cancers including the breast, colon, lung, and liver (reviewed in [108]).

While clinical applications for smoking-related epigenetic marks are currently under investigation, such data ultimately may help to identify individuals at increased susceptibility to cigarette-associated addiction or disease. Of particular interest to the field of personalized epigenetics, in utero exposure is associated with differential methylation of a number of key genes that may underlie the current manifestation of disease. Additional studies are needed to assess the roles differentially expressed miR-NAs and differentially methylated imprinted genes may have in predicting future health outcomes associated with these epigenetic alterations.

8. FURTHER CONSIDERATIONS

Throughout this chapter, literature detailing associations between environmental contaminants/factors and epigenetic alterations has been highlighted. In many cases, genes that are targets of environmental modifications have known potential applications in predicting disease or informing disease prognosis. As many environmentally induced epigenetic modifications are associated with exposure, these marks may ultimately be used for early disease detection where exposure to environmental toxicants is suspected. There is also evidence that many of these contaminants induce epigenetic alterations after prenatal and early life exposures, highlighting the usefulness of such biomarkers to assess critical developmental windows of susceptibility to disease.

While the field of environmental epigenetics is growing substantially and the number of contaminants and their known epigenetic targets is increasing, challenges remain in the incorporation of environmentally induced epigenetic marks into a personalized medicine framework. There are four primary limitations that currently impede their application including: (1) tissue specificity of epigenetic marks, (2) stability of the epigenetic marks, (3) functional implications of the epigenetic modifications, and (4) feasibility of broadscale application in the clinical setting. To detail, the most commonly used human samples in epigenetic studies are adult-derived peripheral blood (i.e., leukocytes), whereas for prenatal exposures umbilical cord blood is used. Although these blood samples are often used as surrogates for target tissue, tissue-specific epigenetic alterations may complicate inferences derived from blood samples. Another issue to consider is that blood comprises various white blood cell types that can change in abundance upon environmental exposures. Consequently, epigenomic studies that simultaneously measure tissue-specific epigenetic alterations are needed to determine the extent to which these convenient tissue samples reflect patterning in the target tissue. As a second limitation, additional research is needed to assess the stability of epigenetic marks. While current studies can measure global or gene-specific epigenetic modifications, little is known about the stability of such alterations during an individual's lifetime and the possibility of transgenerational inheritance. Additionally, further research is needed for quantitative determinations of specific relationships between epigenetic alterations and associated functional effects at the transcriptional and proteomic levels. Lastly, research is needed to determine interindividual differences in epigenetic alterations and various factors that may modulate differences in epigenetic patterning. Future studies are warranted to identify contaminant-specific epigenetic signatures that may be used as either exposure or effect biomarkers. While the advent of high-throughput technologies has enabled researchers to quickly and effectively conduct epigenome-wide studies, these technologies are still relatively costly and have yet to reach widespread adoption in the clinical setting. As a result, clinical use of personalized epigenetic alterations is hindered by the availability and prohibitive cost of sampling. As research continues and the cost of acquiring personalized epigenetic profiles diminishes, it is predicted that gene-specific or epigenome-wide screens may inform clinicians into the dysregulation of specific gene targets with therapeutic significance. However, further testing is necessary to ensure that the results of these individual epigenetic profiles are reproducible, and special attention should be placed on ensuring that the methodology is assessed using interlaboratory validation techniques.

In spite of these limitations, there is tremendous potential that environmentally induced epigenetic alterations may serve as targets for therapeutic interventions in disease prevention. Research continues into the design, testing, and adoption of drugs that specifically reverse or inhibit the actions of epigenetic enzymes. The advent of such therapeutics, as well as the technological advances necessary to efficiently query the epigenome, suggest that individual environmentally associated epigenetic profiling may serve to inform both disease prognosis and treatment.

LIST OF ACRONYMS AND ABBREVIATIONS

5caC	5-carboxylcytosine
5fC	5-formylcytosine
5hmC	5-hydroxymethycytosine
5mc	5-methyl cytosine
ADRB2	adrenoceptor beta 2, surface

AHRR	aryl-hydrocarbon receptor repressor
APC	adenomatous polyposis coli
AS3MT	arsenic (+3 oxidation state) methyltransferase
AXL	AXL receptor tyrosine kinase
BC	black carbon
BDNF	brain-derived neurotrophic factor
BisChIP-seq	bisulfite sequencing of chromatin immunoprecipitated DNA
BRCA1	breast cancer 1, early onset
CCND1	cyclin D1
CDC	center for disease control and prevention
CDKN2A	cyclin-dependent kinase inhibitor 2A
ChIP-BS-seq	chromatin immunoprecipitation followed by bisulfite sequencing
CYP1A1	cytochrome P450, family 1, subfamily A, polypeptide 1
DALYs	disability-adjusted life years
DAPK	death-associated protein kinase 1
DMRs	differentially methylated regions
DNMTi	DNMT inhibitors
DNMTs	DNA methyltransferases
EGFR	epidermal growth factor receptor
FeNO	fractional exhaled nitric oxide
FOXP3	forkhead box transcription factor 3
GSTM1	glutathione S-transferase mu 1
H19	H19, imprinted maternally expressed transcript (non-protein coding)
H3K18ac	acetylation of histone H3 lysine 18
H3K27me3	trimethylation of histone H3 lysine 27
H3K4me3	trimethylation of histone H3 lysine 4
H3K9me2	dimethylation of histone H3 lysine 9
H3Kme2	dimethylation of histone H3 lysine 4
HAT	histone acetyltransferase
HDAC	histone deacetylase
HDMT	histone demethyltransferase
HMT	histone methyltransferases

IFNG	interferon-gamma
IARC	International Agency for Research on Cancer
IGF2	insulin-like growth factor 2
iNOS	inducible nitric oxide synthase
KRAS	Kirsten rat sarcoma viral oncogene homolog
MGMT	O^6-methylguanine-DNA methyltransferase
miRNAs	microRNAs
MLH1	mutL homolog 1
mRNAs	messenger RNAs
oxBS-Seq	oxidative bisulfite sequencing
PBL	peripheral blood lymphocytes
PLAGL1	pleomorphic adenoma gene-like 1
PM	particulate matter
PM_{10}	coarse particulate matter
$PM_{2.5}$	fine particulate matter
ppb	parts per billion
RASSF1A	ras association (RalGDS/AF-6) domain family member 1
RB1	retinoblastoma 1
SAM	S-adenosylmethionine
SNPs	single nucleotide polymorphisms
SLC6A3	Solute Carrier Family 6 (Neurotransmitter Transporter), member 3
TDG	thymine-DNA glycosylase
TP53	tumor protein 53
UTR	untranslated region
WHO	world health organization

References

[1] Pruss-Ustun A, et al. Knowns and unknowns on burden of disease due to chemicals: a systematic review. Environ Health 2011;10:9.
[2] Bollati V, Baccarelli A. Environmental epigenetics. Heredity 2010;105(1):105–12.
[3] Bailey KA, Fry RC. Arsenic-induced changes to the epigenome. Toxicology and epigenetics. West Sussex (United Kingdom): Wiley; 2012. p. 149–190.
[4] Ray PD, Yosim A, Fry RC. Incorporating epigenetic data into the risk assessment process for the toxic metals arsenic, cadmium, chromium, lead, and mercury: strategies and challenges. Front Genet 2014;5:201.

[5] Sun H, Shamy M, Costa M. Nickel and epigenetic gene silencing. Genes 2013;4(4): 583–95.

[6] Syed A, et al. Air pollution and epigenetics. J Environ Prot 2013;4:114–22.

[7] Nieratschker V, Batra A, Fallgatter AJ. Genetics and epigenetics of alcohol dependence. J Mol Psychiatry 2013;1(11):1–6.

[8] Csoka AB, Szyf M. Epigenetic side-effects of common pharmaceuticals: a potential new field in medicine and pharmacology. Med Hypotheses 2009;73(5):770–80.

[9] Ziller MJ, et al. Genomic distribution and inter-sample variation of non-CpG methylation across human cell types. PLoS Genet 2011;7(12).

[10] Kohli RM, Zhang Y. TET enzymes, TDG and the dynamics of DNA demethylation. Nature 2013;502(7472):472–9.

[11] Rodenhiser D, Mann M. Epigenetics and human disease: translating basic biology into clinical applications. CMAJ 2006;174(3):341–8.

[12] Jones PA. Functions of DNA methylation: islands, start sites, gene bodies and beyond. Nat Rev Genet 2012;13(7):484–92.

[13] Esteller M. Epigenetics in cancer. N Engl J Med 2008;358(11):1148–59.

[14] Madhusudan S, Wilson DM. DNA repair and cancer: from bench to clinic. Boca Raton: Taylor & Francis; 2013.

[15] Lyko F, Brown R. DNA methyltransferase inhibitors and the development of epigenetic cancer therapies. J Natl Cancer Inst 2005;97(20):1498–506.

[16] Tan C, Du X. KRAS mutation testing in metastatic colorectal cancer. World J Gastroenterol 2012;18(37):5171–80.

[17] Esteller M, et al. Inactivation of the DNA repair gene O6-methylguanine-DNA methyltransferase by promoter hypermethylation is associated with G to A mutations in K-ras in colorectal tumorigenesis. Cancer Res 2000;60(9):2368–71.

[18] McCabe MT, Brandes JC, Vertino PM. Cancer DNA methylation: molecular mechanisms and clinical implications. Clin Cancer Res 2009;15(12):3927–37.

[19] Bartova E, et al. Histone modifications and nuclear architecture: a review. J Histochem Cytochem 2008;56(8):711–21.

[20] Gunjan A, Singh RK. Epigenetic therapy: targeting histones and their modifications in human disease. Future Med Chem 2010;2(4):543–8.

[21] Seligson DB, et al. Global levels of histone modifications predict prognosis in different cancers. Am J Pathol 2009;174(5):1619–28.

[22] Bushati N, Cohen SM. MicroRNA functions. Annu Rev Cell Dev Biol 2007;23: 175–205.

[23] Cai Y, et al. A brief review on the mechanisms of miRNA regulation. Genomics Proteomics Bioinf 2009;7(4):147–54.

[24] Iorio MV, Croce CM. MicroRNA dysregulation in cancer: diagnostics, monitoring and therapeutics. A comprehensive review. EMBO Mol Med 2012;4(3):143–59.

[25] Schmitz U. MicroRNA cancer regulation: advanced concepts, bioinformatics and systems biology tools. New York: Springer; 2013.

[26] Hou L, Wang D, Baccarelli A. Environmental chemicals and microRNAs. Mutat Res 2011;714(1–2):105–12.

[27] Rivera CM, Ren B. Mapping human epigenomes. Cell 2013;155(1):39–55.

[28] McGraw S, Shojaei Saadi HA, Robert C. Meeting the methodological challenges in molecular mapping of the embryonic epigenome. Mol Hum Reprod 2013;19(12):809–27.

[29] Organization, W.H., World Health Organization (WHO). Burden of disease from household and ambient air pollution for 2012, 2014.

[30] Pope 3rd CA. Epidemiology of fine particulate air pollution and human health: biologic mechanisms and who's at risk? Environ Health Perspect 2000;108 (Suppl. 4):713–23.

[31] Anderson JO, Thundiyil JG, Stolbach A. Clearing the air: a review of the effects of particulate matter air pollution on human health. J Med Toxicol 2012;8(2):166–75.

[32] Brunekreef B, Holgate ST. Air pollution and health. Lancet 2002;360(9341):1233–42.

[33] Valavanidis A, Fiotakis K, Vlachogianni T. Airborne particulate matter and human health: toxicological assessment and importance of size and composition of particles for oxidative damage and carcinogenic mechanisms. J Environ Sci Health C Environ Carcinog Ecotoxicol Rev 2008;26(4):339–62.

[34] Brook RD, et al. Particulate matter air pollution and cardiovascular disease: an update to the scientific statement from the American Heart Association. Circulation 2010;121(21):2331–78.

[35] Breton C, Marutani AN. Air pollution and epigenetics: recent findings. Curr Environ Health Rep 2004;1(1):35–45.

[36] Salam MT, et al. Genetic and epigenetic variations in inducible nitric oxide synthase promoter, particulate pollution, and exhaled nitric oxide levels in children. J Allergy Clin Immunol 2012;129(1):232–9, e1–e7.

[37] Dweik RA, et al. An official ATS clinical practice guideline: interpretation of exhaled nitric oxide levels (FENO) for clinical applications. Am J Respir Crit Care Med 2011;184(5):602–15.

[38] Fry RC, et al. Air toxics and epigenetic effects: ozone altered microRNAs in the sputum of human subjects. Am J Physiol Lung Cell Mol Physiol 2014;306(12):L1129–37.

[39] Kohli A, et al. Secondhand smoke in combination with ambient air pollution exposure is associated with increasedx CpG methylation and decreased expression of IFN-gamma in T effector cells and Foxp3 in T regulatory cells in children. Clin Epigenetics 2012;4(1):17.

[40] Janssen BG, et al. Placental DNA hypomethylation in association with particulate air pollution in early life. Part Fibre Toxicol 2013;10:22.

[41] Herbstman JB, et al. Prenatal exposure to polycyclic aromatic hydrocarbons, benzo[a] pyrene-DNA adducts, and genomic DNA methylation in cord blood. Environ Health Perspect 2012;120(5):733–8.

[42] Fu A, et al. An environmental epigenetic study of ADRB2 5'-UTR methylation and childhood asthma severity. Clin Exp Allergy 2012;42(11):1575–81.

[43] Bind MA, et al. Air pollution and markers of coagulation, inflammation, and endothelial function: associations and epigene-environment interactions in an elderly cohort. Epidemiology 2012;23(2):332–40.

[44] Edenberg HJ. The genetics of alcohol metabolism: role of alcohol dehydrogenase and aldehyde dehydrogenase variants. Alcohol Res Health 2007;30(1):5–13.

[45] Yan J, et al. Using genetic information from candidate gene and genome-wide association studies in risk prediction for alcohol dependence. Addict Biol 2013:708–21.

[46] Taqi MM, et al. Prodynorphin CpG-SNPs associated with alcohol dependence: elevated methylation in the brain of human alcoholics. Addict Biol 2011;16(3):499–509.

[47] Hillemacher T, et al. Promoter specific methylation of the dopamine transporter gene is altered in alcohol dependence and associated with craving. J Psychiatr Res 2009;43(4):388–92.

[48] Varela-Rey M, et al. Alcohol, DNA methylation, and cancer. Alcohol Res 2013;35(1): 25–35.

[49] Mason JB, Choi SW. Effects of alcohol on folate metabolism: implications for carcinogenesis. Alcohol 2005;35(3):235–41.

[50] Ulrich CM, Potter JD. Folate and cancer–timing is everything. JAMA 2007;297(21): 2408–9.

[51] Naujokas MF, et al. The broad scope of health effects from chronic arsenic exposure: update on a worldwide public health problem. Environ Health Perspect 2013;121(3):295–302.

[52] Jomova K, et al. Arsenic: toxicity, oxidative stress and human disease. J Appl Toxicol 2011;31(2):95–107.

[53] Cheung HH, et al. DNA methylation of cancer genome. Birth Defects Res C Embryo Today 2009;87(4):335–50.

[54] Rager JE, et al. Prenatal arsenic exposure and the epigenome: altered microRNAs associated with innate and adaptive immune signaling in newborn cord blood. Environ Mol Mutagen 2014;55(3):196–208.

[55] Rodriguez-Paredes M, Esteller M. Cancer epigenetics reaches mainstream oncology. Nat Med 2011;17(3):330–9.

[56] Jarup L, Akesson A. Current status of cadmium as an environmental health problem. Toxicol Appl Pharmacol 2009;238(3):201–8.

[57] Anetor JI. Rising environmental cadmium levels in developing countries: threat to genome stability and health. Niger J Physiol Sci 2012;27(2):103–15.

[58] ATSDR. Priority list of hazardous substances. 2011. Available from: http://www.atsdr.cdc.gov/spl/.

[59] Waisberg M, et al. Molecular and cellular mechanisms of cadmium carcinogenesis. Toxicology 2003;192(2–3):95–117.

[60] Sanders AP, et al. Cadmium exposure and the epigenome: exposure-associated patterns of DNA methylation in leukocytes from mother-baby pairs. Epigenetics 2014;9(2):212–21.

[61] Kippler M, et al. Sex-specific effects of early life cadmium exposure on DNA methylation and implications for birth weight. Epigenetics 2013;8(5):494–503.

[62] Benbrahim-Tallaa L, et al. Tumor suppressor gene inactivation during cadmium-induced malignant transformation of human prostate cells correlates with overexpression of de novo DNA methyltransferase. Environ Health Perspect 2007;115(10):1454–9.

[63] Person RJ, et al. Chronic cadmium exposure in vitro induces cancer cell characteristics in human lung cells. Toxicol Appl Pharmacol 2013;273(2):281–8.

[64] Zhou ZH, Lei YX, Wang CX. Analysis of aberrant methylation in DNA repair genes during malignant transformation of human bronchial epithelial cells induced by cadmium. Toxicol Sci 2012;125(2):412–7.

[65] Mushak P. Lead and public health: science, risk, and regulation. Trace metals and other contaminants in the environment, vol. ix. Amsterdam; Boston: Elsevier; 2011. p. 980.

[66] World Health Organization (WHO). Childhood lead poisoning. World Health Organization; 2010.

[67] Olympio KP, et al. Neurotoxicity and aggressiveness triggered by low-level lead in children: a review. Rev Panam Salud Publica 2009;26(3):266–75.

[68] Needleman HL, Bellinger D. The health effects of low level exposure to lead. Annu Rev Public Health 1991;12:111–40.

[69] Nevin R. Understanding international crime trends: the legacy of preschool lead exposure. Environ Res 2007;104(3):315–36.

[70] Abadin H, et al. Toxicological profile for lead. Atlanta: Agency for Toxic Substances and Disease Registry; 2007.

[71] Pilsner JR, et al. Influence of prenatal lead exposure on genomic methylation of cord blood DNA. Environ Health Perspect 2009;117(9):1466–71.

[72] Gulson BL, et al. Pregnancy increases mobilization of lead from maternal skeleton. J Lab Clin Med 1997;130(1):51–62.

[73] McShea A, et al. Abnormal expression of the cell cycle regulators P16 and CDK4 in Alzheimer's disease. Am J Pathol 1997;150(6):1933–9.

[74] Hayes RB. The carcinogenicity of metals in humans. Cancer Causes Control 1997;8(3):371–85.

[75] Nordberg G. Handbook on the toxicology of metals. 3rd ed. vol. xlvii. Amsterdam; Boston: Academic Press; 2007. p. 975.

[76] Salnikow K, Zhitkovich A. Genetic and epigenetic mechanisms in metal carcinogenesis and cocarcinogenesis: nickel, arsenic, and chromium. Chem Res Toxicol 2008;21(1):28–44.
[77] Goodrich JM, et al. Mercury biomarkers and DNA methylation among Michigan dental professionals. Environ Mol Mutagen 2013;54(3):195–203.
[78] Yang J, et al. Relationship between urinary nickel and methylation of p15, p16 in workers exposed to nickel. J Occup Environ Med 2014;56(5):489–92.
[79] Arita A, et al. Global levels of histone modifications in peripheral blood mononuclear cells of subjects with exposure to nickel. Environ Health Perspect 2012;120(2):198–203.
[80] Brocato J, Costa M. 10th NTES Conference: nickel and arsenic compounds alter the epigenome of peripheral blood mononuclear cells. J Trace Elem Med Biol 2014.
[81] Jose CC, et al. Epigenetic dysregulation by nickel through repressive chromatin domain disruption. Proc Natl Acad Sci USA 2014;111(40):14631–6.
[82] Ajslev TA, et al. Childhood overweight after establishment of the gut microbiota: the role of delivery mode, pre-pregnancy weight and early administration of antibiotics. Int J Obes 2011;35(4):522–9.
[83] Vidal AC, et al. Associations between antibiotic exposure during pregnancy, birth weight and aberrant methylation at imprinted genes among offspring. Int J Obes 2013;37(7):907–13.
[84] Ferguson-Smith AC. Genomic imprinting: the emergence of an epigenetic paradigm. Nat Rev Genet 2011;12(8):565–75.
[85] Smeester L, et al. Imprinted genes and the environment: links to the toxic metals arsenic, cadmium, lead and mercury. Genes 2014;5(2):477–96.
[86] Varrault A, et al. Zac1 regulates an imprinted gene network critically involved in the control of embryonic growth. Dev Cell 2006;11(5):711–22.
[87] Mitchell AA, et al. Medication use during pregnancy, with particular focus on prescription drugs: 1976–2008. Am J Obstet Gynecol 2011;205(1):51.e1–e8.
[88] Lopez JP, et al. Epigenetic regulation of BDNF expression according to antidepressant response. Mol Psychiatry 2013;18(4):398–9.
[89] Nagahara AH, Tuszynski MH. Potential therapeutic uses of BDNF in neurological and psychiatric disorders. Nat Rev Drug Discov 2011;10(3):209–19.
[90] Soubry A, et al. The effects of depression and use of antidepressive medicines during pregnancy on the methylation status of the IGF2 imprinted control regions in the offspring. Clin Epigenetics 2011;3:2.
[91] Detich N, Bovenzi V, Szyf M. Valproate induces replication-independent active DNA demethylation. J Biol Chem 2003;278(30):27586–92.
[92] Milutinovic S, et al. Valproate induces widespread epigenetic reprogramming which involves demethylation of specific genes. Carcinogenesis 2007;28(3):560–71.
[93] Eyal S, et al. The activity of antiepileptic drugs as histone deacetylase inhibitors. Epilepsia 2004;45(7):737–44.
[94] Lokk K, et al. DNA methylome profiling of human tissues identifies global and tissue-specific methylation patterns. Genome Biol 2014;15(4):r54.
[95] DiFranza JR, Aligne CA, Weitzman M. Prenatal and postnatal environmental tobacco smoke exposure and children's health. Pediatrics 2004;113(4 Suppl.):1007–15.
[96] Tong VT, et al. Trends in smoking before, during, and after pregnancy - pregnancy Risk Assessment Monitoring System (PRAMS), United States, 31 sites, 2000–2005. MMWR Surveill Summ 2009;58(4):1–29.
[97] Godfrey KM, Barker DJ. Fetal programming and adult health. Public Health Nutr 2001;4(2B):611–24.
[98] Guerrero-Preston R, et al. Global DNA hypomethylation is associated with in utero exposure to cotinine and perfluorinated alkyl compounds. Epigenetics 2010;5(6):539–46.

[99] Suter M, et al. Maternal tobacco use modestly alters correlated epigenome-wide placental DNA methylation and gene expression. Epigenetics 2011;6(11):1284–94.

[100] Suter M, et al. In utero tobacco exposure epigenetically modifies placental CYP1A1 expression. Metabolism 2010;59(10):1481–90.

[101] Bartsch H, et al. Genetic polymorphism of CYP genes, alone or in combination, as a risk modifier of tobacco-related cancers. Cancer Epidemiol Biomarkers Prev 2000;9(1):3–28.

[102] Joubert BR, et al. 450K epigenome-wide scan identifies differential DNA methylation in newborns related to maternal smoking during pregnancy. Environ Health Perspect 2012;120(10):1425–31.

[103] Zudaire E, et al. The aryl hydrocarbon receptor repressor is a putative tumor suppressor gene in multiple human cancers. J Clin Invest 2008;118(2):640–50.

[104] Breton CV, Salam MT, Gilliland FD. Heritability and role for the environment in DNA methylation in AXL receptor tyrosine kinase. Epigenetics 2011;6(7):895–8.

[105] Paccez JD, et al. The receptor tyrosine kinase Axl in cancer: biological functions and therapeutic implications. Int J Cancer 2014;134(5):1024–33.

[106] Murphy SK, et al. Gender-specific methylation differences in relation to prenatal exposure to cigarette smoke. Gene 2012;494(1):36–43.

[107] Chao W, D'Amore PA. IGF2: epigenetic regulation and role in development and disease. Cytokine Growth Factor Rev 2008;19(2):111–20.

[108] Maccani MA, Marsit CJ. Exposure and fetal growth-associated miRNA alterations in the human placenta. Clin Epigenetics 2011;2(2):401–4.

11

Nutriepigenomics: Personalized Nutrition Meets Epigenetics

Anders M. Lindroth[1], Joo H. Park[2], Yeongran Yoo[2], Yoon J. Park[2]

[1]Graduate School of Cancer Science and Policy, National Cancer Center, Goyang-si, Republic of Korea; [2]Department of Nutritional Science and Food Management, Ewha Womans University, Seoul, Republic of Korea

1. INTRODUCTION

Food and nutritional intake play a lifelong role in humans as a major environmental health factor. Conventionally, nutritional research and policies have focused on providing guidelines for sufficient daily intake of nutrients to prevent deficiencies and promote human health. In recent decades, the research field has acknowledged major differences in nutrient response between people that call for individual dietary adjustments, referred to as *personalized nutrition*.

Personalized nutrition is conceptually similar to personalized medicine or personalized pharmacology in terms of considering individual needs. The term "personalized nutrition" is often described as nutritional recommendations tailored to individual needs, characteristics, and preferences that can accommodate all stages of individual care. This typically considers growth, health promotion and disease prevention, disease management, and aging. It is now widely accepted that nutrients and bioactive food compounds alter molecular and cellular processes such as gene expression and metabolism, and that the presence or extent of alterations differs with individual genetic variations, physiological conditions, and lifestyle or environment. The recognition of individual differences in response to diets is the result of advances in genome-wide analysis, a field coined nutrigenomics, which allows the entire genome to be queried.

Nutrigenomics is by and large about the complex interplay between diets and the genome. It is sometimes divided into two major fields: *nutrigenetics*, aimed at understanding how an individual's genetic makeup is predisposed to dietary susceptibility, and *nutrigenomics*, focused on understanding how diets and nutrients affect gene expression in the human genome. Since completion of sequencing of the human genome in 2003, nutrigenomics has been intensively studied. The available information about the human genome's structure and sequences opened a new era in biological and medical sciences. It enabled the understanding of gene functions and interactions genome-wide, followed by the explosive development of large-scale molecular analytical technology, commonly labeled with the suffix "-genomics." Genome-wide identification and use of DNA variations, RNA transcripts, proteins, and metabolites have been instrumental in the investigation of complex traits and the responses of these traits to environmental stimulation such as diets. Such systemic analyses have provided powerful tools for understanding the dietary influence of the human homeostatic metabolic system, leading to the prediction of beneficial and adverse effects of nutrients and bioactive food compounds, by connecting dietary response to individual genetic variations. For example, variations in methylenetetrahydrofolate (*MTHFR*) and 1,25-dihydroxyvitamin D3 receptor (*VDR*) show reduced metabolic efficiency or responsiveness to folate and vitamin D/calcium homeostasis, respectively [1–3]. Therefore, it has been recommended that individuals who carry variations in *MTHFR* and *VDR* pursue relatively high intake of those nutrients.

Over the past two decades, epigenetic phenomena have been recognized as driving forces behind development, aging, and various diseases such as cancer, cardiovascular disease, obesity, diabetes, and neurodegenerative disease. Epigenetics is the study of gene regulation mediated by changes in chromatin status via chemical modifications in the absence of changes to the DNA sequence. The modifications can be modulated by nutrients, metabolites, and bioactive food compounds, emphasizing the links between nutrients and epigenetic changes in nutritional research. Thus, incorporating epigenetics into nutrigenomics, hereafter referred to as nutri*epi*genomics, promises to explain the links between nutrients and gene regulation and response (Figure 1). Before implementing nutri-epigenetic biomarkers in a clinical setting, only those markers that report robustly and reproducibly should be used. Since most epigenetic modifications are dynamic and plastic, a detailed understanding of their changing patterns must be achieved by investigations that use multiple cohorts with statistically relevant sizes. In order to identify proper nutriepigenetic biomarkers, the following important questions can be asked: Can diets and nutrients direct epigenetic changes and lead to changes in gene expression? Can a certain epigenetic status explain different responses to the same diet between individuals that are not fully elucidated solely

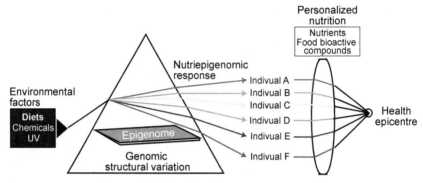

FIGURE 1 **Personalized nutriepigenomics.** In a typical nutrient–genome interaction, diet is fed into the system (referred to as a light-dispersing prism) consisting of individual genomic variation and its epigenomic landscape. The nutrients and bioactive compounds in the diet will exert their functions on the chromatin in a multiparameter fashion (referred to as individual dispersed light). The nutriepigenomic response is a function of the characteristics of the each individual reaction to the diet. The use of diagnostic biomarkers aims to extract these individual responses. When taken through a personal nutrition regime, the response will ultimately be controlled by the personalized diet intervention (referred to as the last converging prism), resulting in an optimal focal health status (the health epicenter) that takes individual characteristics into consideration.

by genetic variations? Here, we review the emerging field of nutriepigenomics with its possibilities and limitations, and how we anticipate that it will significantly contribute to personalized nutrition.

2. CONNECTION BETWEEN CHROMATIN STRUCTURE AND METABOLIC STATUS

2.1 Epigenetic Modifications Determine Chromatin Structure

The genome is not just a collection of genetic codes carrying sequence information. Over the last three decades, detailed molecular research has uncovered key factors in eukaryotes that package DNA into functional units [4]. The chromatin, consisting of core histone proteins forming nucleosomes with the DNA wrapped around them, dictate gene transcriptional activity and maintain chromosome integrity by toggling between condensed and decondensed states. To accomplish this, DNA and histones in the nucleosome are subject to various chemical modifications: cytosine methylation of DNA, as well as a large array of histone modifications—e.g., methylation, acetylation, phosphorylation, ubiquitination, and O-GlcNAcylation. The acquisition or removal of posttranslational modifications of the N-terminal tails of histones is strongly correlated with gene activities such as transcription, elongation, and other transcriptional

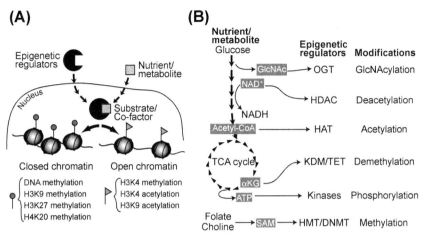

FIGURE 2 **Metabolic regulations of epigenetic modifications.** (A) Nutrients and their metabolites play key roles in chromatin changes as substrates or cofactors for epigenetic regulators whose activities lead to changes in chromatin structure and gene expression. Representative epigenetic modifications for open chromatin (red lollipops) and closed chromatin (green triangles) are shown. (B) Epigenetic enzymes that write or erase modifications require specific metabolites as substrates or cofactors, which are intermediates in glycolysis, TCA cycle, hexosamine biosynthetic pathway, or one-carbon metabolism. N-Acetylglucosamine (GlcNAc) produced from the hexosamine biosynthetic pathway can be used as substrate for histone GlcNAc glycosylation by O-linked GlcNAc transferase. Flux through the glycolytic pathway influences NAD+/NADH ratio in the cytosol, which affects sirtuin histone deacetylase (HDAC) activities. Acetyl-CoA donates acetyl groups to the histone acetyltransferase (HAT). An intermediate molecule from the TCA cycle, α-Ketoglutarate (α-KG), is a required cofactor for histone demethylation enzymes (KDM) and DNA demethylation enzymes (TET). The ATP/AMP ratio influences kinase activity such as AMPK, which phosphorylates histones. The substrate for histone methyltransferase (HMT) and DNA methyltransferase (DNMT) is S-adenosylmethionine (SAM), a product of one-carbon metabolism, involving folate and choline.

regulations. Methylation marks at histones differentially influence transcriptional activity, depending on which lysine residues become methylated and how many methylation groups are attached. For instance, trimethylation of histone H3 lysine 4 (H3K4me3) is associated with transcriptional activation, while H3K9me3, H3K27me3, and H4K20me3 are associated with transcriptional repression [5] (Figure 2(A)). In contrast, acetylation of histone lysines, such as acetylation of histone H3 lysine 9 (H3K9ac) or H3K27ac, is commonly associated with transcriptional activation regardless of which histone lysine it involves.

The most well characterized epigenetic modification is DNA methylation, commonly occurring at cytosines followed by a guanine (together CpG) in mammals. The majority of CpGs in the genome is methylated at all times, except at regulatory elements such as promoters and enhancers, and CpG-rich islands. DNA methylation at promoters and enhancers predominantly results in gene silencing (Figure 2(A)), whereas its presence in the gene body usually leads to gene activation, indicating that DNA methylation has different roles depending on where it appears. Removal of DNA methylation has not been completely elucidated, but recent studies suggest that oxidation of methylated cytosines by the ten-eleven-translocation methylcytosine dioxygenases (TETs) is an initial step for DNA demethylation processes followed by the base excision repair machinery [6,7]. The demethylation is required for proper gene activation, especially during early embryogenesis and differentiation. On many occasions, multiple epigenetic modifications, including DNA methylation and various histone modifications, work coordinately or antagonistically [8,9].

Besides histone modifications and DNA methylation-mediated chromatin changes, noncoding RNA has also emerged as influencing chromatin. Noncoding-RNA-mediated gene regulation occurs both transcriptionally and posttranscriptionally. While noncoding RNA has been recognized as mediating important gene regulatory control and may play important roles in response to nutrition, it still remains poorly investigated in regard to nutrition. For this reason, we have decided to focus on the wealth of information on nutrition and DNA methylation and histone modifications that has accumulated over the past few years.

2.2 Development and Disease Dictated by Epigenetic Modifications

The placement of epigenetic modifications is tightly controlled both spatially and temporally. Each tissue has a unique epigenetic pattern, and changes occur as results of developmental and regenerative processes. Related to development, embryonic stem cells clearly have a unique epigenetic pattern that changes upon differentiation. Extrinsic signals such as hormones and cytokines regulate differentiation via epigenetic remodeling [10]. The epigenetic pattern observed in any particular tissue at any particular time point is a reflection of its related gene expression pattern. Consequently, epigenetic patterns inform about the transcriptional conditions of cells and tissues, which can be used to infer their nutritional responses accurately and reproducibly.

Ever-increasing evidence underscores that an understanding of epigenetic patterns in disease progression will provide invaluable information in the diagnosis and treatment of human disease [11]. Early evidence on epigenetic changes in disease came from cancer research in particular, but has since expanded to include various other diseases such as cardiovascular disease, obesity, diabetes, neurodegenerative disease, and immune disease. Disease-specific epigenetic patterns can be used as biomarkers for diagnosis to evaluate disease progression and therapy choice, and as molecular clues for understanding the pathological process of the disease.

For example, tumor tissues commonly show global DNA hypomethylation and locus-specific DNA hypermethylation mainly at tumor suppressor genes [12].

Genome-wide analyses of epigenetic patterns that undergo defined changes as a result of external stress have identified novel genomic loci that provide insight into how cells respond to external factors [13]. Furthermore, such analyses have offered detailed understanding of how cells cope with the changes and inform about outstanding factors that determine whether the changes can be dealt with or result in lineage-skewed differentiation or cell death. Hence, epigenetic patterns that suggest a predisposition for a particular condition have been identified, and it should thus be possible to utilize these patterns to develop personalized recommendation regimes that prevent future problems.

2.3 Individual Differences Reflected in Epigenetic Patterns

One important characteristic of epigenetic patterns is that they can be inherited through cell divisions and even through germ cell transmission. Their heritability emphasizes the prospect that epigenetic patterns contribute to variations, similar to the well-known fact that heritable genetic variations contribute to individual differences. As evidence for an environmental influence, identical twins share the same DNA sequence, yet still display significant differences in genomic response and disease etiology. These differences can be explained by epigenetic patterns differently inherited or established via environmental exposure.

For example, evidence from animal studies shows that the effects of exposure to toxins or stress during early development can be maintained throughout development or even transmitted to the next generation without genetic mutations [14,15]. Anway et al. showed that transient exposure to endocrine disruptors during pregnancy led to reduced fertility of male progeny up to the fourth generation, by DNA methylation changes in sperm [14]. Also, Weaver et al. showed that the lack of maternal care in early-development-induced DNA hypermethylation at the promoter of the glucocorticoid receptor gene and downregulation of its expression, thus inhibiting stress response [15]. Intriguingly, this hypermethylation persisted even to adulthood. These data indicate that early exposure to environmental factors programs epigenetic patterns in individuals and influences future response variations, even in the offspring of affected individuals.

2.4 Nutrients and Their Metabolites Shape the Epigenetic Landscape

Epigenetic alterations are sensitive to environmental factors including nutrients, stress, and toxins, suggesting that modifications are dynamic.

Chromatin structure is modulated by nutrient availability and meta-bolic status. Most epigenetic modification enzymes require nutrients and metabolites as substrates or cofactors (Figure 2(A)).

Methyl groups are transferred by DNA methyltransferases (DNMTs) or histone methyltransferases (HMTs) from the methyl donor S-adenosylmethionine (SAM) to cytosines in DNA or lysine and arginine residues in histones (Figure 2(B)). During the transfer reaction, SAM is converted to S-adenosyl homocysteine (SAH), which is a potent inhibi-tor of DNMTs and HMTs. SAM availability is influenced by diets, since SAM and SAH are both part of the one-carbon metabolism that requires multiple coenzymes including folate, vitamin B12, vitamin B6, and riboflavin, and key intermediary compounds such as methionine and choline [16]. The removal of methylation is mediated by the TETs and histone lysine demethylases (KDMs) at DNA and histones, respectively. TETs and KDMs are oxygenases requiring Fe(II) and α-ketoglutarate (α-KG) as coenzymes and can be inhibited by succinate or fumarate. The α-ketoglutarate is an intermediate in the tricarboxylic acid cycle (TCA cycle) and can be produced from glutamate by deamination.

Histone acetylation is catalyzed by histone acetyltransferases (HATs) that transfer one acetyl group of acetyl-CoA to histone lysine residues. HATs include P300/CBP, Gcn5/PCAF, MYST, and steroid receptor acti-vator (SRC) families. The level of cytosolic acetyl-CoA or ATP-citrate lyase, catalyzing citrate cleavage to yield the acetyl-CoA, is tightly asso-ciated with histone acetylation level [17]. The removal of acetyl groups in histones is catalyzed by histone deacetylases (HDACs). HDACs can be divided into two groups according to mechanical similarity: classi-cal HDACs and (NAD+)-dependent sirtuins. In classical HDACs such as HDAC1 and HDAC2, the activity can be increased by intermediate metab-olites including acetyl-CoA, malonyl-CoA, and HMG-CoA, and inhibited by free CoA. The second group of HDACs, NAD+ dependent sirtuins, has seven members, of which SIRT1, SIRT6, and SIRT7 have histone deacety-lation activity. Although the deacetylation reaction is energetically favor-able, sirtuins function by using NAD+ and producing nicotinamide and O-acetyl-ADP-ribose. The NAD+ is a cellular sensor of the redox state and energy availability. For example, calorie restriction results in a change in NAD+/NADH ratio and consequently increased sirtuin activity leading to antiaging effects [18].

Besides methylation/demethylation and acetylation/deacetylation, other relatively less studied modifications include phosphorylation in histone H3 and H2B and O-linked N-acetylglucosamine (GlcNAc) in histone H2B. Although those epigenetic modifications are associated with metabolites—e.g., ATP and glucose derivatives—the function of the modifications and the influence of metabolites remain poorly investigated.

3. EVIDENCE FOR DIETARY FACTORS AFFECTING EPIGENETIC PATTERNS

Dietary nutrients and bioactive compounds influence the regulation of development and etiology. While we foresee the development of a detailed understanding of how specific nutrients or bioactive compounds possibly provide dietary effects on the genome, we also acknowledge that individual nutrients or bioactive compounds ingested during regular food intakes will likely provide different responses when ingested in conjunction with daily foods rather than as individual dietary supplements. Furthermore, the compound dietary effects from food will be difficult to recapitulate by combining purified nutrients, largely because of the complexity and bioavailability specifics in most food items.

Despite these potential shortcomings, we continue to build our knowledge base on the mechanistic understanding of individual nutrients and their influence on the genetic material. We know, for instance, that various compounds interfere with DNMTs or HDACs, which profoundly influences transcription globally, as will be discussed in the following paragraphs. We postulate that these mechanistic insights will guide prevention and treatment of various diseases associated with metabolic disorders, cancer, neurodegenerative disease, and immune disease [19]. Below we summarize the effects of nutrients and common bioactive compounds on epigenetic targeting, by focusing on metabolic disorders and cancer (Table 1).

3.1 Diet Compositions and Nutrients

3.1.1 Calorie Restriction and Low-Protein Diets

Emerging evidence in human and animal studies links suboptimal nutrition—of both parents, as well as gestational mothers and their fetuses—with increased risk of metabolic disorders when offspring reach adulthood [20]. Calorie restriction and low-protein diets have been reported to induce epigenetic changes and metabolic alterations that persist later in development, referred to as *metabolic programming* or *metabolic imprinting*. A striking association between early nutrition and epigenetics was observed through the analysis of children to mothers who became pregnant during or just prior to the Dutch Hunger Winter (also known as the Dutch famine) in the years 1944–1945. Individuals born to starved mothers (mainly due to protein restriction) showed increased incidence of cardiovascular disease later in life, possibly due to altered DNA methylation [21]. The methylation patterns differed depending on the individual's sex and the timing of exposure [22]. A recent follow-up study of children from mothers that 9 years earlier were subject to calorie restriction during pregnancy showed altered DNA methylation at the proopiomelanocortin

TABLE 1 Epigenetic Regulation by Diets and Bioactive Food Compounds

Dietary compound	Food source	Epigenetic modifications	Epigenetic targets	Roles in cancer and metabolic disorders	References
DIETARY COMPOSITION AND NUTRIENTS					
Calorie restriction and low-protein diet		Histone modifications	Unknown/SIRT1	Prevention of type 2 diabetes and cancer, and extension of lifespan	[26,31]
		DNA methylation	POMC, IGF2	Acting as an early predictor of late-onset metabolic disorders	[23–25,27]
High-fat diet		Histone modifications	Unknown/HDAC5, HDAC8, HDAC1	Influence on expression of appetite-regulating neuropeptides and energy metabolism	[34,40–42]
		DNA methylation	MC4R, LEP, PGARGC1α, POMC, IRS		
Methyl donors		DNA methylation	MTHFR, agouti, AxinFu	Stimulation of DNA methylation associated with genomic stability	[43,45,46,49]
BIOACTIVE FOOD COMPOUNDS					
Resveratrol	Blueberries, mulberries, cranberries, peanuts, and grapes	Histone modifications	TNFα, IL-8, BRCA-1, RBP/SIRT1,	Reduction of activation of NF kappa B, and activation of antiproliferative and antioxidative pathways	[60,61,114]
		DNA methylation	Unknown/DNMT		
Sulforaphane	Cruciferous vegetables such as broccoli, cabbage, and brussels sprout	Histone modifications	HBD-2, RARβ, HBD-2, p21, BAX/HDAC	Inhibition of the growth of tumor cells and induction of apoptosis in cancer cells	[67,68,70]
		DNA methylation	hTERT/DNMT1, DNMT3a		

EGCG	Green tea	Histone modifications	NF-κB, IL-6, BMI-1, EZH2, SUZ12/HAT, HDAC	Induction of apoptotic cell death and cell cycle arrest in tumor cells	[71,73,77,115]
		DNA methylation	P16^{INK4a}, RNRβ, MGMT, hMLH1, RECK1, hTERT, WIF-1, RXRα, GSTP1, CDKN2A, RXRβ, CDX2/DNMTl, MBD1, MeCP2		
Genistein	Soy beans or soy products	Histone modifications	p21, p16, PTEN, p53, FOXA3, BTG3, RARβ, hTERT, CCLD/HAT, HDAC, (SIRT1)	Induction of apoptotic cell death in tumor cells,	[81,82,83,87, 116,117]
		DNA methylation	P16, RARβ2, MGMT, hTERT, BTG3, GSTP1, EPHB2, HMGN5, CDKN2A/DNMT, MBD1, MBD4, MeCP2		
Curcumin	Turmeric	Histone modifications	p53, GATA4, GZMB, PRF1, EOMES/HAT, HDAC	Prevention of DNA damage and blocking the inflammatory molecule NF kappa B	[52,88–90,118]
		DNA methylation	Unknown/DNMT1		
Quercetin	Onions, citrus fruits, and buckwheat	Histone modifications	IP-10, MIP-2/HAT, SIRT1, HDAC1	Induction of cell cycle arrest	[93–96]
		DNA methylation	CDKN2A/DNMT		

(*POMC*) gene, suggesting that DNA methylation might be an early predictor of late-onset metabolic syndrome [23].

In animal models, calorie restriction during the periconceptional period led to changes in DNA methylation—e.g., hypomethylation of the insulin-like growth factor 2 (*IGF2*) locus in the adrenal glands of ewes [24]. In mice, recent genome-wide bisulfite sequencing revealed that calorie restriction also influenced DNA methylation in the genome of the offspring's germ cells, globally as well as locally [25]. This strongly argues that metabolic alteration can be transmitted to the next generation through calorie-restriction-induced DNA methylation alterations. In addition, analysis of the glucose transporter type 4 (*GLUT4*) promoter revealed that calorie restriction increases histone H4 acetylation in adipose tissue of obese mice, resulting in increased expression of *GLUT4* upon calorie restriction [26]. These studies indicate that both DNA methylation and histone modifications are sensitive to calorie restriction in a locus-specific manner.

Gestational protein restriction in rats can change expression of the *IGF2* locus in liver by altering DNA methylation [27] and also can increase cholesterol in offspring by instigating repressive histone methylation at the promoter of the cholesterol 7 alpha-hydroxylase (*CYP7A1*) locus [28]. These studies suggest that adequate nutrient supply during early development is crucial in metabolic programming via epigenetic modulation and could be responsible for metabolism-related health issues in adults. Intriguingly, stress caused by fluctuating calorie restriction exposure influences appetite-stimulating pathways and alters the reward circuitry in the mouse brain via epigenetic mechanisms, resulting in binge-eating [29]. This may provide a partial explanation for the high failure rate associated with long-term maintenance of dieting.

Together, calorie and protein restrictions modulate epigenetic patterns by influencing the levels of epigenetic enzymes in exposed cells, pertaining limited or physical influence on the absolute activity of the enzymes. Interestingly, calorie restriction is one of the most effective interventions reported for extending lifespan in various animal models by means of a partially conserved pathway, and is conditionally involved in delaying age-related diseases in humans [30]. Although one of the major mechanisms underlying this effect involves the histone deacetylase SIRT1 [31], it mainly acts by alteration of nonhistone proteins rather than by making chromatin changes. This topic will be further discussed below, with special emphasis on *resveratrol*.

3.1.2 High-Fat Diets

A long-term high-fat diet (HFD) is well known to induce metabolic stress that results in obesity, insulin resistance, diabetes, and cardiovascular disease. It affects hypothalamic expression of appetite-regulating neuropeptides and energy metabolism, and consequently expression of metabolic genes in liver and adipose tissue, strongly influenced by epigenetic modifications (reviewed in Milagro et al.) [32]. In animal models, it has been demonstrated that long-term HFDs affect DNA methylation at metabolic-related genes, including melanocortin receptor 4 (*MC4R*) in the brain [33] and leptin (*LEP*) in adipose tissue [34]. In response to fasting or HFD, the hypothalamus shows altered expression of different subsets of HDACs such as HDAC5 and HDAC8, leading to changes in histone modification patterns [35]. Thus, an HFD systemically influences gene expression patterns in tissues via epigenetic alterations.

Early exposure to an HFD during development, as with calorie restriction, leads to metabolic programming that affects metabolic status in adulthood. It has been reported that in animal models maternal HFD intake during pregnancy and lactation induced a markedly obese phenotype, with hyperinsulinemia, hyperleptinemia, and nonalcoholic fatty liver disease in offspring that was independent of postnatal nutrition, possibly due to mitochondrial dysfunction and altered expression of lipogenic genes. Intriguingly, metabolic alterations can be epigenetically transmitted to the next generation [36,37], indicating that heritable factors might be involved in metabolic programming. Maternal HFD intake prior to pregnancy affects energy homeostasis of the offspring by altered hypothalamic gene regulation at *POMC*, anocortin, and leptin receptor (*LEP-R*). Paternal HFD intake, previously neglected as a contributor to altered gene expression compared with maternal nutrition, also influences glucose metabolism in offspring over multiple generations [37,38]. Consequently, transgenerational HFD effects may substantially contribute to the dramatically increased incidence of metabolic disorders worldwide.

Metabolic alterations occur in part due to epigenetic modifications established in early life or even during previous generations. For example, HFD exposure during pregnancy resulted in hypermethylation at the promoter of PGC1a in the muscle of the offspring, and neonatal exposure to overfeeding altered and maintained DNA methylation patterns at promoters of the *POMC* and *insulin receptor* genes, in rodent models [39–41]. Also, HFDs during pregnancy led to increased histone H3 acetylation at multiple lysine residues and decreased HDAC1 expression in the livers of Japanese macaques, resulting in altered expression of genes including *Rdh12* [42]. It is not clear, however, why a subset of genes specifically acquired epigenetic modifications while the global effects were limited. Despite these observations, diagnostic epigenetic biomarkers identified in parental germ cells in response to HFD remain to be discovered. Importantly, a robust epigenetic state must be maintained in the target tissue of the offspring, a criteria not yet been fulfilled.

3.1.3 Folate and Other Vitamin B Compounds Related to One-Carbon Metabolism

Folate, a water-soluble vitamin of the B family, is required for DNA biosynthesis and DNA repair. It plays an important role in the buildup of the pool of SAM that serves as a methyl group donor in DNMT and HMT reactions. According to animal models and human studies, folate deficiency leads to global DNA hypomethylation, in turn leading to genomic instability and chromosomal aberrations (reviewed in Crider et al.) [43]. The vitamin B group, including B2, B6, and B12, are also crucial in one-carbon metabolism [44]. Insufficiency of the cofactors blocks a series of pathways that synthesize methionine from homocysteine, resulting in increased blood homocysteine concentration. Elevated homocysteine induces cell toxicity and alters DNA methylation and gene expression.

Genetic variations in enzymes related to folate metabolism, such as *MTHFR*, have also been reported to result in the lowering of enzymatic activities by diverting the availability of methyl groups from DNA methylation toward DNA biosynthesis pathways [43]. The effects of genetic variation can to some extent be dealt with by folate-fortified diet, a direct evidence for a gene–nutrient interaction. Such interactions can be mediated by not only genetic variation but also epigenetic variations. A compelling example is a link between folate deficiency and DNA hypomethylation in the *agouti* variable yellow (A^{vy}) mouse model, which shows a yellow coat color as a result of folate deficient diet that reduces the amount of DNA methylation at the *agouti* allele [45]. When pregnant mice were fed with methyl donor-rich diets (folate, methionine, vitamin B12, and choline), the offspring had a higher rate of wild-type coat color compared with the control group with normal diet [46]. Similarly, methyl donor-rich diets fed to pregnant mothers resulted in offspring with high incidence of a kinked tail in *AxinFused* mice, which was a direct result of increased DNA methylation at the *AxinFu* promoter [47]. Intriguingly, the effect of methyl donors on DNA methylation in A^{vy} models is associated with an obese phenotype that becomes progressively exacerbated through successive generations [48] mediated by DNA methylation marks in the germ cells [49].

Over the last two decades, the associations between folate, DNA methylation, and cancer risk have been analyzed in numerous studies. The results are incohesive largely because of various study designs that include dose and timing of folate intervention and the severity of folate deficiency [50]. In addition, recently published data indicate that a high dose of folate supplementation might increase cancer risk by accelerating growth of precancerous lesions [51]. Consequently, folate supplement recommendations should be carefully determined considering optimal dosage and time to avoid the adverse effects associated with high dose exposure.

3.2 Bioactive Food Compounds

Numerous studies have aimed at identification of bioactive food compounds with effects on epigenetic modifications that can be applied to metabolic disease prevention and treatment (reviewed in Shankar et al.;

Huang et al.) [52,53]. Some phytochemicals, such as resveratrol, sulfora-phane, and catechin have been shown to have protective, partially thera-peutic effects on various diseases. These phytochemical compounds might interfere with the activities or the targeting processes of epigenetic modi-fication enzymes such as DNMTs or HDACs. Several food extracts and compounds have been analyzed to find uncharacterized molecules that can modulate epigenetic enzymes in vitro and in vivo. While the results from these analyses have identified new substances, much work is left to do in determining compound-specific optimal doses and treatment dura-tions in animal models as well as in humans. The same applies to their bioavailability and stability, which will have to be carefully determined before application to humans, especially when considering individual response differences. In the following paragraphs, we summarize current knowledge on six epigenetically relevant bioactive food compounds (res-veratrol, sulforaphane, epigallocatechin-3-gallate, genistein, curcumin, and quercetin) influencing metabolic disorders and cancer. We address these substances with special emphasis on their influences on epigenetic mechanisms, while intentionally omitting details on dosage and treat-ments that are currently difficult to translate from disparate in vitro and animal model studies to human conditions.

3.2.1 Resveratrol

Resveratrol (3,5,40-trihydroxystilbene) is a natural polyphenolic com-pound in fruits such as grapes (especially skin), mulberries, cranberries, blueberries, and peanuts as well as the grape-based beverage red wine. Resveratrol has been reported to have preventive and therapeutic effects on various diseases including type 2 diabetes, cancer, and cardiovascular disease [54]. In addition, it shows a significant effect in prolonging lifespan in various model organisms from worms to mice [31]. The effects are mediated by its antioxidative, antiproliferative, and antiinflammatory properties [55]. For example, according to in vitro and in vivo analysis in mice, intake leads to increased lipolysis and decreased lipogenesis in adi-pocytes [56]. It also decreases severity of hepatic steatosis [57] and reduces insulin secretion in rats [58].

Although the mechanism by which resveratrol operates remains par-tially elusive, it seems to involve activating SIRT1, sharing some health effects with calorie restriction (see 3.1.1). While SIRT1 is a member of the (NAD(+))-dependent HDAC family, it also influences deacetylation of nonhistone proteins such as the transcriptional factor Forkhead box pro-tein O1 (FOXO1), which is critical for glucose homeostasis [31]. Addition-ally, it was recently shown that AMP-activated protein kinase (AMPK) is stimulated by resveratrol. In C2C12 cells, AMPK is activated through mechanisms that are both dependent and independent of SIRT1, via low and high doses of resveratrol, respectively [59].

The effects provided by resveratrol are partially mediated by chromatin changes, in consequence by altering gene expression. It might for instance inhibit carcinogenesis at multiple stages, possibly through changes in chromatin modifications. In human breast cancer cell lines, pretreatment of resveratrol increased acetylation of H3K9 and reduced methylation of H3K9 and modulated the recruitment of Methyl-CpG-binding domain protein 2 (MBD2) to the breast cancer 1 (*BRCA-1*) promoter [60]. While lysine acetylation of signal transducer and activator of transcription 3 (STAT3) is elevated in tumors, resveratrol reduces STAT3 acetylation [61].

3.2.2 Sulforaphane

Sulforaphane is an isothiocyanate that can be found in cruciferous vegetables such as broccoli, cabbage, and brussels sprout, and has been reported to have anticancer activity in human colon, prostate, and breast cancers (reviewed in Ho et al.) [62]. Sulforaphane has HDAC inhibitory activity that might be explained by its chemical structure, which is similar to that of the potent HDAC inhibitor trichostatin A (TSA), and potentially contributes to its anticancer activity [63]. In the colon cancer cell line HCT116, sulforaphane treatment reduces HDAC activity and represses HDAC2 expression. The HDAC inhibitory effect was confirmed in the Apc^{min} mouse model and resulted in increased histone H3 and H4 acetylation [64].

Broccoli extract sulforaphane-induced expression of phase II detoxification enzymes such as quinone oxidoreductase and glutathione transferase in rodent tissues and cell lines [62], mainly by activating nuclear factor E2-related factor 2 (NRF2) [65], is a key transcription factor in antioxidant, anti-inflammatory, and anticancer pathways. NRF2 induces antioxidant enzymes upon oxidative stress after release from Kelch-like ECH-associated protein 1 (Keap1), which sequesters NRF2 and thus renders it inactive. Subsequent release allows NRF2 to translocate to the nucleus and bind to the antioxidant response elements (AREs) in antioxidant-encoding genes [66]. The sulforaphane-induced activation of NRF2 is contributed by its HDAC inhibitor activity. Treatment with HDAC inhibitors such as TSA and valproic acid has been shown to activate NRF2 by decreasing Keap1 and increasing binding of NRF1 to AREs in prostate cancer cell lines, implying that sulforaphane exerts NRF2 activation in a similar way [62]. A recent study suggested another possibility, that sulforaphane might be influencing NRF2 gene expression through hypomethylation at the promoter followed by reduction of gene expression of DNMTs in mouse skin cell teratoma [67]. Decreased expression of DNMT1 and DNMT3a upon sulforaphane treatment has also been reported in breast cancer cells leading to hypomethylation at the human telomerase reverse transcriptase (*hTERT*) gene and inhibiting TERT expression [68]. Importantly, DNA methylation at the promoter of the *hTERT* locus is associated with transcriptional activation by blocking the binding of the repressor E2F-1 to the promoter.

Still, little is known about the role of sulforaphane in obesity and metabolic syndrome. However, HDAC1, 2, and 5 are downregulated during adipogenesis, resulting in hyperacetylation in the promoter regions of adipogenic genes—such as peroxisome proliferator-activated receptor γ (*PPARγ*), fatty acid synthase, and fatty acid binding protein 4 (*FABP4*) [69]. Despite sulforaphane having an HDAC-inhibitory effect, adipogenesis is expected to be promoted; however, recent evidence suggests that sulforaphane inhibits adipocyte differentiation [70]. Treatment of sulforaphane in 3T3-L1 preadipocytes reduced the accumulation of lipid droplets and inhibited PPARγ and CCAAT/enhancer-binding protein alpha (C/EBPα) expression early in the adipogenesis pathway and treatment of mature adipocytes enhances lipolysis through reduced AMPK signaling followed by hormone sensitive lipase activation. It is not clear, however, whether the effect is independent of its HDAC-inhibiting activity. Noteworthy current evidence on sulforaphane effects on adipogenesis has come from in vitro studies, and is the reason why further investigation is needed under in vivo conditions.

3.2.3 Epigallocatechin-3-gallate

Epigallocatechin-3-gallate (EGCG) is a major catechin found in green tea, with potent antioxidant activity. It has also been shown to have chemopreventive and anticancer effects in, among others, breast, prostate, and stomach cancers [50,71]. In various cancer cell lines, EGCG treatment leads to decreased methylation both locally and globally, possibly by inhibiting DNMT activity (reviewed in Singh et al.) [72]. The inhibitory effects may be mediated by direct binding and blocking the active site of DNMT1 as a nonnucleoside compound or by being processed to hydrogen peroxide via EGCG degradation, thereby rendering DNMT inactive by oxidation [73]. Unlike nucleoside inhibitors such as 5-aza-deoxycytidine (DAC), it does not have the inherent toxicity triggered by covalent trapping of DNMTs.

Moreover, EGCG treatment of an esophageal cancer cell line led to reactivation and demethylation at promoters of the tumor suppressor genes *p16INK4a*, *RARβ*, *MGMT*, and *hMLH1*, all previously shown to be demethylated by DAC treatment [71]. DNA-demethylation-mediated reactivation of the *RARβ* and the glutathione S-transferase-π (*GSTP1*) genes was also observed in breast [74] and prostate cancer cell lines [75], respectively. EGCG treatment also led to downregulation of the *hTERT* gene, supporting its anticancer properties [76]. Besides inhibition of DNMTs, EGCG treatment in skin cancer cells showed decreased HDAC activity that results in increased acetylation at histone H3K9 and H3K14 as well as H4K5, H4K12, and H4K16, reactivating the expression of *p16INK4a* and *Cip1/p21* [77].

EGCG has also been demonstrated to be a preventive and therapeutic agent for obesity and metabolic syndrome [78]. EGCG treatment in mice significantly reduced body weight and insulin resistance, decreased damage and triglyceride levels in the liver, and reduced plasma cholesterol and inflammatory cytokines such as monocyte chemoattractant protein-1 (MCP-1), C-reactive protein, or interleukin 6 (IL-6) in the blood [79]. Similarly, EGCG treatment decreased blood glucose levels and increased glucose tolerance in obese mice in addition to reducing the reactive oxygen species (ROS) in adipocytes and an obese mouse model. Recent evidence suggests that miRNA-induced expression by EGCG, potentially via DNA methylation loss, might be involved in EGCG effects on metabolic disorders. However, further investigation is needed to elucidate the pathways by which EGCG has such diverse epigenetic modulator functions.

3.2.4 Genistein

Genistein is an isoflavone from soy that has been reported to have anticancer properties in various cancers including those of the esophagus, colon, prostate, and breast, by activation of the expression of several aberrantly silenced tumor suppressor genes [80]. Genistein treatment leads to *p16INK4a*, *RARβ*, and *MGMT* reactivation in an esophageal cancer cell line by inhibiting DNMT activity in a substrate and methyl donor-dependent manner [81]. Also, genistein in combination with other inhibitors of DNMTs and HDACs showed a synergistic effect and higher efficacy. The effect was reproduced in prostate cancer cell lines such as LNCaP and PC12, and breast cancer cell lines such as MCF-7 and MDA-BM-231. In experiments with colon cancer cell lines, genistein reactivates the WNT-signaling antagonists WNT5a and sFBP2, potently blocking WNT signaling [80].

Intriguingly, genistein is a weaker inhibitor of DNMT than EGCG. However, it reactivates tumor suppressor genes equally as well or better than EGCG, indicating that genistein may carry additional activities as an epigenetic modulator, perhaps via histone modifications. Genistein-induced reactivation of tumor suppressor genes involved demethylation and acetylation of histone H3K9 at the promoter of phosphatase and tensin homolog (*PTEN*) and cylindromatosis [82]. Also, genistein reduced SIRT1 activity, followed by acetylation of histone H3K9 and derepression at p53 and FOXO3a [82]. In breast cancer cell lines, long-term genistein treatment showed HDAC inhibition, resulting in increased H3 acetylation [83]. It is noteworthy to mention that genistein can have adverse effects in breast cancer due to its estrogenic properties [84]. Some data from in vitro and animal studies suggest that genistein might promote the growth of estrogen-sensitive tumors, but only a limited amount of data supports this claim. Currently, no definitive evidence links genistein with women at high risk for breast cancer or with breast cancer patients who have estrogen-sensitive tumors, calling for further studies that also include long-term intervention regimes.

Considering the estrogenic effect of genistein, it has been suggested that it might play a role in the etiology of obesity, possibly acting as an endocrine-disrupting substance [32]. Genistein may influence lipid deposition in adipose tissue in a dose-dependent and sex-specific manner [85] and further contribute to mild insulin resistance. However, numerous epidemiological studies have demonstrated that high consumption of soy and soy products in Asian countries is highly associated with lower risk for metabolic disorders such as obesity, type 2 diabetes, and cardiovascular disease [86]. Genistein involves the inhibition of adipogenesis in vitro similar to the effect of the female hormone estrogen. In the A^{vy} mouse model, maternal genistein supplementation during early development changes the coat color, indicating shifts in DNA methylation at the *agouti* allele. Early exposure of genistein protects mice from obesity in adulthood by altering the epigenome. The protective effects were also observed in monkeys, showing improved body weight, insulin sensitivity, and lipid profiles by modification of DNA methylation in liver and muscle [87].

3.2.5 *Curcumin*

Curcumin (diferuloylmethane) is a polyphenolic compound extracted from the common Indian spice turmeric, or *Curcuma longa*. It has been used in traditional medicines in India and its functional benefits have been reported for cancer and chronic metabolic disorders. It has been shown to modulate multiple intercellular signaling pathways including proliferation, apoptosis, and inflammation. *In silico* studies suggest that curcumin functions as an inhibitor of DNMT activity by covalently blocking the active binding site [52]. Curcumin treatment has therefore shown to cause global hypomethylation in leukemia cells, but only a few tumor suppressor genes have been identified so far.

Curcumin has been demonstrated to inhibit HAT activity by either covalently binding to the active site or promoting proteasome-dependent degradation. It interferes effectively with p300 HAT activity on histones as well as nonhistone protein; e.g., the tumor suppressor p53 [88], which decreases acetylation of histone H3 and H4 in neural progenitor cells and hepatoma cell lines [89]. Recent studies suggest, however, that it also inhibits expression of HDACs (i.e., HDAC1, 3, and 8), resulting in reduced cell proliferation by increased acetylation of H4 in a leukemia cell line. According to the result from fluorometric analysis, curcumin seems to bind and inhibit HDAC8 activity in addition to repressing expression. However, another study suggests that curcumin activated p53 transcriptionally and translationally, which in turn triggered apoptosis because of increased ROS in prostate cancer cell lines. This process is mediated by increased acetylation of histone H3 and H4 that leads to apoptosis via the involvement of B-cell lymphoma 2 (Bcl-2) family and p53 [90]. The conflicting data regarding acetylation of p53 upon curcumin treatment could be due to targeting of different lysine residues in p53. However, this has not been examined in these studies and therefore further investigation is required.

Curcumin also has potent anti-inflammatory properties that became evident when treating obese individuals suffering from elevated inflammatory symptoms. In vitro studies suggest that curcumin suppresses 3T3-L1 differentiation and leads to apoptosis [91]. Experiments from HFD-induced obese animal models demonstrated that curcumin supplementation reduces body weight gain and adiposity, probably by increasing fatty acid oxidation and decreasing fatty acid esterification in adipose tissue [91]—it reduces lipogenic gene expression in the liver and in adipose tissue during inflammatory response [92]. Despite its strong effect on obesity, it has not been mechanistically connected to epigenetic modulations in contrast to its role in cancer. A recent study provides a clue to the connection by showing that hyperglycemia induces cytokine production in monocytes, which can be suppressed by curcumin treatment via reduction of p300 activity and its target gene expression.

3.2.6 Quercetin

Quercetin is a ubiquitous dietary polyphenolic compound found in onions, citrus fruits, and buckwheat, and is known to have antioxidant and anticancer activity [93]. Quercetin induces cell cycle arrest and apoptosis in hamster buccal pouch tumors that correlates with the inhibition of DNMT1 and HDAC1, resulting in substantial tumor growth delay. Also, it suppresses the growth of colon cancer cells by reactivating *p16INK4a* via hypomethylation of its promoter [94]. In human leukemia HL-60 cells, quercetin could induce FasL-mediated apoptosis through increased H3 acetylation by activation of HAT and inhibition of HDAC [95]. Similar observations were made in murine intestinal epithelial cells, showing inhibition of tumor necrosis factor alpha (*TNFα*)-induced proinflammatory gene expression, such as TNF-induced interferongamma–inducible protein 10 (*IP-10*) and macrophage inflammatory protein-2 (*MIP-2*) that occurred through inhibition of CBP/p300 activity and acetylation and phosphorylation of histone H3 at their respective promoters [96].

Quercetin supplements in animal models fed HFDs decrease metabolic symptoms such as high weight gain and high blood pressure, as well as impairing glucose intolerance and nonalcoholic fatty acid disease. This decrease is mediated by activation of Nrf2 and its target genes including heme oxygenase-1 (*HO-1*) and carnitine palmitoyltransferase (*CPT1*) expression. It is possible that quercetin-facilitated Nrf2 activation acts by providing epigenetic changes, a process reminiscent of sulforaphane activity.

4. EPIGENETIC BIOMARKERS—NOVEL TOOLS TO ASSESS METABOLIC DISORDERS AND DIET RESPONSE

4.1 The Use of Biomarkers in Diagnosis of and Predisposition to Metabolic Disorders

In order to develop a successful strategy for personalized nutrition, use of proper molecular biomarkers is required. Molecular biomarkers are diagnostic tools, typically genetic elements or genes that can readily be analyzed objectively as indicators of biological processes, for disease diagnoses, or as response indicators to pharmacological therapies. Numerous molecular biomarkers have been developed for the diagnosis of disease—e.g., cancer, commonly by utilizing mRNA or protein expression, DNA sequence variations and mutations, and/or epigenetic patterns. Biomarkers indicating individual differences, predisposition to disease, and responses to therapy remain scarce. In the nutritional field, genome-wide association studies have uncovered suitable biomarkers that reflect individual and differential responses to diets or nutrients, but the numbers are still very limited. Typical biomarkers stemming from these studies include the *MTHFR* and *VDR* variations discussed in Section 1 above in the highlighting of genetic variations in metabolically related genes that influence nutrient bioactivity and efficacy.

When considering the plethora of physiological changes throughout life (in part driven by extrinsic factors such as diets), ultimately translating into genetic and chromatin changes, classical genetic biomarkers have provided limited information. In contrast, epigenetic modifications have been recognized as potentially potent diagnostic tools due to their plastic features of both flexibility and stability. Remodeling of epigenetic modifications has been closely correlated with environmental exposures including diets. The modifications can be modulated by nutrients and bioactive food compounds, as discussed in Section 3 above. While generally considered dynamic, some epigenetic modifications are surprisingly solid. For instance, DNA methylation in centromeric or pericentromeric regions is robustly maintained, and genes residing in those areas may prove diagnostic to dramatic global epigenetic changes. Established DNA methylation patterns are also frequently maintained during cell division and development and often inherited through subsequent generations. These inherited patterns may induce long-term changes in gene expression that determine the response to diet or nutrient intakes.

The use of proper nutriepigenetic biomarkers can be divided into three categories: the first consists of biomarkers that accurately and reproducibly report on a specific condition with the epigenetic marker ideally placed in a regulatory region of the gene directly associated with the condition in question. The second category contains biomarkers that quantitatively allow a dose response estimation to be made. A potentially useful third category consists of epigenetic markers that are dynamic and change upon dietary intervention. In addition, a good marker should preferably be diagnostic in the blood or any other easily accessible tissue.

The act of changes of epigenetic modification in response to diets or nutrients is a complex process, and they appear synergistically or antagonistically when establishing an open or closed chromatin configuration. The most reliable approach is therefore to develop combinatorial nutriepigenetic biomarkers that more robustly report on individual dietary responses. While some metabolic disorder-related epigenetic biomarkers have been identified, the influence of nutrients on these marks remains poorly investigated. In the paragraphs to follow, we summarize some emerging evidence of epigenetic biomarkers in metabolic disease and their responses to interventions by low-calorie diets—these may be indicative examples of nutriepigenetic biomarkers.

4.1.1 Obesity

To assess the role of epigenetic variations related to individual risks of developing obesity, a number of studies have analyzed DNA methylation or histone modification patterns to extract epigenetic biomarkers. Global DNA methylation changes are considered a hallmark of many diseases, especially cancer, but associations between global DNA methylation aberrations and obesity are inconsistent. Therefore, locus-specific DNA methylation with direct links to regulation of gene expression of relevant genes could be critical biomarkers for identifying susceptibility to obesity. Several studies report on the identification of specific gene methylation marks related to obesity, providing candidate genes linked to the susceptibility for this condition. The majority of these studies are cross-sectional and focus on the relationships between body mass index (BMI) and DNA methylation levels at a variety of obesity-related genes that influence appetite control, insulin signaling, inflammation, or immunity. Individuals with high BMIs, compared with individuals who have low BMIs, have hypermethylation at the *POMC* locus in whole blood of children [97]. Genes related to circadian genes such as the circadian locomotor output cycles kaput (*CLOCK*) gene, the aryl hydrocarbon receptor nuclear translocator-like (*BMAL1*) gene, and period circadian protein homolog 2 (*PER2*) gene, have also been associated with obesity. Overweight and obese individuals, compared with those of normal body weight, have significantly higher methylation at the *CLOCK* gene and methylation levels of the *CLOCK*, *BMAL1*, and *PER2* genes have been linked to anthropometric parameters such as BMI and adiposity, and also with MetS scores [98]. Women of normal weight with high central adiposity have lower methylation at *TNFα* in peripheral blood leukocyte, compared with women who have low central adiposity [99]. Likewise, in comparison with lean individuals, it has been observed that peripheral blood leukocytes of obese people have hypermethylation in the ubiquitin-associated and SH3 domain–containing protein A (*UBASH3A*) gene and in the tripartite motif containing 3 (*TRIM3*) gene [100].

4.1.2 Diabetes

Type 2 diabetes is one of the most prevalent metabolic conditions across all age groups and results from a combination of age, obesity, and lifestyle. The influence by which epigenetic modifications contribute to higher risk of type 2 diabetes has been addressed in a number of studies, and the majority of them analyzed genes related to pancreatic regulation of glucose metabolism. The methylation level at the promoter of the monocyte chemoattractant protein-1 (*MCP-1*) gene is reduced in patients with type 2 diabetes compared with control groups, resulting in the different MCP-1 levels in the serum [101]. In pancreatic islets of type 2 diabetes patients, distal promoter sites or enhancers of pancreatic duodenal homeobox 1 (*PDX-1*) gene, *PPARGC1A*, and the human insulin gene are hypermethylated compared with those of nondiabetic participants. DNA hypermethylation downregulates those genes and consequently was inversely correlated with the level of insulin secretion [42,102].

4.1.3 Metabolic Syndrome and Cardiovascular Disease

Obesity or diabetes is seen as an independent risk factor for cardiovascular disease, especially during childhood. Consequently, these metabolic diseases act as complications of other diseases. In an attempt to divide obese individuals with susceptibility for metabolic syndrome into high-and low-risk groups, low DNA methylation at the repetitive LINE-1 element was found to correlate well with high risk for metabolic syndrome [103]. This study therefore suggests that hypomethylation is associated with increased risk for metabolic syndrome in obese people. A number of studies in humans have identified epigenetic biomarkers associated with cardiovascular disease risk. The Dutch Hunger Winter study identified several DNA methylation patterns associated with the incidence of metabolic disease [21]. DNA methylation of the *INSIGF* gene was reduced, while elevated at the *IL-10*, *LEP, ABCA1, GNASAS*, and *MEG3* genes among individuals exposed to famine during late gestation [22]. The prevalence of coronary heart disease in these individuals was three times higher than it was in individuals suffering maternal famine during early gestation [104]. Although a number of epigenetic biomarkers specific for cardiovascular disease have been identified, the above-mentioned obesity and diabetes biomarkers may also contribute as risk factors for cardiovascular disease.

4.2 DNA Methylation Changes in Response to Dietary Intervention

Dietary intervention through adjustment of dietary macronutrients has been carried out as a major therapeutic strategy to handle obesity in addition to supplementation of functional compounds. While calorie restriction is the most recommended therapeutic method in patients with obesity [105,106], the response is highly variable among obese people. Therefore, understanding the intrinsic factors leading to variable response is critical in the clinical assessment of metabolic conditions.

Recent studies have reported that DNA methylation patterns could serve as biomarkers to predict responsiveness to calorie restriction in obese people (Table 2). The first evidence of this kind suggested methylation at the promoter of the *TNFα* gene as an epigenetic biomarker

TABLE 2 Epigenetic Biomarkers to Indicate the Responsiveness to Low-Calorie Diet Intervention in Obese People

Response	Tissue	Methylation site	Methylation level in high responder	Reference
Weight loss	PBMC	TNFα promoter (at −170 bp)	Low	[107]
	Adipocyte	TNFα and leptin promoter	Low	[108]
	PBMC	ATP10A promoter (CpG sites, 5-10-16, 18)	Low	[109]
		CD44 promoter (CpG sites, 14, 29)	High	
	Adipocyte	3 CpG regions in RAB3C, HYPK, and DNASEIL2	Low	[110]
		32 CpG regions including KCNA3, GLIS3, ETS, NFIX, and INSM1	High	
	PBMC	HIPK3 (CpG sites, 1,5,7)	High	[111]
		TNNI3 (CpG sites, 1,3)	High	
Weight regain	PBMC	POMC (at +136 bp and 138 bp)	High	[112]
		NPY promoter (CpG sites, 4,5,8-9-10)	Low	

reflecting intervention responsiveness to calorie restriction [107]. In a balanced low-calorie diet where 55% of energy was provided by carbo-hydrates, 15% by proteins, and 30% by fat for 8 weeks in a study group of obese persons (BMI >30), study results indicated different weight-loss responsiveness that was linked to methylation status at the *TNFα* pro-moter. Male responders, defined as losing more than 5% of their initial body weight, had lower levels of methylation at 170 base pairs upstream of the transcriptional starting site of the *TNFα* gene in peripheral blood cells, compared with the levels in nonresponders. A similar result in sub-cutaneous adipose tissue has been reported in responders who showed lower methylation in *Lep* and *TNFα* promoter, and *TNFα* promoter meth-ylation levels were significantly correlated with the systolic and diastolic blood pressures recorded prior to the intervention [108]. Moreover, the baseline methylation levels of the *CLOCK* and *PER2* genes, which serve as obesity risk biomarkers, correlated well with the degree of weight loss after treatment intervention [98].

Studies using genome-wide methylation arrays have identified addi-tional epigenetic biomarkers. For instance, methylation levels at specific CpG sites of the *ATP10A* and *CD44* genes were highly different between high and low responders in peripheral blood mononuclear cells before the intervention [109]. Another study using the methylation array described that high responders, defined as those losing more than 3% body fat mass, showed significantly different DNA methylation patterns at 35 loci prior to the intervention. Of these, three loci (*RAB3C*, *HYPK*, and *DNASEIL2*) were hypomethylated, and 32 loci (e.g., *KCNA3*, *GLIS3*, *ETS*, *NFIX*, and *INSM1*) were hypermethylated in the high compared with low responders [110]. More recently, Moleres et al. revealed that a multidis-ciplinary intervention in overweight and obese adolescents resulted in different weight loss responses depending on basal DNA methylation of CpG sites in the aquaporin 9 (*AQP9*) gene, the dual specificity phospha-tase 22 (*DUSP22*) gene, the homeodomain-interacting protein kinase 3 (*HIPK3*) gene, the troponin T type 1 (*TNNT1*) gene, and the troponin I type 3 (*TNNI3*) gene [111].

Besides weight loss responsiveness, weight-regain responsiveness after weight loss intervention is also associated with DNA methylation bio-markers [112]. Obese men who lost more than 5% of body weight after 8 weeks of weight loss intervention were divided into regainers and non-regainers, depending on whether they gained back more than 10% of their weight loss 32 weeks post-weight loss intervention. Weight regainers had comparably higher methylation levels in the CpGs 136 and 138 base pairs upstream of the transcriptional start site to the *POMC* gene in leukocytes, and comparably lower methylation levels at the promoter of the *NPY* gene, than did nonregainers. Since *POMC* and *NPY* genes are associated with food intake regulation, the epigenetic influence on susceptibility to weight regain via DNA methylation at these loci is obvious. In order to regain weight control, it appears essential to develop calorierestriction regimens that break the epigenetic patterning of *POMC* and *NPY* genes.

5. CONCLUSIONS AND FUTURE PROSPECTS

Noncommunicable diseases are expected to account for 60% of disease burden worldwide and over 73% of global mortality in 2020 [113]. Since up to 80% of chronic disease can be prevented through diet and lifestyle considerations, reevaluation of dietary guidelines is essential to improving their efficacy. These issues highlight the importance of personalized nutrition, which requires detailed individual data collection that includes metabolic and physiological status, even down to the molecular level. Recent advances in studies of nutrigenomics and epigenetics have resulted in substantial progress in our understanding of the complex interactions between diet and the genome, and importantly that these interactions are often mediated by epigenetic changes. The understanding of diet-induced epigenetic alterations and epigenetic determinants in diet response are central to progress in personalized nutrition.

Based on the studies discussed in this review, it is evident that nutrients and bioactive food compounds act on the genome through chromatin changes by means of epigenetic alterations, leading to different gene expression patterns. Assessment of epigenetic patterns provides valuable information on how individuals respond to dietary nutrients and bioactive compounds. Thus, nutriepigenomics carry the fortuitous prospect of contributing to the design of optimized dietary intervention for the promotion of health by guiding prevention and therapy in diet-related or metabolic-related disorders. Identified epigenetic biomarkers suitable for personalized nutriepigenomics remain very limited. Biomarkers currently in use or identified as being diagnostic only provide limited information on how to establish a cutoff prior to intervention that accurately classifies the groups suitable for a particular treatment. Consequently, we are in dire need of further evidence from multiple large-scale intervention studies before launching treatment regimens that provide a robust set of biomarkers for reporting treatment efficacy and implementation. As previously stated, development of combinatorial nutriepigenetic biomarkers seems the best approach toward robust clinical assessments.

Nevertheless, nutriepigenomics is a new area of nutritional science that over the next decade aspires to deliver important insights into genome responses to dietary interventions. Future identification of novel nutriepigenetic biomarkers will primarily emerge from whole genome analysis using various *omics* methods, detailed bioinformatics analysis of existing data, and systems biology evaluations. This requires the use of tools at the frontline of modern molecular biology and genomics. With the right progressive approach and the application of suitable tools, we will expand our understanding of nutriepigenomics as a diagnostic instrument to bring us closer to the goal of effective personalized nutrition.

Glossary

Personalized nutrition Nutritional guidelines that provide the tailoring of nutritional recommendations to the individual needs, characteristics, and preferences of a person during all stages of care including growth, health promotion and disease prevention, disease management, and aging.

Nutrigenomics The research field to study the complex interplay between diets and genes. It contains two major fields: nutrigenetics for understanding how an individual's genetic makeup predisposes for dietary susceptibility, and nutrigenomics for understanding how diets and nutrients affect gene expression in the human genome.

Nutriepigenomics The research field studying the complex interplay between diets and the genome through epigenetic modifications.

Epigenetics The study of heritable and reversible changes in gene expression in the absence of DNA sequence changes, mediated by chromatin changes via chemical modifications to DNA or histones.

LIST OF ACRONYMS AND ABBREVIATIONS

ABCA1	ATP-binding cassette transporter ABCA1
ACL	ATP-citrate lyase
AMPK	AMP-activated protein kinase
AQP9	Aquaporin 9
AREs	Antioxidant response elements
ATP10A	ATPase class V type 10A
A^{vy}	Agouti variable yellow
BCL-2	B-cell lymphoma 2
BMAL1	Aryl hydrocarbon receptor nuclear translocator-like
BMI	Body mass index
BRCA-1	Breast cancer 1
C/EBPα	CCAAT/enhancer-binding protein alpha
CGIs	CpG-rich islands
CLOCK	Circadian locomotor output cycles kaput
CpG	Cytosines followed by guanine
CPT1	Carnitine palmitoyltransferase

CRP	C-reactive protein
CYLD	Cylindromatosis
CYP7A1	Cholesterol 7 alpha-hydroxylase
DAC	5-aza-Deoxycytidine
DNMTs	DNA methyltransferases
DUSP22	Dual specificity phosphatase 22
EGCG	Epigallocatechin-3-gallate
FABP4	Fatty acid binding protein 4
FAS	Fatty acid synthase
FOXO1	Forkhead box protein O1
GlcNAc	O-linked N-acetylglucosamine
GLUT4	Glucose transporter type 4
GR	Glucocorticoid receptor
GSTP1	Glutathione S-transferase-π
GWAS	Genome-wide association studies
HATs	Histone acetyltransferases
HDACs	Histone deacetylases
HFD	High-fat diet
HIPK3	Homeodomain-interacting protein kinase 3
HMTs	Histone methyltransferases
HO-1	Heme oxygenase-1
hTERT	Human telomerase reverse transcriptase
IGF	Insulin-like growth factor
IGF2	Insulin-like growth factor 2
IL-10	Interleukin 10
IL-6	Interleukin 6
INS	Human insulin gene
IP-10	Interferongamma–inducible protein 10
KDMs	Histone lysine demethylases
Keap1	Kelch-like ECH-associated protein 1
α-KG	α-Ketoglutarate
LEP	Leptin

LEP-R	Leptin receptor
lncRNAs	Long noncoding RNAs
MBD2	Methyl-CpG-binding domain protein 2
MC4R	Melanocortin receptor 4
MCP-1	Monocyte chemoattractant protein-1
MEG3	Maternally expressed 3
MIP-2	Macrophage inflammatory protein 2
miRNAs	MicroRNAs
MTHFR	Methylenetetrahydrofolate
NPY	Neuropeptide Y
NRF2	Nuclear factor E2-related factor 2
p53	Tumor suppressor p53
PDX-1	Pancreatic duodenal homeobox 1
PER2	Period circadian protein homolog 2
POMC	Proopiomelanocortin
PPARGC1A	Peroxisome proliferator-activated receptor γ coactivator-1α
PPARγ	Peroxisome proliferator-activated receptor γ
PTEN	Phosphatase and tensin homolog
ROS	Reactive oxygen species
SAH	S-adenosyl homocysteine
SAM	S-adenosylmethionine
SRC	Steroid receptor activator
STAT3	Signal transducer and activator of transcription 3
TCA cycle	Tricarboxylic acid cycle
TETs	Ten-eleven-translocation methylcytosine dioxygenases
TNFα	Tumor necrosis factor alpha
TNNI3	Troponin I type 3
TNNT1	Troponin T type 1
TRIM3	Tripartite motif containing 3
TSA	Trichostatin A
UBASH3A	Ubiquitin-associated and SH3 domain-containing protein A
VDR	1,25-Dihydroxyvitamin D3 receptor

References

[1] Molloy AM, Mills JL, Kirke PN, Weir DG, Scott JM. Folate status and neural tube defects. BioFactors 1999;10(2–3):291–4.
[2] Sharp L, Little J. Polymorphisms in genes involved in folate metabolism and colorectal neoplasia: a HuGE review. Am J Epidemiol 2004;159(5):423–43.
[3] Kim HS, Newcomb PA, Ulrich CM, et al. Vitamin D receptor polymorphism and the risk of colorectal adenomas: evidence of interaction with dietary vitamin D and calcium. Cancer Epidemiol Biomarkers Prev 2001;10(8):869–74.
[4] Bannister AJ, Kouzarides T. Regulation of chromatin by histone modifications. Cell Res 2011;21(3):381–95.
[5] Barski A, Cuddapah S, Cui K, et al. High-resolution profiling of histone methylations in the human genome. Cell 2007;129(4):823–37.
[6] Tahiliani M, Koh KP, Shen Y, et al. Conversion of 5-methylcytosine to 5-hydroxymethylcytosine in mammalian DNA by MLL partner TET1. Science 2009;324(5929):930–5.
[7] Tan L, Shi YG. Tet family proteins and 5-hydroxymethylcytosine in development and disease. Development 2012;139(11):1895–902.
[8] Gal-Yam EN, Egger G, Iniguez L, et al. Frequent switching of Polycomb repressive marks and DNA hypermethylation in the PC3 prostate cancer cell line. Proc Natl Acad Sci USA 2008;105(35):12979–84.
[9] Lindroth AM, Park YJ, McLean CM, et al. Antagonism between DNA and H3K27 methylation at the imprinted Rasgrf1 locus. PLoS Genet 2008;4(8):e1000145.
[10] Cortessis VK, Thomas DC, Levine AJ, et al. Environmental epigenetics: prospects for studying epigenetic mediation of exposure-response relationships. Hum Genet 2012;131(10):1565–89.
[11] Kilpinen H, Dermitzakis ET. Genetic and epigenetic contribution to complex traits. Hum Mol Genet 2012;21(R1):R24–8.
[12] Robertson KD. DNA methylation and human disease. Nat Rev Genet 2005;6(8):597–610.
[13] Bonasio R, Tu S, Reinberg D. Molecular signals of epigenetic states. Science 2010; 330(6004):612–6.
[14] Anway MD, Cupp AS, Uzumcu M, Skinner MK. Epigenetic transgenerational actions of endocrine disruptors and male fertility. Science 2005;308(5727):1466–9.
[15] Weaver IC, Cervoni N, Champagne FA, et al. Epigenetic programming by maternal behavior. Nat Neurosci 2004;7(8):847–54.
[16] Mason JB. Biomarkers of nutrient exposure and status in one-carbon (methyl) metabolism. J Nutr 2003;133(Suppl. 3):941S–7S.
[17] Wellen KE, Hatzivassiliou G, Sachdeva UM, Bui TV, Cross JR, Thompson CB. ATP-citrate lyase links cellular metabolism to histone acetylation. Science 2009;324(5930):1076–80.
[18] Guarente L. Calorie restriction and sirtuins revisited. Genes Dev 2013;27(19):2072–85.
[19] Steinmetz KA, Potter JD. Vegetables, fruit, and cancer prevention: a review. J Am Diet Assoc 1996;96(10):1027–39.
[20] Martin-Gronert MS, Ozanne SE. Mechanisms linking suboptimal early nutrition and increased risk of type 2 diabetes and obesity. J Nutr 2010;140(3):662–6.
[21] Heijmans BT, Tobi EW, Stein AD, et al. Persistent epigenetic differences associated with prenatal exposure to famine in humans. Proc Natl Acad Sci USA 2008;105(44):17046–9.
[22] Tobi EW, Lumey LH, Talens RP, et al. DNA methylation differences after exposure to prenatal famine are common and timing- and sex-specific. Hum Mol Genet 2009;18(21):4046–53.
[23] Yoo JY, Lee S, Lee HA, et al. Can proopiomelanocortin methylation be used as an early predictor of metabolic syndrome? Diabetes Care 2014;37(3):734–9.
[24] Zhang S, Rattanatray L, MacLaughlin SM, et al. Periconceptional undernutrition in normal and overweight ewes leads to increased adrenal growth and epigenetic changes in adrenal IGF2/H19 gene in offspring. FASEB J: Off Publ Fed Am Soc Exp Biol 2010;24(8):2772–82.

[25] Radford EJ, Ito M, Shi H, et al. In utero effects. In utero undernourishment perturbs the adult sperm methylome and intergenerational metabolism. Science 2014;345(6198):1255903.

[26] Wheatley KE, Nogueira LM, Perkins SN, Hursting SD. Differential effects of calorie restriction and exercise on the adipose transcriptome in diet-induced obese mice. J Obes 2011;2011:265417.

[27] Gong L, Pan YX, Chen H. Gestational low protein diet in the rat mediates Igf2 gene expression in male offspring via altered hepatic DNA methylation. Epigenet: Off J DNA Methylation Soc 2010;5(7):619–26.

[28] Sohi G, Marchand K, Revesz A, Arany E, Hardy DB. Maternal protein restriction elevates cholesterol in adult rat offspring due to repressive changes in histone modifications at the cholesterol 7alpha-hydroxylase promoter. Mol Endocrinol 2011;25(5):785–98.

[29] Pankevich DE, Teegarden SL, Hedin AD, Jensen CL, Bale TL. Caloric restriction experience reprograms stress and orexigenic pathways and promotes binge eating. J Neurosci: Off J Soc Neurosci 2010;30(48):16399–407.

[30] Fontana L, Partridge L, Longo VD. Extending healthy life span–from yeast to humans. Science 2010;328(5976):321–6.

[31] Camins A, Sureda FX, Junyent F, et al. Sirtuin activators: designing molecules to extend life span. Biochim Biophys Acta 2010;1799(10–12):740–9.

[32] Milagro FI, Mansego ML, De Miguel C, Martinez JA. Dietary factors, epigenetic modifications and obesity outcomes: progresses and perspectives. Mol Aspects Med 2013;34(4):782–812.

[33] Widiker S, Karst S, Wagener A, Brockmann GA. High-fat diet leads to a decreased methylation of the Mc4r gene in the obese BFMI and the lean B6 mouse lines. J Appl Genet 2010;51(2):193–7.

[34] Milagro FI, Campion J, Garcia-Diaz DF, Goyenechea E, Paternain L, Martinez JA. High fat diet-induced obesity modifies the methylation pattern of leptin promoter in rats. J Physiol Biochem 2009;65(1):1–9.

[35] Funato H, Oda S, Yokofujita J, Igarashi H, Kuroda M. Fasting and high-fat diet alter histone deacetylase expression in the medial hypothalamus. PloS One 2011;6(4):e18950.

[36] Dunn GA, Bale TL. Maternal high-fat diet effects on third-generation female body size via the paternal lineage. Endocrinology 2011;152(6):2228–36.

[37] Ng SF, Lin RC, Laybutt DR, Barres R, Owens JA, Morris MJ. Chronic high-fat diet in fathers programs beta-cell dysfunction in female rat offspring. Nature 2010;467(7318):963–6.

[38] Fullston T, Ohlsson Teague EM, Palmer NO, et al. Paternal obesity initiates metabolic disturbances in two generations of mice with incomplete penetrance to the F2 generation and alters the transcriptional profile of testis and sperm microRNA content. FASEB J: Off Publ Fed Am Soc Exp Biol 2013;27(10):4226–43.

[39] Laker RC, Lillard TS, Okutsu M, et al. Exercise prevents maternal high-fat diet-induced hypermethylation of the Pgc-1alpha gene and age-dependent metabolic dysfunction in the offspring. Diabetes 2014;63(5):1605–11.

[40] Plagemann A, Harder T, Brunn M, et al. Hypothalamic proopiomelanocortin promoter methylation becomes altered by early overfeeding: an epigenetic model of obesity and the metabolic syndrome. J Physiol 2009;587(Pt 20):4963–76.

[41] Plagemann A, Roepke K, Harder T, et al. Epigenetic malprogramming of the insulin receptor promoter due to developmental overfeeding. J Perinat Med 2010;38(4):393–400.

[42] Aagaard-Tillery KM, Grove K, Bishop J, et al. Developmental origins of disease and determinants of chromatin structure: maternal diet modifies the primate fetal epigenome. J Mol Endocrinol 2008;41(2):91–102.

[43] Crider KS, Yang TP, Berry RJ, Bailey LB. Folate and DNA methylation: a review of molecular mechanisms and the evidence for folate's role. Adv Nutr 2012;3(1):21–38.

[44] Rush EC, Katre P, Yajnik CS. Vitamin B12: one carbon metabolism, fetal growth and programming for chronic disease. Eur J Clin Nutr 2014;68(1):2–7.

[45] Michaud EJ, van Vugt MJ, Bultman SJ, Sweet HO, Davisson MT, Woychik RP. Differential expression of a new dominant agouti allele (Aiapy) is correlated with methylation state and is influenced by parental lineage. Genes Dev 1994;8(12):1463–72.

[46] Wolff GL, Kodell RL, Moore SR, Cooney CA. Maternal epigenetics and methyl supplements affect agouti gene expression in Avy/a mice. FASEB J: Off Publ Fed Am Soc Exp Biol 1998;12(11):949–57.

[47] Waterland RA, Dolinoy DC, Lin JR, Smith CA, Shi X, Tahiliani KG. Maternal methyl supplements increase offspring DNA methylation at Axin Fused. Genesis 2006;44(9):401–6.

[48] Waterland RA, Travisano M, Tahiliani KG, Rached MT, Mirza S. Methyl donor supplementation prevents transgenerational amplification of obesity. Int J Obes 2008;32(9):1373–9.

[49] Cropley JE, Suter CM, Beckman KB, Martin DI. Germ-line epigenetic modification of the murine A vy allele by nutritional supplementation. Proc Natl Acad Sci USA 2006;103(46):17308–12.

[50] Li Y, Tollefsbol TO. Impact on DNA methylation in cancer prevention and therapy by bioactive dietary components. Curr Med Chem 2010;17(20):2141–51.

[51] Song J, Medline A, Mason JB, Gallinger S, Kim YI. Effects of dietary folate on intestinal tumorigenesis in the apcMin mouse. Cancer Res 2000;60(19):5434–40.

[52] Shankar S, Kumar D, Srivastava RK. Epigenetic modifications by dietary phytochemicals: implications for personalized nutrition. Pharmacol Ther 2013;138(1):1–17.

[53] Huang J, Plass C, Gerhauser C. Cancer chemoprevention by targeting the epigenome. Curr Drug Targets 2011;12(13):1925–56.

[54] Chung JH, Manganiello V, Dyck JR. Resveratrol as a calorie restriction mimetic: therapeutic implications. Trends Cell Biol 2012;22(10):546–54.

[55] Baur JA, Sinclair DA. Therapeutic potential of resveratrol: the in vivo evidence. Nat Rev Drug Discov 2006;5(6):493–506.

[56] Baile CA, Yang JY, Rayalam S, et al. Effect of resveratrol on fat mobilization. Ann NY Acad Sci 2011;1215:40–7.

[57] Bujanda L, Hijona E, Larzabal M, et al. Resveratrol inhibits nonalcoholic fatty liver disease in rats. BMC Gastroenterol 2008;8:40.

[58] Szkudelski T. Resveratrol-induced inhibition of insulin secretion from rat pancreatic islets: evidence for pivotal role of metabolic disturbances. Am J Physiol Endocrinol Metab 2007;293(4):E901–7.

[59] Price NL, Gomes AP, Ling AJ, et al. SIRT1 is required for AMPK activation and the beneficial effects of resveratrol on mitochondrial function. Cell Metab 2012;15(5):675–90.

[60] Papoutsis AJ, Lamore SD, Wondrak GT, Selmin OI, Romagnolo DF. Resveratrol prevents epigenetic silencing of BRCA-1 by the aromatic hydrocarbon receptor in human breast cancer cells. J Nutr 2010;140(9):1607–14.

[61] Lee H, Zhang P, Herrmann A, et al. Acetylated STAT3 is crucial for methylation of tumor-suppressor gene promoters and inhibition by resveratrol results in demethylation. Proc Natl Acad Sci 2012;109(20):7765–9.

[62] Ho E, Clarke JD, Dashwood RH. Dietary sulforaphane, a histone deacetylase inhibitor for cancer prevention. J Nutr 2009;139(12):2393–6.

[63] Myzak MC, Karplus PA, Chung FL, Dashwood RH. A novel mechanism of chemoprotection by sulforaphane: inhibition of histone deacetylase. Cancer Res 2004;64(16):5767–74.

[64] Myzak MC, Dashwood WM, Orner GA, Ho E, Dashwood RH. Sulforaphane inhibits histone deacetylase in vivo and suppresses tumorigenesis in Apc-minus mice. FASEB J: Off Publ Fed Am Soc Exp Biol 2006;20(3):506–8.

[65] McWalter GK, Higgins LG, McLellan LI, et al. Transcription factor Nrf2 is essential for induction of NAD(P)H:quinone oxidoreductase 1, glutathione S-transferases, and glutamate cysteine ligase by broccoli seeds and isothiocyanates. J Nutr 2004; 134(12 Suppl):3499S–506S.

[66] Zhang DD, Hannink M. Distinct cysteine residues in Keap1 are required for Keap1-dependent ubiquitination of Nrf2 and for stabilization of Nrf2 by chemopreventive agents and oxidative stress. Mol Cell Biol 2003;23(22):8137–51.

[67] Su ZY, Zhang C, Lee JH, et al. Requirement and epigenetics reprogramming of Nrf2 in suppression of tumor promoter TPA-induced mouse skin cell transformation by sulforaphane. Cancer Prev Res 2014;7(3):319–29.

[68] Meeran SM, Patel SN, Tollefsbol TO. Sulforaphane causes epigenetic repression of hTERT expression in human breast cancer cell lines. PloS One 2010;5(7):e11457.

[69] Yoo EJ, Chung JJ, Choe SS, Kim KH, Kim JB. Down-regulation of histone deacetylases stimulates adipocyte differentiation. J Biol Chem 2006;281(10):6608–15.

[70] Choi KM, Lee YS, Kim W, et al. Sulforaphane attenuates obesity by inhibiting adipogenesis and activating the AMPK pathway in obese mice. J Nutr Biochem 2014;25(2):201–7.

[71] Fang MZ, Wang Y, Ai N, et al. Tea polyphenol (-)-epigallocatechin-3-gallate inhibits DNA methyltransferase and reactivates methylation-silenced genes in cancer cell lines. Cancer Res 2003;63(22):7563–70.

[72] Singh BN, Shankar S, Srivastava RK. Green tea catechin, epigallocatechin-3-gallate (EGCG): mechanisms, perspectives and clinical applications. Biochem Pharmacol 2011;82(12):1807–21.

[73] Lyko F, Brown R. DNA methyltransferase inhibitors and the development of epigenetic cancer therapies. J Natl Cancer Inst 2005;97(20):1498–506.

[74] Lee WJ, Shim JY, Zhu BT. Mechanisms for the inhibition of DNA methyltransferases by tea catechins and bioflavonoids. Mol Pharmacol 2005;68(4):1018–30.

[75] Pandey M, Shukla S, Gupta S. Promoter demethylation and chromatin remodeling by green tea polyphenols leads to re-expression of GSTP1 in human prostate cancer cells. Int J Cancer 2010;126(11):2520–33.

[76] Berletch JB, Liu C, Love WK, Andrews LG, Katiyar SK, Tollefsbol TO. Epigenetic and genetic mechanisms contribute to telomerase inhibition by EGCG. J Cell Biochem 2008;103(2):509–19.

[77] Nandakumar V, Vaid M, Katiyar SK. (-)-Epigallocatechin-3-gallate reactivates silenced tumor suppressor genes, Cip1/p21 and p16INK4a, by reducing DNA methylation and increasing histones acetylation in human skin cancer cells. Carcinogenesis 2011;32(4):537–44.

[78] Wang S, Moustaid-Moussa N, Chen L, et al. Novel insights of dietary polyphenols and obesity. J Nutr Biochem 2014;25(1):1–18.

[79] Chen YK, Cheung C, Reuhl KR, et al. Effects of green tea polyphenol (-)-epigallocatechin-3-gallate on newly developed high-fat/Western-style diet-induced obesity and metabolic syndrome in mice. J Agric Food Chem 2011;59(21):11862–71.

[80] Zhang Y, Chen H. Genistein, an epigenome modifier during cancer prevention. Epigenet : Off J DNA Methylation Soc 2011;6(7):888–91.

[81] Fang MZ, Chen D, Sun Y, Jin Z, Christman JK, Yang CS. Reversal of hypermethylation and reactivation of p16INK4a, RARbeta, and MGMT genes by genistein and other isoflavones from soy. Clin Cancer Res : Off J Am Assoc Cancer Res 2005;11(19 Pt 1):7033–41.

[82] Kikuno N, Shiina H, Urakami S, et al. Genistein mediated histone acetylation and demethylation activates tumor suppressor genes in prostate cancer cells. Int J Cancer 2008;123(3):552–60.

[83] Jawaid K, Crane SR, Nowers JL, Lacey M, Whitehead SA. Long-term genistein treatment of MCF-7 cells decreases acetylated histone 3 expression and alters growth responses to mitogens and histone deacetylase inhibitors. J Steroid Biochem Mol Biol 2010;120(4–5):164–71.

[84] Messina M, McCaskill-Stevens W, Lampe JW. Addressing the soy and breast cancer relationship: review, commentary, and workshop proceedings. J Natl Cancer Inst 2006;98(18):1275–84.

[85] Penza M, Montani C, Romani A, et al. Genistein affects adipose tissue deposition in a dose-dependent and gender-specific manner. Endocrinology 2006;147(12):5740–51.

[86] Behloul N, Wu G. Genistein: a promising therapeutic agent for obesity and diabetes treatment. Eur J Pharmacol 2013;698(1–3):31–8.

[87] Howard TD, Ho SM, Zhang L, et al. Epigenetic changes with dietary soy in cynomolgus monkeys. PloS One 2011;6(10):e26791.

[88] Balasubramanyam K, Varier RA, Altaf M, et al. Curcumin, a novel p300/CREB-binding protein-specific inhibitor of acetyltransferase, represses the acetylation of histone/nonhistone proteins and histone acetyltransferase-dependent chromatin transcription. J Biol Chem 2004;279(49):51163–71.

[89] Kang SK, Cha SH, Jeon HG. Curcumin-induced histone hypoacetylation enhances caspase-3-dependent glioma cell death and neurogenesis of neural progenitor cells. Stem Cells Dev 2006;15(2):165–74.

[90] Shankar S, Srivastava RK. Involvement of Bcl-2 family members, phosphatidylinositol 3′-kinase/AKT and mitochondrial p53 in curcumin (diferulolylmethane)-induced apoptosis in prostate cancer. Int J Oncol 2007;30(4):905–18.

[91] Ejaz A, Wu D, Kwan P, Meydani M. Curcumin inhibits adipogenesis in 3T3-L1 adipocytes and angiogenesis and obesity in C57/BL mice. J Nutr 2009;139(5):919–25.

[92] Shao W, Yu Z, Chiang Y, et al. Curcumin prevents high fat diet induced insulin resistance and obesity via attenuating lipogenesis in liver and inflammatory pathway in adipocytes. PloS One 2012;7(1):e28784.

[93] Priyadarsini RV, Vinothini G, Murugan RS, Manikandan P, Nagini S. The flavonoid quercetin modulates the hallmark capabilities of hamster buccal pouch tumors. Nutr Cancer 2011;63(2):218–26.

[94] Tan S, Wang C, Lu C, et al. Quercetin is able to demethylate the p16INK4a gene promoter. Chemotherapy 2009;55(1):6–10.

[95] Lee W-J, Chen Y-R, Tseng T-H. Quercetin induces FasL-related apoptosis, in part, through promotion of histone H3 acetylation in human leukemia HL-60 cells. Oncol Rep 2011;25(2):583–91.

[96] Ruiz PA, Braune A, Holzlwimmer G, Quintanilla-Fend L, Haller D. Quercetin inhibits TNF-induced NF-kappaB transcription factor recruitment to proinflammatory gene promoters in murine intestinal epithelial cells. J Nutr 2007;137(5):1208–15.

[97] Kuehnen P, Mischke M, Wiegand S, et al. An Alu element-associated hypermethylation variant of the POMC gene is associated with childhood obesity. PLoS Genet 2012;8(3):e1002543.

[98] Milagro FI, Gomez-Abellan P, Campion J, Martinez JA, Ordovas JM, Garaulet M. CLOCK, PER2 and BMAL1 DNA methylation: association with obesity and metabolic syndrome characteristics and monounsaturated fat intake. Chronobiol Int 2012;29(9):1180–94.

[99] Hermsdorff HH, Mansego ML, Campion J, Milagro FI, Zulet MA, Martinez JA. TNF-alpha promoter methylation in peripheral white blood cells: relationship with circulating TNFalpha, truncal fat and n-6 PUFA intake in young women. Cytokine 2013;64(1):265–71.

[100] Wang X, Zhu H, Snieder H, et al. Obesity related methylation changes in DNA of peripheral blood leukocytes. BMC Med 2010;8:87.

[101] Liu ZH, Chen LL, Deng XL, et al. Methylation status of CpG sites in the MCP-1 promoter is correlated to serum MCP-1 in type 2 diabetes. J Endocrinol Invest 2012;35(6):585–9.

[102] Yang BT, Dayeh TA, Volkov PA, et al. Increased DNA methylation and decreased expression of PDX-1 in pancreatic islets from patients with type 2 diabetes. Mol Endocrinol 2012;26(7):1203–12.

[103] Turcot V, Tchernof A, Deshaies Y, et al. LINE-1 methylation in visceral adipose tissue of severely obese individuals is associated with metabolic syndrome status and related phenotypes. Clin Epigenet 2012;4(1):10.

[104] Painter RC, Roseboom TJ, Bleker OP. Prenatal exposure to the Dutch famine and disease in later life: an overview. Reprod Toxicol (Elmsford, NY) 2005;20(3):345–52.

[105] Lecheminant JD, Gibson CA, Sullivan DK, et al. Comparison of a low carbohydrate and low fat diet for weight maintenance in overweight or obese adults enrolled in a clinical weight management program. Nutr J 2007;6:36.

[106] Dalle Grave R, Calugi S, Gavasso I, El Ghoch M, Marchesini G. A randomized trial of energy-restricted high-protein versus high-carbohydrate, low-fat diet in morbid obesity. Obes (Silver Spring) 2013;21(9):1774–81.

[107] Campion J, Milagro FI, Goyenechea E, Martinez JA. TNF-alpha promoter methylation as a predictive biomarker for weight-loss response. Obes (Silver Spring) 2009;17(6):1293–7.

[108] Cordero P, Campion J, Milagro FI, et al. Leptin and TNF-alpha promoter methylation levels measured by MSP could predict the response to a low-calorie diet. J Physiol Biochem 2011;67(3):463–70.

[109] Milagro FI, Campion J, Cordero P, et al. A dual epigenomic approach for the search of obesity biomarkers: DNA methylation in relation to diet-induced weight loss. FASEB J: Off Publ Fed Am Soc Exp Biol 2011;25(4):1378–89.

[110] Bouchard L, Rabasa-Lhoret R, Faraj M, et al. Differential epigenomic and transcriptomic responses in subcutaneous adipose tissue between low and high responders to caloric restriction. Am J Clin Nutr 2010;91(2):309–20.

[111] Moleres A, Campion J, Milagro FI, et al. Differential DNA methylation patterns between high and low responders to a weight loss intervention in overweight or obese adolescents: the EVASYON study. FASEB J: Off Publ Fed Am Soc Exp Biol 2013;27(6):2504–12.

[112] Crujeiras AB, Campion J, Diaz-Lagares A, et al. Association of weight regain with specific methylation levels in the NPY and POMC promoters in leukocytes of obese men: a translational study. Regul Pept 2013;186:1–6.

[113] WHO. WHO global report - preventing chronic diseases: a vital investment 2005.

[114] Tili E, Michaille JJ, Alder H, et al. Resveratrol modulates the levels of microRNAs targeting genes encoding tumor-suppressors and effectors of TGFbeta signaling pathway in SW480 cells. Biochem Pharmacol 2010;80(12):2057–65.

[115] Tang SN, Singh C, Nall D, Meeker D, Shankar S, Srivastava RK. The dietary bioflavonoid quercetin synergizes with epigallocathechin gallate (EGCG) to inhibit prostate cancer stem cell characteristics, invasion, migration and epithelial-mesenchymal transition. J Mol Signaling 2010;5:14.

[116] King-Batoon A, Leszczynska JM, Klein CB. Modulation of gene methylation by genistein or lycopene in breast cancer cells. Environ Mol Mutagen 2008;49(1):36–45.

[117] Majid S, Dar AA, Ahmad AE, et al. BTG3 tumor suppressor gene promoter demethylation, histone modification and cell cycle arrest by genistein in renal cancer. Carcinogenesis 2009;30(4):662–70.

[118] Mudduluru G, George-William JN, Muppala S, et al. Curcumin regulates miR-21 expression and inhibits invasion and metastasis in colorectal cancer. Biosci Rep 2011;31(3):185–97.

PHARMACOLOGY AND DRUG DEVELOPMENT OF PERSONALIZED EPIGENETICS

12

Personalized Pharmacoepigenomics

Jacob Peedicayil

Department of Pharmacology and Clinical Pharmacology, Christian Medical
College, Vellore, India

1. INTRODUCTION

The extensive research going on in epigenetics worldwide has impacted pharmacology giving rise to a new subdiscipline in pharmacology called pharmacoepigenetics [1–3]. Pharmacoepigenetics has been defined as the study of the epigenetic basis of variation in response to drugs [1]. The study of pharmacoepigenetics on a genome-wide basis is referred to as pharmacoepigenomics [1]. The epigenetic basis of variation in response to drugs can involve pharmacodynamics (the study of the mechanisms of actions of drugs). Thus, a new area in pharmacotherapy, epigenetic therapy is the use of drugs to correct epigenetic defects [4,5]. Although several classes of epigenetic drugs are being investigated, at present most work is focused on the DNA methyltransferase inhibitor (DNMTi) and histone deacetylase inhibitor (HDACi) classes of epigenetic drugs. Indeed, a few DNMTis and HDACis have been approved for clinical use by US Food and Drug Administration. The epigenetic basis of variation in drug response can also involve pharmacokinetics—the study of the absorption, distribution, metabolism, and excretion (ADME) of drugs. Thus, epigenetic mechanisms of gene expression can alter the activity of membrane transporters (membrane proteins present in all organisms that control the influx of essential nutrients and ions as well as the efflux of cellular waste, environmental toxins, drugs, and foreign chemicals) and hepatic enzymes involved in the metabolism of drugs [2,6]. The epigenetic basis of variation in response to drugs is also important in the development of any adverse drug reaction (ADR) that can involve epigenetics related to pharmacodynamics and pharmacokinetics [7].

Epigenetic patterns are known to vary from individual to individual and all persons have their own individual epigenetic pattern [8,9]. Hence, it has been suggested that epigenetics will help pave the way for the subject matter of this book, personalized medicine [10]. In the application of epigenetics to personalized medicine, it has been suggested that pharmacoepigenomics is likely to play a major role [11–13]. This chapter discusses the application and relevance of pharmacoepigenomics to personalized medicine.

2. GENOME-WIDE EPIGENETIC STUDIES

For personalized pharmacoepigenomics, genome-wide epigenetic studies are essential in order to reveal the epigenetic pattern in different body tissues in normal as well as diseased states. These epigenetic studies could involve the study of various epigenetic patterns such as DNA methylation, histone modifications, and the expressions of noncoding RNAs such as microRNAs (miRNAs). Epigenetic patterns of gene expression are known to vary depending on the tissue and cell type in question, age, sex, and race of the individual, and of course the type of disease in question. Hence, in order to apply personalized and individualized pharmacoepigenomics in clinical practice, it would be necessary to compare the epigenetic pattern associated with the pathogenesis of a disease in the tissue of an individual patient with that of the normal epigenetic pattern in the same tissue. Since there is significant variability in "normal" epigenetic patterns, it would also be useful to compare the epigenetic pattern of the individual patient with the epigenetic pattern in the same tissue in the diseased state. Ideally, when investigating the individual patient, it would be best to study the epigenetic pattern in the tissue affected and involved in the disease process. For example, this would be brain tissue in patients with psychiatric and neurological disorders, and the breast tissue in a patient with cancer of the breast. However, due to practical reasons such as tissue accessibility, this would not always be possible, and one might have to study the epigenetic pattern of suitable surrogate tissues/cells such as leukocytes instead.

The Human Genome Project (HGP) was successfully completed during the first decade of this century [14–16]. The HGP has established a wonderful platform on which further work on the genome can be performed, including the mapping of epigenetic patterns across the genome. Although things will undoubtedly change in the future, at present, genome-wide epigenetic studies comprise the following (Table 1): (1) Genome-wide DNA methylation scans. These scans examine genome-wide DNA methylation patterns using various techniques such as luminometric methylation assay, restriction landmark genome scanning, and cytosine extension assay [17,18]. These techniques enable the determination of changes in global DNA methylation patterns and regional changes in DNA methylation throughout the genome. (2) Genome-wide 5-hydroxymethylcytosine (5-hmC) scans [19,20].

TABLE 1   Techniques Used for Genome-Wide Epigenetic Studies

GENOME-WIDE DNA METHYLATION SCANS
• Luminometric methylation assay • Restriction landmark genome scan • Cytosine extension assay
GENOME-WIDE 5-HYDROXYMETHYLCYTOSINE SCANS
• Tet-assisted bisulfite sequencing • Glucosylation coupled with restriction enzyme digestion and microarray analysis
GENOME-WIDE ChIP SCANS
GENOME-WIDE MICROARRAYS FOR microRNAs

These scans scan the genome for the presence of 5-hmC, which is a nitrogenous base formed from cytosine by adding a methyl group and then a hydroxyl group, and it is abundant in the human brain and in embryonic stem cells. In mammals, it can be formed from 5-methylcytosine, a reaction catalyzed by the Tet family of enzymes. The 5-hmC can be scanned across the genome by using techniques such as Tet-assisted bisulfite sequencing and glucosylation coupled with restriction enzyme digestion and microarray analysis. (3) Genome-wide chromatin immunoprecipitation (ChIP) scans. These techniques involve genome-wide profiling using ChIP followed by sequencing and other ChIP-based techniques to investigate DNA-binding proteins and histone modifications. ChIP can be used to study histone modifications across the genome. Mass spectrometry-based proteomics is also being used to study histone modifications [21,22].

In addition to the above scans, many of which are performed on a private basis, government-funded projects include the Human Epigenome Project conducted by the International Human Epigenome Consortium [23], which aims to catalogue 1000 human reference epigenomes, including those in 250 types of human cells; the Cancer Genome Atlas [24], which aims to accelerate our understanding of the molecular basis of cancer through the analysis of genome analysis technologies including large-scale genome sequencing; the National Institutes of Health (NIH) Roadmap Epigenomics Project [25], which aims to produce a public resource of human epigenomic data to catalyze basic biology and disease-oriented research; and the Encyclopaedia of DNA Elements (ENCODE) project, which has provided voluminous data on the role of regulatory elements such as enhancers and insulators and how they are related to DNA methylation and histone modifications [26–28]. Genome-wide epigenome scans are conducted in normal as well as diseased tissues. From the point of view of personalized pharmacoepigenomics, it is obvious that genome-wide epigenome scans from diseased tissues are more useful, although those from normal tissues are also a useful reference.

Noncoding RNAs (ncRNAs) are RNAs transcribed from genes but not translated into protein [29]. Genes encoding ncRNAs produce a functional RNA product rather than a translated protein, and there are no clear-cut differences between these RNAs. However, they can be divided into three classes, based on their sizes: short, medium, and long ncRNAs. Among all ncRNAs, the ncRNA that has been best and most characterized is one type of small ncRNAs called miRNAs, which are 21–23 nucleotides in length. In humans, more than 2500 mature miRNAs have been reported and deposited into the online repository miRNA database (v20, www.mirbase.org) [30,31]. ncRNAs are known to contribute to the pathogenesis of disease and miRNAs are being used as possible targets for drug therapy [32]. In addition, ncRNAs can be investigated across the genome using microarrays [33].

3. PHARMACOEPIGENOMICS IN RELATION TO PHARMACODYNAMICS

Presently, extensive research is going on worldwide on various classes of epigenetic drugs in academic centers and at pharmaceutical companies, as there is considerable expectation that epigenetic drugs will fulfill many unmet needs of patients in several disease categories, especially common diseases such as cancer, psychiatric diseases, and type II diabetes [5,34,35]. To date, about 380 proteins in the epigenetic machinery of cells have been identified as potential targets for epigenetic drugs [36].

Although several classes of epigenetic drugs are being investigated, presently most research is focused on the DNMTi and HDACi classes of epigenetic drugs, and a few of these drugs have been approved for clinical use (Table 2). Many other DNMTis and HDACis are undergoing clinical trials, some of which are listed in Tables 3 and 4, respectively. In addition to DNMTis and HDACi, other epigenetic drugs being investigated and developed include inhibitors of: histone acetyltransferase (HAT) [41], histone methyltransferase (HMT) [42], and histone demethylases [43]. In addition to these epigenetic drugs, several drugs are already in clinical use and alter epigenetic mechanisms of gene expression. Effects on epigenetic mechanisms of gene expression can contribute to the pharmacological actions of these drugs, although they may not be the major mechanisms of action for these drugs. Examples of such drugs are valproic acid in the treatment of bipolar disorder [44] and theophylline in the treatment of asthma [45]. At present the main disorder for which epigenetic drugs are being investigated is cancer, although preclinical and clinical trials of these drugs for several other disorders are under way, such as psychiatric [34], neurodegenerative [46], obesity-related [47], inflammatory [48], and immunological [49], as well as cardiovascular disorders such as pulmonary hypertension [50] and heart failure [51], and also infectious disorders [52].

TABLE 2   US Food and Drug Administration–Approved Epigenetic Drugs for Clinical Use

Epigenetic drug	Drug class	Therapeutic use
Azacytidine	DNMTi	MDS
Decitabine	DNMTi	MDS
Vorinostat	HDACi	CTCL
Romidepsin	HDACi	CTCL

Abbreviations: CTCL = cutaneous T-cell lymphoma; DNMTi = DNA methyltransferase inhibitor; HDACi = histone deacetylase inhibitor; MDS = myelodysplastic syndrome.

TABLE 3   Examples of DNA Methyltransferase Inhibitors under Development

Drug	Class	Comments
Zebularine	Nucleoside analog	Undergoing clinical trials
5, 6-Dihydroazacytidine	Nucleoside analog	Undergoing clinical trials
5-Fluoro-2′-Deoxycytidine	Nucleoside analog	Undergoing clinical trials
Hydralazine	Nonnucleoside analog	In clinical use as antihypertensive
Procainamide	Nonnucleoside analog	In clinical use as antiarrhythmic
Procaine	Nonnucleoside analog	In clinical use as local anaesthetic
EGCG	Nonnucleoside analog	Natural compound
Psammaplin A	Nonnucleoside analog	Natural compound
Curcumin	Nonnucleoside analog	Natural compound

Abbreviation: EGCG: epigallocatechin gallate.
Data compiled from references [37,38].

TABLE 4   Examples of Histone Deacetylase Inhibitors under Development

Drug	Class	HDAC specificity
Vorinostat	Hydroxamic acid	Pan-inhibitor
Trichostatin A	Hydroxamic acid	Classes I and II
Panabinostat	Hydroxamic acid	Classes I and II
Belinostat	Hydroxamic acid	Pan-inhibitor
Depsipeptide	Cyclic tetrapeptide	Class I
Entinostat (MS-275)	Benzamide	Class I
Phenyl butyrate	Short-chain aliphatic acid	Classes I and IIa

Data compiled from references [39,40].

3.1 DNMTis Approved for Clinical Use

To date, two DNMTis have been approved by the US Food and Drug Administration for clinical use. Both of these DNMTis are nucleoside analog inhibitors—azacytidine has been approved for the treatment of myelodysplastic syndrome (MDS) [53] and decitabine has also been approved for the treatment of MDS [53]. These drugs are thought to act in cancer by being phosphorylated in the cancer cell to active triphosphorylated nucleotides that are incorporated into DNA or RNA as cytosine substitutes. Under normal circumstances, DNMTs establish a covalent bond with the carbon-6 atom, which is resolved by β-elimination via the carbon atom. The presence of a nitrogen atom in the carbon-5 position of a DNMTi inhibits release, leading to covalent trapping of the DNMT to DNA, resulting in DNMT pool depletion. DNMTis are also associated with the formation of adducts between DNMTs and DNA that cause breaks in DNA double strands and lead to activation of DNA damage pathways [53]. Many other DNMTis are being investigated for possible clinical use (Table 3).

3.2 HDACi Approved for Clinical Use

To date, two HDACis have been approved by the US Food and Drug Administration for clinical use. One is suberoylanilide hydroxamic acid (SAHA, vorinostat), which has been approved for the treatment of refractory cutaneous T-cell lymphoma (CTCL) [54]. The other is romidepsin, which has been approved for the treatment of CTCL and peripheral T-cell lymphoma [54]. HDACis promote growth arrest, differentiation, and apoptosis of tumor cells with minimal effects on normal cells [55]. In addition, they interfere with the DNA damage response, exert antiangiogenic effects, have immunosuppressive and antiinflammatory effects, and interfere with protein quality control mechanisms [56]. Many other HDACis are under investigation for possible clinical use (Table 4).

3.3 Targeting of miRNAs

As mentioned above, miRNAs are being used as possible drug targets in two ways [57]. The first is by using antisense oligonucleotides (ASOs), which are short single-stranded RNA or DNA molecules that bind other nucleic acids by base pairing. ASOs are thought to function at least in part by competitively inhibiting miRNAs. ASOs that are used in targeting of miRNAs are called antagomirs [58]. The use of antagomirs is presently the most researched way of targeting miRNAs and several groups have used different backbone modifications of antagomirs to successfully antagonize miRNAs in cell culture. Antagomirs are potentially of use in the treatment of clinical conditions such as cancer [59], heart failure [60], and stroke [61]. The second way that miRNAs are being targeted for drugs is by adding back a miRNA whose expression has been lost or decreased, a procedure called miRNA replacement therapy. This type of therapy could be useful in the treatment of diseases where the expression of a miRNA has been lost or reduced. Adding the miRNA in a single dose may not allow sustained miRNA target regulation due to inadequate delivery or degradation. Instead, three to five doses of miRNA modified or formulated for optimal delivery could provide adequate miRNA for 20–30 days [58].

4. PERSONALIZED PHARMACOEPIGENOMICS IN RELATION TO PHARMACODYNAMICS

As is well known in pharmacology, the major way that drugs act is via receptors. Most receptors in the body are proteins, such as membrane-bound receptors, ion channels, enzymes, and cytosolic receptors. Some drugs, such as some of those used to treat cancer, act by binding to nucleic acids such as DNA rather than to proteins. Among epigenetic drugs, most target enzymes such as DNA methyltransferase and histone deacetylase, although a few target nucleic acids such as miRNAs.

Epigenetic drugs correct abnormal epigenetic patterns of gene expression that contribute to the pathogenesis of disease. In light of the preceding paragraph, epigenetic drugs are likely to act via proteins or nucleic acids linked to the pathogenesis of disease. In order to conduct personalized pharmacoepigenomics, i.e., to use the appropriate epigenetic drug for the appropriate patient, knowledge of the aberrant epigenetic pattern contributing to the pathogenesis of disease in the individual patient is required. For such knowledge, it would be necessary to compare the epigenetic pattern of the individual patient with the epigenetic pattern in the tissue of concern in the diseased state as well as the epigenetic pattern in the same tissue in the normal state. Epigenetic patterns on a genome-wide basis in different tissues in normal and diseased states are being conducted and published. At present, the epigenetic patterns mainly being studied are those of DNA methylation. Such tissues and diseases include the following: pancreatic adenocarcinoma [62], testicular cancer [63], colorectal cancer [64], hepatocellular carcinoma [65], lung cancer [66], acute lymphocytic leukemia [67], acute myeloid leukemia [68], schizophrenia [69], and posttraumatic stress disorder [70]. Personalized pharmacoepigenomics in relation to pharmacodynamics has already started and made some progress. For example, related to DNA methylation patterns, Feinberg's group [71] performed a genome-wide analysis of about 4 million CpG sites in 74 individuals with comprehensive array-based relative methylation analysis. They found 227 regions that showed extreme interindividual variability (VMRs) across the genome, which are enriched for developmental genes based on gene ontology analysis. Half of these VMRs were stable within individuals over an average of 11 years, and the VMRs defined a personalized epigenomic signature. Four of these VMRs showed covariation with body mass index consistently at two study visits and were located in or near genes previously implicated in regulating body weight or diabetes mellitus. These findings suggested to the authors a strategy for identifying a personalized epigenomic signature that may correlate with common disease. More recently, Liao et al. [72] analyzed the methylomic profiles of ovarian tumor-initiating cells (OTICs) and their derived progeny using a human methylation array. In that case, qRT-PCR, quantitative methylation-specific PCR, and pyrosequencing were used to verify gene expression and DNA methylation in cancer cell lines. The methylation statuses of genes were validated quantitatively in cancer tissues and correlated with clinicopathological findings. It was found that two genes, *ATGA* and *HISTIH2BN*, were hypomethylated in OTICS and that the hypomethylation predicted a poor prognosis. The authors suggested that their findings could help in personalized medicine for patients with ovarian cancer. Figueroa and coworkers [73] examined the methylation profiles of 344 patients with acute myeloid leukemia. Clustering of the patients by DNA methylation data enabled a 15 gene DNA methylation classifier predictive of overall survival.

Personalized pharmacoepigenomics related to histone modifications has made some progress with regard to cancers. For example, Seligson et al. [74], using immunohistochemical staining of prostate cancer tissue samples, showed that changes in global levels of individual histone modification are associated with cancer, and these changes are predictive of clinical outcomes including prostate cancer recurrence. Cuomo et al. [75] comprehensively analyzed lysine acetylation and methylation on histones H3 and H4 from four breast cancer cell lines in comparison with normal epithelial breast cells. The authors performed high-resolution mass spectrometry analysis of histones, in combination with stable isotope labeling with amino acids in cell culture, to quantitatively track the histone modifications in cancer cells as compared with their normal counterparts. The authors found several novel histone modifications. Overall, they suggested that their data could represent a breast cancer-specific epigenetic signature with classification of breast cancer into subtypes.

Personalized pharmacoepigenomics related to miRNAs has also made some progress. For example, Yu et al. [76], using real-time PCR, investigated whether miRNA expression profiles can predict clinical outcomes in 112 patients with non-small-cell lung cancer. They identified a fine miRNA signature that predicts treatment outcomes like cancer relapse and survival. Calin et al. [77] found, in patients with chronic lymphocytic leukemia, a unique miRNA signature comprising 13 genes (of 190 analyzed) that differentiated cases between those with the presence or absence of disease progression.

As discussed above, at present only four epigenetic drugs have been FDA-approved for the treatment of disease. However, preclinical and clinical trials of several other epigenetic drugs are under way. At present, the study of the role of epigenetics in the pathogenesis of disease is in its very early stages, and the knowledge of the role of specific epigenetic abnormalities in the pathogenesis of specific diseases is limited. Much more research needs to be done in elucidating the epigenetic basis of disease. When the epigenetic causes of diseases have been determined, more appropriate epigenetic drugs can be developed, enhancing the scope of personalized pharmacoepigenomics.

5. PERSONALIZED PHARMACOEPIGENOMICS IN RELATION TO PHARMACOKINETICS

Pharmacokinetics involves the ADME of drugs. Several proteins such as hepatic enzymes and membrane transporters are involved in the ADME of drugs. In humans, about 300 genes that encode proteins are involved in the ADME of drugs [7]. Of these, about 60 are known to be epigenetically regulated. This epigenetic regulation has been mainly described in cancer cells and after the administration of decitabine to cancer cell lines.

The considerable variations in ADME in humans [78–82] can alter drug efficacy and predispose to ADRs. Interindividual variations in ADME are due to several influences including genetic, epigenetic, physiological, and environmental factors [81]. Over the years, much information regarding the role of genetic polymorphisms in such variation has been learned. The number of genetic biomarkers for the prediction of drug dosage and choice based on such polymorphisms keeps increasing. However, the epigenetic causes of interindividual variations in ADME have been much less studied [81].

As is the case in pharmacodynamics, genome-wide epigenetic studies are required for personalized pharmacoepigenomics related to pharmacokinetics. Efforts are being made in this regard, but further work and improvements in techniques are required. Ivanov and colleagues [83] have reported the successful use of a custom Agilent SureSelect Target Enrichment System for the hybrid capture of bisulfite-converted DNA. They prepared bisulfite-converted next-generation sequencing libraries enriched for the coding and regulatory regions of 174 ADME genes. Sequencing of these libraries on an Illumina HiSeq 2000 revealed that the method allows a reliable quantification of methylation levels of CpG sites in the selected genes, and validation of the method using pyrosequencing and the Illumina Infinium HumanMethylation450 BeadChip Kit revealed good concordance. More recently, Bonder et al. [84] analyzed the methylomes and transcriptomes of fetal and adult human livers and generated a comprehensive resource of factors involved in the regulation of hepatic gene expression, and also estimated the proportion of variation in gene expression that can be attributed to genetic and epigenetic variation. The availability of genome-wide epigenetic studies may enable the tailoring of drug dosages, drug dosing schedules, and routes of drug administration to individual patients, since the knowledge of an individual's epigenetic pattern could give information on how that individual's body would absorb, distribute, metabolize, and excrete a particular drug.

5.1 DNA Methylation and ADME Genes

Hypermethylation of the promoter of the gene *GSTP1* has been observed in several cancers like prostate cancer and breast cancer. This epigenetic change has been associated with reduced expression of GSTP1 protein and changes in prognosis of cancer patients. DNMTis could be useful for treating prostate cancer in order to reverse the hypermethylation of the *GSTP1* gene. Chiam et al. [85] have suggested that assessing *GSTP1* promoter methylation has the potential for determining the efficacy of DNMTis for the treatment of prostate cancer. In breast cancer, the promoter of *GSTP1* shows variable methylation levels and it is thought that the extent of methylation correlates with improved overall survival. Dejeux and colleagues [86] showed a significant association between the variable DNA methylation of the genes *GSTP1* and *ABCB1* and the therapeutic effect of the anticancer drug doxorubicin in breast cancer samples. GPx3 is an enzyme that catalyzes the reduction of hydrogen and lipid peroxides by oxidizing glutathione. It is thought to protect cells from oxidative damage caused by these reactive oxygen species. Hypermethylation of *GPx3* has been found in cancers such as those of the esophagus, prostate, and stomach. There is preliminary evidence that there is a correlation between methylation of the promoter of *GPx3* and chemotherapy with cisplatin in patients with cancers of the head and neck. In colon, endometrial, and breast cancer, aberrant methylation of the *VDR* gene has been reported. There is experimental evidence that methylation of the promoter of this gene is responsible for the resistance to the antiproliferative and differentiating effects of the drug calcitriol (1,25-dihdroxyvitamin D) in breast cancer cells [87].

5.2 MicroRNAs and ADME Genes

It is believed that miRNAs play major roles in the regulation of the expression of ADME genes [79]. Moreover, based on data from in silico studies, several miRNAs are thought to target many genes that encode proteins involved in the ADME of drugs. However, the number of such genes experimentally validated as being targets of specific miRNAs is much smaller. Recently, it was estimated by He and colleagues [31] that about 261 ADME genes are regulated by about 120 miRNAs.

6. PERSONALIZED PHARMACOEPIGENOMICS IN RELATION TO ADVERSE DRUG REACTIONS

ADRs are common and important problems in drug therapy and drug development [7]. ADRs can be due to pharmacodynamic and pharmacokinetic reasons [88,89]. With regard to pharmacoepigenomics, pharmacodynamic reasons for ADRs will be the use of epigenetic drugs. Since all drugs are known to cause ADR, the same can be expected from currently approved epigenetic drugs as well as those epigenetic drugs that will be approved for clinical use in the future. Pharmacokinetic reasons for ADRs in relation to pharmacoepigenomics would be interindividual variations in ADME due to variations in epigenetic mechanisms of gene expression of genes encoding proteins involved in the ADME of drugs.

Once genome-wide epigenome scans are available, and interindividual variations in epigenetic patterns of gene expression are known, it could be possible to give the right dose of the right epigenetic drug to the right individual. This would minimize the likelihood of ADRs in individual patients from a pharmacodynamic point of view, and the availability of interindividual variations in epigenetic patterns of gene expression would allow the adjustment of administration of epigenetic drugs as well as of other drugs used, based on the patient's individual epigenetic pattern, in order to minimize the chances of ADR.

7. CONCLUSIONS

Pharmacoepigenomics is a relatively new subject and this field is in its very early stages. Data in this area can be applied for use in personalized medicine in relation to pharmacodynamics, pharmacokinetics, and ADRs. Genome-wide studies of epigenetic patterns in tissues in both normal healthy and diseased states will help in the application of pharmacoepigenomics to personalized medicine. Personalized pharmacoepigenomics could help usher in the era of the five rights of safe and appropriate pharmacotherapy [90]: "Right drug, right patient, right dose, right route, and right time."

8. FUTURE DIRECTIONS

In the future, more genome-wide studies of epigenetic patterns of gene expression in additional body tissues in normal healthy and diseased states are needed. Such studies should include other epigenetic patterns of gene expression such as histone modification in addition to DNA methylation. A greater number of epigenetic drugs to treat a wider range of diseases need to be developed. More work needs to be done to determine how changes in the epigenetic patterns of gene expression of genes that encode proteins involved in the ADME of drugs lead to altering of the pharmacokinetics of drugs.

Glossary

Adverse drug reaction Undesired, often unpleasant, effect of a drug
Membrane transporter Membrane-bound protein that controls the influx of essential nutrients and ions and the efflux of cellular waste, environmental toxins, drugs, and foreign chemicals
Pharmacodynamics The branch of pharmacology dealing with the mechanisms of action of drugs
Pharmacokinetics The branch of pharmacology dealing with the absorption, distribution, metabolism, and excretion (or elimination) (ADME) of drugs

LIST OF ACRONYMS AND ABBREVIATIONS

ADME	Absorption, distribution, metabolism, and excretion (or elimination) of drugs
ADR	Adverse drug reaction
ChIP	Chromatin immunoprecipitation
DNMTi	DNA methyltransferase inhibitor
HDACi	Histone deacetylase inhibitor
5-hmC	Hydroxymethylcytosine
miRNA	microRNA
ncRNA	Noncoding RNA

References

[1] Peedicayil J. Pharmacoepigenetics and pharmacoepigenomics. Pharmacogenomics 2008;9:1785–6.

[2] Gomez A, Ingelman-Sundberg M. Pharmacoepigenetics: Its role in interindividual differences in drug response. Clin Pharmacol Ther 2009;85:426–30.

[3] Baer-Dubowska W, Majchrzak-Celińska A, Cichocki M. Pharmacoepigenetics: a new approach to predicting individual drug responses and targeting new drugs. Pharmacol Rep 2011;63:293–304.

[4] Egger G, Liang G, Aparicio A, Jones PA. Epigenetics in human disease and prospects for epigenetic therapy. Nature 2004;429:457–63.

[5] Peedicayil J. Epigenetic therapy – a new development in pharmacology. Indian J Med Res 2006;123:17–24.

[6] Zanger UM, Klein K, Thomas M, Rieger JK, Tremmel R, Kandel BA, et al. Genetics, epigenetics, and regulation of drug-metabolizing cytochrome P450 enzymes. Clin Pharmacol Ther 2014;95:258–61.

[7] Kacevska M, Ivanov M, Ingelman-Sundberg M. Perspectives on epigenetics and its relevance to adverse drug reactions. Clin Pharmacol Ther 2011;89:902–7.

[8] Bock C, Walter J, Paulsen M, Lengauer T. Inter-individual variation of DNA methylation and its implications for large-scale epigenome mapping. Nucleic Acids Res 2008;36:e55.

[9] Massicotte R, Whitelaw E, Angers B. DNA methylation: a source of random variation in natural populations. Epigenetics 2011;6:421–7.

[10] Joss-Moore LA, Lane RH. Epigenetics and the developmental origins of disease: the key to unlocking the door of personalized medicine. Epigenomics 2012;4:471–3.

[11] Weber WW. The promise of epigenetics in personalized medicine. Mol Interv 2010;10:363–70.

[12] Hunter P. A new target for personalized medicine. EMBO Rep 2011;12:1229–32.

[13] Peedicayil J. The epigenome in personalized medicine. Clin Pharmacol Ther 2013;93:149–50.

[14] Venter C, Adams MD, Myers EW, Li PW, Mural RJ, Sutton GG, et al. The sequence of the human genome. Science 2001;291:1304–51.

[15] Lander ES, Linton LM, Birren B, Nusbaum C, Zody MC, Baldwin J, et al. Initial sequencing and analysis of the human genome. Nature 2001;409:860–921.

[16] International Human Genome Sequencing Consortium. Finishing the euchromatic sequence of the human genome. Nature 2004;431:931–45.
[17] Butcher LM, Beck S. Future impact of integrated high-throughput methylome analyses on human health and disease. J Genet Genomics 2008;35:391–401.
[18] Unterberger A, Dubuc AM, Taylor MD. Genome-wide methylation analysis. Methods Mol Biol 2012;863:303–17.
[19] Khare T, Pai S, Koncevicius K, Pal M, Kriukiene E, Liutkeviciute Z, et al. 5-hmC in the brain is abundant in synaptic genes and shows differences at the exon-intron boundary. Nat Struct Mol Biol 2012;19:1037–43.
[20] Wen L, Li X, Yan L, Tan Y, Li R, Zhao Y, et al. Whole-genome analysis of 5-hydroxymethylcytosine and 5-methylcytosine at base resolution in the human brain. Genome Biol 2014;15:R49.
[21] Kimura H. Histone modifications for human epigenome analysis. J Hum Genet 2013;58:439–45.
[22] Han Y, Garcia BA. Combining genomic and proteomic approaches for epigenetics research. Epigenomics 2013;5:439–52.
[23] Bae J-B. Perspectives of international human epigenome consortium. Genomics Inform 2013;11:7–14.
[24] Weisenberger DJ. Characterizing DNA methylation alterations from the Cancer genome atlas. J Clin Invest 2014;124:17–23.
[25] Bernstein BE, Stamatoyannopoulos JA, Costello JF, Ren B, Milosavljevic A, Meissner A, et al. The NIH roadmap epigenomics mapping consortium. Nat Biotechnol 2010;28:1045–8.
[26] The ENCODE Project Consortium. The encode (ENCyclopedia of DNA elements) project. Science 2004;306:636–40.
[27] The ENCODE Project Consortium. A user's guide to the Encyclopedia of DNA Elements (ENCODE). PLoS Biol 2011;9:e1001046.
[28] Topol EJ. Individualized medicine from prewomb to tomb. Cell 2014;157:241–53.
[29] Collins LJ, Schönfeld B, Chen XS. The epigenetics of non-coding RNA. In: Tollefsbol T, editor. Handbook of epigenetics: the new molecular and Medical genetics. Burlington, MA: Elsevier; 2011. p. 49–61.
[30] Griffith Jones S. miRBase www.mirbase.org/
[31] He Y, Chevillet JR, Liu G, Kim TK, Wang K. The effects of microRNA on the absorption, distribution, metabolism and excretion of drugs. Br J Pharmacol 2014. http://dx.doi.org/10.1111/bph.12968.
[32] Taft RJ, Pang KC, Mercer TR, Dinger M, Mattick JC. Non-coding RNAs: regulators of disease. J Pathol 2010;220:126–39.
[33] Thomson JM, Parker J, Perou CM, Hammond SM. A custom microarray platform for analysis of microRNA gene expression. Nat Methods 2004;1:47–53.
[34] Peedicayil J. The role of epigenetics in mental disorders. Indian J Med Res 2007;126:105–11.
[35] Reddy MA, Natarajan R. Role of epigenetic mechanisms in the vascular complications of diabetes. Subcell Biochem 2013;61:435–54.
[36] Arrowsmith CH, Bountra C, Fish PV, Lee K, Schapira M. Epigenetic protein families: a new frontier for drug discovery. Nat Rev Drug Discov 2012;11:384–400.
[37] Medina-Franco JL, Caulfield T. Advances in the computational development of DNA methyltransferase inhibitors. Drug Disc Today 2011.
[38] Foulks JM, Parnell KM, Nix RN, Chau S, Swierczek K, Saunders M, et al. Epigenetic drug discovery: targeting DNA methyltransferases. J Biomol Screen 2012;17:2–17.
[39] Wagner JM, Hackanson B, Lübbert M, Jung M. Histone deacetylase (HDAC) inhibitors in recent clinical trials for cancer therapy. Clin Epigenet 2010;1:117–36.
[40] New M, Olzscha H, La Thangue NB. HDAC inhibitor-based therapies: can we interpret the code? Mol Oncol 2012;6:637–56.

[41] Maddox SA, Watts CS, Doyère V, Schafe GE. A naturally-occurring histone acetyltransferase inhibitor derived from *Garcinia indica* impairs newly acquired and reactivated fear memories. PLoS One 2013;8:e54463.

[42] Pang AL, Title AC, Rennert OM. Modulation of microRNA expression in human lung cancer cells by the G9a histone methyltransferase inhibitor BIX01294. Oncol Lett 2014;7:1819–25.

[43] Burridge S. Drugging the epigenome. Nat Rev Drug Discov 2013;12:92–3.

[44] Peedicayil J. Epigenetic management of major psychosis. Clin Epigenetics 2011;2:249–56.

[45] Barnes PJ. Corticosteroid resistance in patients with asthma and chronic obstructive pulmonary disease. J Allergy Clin Immunol 2013;131:636–45.

[46] Peedicayil J. Epigenetic drugs in cognitive disorders. Curr Pharm Des 2014;20:1840–6.

[47] Martin SL, Hardy TM, Tollefsbol TO. Medicinal chemistry of the epigenetic diet and caloric restriction. Curr Med Chem 2013;20:4050–9.

[48] Lin HS, Hu CY, Chan HY, Liew YY, Huang HP, Lepescheux L, et al. Anti-rheumatic activities of histone deacetylase (HDAC) inhibitors in vivo in collagen-induced arthritis in rodents. Br J Pharmacol 2007;150:862–72.

[49] Koch MW, Metz LM, Kovalchuk O. Epigenetics and miRNAs in the diagnosis and treatment of multiple sclerosis. Trends Mol Med 2013;19:23–30.

[50] Saco TV, Parthasarathy PT, Cho Y, Lockey RF, Kolliputi N. Role of epigenetics in pulmonary hypertension. Am J Physiol Cell Physiol 2014;306:C1101–5.

[51] Baccarelli A, Ghosh S. Environmental exposures, epigenetics and cardiovascular disease. Curr Opin Clin Nutr Metab Care 2012;15:323–9.

[52] Zumla A, Maeurer M. Rational development of adjunct immune-based therapies for drug-resistant tuberculosis: hypotheses and experimental designs. J Infect Dis 2012;205:S335–9.

[53] Peedicayil J. The role of DNA methylation in the pathogenesis and treatment of cancer. Curr Clin Pharmacol 2012;7:333–40.

[54] Chabner BA, Bertino J, Cleary J, Ortiz T, Lane A, Supko JG, et al. Cytotoxic agents. In: Brunton LL, Chabner BA, Knollman BC, editor. The pharmacological basis of therapeutics. New York: McGraw-Hill; 2011. p. 1677–730.

[55] Lane AA, Chabner BA. Histone deactylase inhibitors in cancer therapy. J Clin Oncol 2009;32:5459–68.

[56] Dickinson M, Johnstone RW, Prince HM. Histone deacetylase inhibitors: potential targets responsible for their anticancer effect. Invest New Drugs 2010;28:S3–20.

[57] Broderick JA, Zamore PD. MicroRNA therapeutics. Gene Ther 2011;18:1104–10.

[58] Soifer HS, Rossi JS, Sætrom P. MicroRNAs in disease and potential therapeutic applications. Mol Ther 2007;15:2070–9.

[59] Sethi S, Ali S, Sethi S, Sarkar FH. MicroRNAs in personalized cancer therapy. Clin Genet 2014;86:68–73.

[60] Oliveira-Carvalho V, da Silva MM, Guimarães GV, Bacal F, Bocchi EA. MicroRNAs: new players in heart failure. Mol Biol Rep 2013;40:2663–70.

[61] Koutsis G, Siasos G, Spengos K. The emerging role of microRNAs in stroke. Curr Top Med Chem 2013;13:1573–88.

[62] Omura N, Li C-P, Li A, Hong S-M, Walter K, Jimeno A, et al. Genome-wide profiling of methylated promoters in pancreatic adenocarcinoma. Cancer Biol Ther 2008;7:1146–56.

[63] Cheung HH, Lee TL, Davis AJ, Taft DH, Rennert OM, Chan WY. Genome-wide DNA methylation profiling reveals novel epigenetically regulated genes and non-coding RNAs in human testicular cancer. Br J Cancer 2010;102:419–27.

[64] Rodriguez J, Frigola J, Vendrell E, Risques R-A, Fraga MF, Morales C, et al. Chromosomal instability correlates with genome-wide DNA demethylation in human primary colorectal cancers. Cancer Res 2006;66:8462–8.

[65] Lin C-H, Hsieh SY, Sheen IS, Lee WC, Chen TC, Shyu WC, et al. Genome-wide hypomethylation in hepatocellular carcinogenesis. Cancer Res 2001;61:4238–43.

[66] Shames DS, Girard L, Gao B, Sato M, Lewis CM, Shivapurkar N, et al. A genome-wide screen for promoter methylation in lung cancer identifies novel methylation markers for multiple malignancies. PLoS Med 2006;3:e486.

[67] Kuang S-Q, Tong WG,Yang H, Lin W, Lee MK, Fang ZH, et al. Genome-wide identification of aberrantly methylated promoter associated CpG islands in acute lymphocytic leukemia. Leukemia 2008; 22:1529–38.

[68] Gebhard C, Schwarzfischer L, Pham TH, Schilling E, Klug M, Andreesen R, et al. Genome-wide profiling of CpG methylation identifies novel targets of aberrant hypermethylation in myeloid leukemia. Cancer Res 2006;66:6118–28.

[69] Aberg KA, McClay JL, Nerella S, Clark S, Kumar G, Chen W, et al. Methylome-wide association study of schizophrenia: Identifying blood biomarker signatures of environmental insults. JAMA Psychiatry 2014;71:255–64.

[70] Mehta D, Klengel T, Conneely KN, Smith AK, Altmann A, Pace TW, et al. Childhood maltreatment is associated with distinct genomic and epigenetic profiles in posttraumatic stress disorder. Proc Natl Acad Sci USA 2013;110:8302–7.

[71] Feinberg AP, Irizarry RA, Fradin D, Aryee MJ, Murakami P, Aspelund T, et al. Personalized epigenomic signatures that are stable over time and covary with body mass index. Sci Trans Med 2010;2:49ra67.

[72] Liao YP, Chen LY, Huang RL, Su PH, Chan MW, Chang CC, et al. Hypomethylation signature of tumor-initiating cells predicts poor prognosis of ovarian cancer patients. Hum Mol Genet 2014;23:1894–906.

[73] Figueroa ME, Lugthart S, Li Y, Erpelinck-Verschueren C, Deng X, Christos PJ, et al. DNA methylation signatures identify biologically distinct subtypes in acute myeloid leukemia. Cancer Cell 2010;17:13–27.

[74] Seligson DB, Horvath S, Shi T, Yu H, Tze S, Grunstein M, Kurdistani SK. Global histone modification patterns predict risk of prostate cancer recurrence. Nature 2005;435:1262–6.

[75] Cuomo A, Moretti S, Minucci S, Bonaldi T. SILAC-based proteomic analysis to dissect the "histone modification signature" of breast cancer cells. Amino Acids 2011;41:387–99.

[76] Yu SL, Chen HY, Chang GC, Chen CY, Chen HW, Singh S, et al. MicroRNA signature predicts survival and relapse in lung cancer. Cancer Cell 2008;13:48–57.

[77] Calin GA, Ferracin M, Cimmino A, Di Leva G, Shimizu M, Wojcik SE, et al. A microRNA signature associated with prognosis and progression in chronic lymphocytic leukemia. N Engl J Med 2005;353:1793–801.

[78] Glubb DM, Innocenti F. Mechanisms of genetic regulation in gene expression: examples from drug metabolizing enzymes and transporters. Wiley Interdiscip Rev Syst Biol Med 2011;3:299–303.

[79] Kacevska M, Ivanov M, Ingelman-Sundberg M. Epigenetic-dependent regulation of drug transport and metabolism: an update. Pharmacogenomics 2012;13:1373–85.

[80] Zhong X-B, Leeder JS. Epigenetic regulation of ADME-related genes: focus on drug metabolism and transport. Drug Metab Dispos 2013;41:1721–4.

[81] Ingelman-Sundberg M, Zhong X-B, Hankinson O, Beedanagari S, Yu AM, Peng L, et al. Potential role of epigenetic mechanisms in the regulation of drug metabolism and transport. Drug Metab Dispos 2013;41:1725–31.

[82] Rieger JK, Klein K, Winter S, Zanger UM. Expression variability of absorption, distribution, metabolism, excretion – related microRNAs in human liver: Influence of nongenetic factors and association with gene expression. Drug Metab Dispos 2013;41:1752–62.

[83] Ivanov M, Kals M, Kacevska M, Metspalu A, Ingelman-Sundberg M, Milani L. In-solution hybrid capture of bisulfite-converted DNA for targeted bisulfite sequencing of 174 ADME genes. Nucleic Acids Res 2013;41:e72.

[84] Bonder MJ, Kasela S, Kals M, Tamm R, Lokk K, Barragan I, et al. Genetic and epigenetic regulation of gene expression in fetal and adult human livers. BMC Genomics 2014;15:860.

[85] Chiam K, Centenera MM, Butler LM, Tilley WD, Bianco-Miotto T. GSTP1 DNA methylation and expression status is indicative of 5-aza-2′-deoxycytidine efficacy in human prostate cancer cells. PLoS One 2011;6:e25634.

[86] Dejeux E, Rønneberg JA, Solvang H, Bukholm I, Geisler S, Aas T, et al. DNA methylation profiling in doxorubicin treated primary locally advanced breast tumours identifies novel genes associated with survival and treatment response. Mol Cancer 2010;25(9):68.

[87] Marik R, Fackler M, Gabrielson E, Zeiger MA, Sukumar S, Stearns V, et al. DNA methylation-related vitamin D receptor insensitivity in breast cancer. Cancer Biol Ther 2010;10:44–53.

[88] Bates DW, Cullen DJ, Laird N, Petersen LA, Small SD, Servi D, et al. Incidence of adverse drug events and potential adverse drug events. Implications for prevention. JAMA 1995;274:29–34.

[89] Rang HP, Dale MM, Ritter JM, Flower RJ, Henderson G. Rang and Dale's pharmacology. Edinburgh: Elsevier; 2012. pp. 698–709.

[90] Osterhoudt KC, Penning TM. Drug toxicity and poisoning. In: Brunton LL, Chabner BC, Knollmann BC, editors. The pharmacological basis of therapeutics. New York: McGraw-Hill; 2011. p. 73–87.

CHAPTER

13

Personalized Medicine and Epigenetic Drug Development

Kenneth Lundstrom

PanTherapeutics, Lutry, Switzerland

1. INTRODUCTION

Modern drug discovery has encountered serious problems in developing more efficient and safer medicines. In this context, drug developers have started to seriously look for new drug targets and new mechanisms of drug action. One area that has attracted much attention is epigenetics, which can be defined as the study of heritable but reversible changes in the expression of genes without modifications of an individual's primary DNA sequence [1]. Epigenetics does not therefore involve any modifications of the primary DNA sequence and the potentially reversible nature makes it an attractive alternative for therapeutic applications. DNA methylation, histone modifications, and RNA interference are the most common epigenetic changes. Related to DNA methylation, typically a methyl group (CH_3) is covalently added to the 5′-position of cytosine upstream of guanosine and contributes to the regulation of gene expression [2]. This will affect differentiation, genomic imprinting, and DNA repair. Methylated CpG dinucleotides form clusters called CpG islands, which are located prominently within the promoter regions of genes. DNA methylation affects mRNA transcription, which may result in reduction or cessation, but also in upregulation, of transcription. For this reason, some association has been established between methylation patterns and cancer [3,4]. For instance, inactivation of the HIC1, INK4b, and TIMP3 tumor suppressor genes has been related to hypermethylation in promoter regions [5]. Likewise, the process of DNA demethylation has inspired the development of inhibitors of DNA methylation for cancer therapy [6]. Histones play important roles in packaging DNA in chromatin structures and have also been suggested to have an essential function in epigenetics [1]. For instance, acetylation, methylation, ubiquitination, and phosphorylation can modify histones H3 and H4 [7]. In this context, increased acetylation can lead to transcription activation, and histone methylation can result in either repression or activation of transcription [8]. Furthermore, when histone modifications are deregulated, the outcome can be mutations in oncogenes, tumor suppressor genes, and DNA repair genes. Histone deacetylation also plays an important role in the regulation of gene expression and therefore histone deacetylase inhibitors present attractive molecules for therapeutic interventions [9]. Last, the mechanism of RNA interference particularly involves splicing of microRNA (miRNA), which also contributes strongly to the regulation of gene expression [10]. The phenomenon relies on 21- to 23-nucleotide single-stranded RNA molecules, which interfere with messenger RNA and lead to downregulation of gene expression [11]. An alternative mechanism for miRNA has increased transcription and enhanced gene expression [12]. Today, more than 1000 human miRNA sequences have been isolated and it is postulated that approximately one-third of human mRNAs are regulated by miRNAs [13].

As described above, epigenetics has been demonstrated to have a profound effect on disease development and therefore potentially presents attractive opportunities for finding novel targets for drug discovery. This chapter reviews the relationship between epigenetics and disease through examples and presents aspects of the association of personalized medicine with epigenetics and nutrition.

2. EPIGENETICS AND DRUG DISCOVERY

In addition to the study of the biological aspects of epigenetics, this area has seen more and more applications in drug discovery (Table 1). Interestingly, epigenetic targets have been discovered for a variety of diseases. These include, for example, cancer, neurological disorders, and liver, cardiovascular, and skin diseases. Here, it is possible to illustrate the development only through examples from various disease areas.

TABLE 1   Examples of Epigenetic Drugs

Disease	Epigenetic target	Regulatory function	Documented effect	Reference
AUTOIMMUNE				
Inflammation	MCP-1	miR-124 dysregulation	Pathogenesis	[58]
RA	Let-7a, miR-132	miRNA dysregulation	RA pathogenesis	[59]
MS	miR-21	miRNA dysregulation	Disease indicator for MS	[62]
CANCER				
AML	Azacitidine	DNA MT inhibition	AML treatment	[22]
	Decitabine	DNA MT inhibition	AML treatment	[22]
Bone	miR-143, miR-145	miRNA downregulation	Bone metastasis	[18]
Brain	Methylation drugs	DNA, H3K27 methylation	Ependymoma therapy	[85]
CTCL	Vorinostat	Histone deacetylase	CTCL treatment	[23]
	Romidepsin	Histone deacetylase	CTCL treatment	[24]
Gastric	CDKN2A, CHFR	DNA methylation	Patient selection	[65]
Glioma	PTPN6	DNA methylation	Reduced survival	[25]
			Chemotherapy response	[25]
Kidney	FBN2, SFRP1	Hypermethylation	Biomarkers	[86]

Continued

TABLE 1　Examples of Epigenetic Drugs—cont'd

Disease	Epigenetic target	Regulatory function	Documented effect	Reference
Liver	Deacetylase inhibitor	miRNA upregulation	Reduced cell death	[44]
Ovarian	Ruthenium compounds	Histone modifications	Anticancer activity	[26]
Prostate	miR-34a	Regulation of CSCs	Tumor regression	[17]
	miR-200	miRNA downregulation	Tumor development	[19]
	miR-708	Tumor suppressor	Tumor regression	[20]
CARDIOVASCULAR				
Stroke	TNF-α, PON	DNA methylation	Association with stroke	[56]
	HHCy	DNA methylation, miRNAs	Role in stroke	[90]
PH	miR-21	Epigenetic regulation	Control of PH	[91]
CH	CBP	HDACs	Suppression of CH	[92]
Heart failure	Multiple miRNAs	Differential expression	Heart disease association	[93]
LIVER				
NAFLD	DNMT1 inhibitor	Differential methylation	Inhibition of HSC	[42]
Liver fibrosis	TGF-β1	miRNA	HSC phenotype	[43]
NASH	GALNTL4, ACLY	DNA methylation	Correlation to NASH	[47]
	PTPRE	miR-122	Circadian rhythm linkage	[48]
METABOLIC				
Diabetes	PLAGL1, HYMAI	DNA methylation	Diabetes association	[87]
	TSA	Histone modification	Diabetes protection	[88]
	IGFBP1	DNA methylation	DT2 association	[89]

TABLE 1   Examples of Epigenetic Drugs—cont'd

Disease	Epigenetic target	Regulatory function	Documented effect	Reference
NEUROLOGICAL				
Alzheimer	p25	HDAC inhibitors	Reinstated learning	[29]
	Folic acid	DNA methylation	Reduced pathology	[35]
	BACE1	miR-29	Pathology promotion	[38]
	BACE1	miR-107	Link to Alzheimer	[39]
	BACE1	miR-298, miR-328	Link to Alzheimer	[40]
SKIN				
Psoriasis	Multiple genes	DNA methylation	Psoriasis pathogenesis	[52]
		Histone modifications	PASI correlation	[52]
		miR.143, miR-223	Biomarkers for psoriasis	[53]

AML, acute myeloid leukemia; BACE1, β-secretase 1 CBP, CREB-binding protein; CH, cardiac hypertrophy; CSC, cancer stem cell; CTCL, cutaneous T-cell lymphoma; DNA MT, DNA methyltransferase; FBN2, fibrillin 2; HDAC, histone deacetylase; HHCy, hyperhomocysteinemia; HSC, hepatic stellate cell; IGFBP1, insulin growth factor-binding protein 1; MCP-1, monocyte chemoattractant protein 1; MS, multiple sclerosis; NAFLD, nonalcoholic fatty liver disease; NASH, nonalcoholic steatohepatitis; PASI, psoriasis area severity index; PH, pulmonary hypertension; PON, paraoxonase; PTPN6, protein tyrosine phosphatase nonreceptor 6; PTPRE, protein tyrosine phosphatase ε; RA, rheumatoid arthritis; SFRP1, secreted frizzled-related protein 1; TGF-β1, tumor growth factor β1; TNF-α, tumor necrosis factor α.

2.1 Cancer

There is growing evidence of an epigenetic basis of cancer based on altered DNA methylation [14]. Studies of primary cancers and chromatin development and epithelial–mesenchymal transition suggest a role for epigenetic stochasticity as the driving force of cancer, which seems to be mediated by a large-scale change in DNA methylation. In this context, epigenetic analysis detected widespread gene-body DNA hypomethylation in chronic lymphocytic leukemia [15].

Relative to prostate cancer, master regulators of prostate cancer stem cells have been identified, including epigenetic factors [16]. For instance, several miRNAs have been implicated. In this context, expression of the p53-specific miR-34a can inhibit clonogenic expansion and tumor regression [17]. In contrast, downregulation of expression by miR-34a antagomirs promoted tumor development and metastasis. Additionally, when tumor-bearing mice were injected with miR-34a, inhibition of metastasis and survival prolongation was observed. Furthermore, reduced expression of miR-143 and miR-145 has been associated with metastasis of the bone [18]. Likewise, downregulation of miR-200 has been demonstrated in clinical prostate tumors and prostate cancer cell lines [19]. Finally, miR-708 acts as a tumor suppressor, as delivery of synthetic miR-708 oligonucleotides resulted in regression of tumor xenografts, and silencing of the expression promoted tumor growth [20].

At least four epigenome-targeted drugs have been approved by the U.S. Food and Drug Administration [21]. Azacitidine and decitabine inhibit DNA methyltransferases (DNMTs) and have been used for the treatment of acute myeloid leukemia [22]. The histone deacetylase inhibitors vorinostat [23] and romidepsin [24], on the other hand, have been approved for the treatment of cutaneous T-cell lymphoma.

In attempts to treat glioma, the role of the epigenetically regulated protein tyrosine phosphatase nonreceptor type 6 (PTPN6) was analyzed by immunohistochemistry in 89 high-grade glioma patients [25]. Furthermore a panel of 16 chemotherapeutic drugs was subjected to drug resistance evaluation in PTPN6-overexpressing clones. Clearly, PTPN6 expression correlated with poor survival for anaplastic glioma patients. Moreover, PTPN6 overexpression increased resistance to bortezomib, cisplatin, and melphalan. In glioma-derived cells decreased PTPN6 promoter methylation correlated with protein expression, which was increased after demethylation. Therefore, PTPN6 expression may contribute to the poor survival of anaplastic glioma patients and may influence the response to chemotherapy because of its epigenetic nature of regulation.

A promising approach to developing novel epigenetic cancer drugs is presented by ruthenium compounds [26]. The cytotoxic antiprimary tumor compound $[(\eta^6\text{-}p\text{-cymene})\text{Ru(ethylenediamine)Cl}]\text{PF}_6$ and the relatively noncytotoxic antimetastatic compound $(\eta^6\text{-}p\text{-cymene})\text{Ru}(1,3,5\text{-triaza-7-phosphaadamantane})\text{Cl}_2$ target the DNA chromatin and histone proteins, respectively. A novel "atom-to-cell" approach demonstrated the basis for the surprisingly site-selective adduct formation and distinct cellular impact of the two chemically similar anticancer compounds. It suggested that the cytotoxic effect was based largely on DNA lesions, whereas the histone protein adducts may be due to other therapeutic activities.

2.2 Neurological Disorders

Several drugs that target the epigenetic machinery have been found to enhance memory function in rodents and have shown therapeutic efficacy in animal models for Alzheimer disease, schizophrenia, and depression [27]. In the context of Alzheimer disease, HDAC inhibitors have been evaluated in a mouse model overexpressing the p25 protein, a pathological activator of cyclin-dependent kinase 5, which causes amyloid and tau pathology with severe neurodegeneration and memory impairments [28]. Intraperitoneal administration of HDAC inhibitors for 4 weeks induced sprouting of dendrites, increased the number of synapses, and reinstated learning behavior and synaptic plasticity even after onset of severe neuronal loss [29]. The rationale for applying HDAC inhibitors was based on environmental enrichment mediated via changes in neuronal histone acetylation. A number of studies have demonstrated differences in DNA methylation between control and Alzheimer patients or animal models for Alzheimer disease [30–34]. For instance, altered methylation/demethylation patterns in vulnerable brain regions were observed before the onset of clinical symptoms of Alzheimer disease [30]. Moreover, analysis of 12 distinct mouse brain regions according to their CpG 5'-end gene methylation patterns that were translated to findings from Alzheimer disease patients identified DNA methylation-associated silencing of the thromboxane A2 receptor (TBXA2R), sorbin and SH3 domain-containing 3 (SORBS3), and spectrin β 4 (SPTBN4) genes [31]. In another study, the levels of the two most important DNA methylation (5-methylcytidine, 5-mC) and hydroxymethylation (5-hydroxymethylcytidine, 5-hmC) markers were analyzed in the hippocampus of Alzheimer disease patients [33]. Quantitative immunohistochemistry showed a significant reduction in 5-mC and 5-hmC levels, and also a negative correlation with amyloid plaque load in the hippocampus was observed. Furthermore, dietary factors have a role; for instance, a diet rich in folic acid ensures high levels of the methyl donor S-adenosylmethionine, which was demonstrated to reduce disease pathology in a mouse model for amyloid pathology [35]. A similar effect was discovered for vitamin-deficient mice suffering from memory impairment [36]. Noncoding RNAs have been demonstrated to have an effect on Alzheimer disease development. For instance, the miR-29 cluster targeting the β-secretase 1 (BACE1) is downregulated in Alzheimer brains, which leads to elevated BACE1, promoting amyloid pathology [37]. Furthermore, single-nucleotide polymorphisms (SNPs) within miR-29 binding sites of the BACE1 gene have shown a correlation with sporadic Alzheimer disease [38]. In addition to miR-29, other miRNAs, such as miR-107, miR-298, and miR-328, target BACE1 and have also been linked to Alzheimer disease [39,40]. The loss of miRNA control of BACE1 expression may lead to increased Aβ formation and disease progression. A systematic computational approach to the construction of the small molecule and miRNA association network in Alzheimer disease (SmiRN-AD) has allowed the therapeutic targeting of miRNAs based on the gene expression signatures of bioactive small-molecule perturbation and Alzheimer disease-related miRNA regulation [41]. The approach allowed recognition of 25 Alzheimer disease-related miRNAs and 275 small molecules. Analysis of positive connections (quinostatin and miR-148b and amantadine and miR-15a) as well as the negative connection, melatonin and miR-30e-5p, provided specific biological insights into Alzheimer disease pathogenesis and therapy.

2.3 Liver Diseases

Epigenetics has also had some significant impact on drug development for liver diseases [47]. For example, small-molecule epigenetic inhibitors such as the DNMT1 inhibitor 5-azadeoxycytidine and the EZH2 inhibitor 3-deazaneplanocin A potently inhibit hepatic stellate cell (HSC) activation [43]. Moreover, 69,247 differentially methylated CpG sites in liver biopsy material from nonalcoholic fatty liver disease (NAFLD), patients stratified into advanced (F3–F4) versus mild (F0–F1) disease [44]. Among the differentially modified CpG sites 76% became hypomethylated in advanced disease, whereas 24% underwent hypermethylation. Furthermore, noncoding RNAs most likely play a fundamental role in the determination of HSC phenotype and liver fibrosis. Several miRNAs regulating proliferation, apoptosis, TGF-β1 signaling, and collagen expression can function as regulators of HSC phenotype and progression of fibrosis [45]. In this context, deacetylase inhibitors (DACis) have been shown to interfere with gene expression via miRNA pathways in addition to histone acetylation [46]. For instance, the novel DACi panobinostat resulted in highly aberrant modulation of several miRNAs. In hepatocellular carcinoma cell lines, miR-19a and miR-19b1 were downregulated, while the miRNA targets APAF1 and Beclin 1 were upregulated.

Epigenetics has been firmly connected to nutrition and metabolic disease [42]. For instance, in rodents diets depleted of methyl donors promote DNA hypomethylation and the development of steatosis. However, when high-calorie diets were supplemented with methyl donors, NAFLD was prevented, which suggested that epigenetic changes affecting hepatic fat metabolism may be related to dynamic alterations in DNA methylation [47]. Furthermore, a close association between lipid metabolism and circadian rhythm has been established. The CLOCKBMAL1 circadian transcription factors regulate hundreds of genes, and mice lacking the expression of Clock are hyperphagic and obese and develop nonalcoholic steatohepatitis (NASH) [48]. Moreover, it was discovered that Clock has intrinsic histone acetyl transferase (HAT) activity, which revealed the molecular link between epigenetic control and the circadian clock [49]. Owing to the presence of several homologous regions to the HAT activator in Clock, it has been suggested that Clock is more than a transcription factor and a new type of DNA-binding HAT. DNA methylation and transcriptome analysis of 45,000 CpG sites from liver biopsies from lean control healthy, obese, and NASH patients revealed 467 dinucleotides with differences in methylation. Eight genes (GALNTL4, ACLY, IGFBP2, PLCG1, PRKCE, IGF1, IP6K3, and PC) with obesity-related expression alterations showed inverse correlation with altered CpG methylation [50]. Analysis of paired liver biopsies 5–9 months after bariatric surgery showed hypermethylation and downregulation of protein tyrosine phosphatase ε, a negative regulator of insulin signaling. Its correlation with weight loss suggests an interesting mechanistic link between weight loss and control of hepatic insulin sensitivity. As of this writing, approximately 100 miRNAs have been reported to be differentially expressed in NASH [51]. They show a wide functional diversity, such as control of lipid and glucose metabolism. For instance, miR-122 plays important regulatory functions in lipid and cholesterol metabolism and further has shown a close linkage to the circadian rhythm. Moreover, miR-122 expression is abundant in healthy liver, but downregulated in NASH patients. In mice, miR-122 has been functionally implicated in NAFLD pathology [52].

2.4 Other Diseases

Epigenetics has also been suggested to provide an explanation for the link between psoriasis risk alleles and disease development [53]. The methylation pattern is abnormal for multiple genes involved in psoriasis pathogenesis. Moreover, the peripheral blood mononuclear cells from psoriasis patients show hypoacetylation of histone H4, which is inversely correlated to the psoriasis area severity index. Also, increased and decreased expression of miRNAs in psoriasis patients has been correlated to roles in pathogenesis [54]. For instance, miR-223 and miR-143 have become important systemic biomarkers for psoriasis activity.

Epigenetics has also been linked to diabetes. For instance, obese and diabetic individuals have a different epigenetic pattern compared to nonobese and nondiabetic persons [55]. Similarly, a connection between epigenetics and stroke was discovered, in which the total tumor necrosis factor-α (TNF-α) promoter DNA methylation was lower in stroke patients and showed no interaction with body composition [56]. On the other hand, the +309 CpG methylation site of the TNF-α promoter was related to body weight in nonstroke individuals. Moreover, the methylation of the paraoxonase (PON) promoter correlated with body weight, waist circumference, and energy intake. Therefore, TNF-α and PON promoter methylation is suggested to be involved in the susceptibility to the outcome of stroke and obesity.

Several miRNAs have been associated with autoimmune disease and rheumatoid arthritis [57]. For instance, miR-124 regulates the proliferation and secretion of monocyte chemoattractant protein-1, and its dysregulation might lead to inflammatory pathogenesis [58]. Furthermore, let-7a and miR-132 have been shown to be involved in rheumatoid arthritis pathogenesis [59]. Similarly, circulating miR-21 has been suggested as an indicator of the disease course of multiple sclerosis [60]. Furthermore, a comparison between cancer patients and healthy individuals revealed that certain types of circulating miRNAs are associated with initiation and progression of cancer [61] and multiple sclerosis [62]. The noninvasive nature of miRNA-based screening assays and their sensitivity and specificity in detecting cancer make them attractive biomarkers.

3. PERSONALIZED MEDICINES

The vast accumulated studies on drug function, especially in clinical trials, have revealed the individual differences in drug response seen from one patient to another. For this reason, the approach of developing personalized medicines has become highly attractive. Typically, it is achieved by applying diagnostic biomarkers or biomolecular signatures, which can predict clinical responses in patients before treatment [63]. Particularly in oncology the tumors and biofluids from patients are isolated and analyzed for genetic, biochemical, and immunohistochemical markers to define a selective treatment [64]. The prescreening allows for disease-specific therapy without the need of subjecting patients to drugs unlikely to provide clinical benefit.

3.1 Personalized Drugs

Epigenetic modifications have had an important function in the evaluation of potential biomarkers for patient cancer diagnostics [18]. The goals have been to identify patient populations likely to respond to specific anticancer therapies and to define reasonable dosages for investigational anticancer drugs. The profiles of genetic and epigenetic alterations were characterized for gastric cancer [65]. Analysis of DNA methylation by a bead array with 485,512 probes revealed that genes involved in cancer-related pathways were more frequently affected by epigenetic modifications than by genetic alterations. This knowledge will be useful for selection of patients likely to benefit from specific drug treatment.

Application of stem cell technology for epigenetic reprogramming of tumor cells isolated from cancer patients can aid in the creation of self-evolving personalized translational platforms [66]. The generated stem cell population is immediately available for experimental manipulations including pharmacological screening for personalized "stemotoxic" cancer drugs. A cell-based epigenetic assay system (EPISSAY) has been designed based on a silenced triple-mutated bacterial nitroreductase, TMnfsB, fused with red fluorescent protein expressed in the nonmalignant human breast cell line MCF10A [67]. The potency of the epigenetic drug decitabine was evaluated for the naked drug and a PEGylated liposome-encapsulated formulation showing a 50% higher potency for the encapsulated version. The EPISSAY assay system provided rapid comparison of novel drug formulations to existing drugs.

Epigenetic aberrations can influence drug treatment by key gene expression modulations related to metabolism and distribution of drugs and drug targets [68]. The interindividual variation in drug response, the epigenetic alterations, and the epigenetic profiles of circulating nucleic acids present great potential for the application of biomarkers for personalized drugs. Moreover, the pharmacoepigenetic results provide information for the regulation of ADME (adsorption, distribution, metabolism, and excretion) genes and drug targets, which can be used for interindividual variations in drug response.

Furthermore, in addition to inactivation of tumor suppressor genes, epigenetics may play an important role in the development of drug resistance [69]. Epigenetic mechanisms triggering resistance to the colorectal cancer drugs 5-fluorouracil, irinotecan, and oxaliplatin have been investigated to aid in the stratification of cancer patients and to develop tailored personalized therapy. Also, preclinical studies on DNA methyltransferase and histone deacetylase inhibitors have suggested that they may provide reversible chemoresistance in colorectal tumors.

DNA methylation of a number of genes involved in DNA repair (MGMT, hMLH1, WRN) and cell cycle regulation (CHFR, CDK10, p73) influence the sensitivity to chemotherapeutic drugs [70]. Therefore, DNA methylation should be explored to discover molecular markers for the prediction of the tumor responsiveness of chemotherapy by applying genome-wide analysis of DNA methylation using microarrays and next-generation sequencing. Furthermore, epigenetic-based agents may help in ameliorating chemoresistance or enhancing tumor cytotoxicity.

3.2 Personalized Nutrition

Not only drugs play an important role in disease treatment. The diet has an enormous effect on well-being, especially as a preventive factor [71]. Studies have indicated that a low-fat diet has been preferable, but mounting evidence has supported other alternatives such as the Mediterranean diet. Studies have also demonstrated that individual responses to similar environmental challenges can be dramatically different, which is due to genomic variability. Nutrigenetics can therefore play an important role in personalized nutrition and the potential prevention of chronic diseases. For example, the transcription factor 7-like 2 (TCF7L2) gene locus is strongly associated with type 2 diabetes [72]. Individuals with the TCF7L2 rs7903146T allele present a higher risk of diabetes. However, in the PRE-DIMED study, when subjected to a Mediterranean diet, they did not show the higher risk of diabetes and stroke [73]. Another example demonstrated the differential regulation of miRNAs in lipoprotein lipase (LPL) polymorphisms associated with plasma lipid and dietary response [74]. In this context, it was shown that the LPL rs13702 minor allele disrupted an miRNA recognition element seed site for miR-410, which induced allele-specific regulation of LPL and explained reported LPL variants associated with plasma lipid phenotypes and gene–diet interactions.

Furthermore, newly discovered candidates for gene–diet interactions such as the peroxisome proliferator-activated receptor transcription factors [75] and genes involved in inflammation-related pathways [76] have been studied. Additionally, the interaction between the functional APOA2 rs5082 gene variant, saturated fat (SFA), and the body mass index (BMI) showed a 6.2% increase in BMI between genotypes when the SFA consumption was high. Furthermore, the CC genotype was associated with enhanced prevalence of obesity only in the case of high SFA [77]. More recently, this association was shown to be applicable to specific food groups such as dairy products [78].

A most interesting aspect of personalized nutrition relates to the circadian rhythm and the association of chronodisruption with chronic disease development. Genome-wide association studies have demonstrated that there exists a relationship between the Clock circadian regulator genes and fasting glucose concentrations, obesity, and metabolic syndrome [79]. For instance, the Clock SNPs rs4580704 and rs1801260 showed an association with BMI and other variables related to glucose and insulin resistance. An interaction between SNPs and fat intake (SFA and monosaturated fat) could also be established [80]. Additionally, it was demonstrated that Clock locus variations were associated with energy intake [81].

Similarly, dietary intervention can have an impact on epigenetic mechanisms such as DNA methylation, which means that, for example, not all Alzheimer disease patients would benefit from the same therapy [34]. Additional knowledge of epigenetics transferred into suitable biomarker information would allow the selection of the right patients for the right treatment and help explain how diet affects epigenetics. However, in the case of weight gain and insulin resistance/diabetes, the development of drugs or dietary interventions aiming at delaying or reversing epigenetic changes is hampered by the influence of other factors [55]. For instance, the impact of dietary and environmental factors may be small but cumulative. Furthermore, factors such as age and genetic and ethnic backgrounds should be considered. As mentioned above, the epigenetic function of TNF-α and PON promoter methylation presents a link to susceptibility for stroke and obesity, which can be strongly influenced by changes in food intake [56].

As many nutrients present a wide range of activities in prevention and alleviation of various diseases, their potential roles in the epigenetic regulation of human health have become important, leading to the establishment of the new field of nutriepigenetics [82]. Plant-derived polyphenols particularly are well known as bioactive food components, which contribute to the improvement of disease and promotion of health [83]. Epigenetic regulation of gene expression has also been associated with the consumption of polyphenols, suggesting the relevance of human nutrition to human health and disease. Furthermore, nutritional epigenetics has emerged as a new field in aging and age-related disease in relation to the interaction between genes and diet [84]. For example, nutrients involved in one-carbon metabolism such as folate, vitamin B12, vitamin B6, riboflavin, methionine, choline, and betaine regulate the levels of the universal methyl donor S-adenosylmethionine and methyltransferase inhibitor S-adenosylhomocysteine affecting DNA methylation. Additionally, retinoic acid, resveratrol, curcumin, sulforaphane, and tea polyphenols are able to modulate epigenetic patterns, which catalyze DNA methylation and histone modifications. Findings indicate that aging and age-related diseases are associated with changes in epigenetic patterns. Importantly, the bioavailability of plant-derived bioactive food compounds needs to be addressed to determine required concentrations for achieving epigenetic activity and therapeutic effect.

4. CONCLUSIONS

Epigenetics has received much attention in relation to drug discovery lately. DNA methylation, histone modifications, and gene silencing can all contribute significantly to the discovery of new mechanisms of drug function and new drug targets. This approach has convincingly revealed particularly the interaction of a number of miRNAs in the progress of various diseases such as cancer, neurological disorders, diabetes, and liver and other diseases. The reversible nature of miRNA activity has made the approach even more attractive.

Emerging technologies in genomics and diagnostics have strongly influenced the area of personalized medicine. Epigenetics provides great opportunities especially to screen for biomarkers indicating potential drug resistance to avoid any unnecessary drug treatment. Moreover, the influence of diet intake on disease development has caught much attention and has generated substantial data on how radical dietary changes can reduce the risk of chronic disease. This has triggered the increased development of the design of personalized nutrition.

In conclusion, epigenetics will obviously not replace past and present traditional drug discovery approaches. However, developments in a number of areas such as genomics, bioinformatics, and delivery technologies have enhanced the application range of epigenetics and might provide a novel "niche" for epigenetics in the search for novel and better drugs in the future.

References

[1] Su LJ, Mahabir S, Ellison GL, McGuinn LA, McGuinn LA, Reid BC. Epigenetic contributions to the relationship between cancer and dietary intake of nutrients, bioactive food components and environmental toxicants. Front Genet 2012;2:1–11.
[2] Fang M, Chen D, Yang CS. Dietary polyphenols may affect DNA methylation. J Nutr 2007;137:223S–8S.
[3] Boehm TL, Drahovsky D. Alteration of enzymatic methylation of DNA cytosines by chemical carcinogens; a mechanism involved in the initiation of carcinogenesis. J Natl Cancer Inst 1983;71:429–33.
[4] Costello JF, Plass C. Methylation matters. J Med Genet 2001;38:285–303.
[5] Jones PA, Baylin SB. The epigenomics of cancer. Cell 2007;128:683–92.
[6] Cheishvili D, Boureau L, Szyf M. DNA demethylation and invasive cancer: implications for therapeutics. Br J Pharmacol August 18, 2014. http://dx.doi.org/10.1111/bph.12885. [Epub ahead of print].
[7] Baccarelli A, Bollati V. Epigenetics and environmental chemicals. Curr Opin Pediatr 2009;21:243–51.
[8] Bollati V, Baccarelli A. Environmental epigenetics. Heredity 2010;105:105–12.
[9] Yang PH, Zhang L, Zhang YJ, Zhang J, Xu WF. HDAC6: physiological function and its selective inhibitors for cancer treatment. Drug Discov Ther 2013;7:233–42.
[10] Lundstrom K. MicroRNA in disease and gene therapy. Curr Drug Discov Technol 2011;8:76–86.
[11] Mathers JC, Strathdee G, Relton CL. Induction of epigenetic alterations by dietary and other environmental factors. Adv Genet 2010;71:3–39.
[12] Barros SP, Offenbacher S. Epigenetics: connecting environment and genotype to phenotype and disease. J Dent Res 2009;88:400–8.
[13] Esquela-Kerscher A, Slack FJ. Oncomirs – microRNAs with a role in cancer. Nat Rev Cancer 2006;6:259–69.
[14] Feinberg AP. Epigenetic stochasticity, nuclear structure and cancer: the implications. J Intern Med March 17, 2014;276:5–11. http://dx.doi.org/10.1111/joim.12224. [Epub ahead of print].
[15] Kulis M, Heath S, Bibikova M, Queiros AC, Navarro A, Clot G, et al. Epigenomic analysis detects widespread gene-body DNA hypomethylation in chronic lymphocytic leukemia. Nat Genet 2012;44:1236–42.
[16] Kerr CL, Hussain A. Regulators of prostate cancer stem cells. Curr Opin Oncol 2014;26:328–33.
[17] Liu C, Kelnar K, Liu B, Chen X, Calhoun-Davis T, Li H, et al. The microRNA miR-34a inhibits prostate cancer stem cells and metastasis by directly repressing CD44. Nat Med 2011;17:211–5.
[18] Huang S, Guo W, Tang Y, Ren D, Zou X, Peng X. miR-143 and miR-145 inhibit stem cell characteristics of PC-3 prostate cancer cells. Oncol Rep 2012;28:1831–7.
[19] Kong D, Li Y, Wang Z, Banerjee S, Ahmad A, Kim HR, et al. miR-200 regulates PDGF-D-mediated epithelial-mesenchymal transition, adhesion, and invasion of prostate cancer cells. Stem Cells 2009;27:1712–21.
[20] Saini S, Majid S, Shahryari V, Arora S, Yamamura S, Chang I, et al. miRNA-708 control CD44(+) prostate cancer-initiating cells. Cancer Res 2012;72:3618–30.

[21] Mummaneni P, Shord SS. Epigenetics and oncology. Pharmacotherapy 2014;34:495–505.

[22] Estey EH. Epigenetics in clinical practice: the examples of azicitidine and decitabine in myelodysplasia and acute myeloid leukemia. Leukemia 2013;27:1803–12.

[23] Iwamoto M, Friedman EJ, Sandhu P, Agrawal NG, Rubin EH, Wagner JA. Clinical pharmacology profile of vorinostat, a histone deacetylase inhibitor. Cancer Chemother Pharmacol 2013;72:493–508.

[24] Prince HM, Dickinson M, Khol A. Romidepsin for cutaneous T-cell lymphoma. Future Oncol 2013;9:1819–27.

[25] Sooman L, Ekman S, Tsakonas G, Jaiswal A, Navani S, Edqvist PH, et al. PTPN6 expression is epigenetically regulated and influences survival and response to chemotherapy in high-grade gliomas. Tumour Biol February 9, 2014;35:4479–88. [Epub ahead of print].

[26] Adhireksan Z, Davey GE, Campomanes P, Groessl M, Clavel CM, Yu H, et al. Ligand substitutions between ruthenium-cymene compounds can control protein versus DNA targeting and anticancer activity. Nat Commun 2014;5:3462.

[27] Fisher A. Epigenetic memory: the Lamarckian brain. Embo J 2014;33:945–67.

[28] Fischer A, Sananbenesi F, Pang PT, Lu B, Tsai LH. Opposing roles of transient and prolonged expression of p25 in synaptic plasticity and hippocampus-dependent memory. Neuron 2005;48:825–38.

[29] Fischer A, Sananbenesi F, Wang X, Dobbin M, Tsai LH. Recovery of learning and memory is associated with chromatin remodelling. Nature 2007;447:178–82.

[30] Bradley-Whitman MA, Lovell MA. Epigenetic changes in the progression of Alzheimer's disease. Mech Ageing Dev 2013;134:486–95.

[31] Sanchez-Mut JV, Aso E, Panayotis N, Lott I, Dierssen M, Rabano A, et al. DNA methylation map of mouse and human brain identifies target genes in Alzheimer's disease. Brain 2013;136:3018–27.

[32] Wang J, Yu JT, Tan MS, Jiang T, Tan L. Epigenetic mechanisms in Alzheimer's disease: implications for pathogenesis and therapy. Ageing Res Rev 2013;12:1024–41.

[33] Chouliaras L, Mastroeni D, Delvaux E, Grover A, Kenis G, Hof PR, et al. Consistent decrease in global DNA methylation and hydroxymethylation in the hippocampus of Alzheimer's disease patients. Neurobiol Aging 2013;34:2091–9.

[34] Rogers J, Mastroeni D, Grover A, Delvaux E, Whiteside C. Coleman PD. The epigenetics of Alzheimer's disease–additional considerations. Neurobiol Aging 2011;32:1196–7.

[35] Chen TF, Huang RF, Lin SE, Lu JF, Tang MC, Chiu MJ. Folic acid potentiates the effect of memantine on spatial learning and neuronal protection in an Alzheimer's disease transgenic model. J Alzheimers Dis 2010;20:607–15.

[36] Fuso A, Nicolia V, Ricceri L, Cavallaro RA, Isopi E, Mangia F, et al. S-adenosylmethionine reduces the progress of the Alzheimer-like features induced by B-vitamin deficiency in mice. Neurobiol Aging 2012;33:1–16.

[37] Hébert SS, Horré K, Nicolaï L, Papadopoulou AS, Mandemakers W, Silahtaroglu AN, et al. Loss of microRNA cluster miR-29a/b-1 in sporadic Alzheimer's disease correlates with increased BACE1/beta-secretase expression. Proc Natl Acad Sci USA 2008;205:6415–20.

[38] Bettens K, Brouwers N, Engelborghs S, Van Miegroet H, De Deyn PP, Theuns J, et al. APP and BACE1 miRNA genetic variability has no major role in risk for Alzheimer disease. Hum Mutat 2009;30:1207–13.

[39] Wang WX, Rajeev BW, Stromberg AJ, Ren N, Tang G, Huang Q, et al. The expression of microRNA miR-107 decreases early in Alzheimer's disease and may accelerate disease progression through regulation of beta-site amyloid precursor protein-cleaving enzyme 1. J Neurosci 2008;28:1213–25.

[40] Boissonneault V, Plante I, Rivest S, Provost P. MicroRNA-298 and microRNA-328 regulate expression of mouse beta-amyloid precursor protein-converting enzyme 1. J Biol Chem 2009;284:1971–81.

[41] Meng F, Dai E, Yu X, Zhang Y, Chen X, Liu X, et al. Constructing and characterizing a bioactive small molecule and microRNA association network for Alzheimer's disease. J R Soc Interface 2013;11:20131057.

[42] Mann DA. Epigenetics in liver disease. Hepatology March 16, 2014. http://dx.doi.org/ 10.1002/hep.27131. [Epub ahead of print].

[43] Mann J, Chu DC, Maxwell A, Oakley F, Zhu NL, Tsukamoto H, et al. MeCP2 controls an epigenetic pathway that promotes myofibroblast transdifferentiation and fibrosis. Gastroenterology 2010;138:705–14.

[44] Murphy SK, Yang H, Moylan CA, Pang H, Dellinger A, Abdelmalek MF, et al. Relationship between the methylome and transcriptome in patients with non-alcoholic fatty liver disease. Gastroenterology 2013;145:1076–87.

[45] Chen SL, Zheng MH, Shi KQ, Yang T, Chen YP. A new strategy for treatment of liver fibrosis: letting MicroRNAs do the job. BioDrugs 2013;27:25–34.

[46] Henrici A, Montalbano R, Neureiter D, Krause M, Stiewe T, Slater EP, et al. The pan-deacetylase inhibitor panobinostat suppresses the expression of oncogenic miRNAs in hepatocellular carcinoma cell lines. Mol Carcinog December 23, 2013. http://dx.doi.org/10.1002/mc.22122. [Epub ahead of print].

[47] Pogrigbny IP, Tryndyak VP, Bagnyukova TV, Melnyk S, Montgomery B, Ross SA, et al. Hepatic epigenetic phenotype predetermines individual susceptibility to hepatic steatosis in mice fed a lipogenic methyldeficient diet. J Hepatol 2009;51: 176–86.

[48] Bellet MM, Sassone-Corsi P. Mammalian circadian clock and metabolism — the epigenetic link. J Cell Sci 2010;123:3837–48.

[49] Doi M, Hirayama J, Sassone-Corsi P. Circadian regulator CLOCK is a histone acetyltransferase. 2006 Cell 2006;125:497–508.

[50] Ahrens M, Ammerpohl O, von Schonfels W, Kolarova J, Bens S, Itzel T, et al. DNA methylation analysis in nonalcoholic fatty liver disease suggests distinct disease-specific and remodeling signatures after bariatric surgery. Cell Metab 2013;18:296–302.

[51] Yu-Yuan L. Genetic and epigenetic variants influencing the development of nonalcoholic fatty liver disease. World J Gastroenterol 2012;18:6546–51.

[52] Esau C, Davis S, Murray SF, Yu XX, Pandey SK, Pear M, et al. miR-122 regulation of lipid metabolism revealed by in vitro antisense targeting. Cell Metab 2006;3:87–98.

[53] Trowbridge RM, Pittelkow MR. Epigenetics in the pathogenesis and pathophysiology of psoriasis vulgaris. J Drugs Dermatol 2014;13:111–8.

[54] Løvendorf MB, Zibert JR, Gyldenløve M, Røpke MA, Skov L. MicroRNA-223 and miR-143 are important systemic biomarkers for disease activity in psoriasis. J Dermatol Sci 2014;75:133–9.

[55] Martínez JA, Milagro FI, Claycombe KJ, Schalinske KL. Epigenetics in adipose tissue, obesity, weight loss, and diabetes. Adv Nutr 2014;5:71–81.

[56] Gómez-Uriz AM, Goyenechea E, Campión J, de Arce A, Martinez MT, Puchau B, et al. Epigenetic patterns of two gene promoters (TNF-α and PON) in stroke considering obesity condition and dietary intake. J Physiol Biochem 2014;70:603–14.

[57] Qu Z, Li W, Fu B. MicroRNAs in autoimmune diseases. Biomed Res Int 2014;2014:1–8.

[58] Nakamachi Y, Kawano S, Takenokuchi M, Nishimura K, Sakai Y, Chin T, et al. MicroRNA-124a is a key regulator of proliferation and monocyte chemoattractant protein 1 secretion in fibroblast-like synoviocytes from patients with rheumatoid arthritis. Arthritis Rheumatism 2009;60:1294–304.

[59] Pauley KM, Satoh M, Chan AL, Bubb MR, Reeves WH, Chan EKL. Upregulated miR-146a expression in peripheral blood mononuclear cells from rheumatoid arthritis patients. Arthritis Res Ther 2008;10:R101.

[60] Cheng G. Circulating miRNAs: roles in cancer diagnosis, prognosis and therapy. Adv Drug Deliv Rev 2014:S0169–409X(14)00199-9.

[61] Lindner K, Haier J, Wang Z, Watson DI, Hussey DJ, Hummel R. Circulating microR-NAs: emerging biomarkers for diagnosis and prognosis in patients with gastrointestinal cancers. Clin Sci (Lond) 2015;128:1–15.

[62] Lindberg RLP, Hoffmann F, Kuhle J, Kappos L. Circulating microRNAs as indicators for disease course of multiple sclerosis. Mult Scler 2010;16:S41–196.

[63] Woodcock J. Assessing the clinical utility of diagnostics used in drug therapy. Clin Pharmacol Ther 2010;88:765–73.

[64] Nicolaides NC, O'Shanessy DJ, Albone E, Grasso L. Co-development of diagnostics vectors to support targeted therapies and theranostics: essential tools in personalized cancer therapy. Front Oncol 2014;4:1–14.

[65] Yoda Y, Takeshima H, Niwa T, Kim JG, Ando T, Kushima R, et al. Integrated analysis of cancer-related pathways affected by genetic and epigenetic alterations in gastric cancer. Gastric Cancer February 9, 2014. [Epub ahead of print].

[66] Menendez JA, Alarcón T, Corominas-Faja B, Cuyàs E, López-Bonet E, Martin AG, et al. Xenopatients 2.0: reprogramming the epigenetic landscapes of patient-derived cancer genomes. Cell Cycle 2014;13:358–70.

[67] Lim SP, Kumar R, Akkamsetty Y, Wang W, Ho K, Neilsen PM, et al. Development of a novel cell-based assay system EPISSAY for screening epigenetic drugs and liposome formulated decitabine. BMC Cancer 2013;13:113.

[68] Ivanov M, Kacevska M, Ingelman-Sundberg M. Epigenomics and interindividual differences in drug response. Clin Pharmacol Ther 2012;92:727–36.

[69] Crea F, Nobili S, Paolicchi E, Perrone G, Napoli C, Landini I, et al. Epigenetics and chemoresistance in colorectal cancer: an opportunity for treatment tailoring and novel therapeutic strategies. Drug Resist Updat 2011;14:280–96.

[70] Toyota M, Suzuki H, Yamashita T, Hirata K, Imai K, Tokino T, et al. Cancer epigenomics: implications of DNA methylation in personalized cancer therapy. Cancer Sci 2009;100:787–91.

[71] Konstantinidou V, Daimiel-Ruiz LA, Ordivas JM. Personalized nutrition and cardiovascular disease: from Framingham to PREDIMED. Adv Nutr 2014;5:3675–815.

[72] Grant SF, Thorleifsson G, Reynisdottir I, Benediktsson R, Manolescu A, Sainz J, et al. Variant of transcription factor 7-like 2 (TCF7L2) gene confers risk of type 2 diabetes. Nat Genet 2006;38:320–3.

[73] Corella D, Carrasco P, Sorlí JV, Estruch R, Rico-Sanz J, Martínez-González MÁ, et al. Mediterranean diet reduces the adverse effect of the TCF7L2-rs7903146 polymorphism on cardiovascular risk factors and stroke incidence: a randomized controlled trial in a high-cardiovascular-risk population. Diabetes Care 2013;36:3803–11.

[74] Richardson K, Nettleton JA, Rotllan N, Tanaka T, Smith CE, Lai CQ, et al. Gain-of-function lipoprotein lipase variant rs13702 modulates lipid traits through disruption of a microRNA-410 seed site. Am J Hum Genet 2013;92:5–14.

[75] Vohl MC, Lepage P, Gaudet D, Brewer CG, Bétard C, Perron P, et al. Molecular scanning of the human PPARa gene: association of the L162v mutation with hyperapobetalipoproteinemia. J Lipid Res 2000;41:945–52.

[76] Shen J, Arnett DK, Peacock JM, Parnell LD, Kraja A, Hixson JE, et al. Interleukin1beta genetic polymorphisms interact with polyunsaturated fatty acids to modulate risk of the metabolic syndrome. J Nutr 2007;137:1846–51.

[77] Corella D, Peloso G, Arnett DK, Demissie S, Cupples LA, Tucker K, et al. APOA2, dietary fat, and bodymass index: replication of a gene-diet interaction in 3 independent populations. Arch Intern Med 2009;169:1897–906.

[78] Smith CE, Tucker KL, Arnett DK, Noel SE, Corella D, Borecki IB, et al. Apolipoprotein A2 polymorphism interacts with intakes of dairy foods to influence body weight in 2 U.S. populations. J Nutr 2013;143:1865–71.

[79] Garaulet M, Lee YC, Shen J, Parnell LD, Arnett DK, Tsai MY, et al. CLOCK genetic variation and metabolic syndrome risk: modulation by monounsaturated fatty acids. Am J Clin Nutr 2009;90:1466–75.

[80] Garaulet M, Lee YC, Shen J, Parnell LD, Arnett DK, Tsai MY, et al. Genetic variants in human CLOCK associate with total energy intake and cytokine sleep factors in overweight subjects (GOLDN population). Eur J Hum Genet 2010;18:364–9.

[81] Remely M, Lovrecic L, de la Garza AL, Migliore L, Peterlin B, Milagro FI, et al. Therapeutic perspectives of epigenetically active nutrients. Br J Pharmacol July 22, 2014. http://dx.doi.org/10.1111/bph.12854. [Epub ahead of print].

[82] Lundstrom K. Past, present and future of nutrigenomics and its influence on drug development. Curr Drug Discov Technol 2013;10:35–46.

[83] Joven J, Micol V, Segura-Carretero A, Alonso-Villaverde C, Menéndez JA, Bioactive food components platform. Polyphenols and the modulation of gene expression pathways: can we eat our way out of the danger of chronic disease? Crit Rev Food Sci Nutr 2014;54:985–1001.

[84] Park LK, Friso S, Choi SW. Nutritional influences on epigenetics and age-related disease. Proc Nutr Soc 2012;71:75–83.

[85] Mack SC, Witt H, Piro RM, Gu L, Zuyderduyn S, Stütz AM, et al. Epigenomic alterations define lethal CIMP-positive ependymomas of infancy. Nature 2014;506:445–50.

[86] Ricketts CJ, Hill VK, Linehan WM. Tumor-specific hypermethylation of epigenetic biomarkers, including SFRP1, predicts for poorer survival in patients from the TCGA Kidney Renal Clear Cell Carcinoma (KIRC) project. PLoS One 2014;9:e85621.

[87] Jayaraman S. Novel methods of diabetes 1 treatment. Discov Med 2014;17:347–55.

[88] Patel T, Patel V, Sing R, Jayaraman S. Chromatin remodeling resets the immune system to protect against autoimmune diabetes in mice. Immunol Cell Biol 2011;89:640–9.

[89] Gu T, Gu HF, Hilding A, Sjöholm LK, Ostenson CG, Ekström TJ, et al. Increased DNA methylation levels of the insulin-like growth factor binding protein 1 gene are associated with type 2 diabetes in Swedish men. Clin Epigenetics 2013;5:21.

[90] Kalani A, Kamat PK, Tyagi SC, Tyagi N. Synergy of homocysteine, microRNA, and epigenetics: a novel therapeutic approach for stroke. Mol Neurobiol 2013;48:157–68.

[91] Parikh VN, Jin RC, Rabello S, Gulbahce N, White K, Hale A, et al. MicroRNA-21 integrates pathogenic signaling to control pulmonary hypertension: results of a network bioinformatics approach. Circulation 2012;125:1520–32.

[92] Abi Khali C. The emerging role of epigenetics in cardiovascular disease. Ther Adv Chronic Dis 2014;5:178–87.

[93] Ikeda S, Kong SW, Lu J, Bisping E, Zhang H, Allen PD, et al. Altered microRNA expression in human heart disease. Physiol Genomics 2007;31:367–73.

PERSONALIZED EPIGENETICS OF DISORDERS AND DISEASE MANAGEMENT

14

Epigenetics and Personalized Pain Management

Seena K. Ajit

Department of Pharmacology & Physiology, Drexel University College of
Medicine, Philadelphia, PA, USA

OUTLINE

1. INTRODUCTION

According to the American Academy of Pain Medicine, pain affects more Americans than diabetes, heart disease, and cancer combined. Common chronic pain conditions include headache, lower back pain, cancer pain, arthritis pain, and neuropathic pain. Neuropathic pain resulting from injury or malfunction of the nervous system affects millions of people worldwide and is generally acknowledged to be extremely difficult to treat. Several conditions, including postherpetic neuralgia, phantom limb pain, diabetic peripheral neuropathy, and trigeminal neuralgia, are associated with neuropathic pain. Inflammatory diseases, as well as cancer-related pain and chemotherapy, likewise result in acute and chronic pain that is often debilitating, adding to the burden of an already significant diagnosis. In addition to adversely affecting quality of life, chronic pain imposes a large burden on health care systems because the primary reason for visiting a doctor is related to the onset and/or persistence of pain. Pain is highly prevalent among the elderly. In the United States, more than 38 million individuals are 65 years of age or older. Neuropathic pain is disproportionately high in older patients because many diseases associated with neuropathic pain, such as cancer and musculoskeletal degeneration, increase in incidence with age [1]. It is important to understand the impact of advancing age on pain to provide more effective translational approaches to achieve better treatment outcomes for the elderly.

Regardless of age, however, most types of pain are treated with a limited number of standard drugs, most of which are far from optimal and also produce undesirable side effects. Present treatment options are limited to nonsteroidal anti-inflammatory drugs, opioids, anticonvulsants, and antidepressants, which provide pain relief to only about 50% of patients, clearly highlighting the unmet medical need for such a common ailment [2]. One reason for the low efficacy of existing treatments for chronic pain is the interindividual variability in pain sensitivity and response to analgesic drugs [3]. Several studies have addressed the genetic determinants of pain and analgesia in rodents and humans [3–5]. Excitement in the pain field concerning the plausibility of revolutionizing our understanding of genetic risk factors for the development of chronic pain has been tempered by slow progress and contradictory data, as well as the modest percentage of the trait variance accounted for by the associated genes [3]. The pain field can benefit tremendously from entirely new approaches rendering novel perspectives on how we treat pain. A fresh perspective in understanding the molecular mechanisms underlying chronic pain is needed to obtain insight, thereby leading to more effective therapeutic intervention strategies devoid of side effects.

The goal of personalized medicine is to establish patient-specific diagnosis and treatment strategies based on the unique profile of the

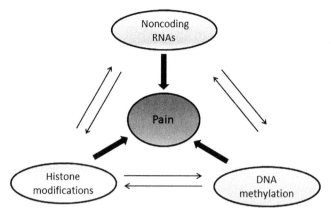

FIGURE 1 Epigenetic mechanisms attributed to pain. Evidence linking epigenetic mechanisms to the development and maintenance of chronic pain conditions and treatment response is beginning to emerge. Mechanisms commonly attributed to pain include DNA methylation, posttranslational modifications of histone proteins, and noncoding RNA-mediated gene regulation. The hierarchical order of epigenetic events and dependencies is not known but these mechanisms can occur simultaneously or sequentially, influencing gene expression changes mediating pain.

individual [6]. The impetus for this is the enormous differences between individuals, combined with the technical advances in high-throughput methodologies providing insights into DNA, RNA, proteins, and metabolites. These advances have increasingly made it feasible to characterize the disease process. Personalized or individualized medicine also aspires to recommending preventive measures tailored to individuals based on available biomolecular signature and risk or susceptibility factors, including lifestyle, to include epigenetic traits. Awareness is growing regarding the need to integrate biological and biographical factors, including exposure to physical stressors such as pollutants, toxins, and radiation. Nutritional deprivation during a child's growing years can influence later susceptibility to certain diseases [7]. The progress achieved in elucidating the role of epigenetic regulation in the context of pain is discussed here (Figure 1).

2. EPIGENETIC REGULATION

Factors other than DNA can affect the way our genes are expressed, a phenomenon referred to as epigenetics [8]. Epigenetics encompasses heritable alterations in gene expression and chromatin without accompanying changes in the DNA sequence [8]. Though disease states are often the result of altered gene expression caused by mutations within a gene, aberrant epigenetic modifications of the chromatin surrounding the gene can also result in profound changes in gene expression. Molecular and genetics studies have shown the fundamental role of epigenetics in key biological processes and have elucidated how epigenetic aberrations result in disease. DNA methylation and histone modification are two well-documented epigenetic modifications that influence access to DNA and thus protein expression [7,9]. RNA-mediated gene silencing is another mechanism that has been linked to posttranscriptional gene silencing [10,11]. A common comparison is that the keys of the piano represent the genome, and the pianist's fingers represent the epigenome. While it is hard to change an error in DNA sequence, it has been demonstrated that modification of epigenetic tags on DNA, such as methyl groups or histones, can be reversed with drug treatments. The underlying principle that is emerging from the study of epigenetics is that DNA is not destiny, and there may be ways to alter gene expression and hence disease states. Studies on identical twins have shown that over the years, measurable epigenetic variations accumulate, indicating that an individual's lifestyle choices and personal history, such as diet, smoking, and stress, can influence epigenetic factors and their buildup in the body [7].

3. DNA METHYLATION

Methylation of cytosine in the DNA at the carbon 5 position is a common epigenetic modification observed in many eukaryotes [12]. It is a major covalent modification of the DNA and considered to be one of the most stable epigenetic marks. It is often found in the sequence context of CpG or CpHpG (H = A, T, C); the "p" in CpG refers to the phosphodiester bond between the cytosine and the guanine. Epigenetic modification resulting from DNA methylation plays a critical role in embryonic development, cellular differentiation, and disease [13,14]. DNA methylation plays a fundamental role in transcriptional silencing of imprinted genes in normal cells, X-chromosome inactivation in females (important for dosage compensation), and suppression of retrotransposon elements [12,15]. DNA methylation is thus a nonrandom process that is well regulated and tissue specific [16–18]. In addition to CpG islands, alterations in DNA methylation can occur in "CpG shores," which are genomic regions with lower CpG density than in conventional CpG islands and occur in proximity (~2 kb) to CpG islands [19]. Methylation of CpG shores is closely associated with transcriptional inactivation, and their differential methylation is sufficient to distinguish tissue-specific DNA methylation [20].

DNA methylation is catalyzed by the enzyme DNA methyltransferase (DNMT). DNMTs mediate both propagation of methylation patterns and de novo methylation. In mammals, three catalytically active DNMTs have been identified: DNMT1, DNMT3A, and DNMT3B.

DNMT1 is considered to be the maintenance methyltransferase because of its preference for hemimethylated CpG sites. Hemimethylated DNA is generated during the process of DNA replication and DNMT1 copies the preexisting methylation patterns to the newly synthesized DNA strand. DNMT3A and DNMT3B are referred to as the de novo methyltransferases because of their ability to establish methylation patterns. They do not have a preference for hemimethylated CpG substrates in vitro. Methylation of cytosine is usually associated with stable transcriptional repression when present in promoters and enhancers. However, methylation of gene bodies is positively correlated with transcription [20]. Three possible ways in which gene silencing is caused by DNA methylation include (1) changing the DNA accessibility by altering the higher order structure of chromatin; (2) interfering with or inhibiting the interaction between particular DNA-binding proteins, such as a transcriptional activator and its cognate binding site on the DNA; and (3) attracting proteins, such as methyl–CpG-binding proteins (MeCPs and MBDs), together with corepressor molecules that have an affinity for methylated DNA, which in turn mediate downstream biological effects [21]. Epigenetic regulators are broadly classified as readers, writers, or erasers depending on their role in interpreting the epigenetic marks [22]. For example, DNMTs responsible for cytosine methylation are considered writers. MeCP2 protein is capable of deciphering methylation patterns across the genome before binding to methylated DNA. Hence MeCP2 is commonly referred to as a reader.

Though the enzymes catalyzing DNA methylation and proteins interpreting the methylation pattern are well characterized, factors responsible for the removal of methyl groups from cytosine have remained elusive until recently. A novel discovery in the field of DNA methylation is the elucidation of the role of ten–eleven translocation (TET) family enzymes. TET1, one of the three enzymes of the TET family, was found to oxidize 5-methylcytosine (5mC) through oxidation to 5-hydroxymethylcytosine (5hmC) and has tremendously advanced our understanding of DNA demethylation mechanisms. 5hmC serves as a key intermediate in active demethylation pathways [23,24]. Through iterative oxidation, TET can generate 5-formylcytosine (5fC) and 5-carboxylcytosine (5caC). In other words, 5mC bases introduced by DNMT enzymes can be oxidized iteratively to 5hmC, 5fC, and 5caC. Thus 5hmC, 5fC, and 5caC are chemically distinct modifications of cytosine that could be specifically recognized by different DNA-binding proteins [25]. 5hmC is the main link in demethylation and can be either passively depleted through DNA replication or actively reverted to cytosine. In the pathway of active modification followed by passive dilution, 5hmC is diluted in a replication-dependent manner, resulting in the regeneration of unmodified cytosine. In the pathway of active modification followed by active restoration, 5fC or 5caC is excised by a thymine DNA glycosylase-mediated base excision repair process that regenerates unmodified cytosine [25]. Distribution of 5hmC in the genome differs from that of 5mC and is associated with promoters mediating gene expression, as well as with polycomb-mediated silencing. All three TETs oxidize 5mC to 5hmC in vitro and in vivo and the presence of 5hmC depends on preexisting 5mC in vivo, suggesting that this is the only course for the synthesis of genomic 5hmC [26].

A 2012 quantitative, genome-wide analysis of 5hmC, 5mC, and gene expression in differentiated central nervous system (CNS) cell types in vivo showed that 5hmC is enriched in active genes. There was a strong depletion of 5mC over these regions. The contribution of these epigenetic marks to gene expression depends on the cell type, suggesting a cell-specific epigenetic mechanism for regulation of chromatin structure and gene expression. These investigators also demonstrated that MeCP2 can bind 5hmC- and 5mC-containing DNA with similar high affinities [27]. In adult tissues 5mC levels are fairly constant (4–5% of all cytosines) but levels of 5hmC vary between <0.1% and ~0.7% and are highest in CNS tissues [28]. Although as of this writing none of the studies have addressed the role of 5hmC in pain, its abundance in the CNS suggests that these epigenetic marks could play a role in mediating pain. Further studies delineating the contribution of 5hmC and 5mC will provide insight into the gene-regulatory consequences resulting from these modifications.

4. DNA METHYLATION AND PAIN

Direct evidence-linking DNA methylation to the development and/or maintenance of chronic pain is beginning to emerge [29–32]. Epigenetics has been predicted to play a key role in pain and analgesia by influencing pro- and antinociceptive gene expression and by modulating pharmacodynamics or pharmacokinetic properties of analgesics [33–40]. A few studies have directly linked methylation-induced changes to pain. Increased methylation of the SPARC (secreted protein acidic rich in cysteine) promoter and decreased SPARC expression have been linked to chronic low back pain and degeneration of the intervertebral discs in human subjects and in aged mice [31,41]. SPARC is an extracellular matrix protein, and SPARC-null mice exhibit symptoms consistent with chronic low back and radicular pain, attributable to intervertebral disc degeneration [41]. In human subjects with back pain, disc degeneration was observed along with increased methylation of the SPARC promoter. Transient transfection assays were employed to demonstrate that methylation of either the human or the mouse SPARC promoter can induce silencing of its activity, providing the first evidence that DNA methylation of a single gene can play a crucial role in chronic pain in both humans and animal models [31].

Methylation-induced gene regulation has been shown to have a role in mediating cancer pain. The endothelin-1B receptor (EDNRB) was methylated in painful human oral squamous cell carcinoma lesions but not in nonpainful oral dysplasias. The four endothelin peptides that constitute the endothelin system exert their function by binding to their G-protein-coupled receptors A and B. Members of the endothelin system play roles in the growth and progression of multiple tumors. Hypermethylation of EDNRB resulted in reduction of its mRNA in human oral squamous cell carcinoma lesions. Pain attenuation was achieved by reexpression of EDNRB in a mouse model of cancer pain, suggesting EDNRB methylation as a novel regulatory mechanism in cancer-induced pain [29].

Several methods are available for global mapping of the DNA methylome. Sodium bisulfite treatment of DNA, which converts unmethylated but not methylated cytosine to uracil, is used in reduced representation bisulfite sequencing (RRBS), methylC-seq, and the Infinium-27K bead-array technology (Illumina, San Diego, CA, USA). The other three methods, including methylated DNA immunoprecipitation sequencing (Me-DIP-seq), methylated DNA capture by affinity purification (MethylCap-seq), and methylated DNA binding domain sequencing (MBD-seq), rely on capture of methylated DNA by a monoclonal antibody or by the recombinant methyl-binding domains of MeCP2 or MBD2, respectively [42]. Comparison of these methods has confirmed that all of these approaches can provide accurate DNA methylation measurements and can be used to detect differential DNA methylation [42,43]. Patterns of DNA methylation in promoters, CpG islands, and gene bodies have been discovered using methodologies available. However, no single method can address all aspects of DNA methylation detection. For example, selective immunoprecipitation of methylated DNA fragments using antibodies to 5-methylcytosine or methylated DNA binding proteins may lack sensitivity in genomic regions with a relatively low density of CpG sites. It may be difficult to determine exactly how many and which CpG among multiple CpGs within a given fragment are methylated. Thus evaluating, the type of information that can be obtained from different methods and shortcomings such as inherent potential bias and artifacts can be helpful in determining the most appropriate methodology to pursue for a specific research question. A 2014 review provides an excellent comparison of strengths and limitations of DNA methylation assays, cost, minimum sample input requirements, accuracy, and throughput and provides general guidelines for selecting appropriate methods for specific experimental contexts [44]. PCR-based DNA methylation studies permit detailed analysis of the methylation status of specific regions of the genome and critical experimental parameters have been reviewed [45].

The 450K HumanMethylation450 BeadChip kit allows researchers to interrogate >485,000 methylation sites per sample at single-nucleotide resolution and covers 96% of CpG islands, whereas the Infinium-27K bead-array mentioned earlier probes for 27,578 individual CpG sites.

In a genome-wide methylation study using the 450K HumanMethylation chip (Illumina), higher levels of methylation were observed in women with fibromyalgia compared with control subjects [46]. This study, conducted in 10 patients and 8 controls, attributed 91% of differentially methylated sites to an increase in methylation in fibromyalgia patients. The finding of differential methylation of genes with functional relevance to fibromyalgia [46] suggests a role for DNA methylation in this chronic pain disorder.

In the largest and most comprehensive study as of this writing investigating methylation and pain, differential methylation in the genomic DNA from whole blood was performed using Me-DIP-seq. This study enrolled 100 people, including 25 pairs of identical twins with different heat pain thresholds [47]. Inclusion of twins and an unrelated, independent cohort enabled the investigators to differentiate regions of the genome at which genetic determinants of pain sensitivity may be mediated through DNA methylation from regions where differential methylation resulted from alterations in epigenetic regulation, both of which may contribute to variations in pain sensitivity. They identified nine differentially methylated regions including the transient receptor potential cation channel, subfamily A, member 1 (TRPA1). The promoter region of TRPA1 was hypermethylated in individuals with lower pain thresholds. DNA methylation in differentially methylated regions was validated using bisulfite conversion methods. These regions also showed DNA methylation stability over time, consistent with susceptibility effects on pain sensitivity. In other words, changes in heat pain tolerance were lower in individuals when DNA methylation remained stable over time and higher when DNA methylation levels significantly changed over time. This observation indicates a role for epigenetic regulation in mediating pain sensitivity. A member of the transient receptor potential channel family, TRPA1 plays an important role in chemonociception. In addition to contributing to injury-evoked cold hypersensitivity, this ligand-gated ion channel is activated by a variety of plant-derived pungent agents, toxic and volatile air pollutants, metabolic by-products of chemotherapeutic agents, endogenous products of oxidative or nitrative stress, and calcium released from intracellular stores [48]. However, TRPA1 has not been reported as a heat sensor in mammals and this study was designed to investigate various heat pain thresholds. TRPA1 is expressed in peptidergic C-fibers of mammalian sensory ganglia that project to a variety of peripheral targets, including the skin and viscera [49], and all TRPA1-positive neurons coexpress transient receptor potential cation channel subfamily V member 1 (TRPV1), a heat-activated ion channel that mediates acute sensation of noxious heat, as well as thermal hyperalgesia. TRPA1 can be regulated by and interact with the thermosensor TRPV1, promoting sensitization to thermal stimuli, and the authors suggest that this interaction is the link to heat pain threshold. They then investigated differential methylation and expression of TRPA1 and the three thermosensors TRPV1, TRPV2, and TRPV3 using skin biopsies. Expression of TRPA1 was higher in the skin of subjects with higher pain thresholds, and this increase was described as nominally significant. This finding is in agreement with the observation of hypermethylation in individuals with lower pain thresholds and the downregulation in gene expression that can ensue from a higher methylation in TRPA1 promoter. TRPV1, TRPV2, and TRPV3 were differentially methylated but the differences did not reach statistical significance. In addition to providing important insight into epigenetic regulation in pain states, this study also investigated the suitability of using whole blood for epigenome-wide studies for pain. Variability in white blood cell subtypes did not appear to have a major effect on differentially methylated regions of the top-ranked loci linked to pain. To further study whether blood is the most appropriate source, representing the tissue most relevant to disease such as nociceptors, dorsal root ganglia (DRG), or components of the CNS, the investigators used whole blood, multiple brain tissues, and skin to test whether highly differentially methylated regions had consistent levels of DNA methylation and gene expression. They concluded that many of the top-ranked differentially methylated regions showed consistent methylation or expression changes across blood and tissues relevant to pain [47]. Although this finding must be further validated for all genomic regions, the study is an important step forward for the pain field because blood is the source researchers are most likely to be able to obtain for these types of studies. This topic is discussed further under "Conclusions and future perspectives."

A few studies have investigated methylation-induced alterations in gene expression using rodent models of pain. Complete Freund's adjuvant (CFA)-induced inflammation resulted in significant upregulation of cystathionine-β-synthetase expression in DRG and was associated with the demethylation of its promoter [32]. In the peripheral nervous system, the DRG are major players responsible for conveying noxious stimuli from the peripheral to the CNS. RRBS of rat DRG and a genome-wide integrated methylome–transcriptome analysis showed that genes with low CpG content promoters were markedly hypermethylated when repressed and hypomethylated when active. Low and high CpG levels in the promoters showed opposite methylome–transcriptome associations [50]. Determination of global DNA methylation using luminometric methylation assay 6 months following peripheral nerve injury showed a decrease in DNA methylation in mouse prefrontal cortex and amygdala, and global methylation was significantly correlated with the severity of mechanical and cold sensitivity [51]. This study is important because aging-induced epigenetic changes may influence development or maintenance of pain.

Mutations in *DNMT1*, the maintenance methyltransferase, are now associated with one form of hereditary sensory neuropathy causing both central and peripheral neurodegeneration, associated with dementia and hearing loss [52]. All mutations were within the targeting-sequence domain of *DNMT1*, causing premature degradation of mutant proteins, reduced methyltransferase activity, and impaired heterochromatin binding. This mutation also resulted in global hypomethylation, with local, site-specific hypermethylation. This study demonstrated a direct link between DNMT1 alteration and a neurodegenerative disorder of both the central and the peripheral nervous systems [52].

Evidence now suggests a role for MeCP2 in pain sensitivity. Mutations within the *MECP2* gene cause Rett syndrome (RTT), a neurodevelopmental disorder that predominantly affects girls. Children with RTT syndrome display decreased awareness of pain and temperature [53,54]. RTT is a progressive postnatal neurological disorder the clinical hallmarks of which include a period of apparently normal early development followed by a plateau or stagnation and subsequent regression. Earlier reports of RTT described afflicted girls touching candle flames, and one study noted that parents and caregivers of these children observe decreased and delayed responses to injections, falls, trauma, and burns. Abnormal pain response in 75.2% of subjects and decreased sensitivity to pain in 65% were reported in a study comprising subjects with a pathogenic *MECP2* mutation [53]. In another study characterizing the relevance of *MECP2* overexpression to autism spectrum disorder (ASD)-related behaviors, boys with idiopathic ASD were compared with 10 age-matched boys with *MECP2* duplication syndrome [54]. Boys with *MECP2* duplication syndrome share the core behavioral features of ASD. Hyposensitivity to pain and temperature were part of the behavioral phenotype in children with *MECP2* duplication syndrome [54].

Mouse models have been pivotal for in-depth studies of the biological basis of *Mecp2* gene function. Several mouse models have been generated [55] based on manipulations of the endogenous gene or on targeted introduction of the human *MECP2* gene, either wild type or carrying RTT-associated mutations, and some mutant mouse studies have assessed pain behavior [56,57]. Increased paw withdrawal latency in a hot-plate assay was observed in mice expressing 50% of the wild-type level of MeCP2 and female Mecp2$^{+/-}$ mice [57,58]. Several *Mecp2* target genes were unregulated in the dorsal horn of the spinal cord when CFA was injected into the rat ankle joint. MeCP2 binding can lead to gene repression, and the release of MeCP2 due to phosphorylation in lamina I projection neurons results in upregulation of its target genes [59]. Further studies showed that the activity of MeCP2 in spinal cord was regulated by descending serotonergic pathways that are crucial in regulating gene expression in the dorsal horn and mechanical sensitivity associated with inflammatory pain states [34,60]. Another study investigating the role of MeCP2 in regulating pain and morphine reward showed that both persistent pain and repeated morphine administration upregulated MeCP2 in the central nucleus of mouse amygdala [61]. They confirmed that the expression of the histone dimethyltransferase G9a is decreased in the amygdala of transgenic mice overexpressing MeCP2, as was previously observed in hypothalamus [62]. G9a catalyzes the dimethylation of histone 3 at lysine 9 (H3K9me2), an epigenetic mark for transcriptional repression, and plays an important role in MeCP2-related mechanism of gene repression in cocaine-induced plasticity [63]. G9a protein was significantly reduced in mice with persistent pain [61]. The repressive role of MeCP2 on G9a expression was further confirmed by MeCP2 knockdown, resulting in increased levels of G9a protein, H3K9me2 marks, and G9a mRNA. Repression of G9a increased expression of brain-derived neurotrophic factor (BDNF), suggesting that MeCP2 may regulate BDNF indirectly through G9a. Thus MeCP2 and G9a induce a transcriptional derepression of *Bdnf* and thus play a role in promoting pain behavior through BDNF upregulation [61].

5. METHYLATION CHANGES IN RESPONSE TO DRUGS

A rapid increase in the number of genes that show epigenetic alterations in disease states indicates their importance in diagnosis, prognosis, and prediction of response to therapies [64]. Learning how aberrant placement of these epigenetic marks leads to disease will be crucial in developing better therapeutic intervention strategies [14]. The different types of epigenetics marks—erasing, writing (adding), or reading (interpreting)—can contribute to transcriptional changes observed in chronic pain. DNA methylation has been suggested to play a crucial role in modulating chronic pain and may underlie interindividual differences in the susceptibility to developing chronic pain [37]. Interest has increased in exploring the utility of DNMT inhibitors targeting enzymes responsible for inducing epigenetic changes, beyond the field of oncology [65]. Diet is known to affect pain sensitivity in rats [66] and a methylation-rich diet can modify behavior in rodents [7]. Although diet modulation cannot cure pain, it is interesting to ask whether diet can influence the efficacy of existing treatment by altering methylation marks in the DNA. MeCP2 plays an important role in the reading type of methylation marks in the DNA. Depending on its interacting protein partners and target genes, MeCP2 can act as either an activator or a repressor [67]. A role in dampening genome-wide transcriptional noise in a DNA methylation-dependent manner has also been attributed to MeCP2 [68]. DNA methylation is a dynamic modification, a reversible biological signal. As we noted, the study of epigenetics suggests that DNA is not destiny and that we may be able to alter gene expression and hence disease states by modulating these marks and thus their interpretation. Though global modulation of the epigenome is challenging because of the lack of selectivity and specificity, this approach can be thought of as a means of resetting the epigenetic landscape, rendering a more conducive environment for current therapies or complementing them to increase effectiveness.

Methylation changes can result from exposure to drugs. Chronic opioid use is associated with increased DNA methylation in methadone-substituted addicts and in pain patients on long-term opioid treatment. This study, performed in two independent cohorts, implied a causal relationship between chronic opioid exposure and increased DNA methylation [69]. However, direct proof of causality could not be obtained, as the assessment was cross-sectional [70]. Surprisingly, methylation of the *OPRM1* gene, coding for mu-opioid receptors, had no immediate effect on transcription [69]. Thus methylation status may not be the sole determinant of transcriptional activity. The discrepancy could also be due to differences in source material, as mRNA expression analyses were conducted in postmortem brain tissue, whereas DNA methylation was determined in peripheral blood cells.

In the same study, the investigators found that global DNA methylation at LINE1 (L1) was significantly correlated with increased chronic pain. L1 retrotransposons are repetitive elements in mammalian genomes capable of synthesizing DNA on their own RNA templates using reverse transcriptase that they encode. Abundantly expressed full-length L1's globally influence gene expression profiles and cellular differentiation, and their methylation changes are associated with cancer [71]. It has been hypothesized that increased methylation of L1 led to a decreased expression of nocifensive genes or to an increased expression of nociceptive genes [69], possibly explaining the correlation of DNA methylation with increased pain.

Another study investigated the effects of dietary methyl donor content on opioid responses in mice. Mice treated with high- and low-methyl-donor diets either in the perinatal period or after weaning were tested for their analgesic responses and tolerance to morphine and for opioid-induced hyperalgesia. Mice on a high-methyl-donor diet in the perinatal period had higher physical dependence. Analgesic responses to low doses of morphine were altered when the dietary treatments were given to the mice after weaning. Opioid-induced hyperalgesia was unaltered by dietary methyl donor content. The authors concluded that the effect of high- and low-methyl-donor diet treatment on opioid responses may vary depending on the timing of exposure [72]. This study provides yet another example of how diet may influence individual response to treatment.

Epigenetic aberrations can also affect drug treatment by modulating the expression of key genes involved in the metabolism and distribution of drugs as well as drug targets, thereby contributing to interindividual variation in drug response [73]. Thus in addition to genetic polymorphisms, epigenetic regulation of the ADME genes (related to absorption, distribution, metabolism, and excretion) and of drug targets can cause different responses to the same treatment. Changes in the DNA methylation of ATP-binding cassette transporter member 1, a major regulator of cellular cholesterol and phospholipid homeostasis, were observed in individuals who were exposed prenatally to wartime famine. These authors concluded that persistent changes in DNA methylation may be a common consequence of prenatal famine exposure and that these changes varied by sex and by the gestational timing of the exposure [74]. Thus the fact that our epigenome is influenced by our parents and grandparents, in addition to our own lifestyle choices, holds true and continues to be validated in different contexts of disease and therapy.

6. HISTONE MODIFICATIONS

Another widely described epigenetic mechanism involves histone modifications [75]. Nucleosomes, the building blocks of chromatin, comprise an octamer of four histone proteins, H2A, H2B, H3, and H4, around which are wrapped approximately 147 base pairs of DNA. Because the N-terminal tails of histone proteins protrude from the nucleosomal core, they are freely accessible to enzymes for the addition or removal of posttranslational modifications. A wide variety of posttranslational modifications occur on their N-terminal tails, including acetylation, methylation, phosphorylation, sumoylation, ubiquitination, and ADP ribosylation [75]. A number of enzymes are responsible for bringing about the posttranslational modifications, among them the histone acetyltransferases (HATs) and deacetylases (HDACs), the methyltransferases and demethylases, and the kinases and phosphatases. These modifications can contribute to the "open" or "closed" conformation of the chromatin. For example, HATs create a more open chromatin, whereas deacetylation induced by HDACs results in transcriptional repression. Aberrant chromatin-modifying enzymes can thus lead to altered histone modifications and disease states. HDAC inhibitors increase histone acetylation, and the resulting open chromatin conformation can allow access of transcription factors to DNA [76].

The multitudes of histone modifications, influencing either activation or repression of gene regulation, led to the hypothesis that the modifications constitute a code that is deciphered by the transcription factors, thus determining the transcriptional state of a gene. These histone modifications have context-dependent effects, and thus some consider their interplay to be more like a language within the chromatin signaling pathway than a code [77]. Thus whether histone modifications constitute a "code" has been debated, and some propose that key histone modifications are better understood as cogs in the machinery that regulates transcriptional elongation and heterochromatic silencing [78]. It has also been argued that the histone mark is neither "activating" nor "repressive" but rather facilitates nucleosome dynamics. The average genomic location of the midpoint of a nucleosome in a population of cells defines the nucleosome position, and nucleosome occupancy may exhibit very little variation or may be random. Thus nucleosome occupancy defines the frequency with which a nucleosome is found at a given position in a population of cells; incomplete occupancy suggests that a nucleosome moved or changes position, thus contributing to the variability in different cell types or in response to signaling events. The dynamic processes affecting nucleosomes result in patterns of histone modifications, which in turn affect the physical properties of nucleosomes and help to maintain the active or silent state of chromatin [78]. Modification of the terminologies used to define different epigenetic players has also been proposed. Some authors argue that the histone code terms "writers" and "readers" are poor metaphors for what these proteins do, because the "writers" do not write, but only modify one amino acid residue at a time, and "readers" do not read but only bind, also one residue at a time. "Modifiers" and "binders," respectively, have thus been proposed as more accurate terminology [78].

In mammals there are 18 HDAC genes that are grouped into four classes: class I (HDAC1, 2, 3, and 8), class II (HDAC4, 5, 7, 9 in IIa and HDAC6 and 10 in IIb), class III (sirtuins 1–7), and class IV (HDAC11) [79]. Given the role of histone acetylation and deacetylation in multiple facets of development and normal physiology, systemic inhibition of HDACs with pharmacologic inhibitors could result in nonspecific and harmful side effects as a consequence of global modulation of gene expression. However, compounds that broadly inhibit most or all HDACs are reasonably well tolerated in vivo and block numerous disease states, including cancer, cognitive dysfunction, immunological disorders, and neurodegeneration [80]. Several small-molecule HDAC inhibitors have been developed, and the majority of HDAC inhibitors that are currently either in clinical trials or on the market target multiple isoforms of the classic HDAC family (classes I, II, and IV). Although how reduced deacetylase activity can be beneficial in diverse pathophysiological conditions is unknown, this fact suggests that most of these diseases have an underlying epigenetic component of aberrant histone acetylation and that treatment with HDAC inhibitors resets the epigenetic memory of the cell to a predisease state [79].

7. HISTONE MODIFICATIONS IN PAIN STATES AND IN RESPONSE TO DRUGS

Investigations on whether posttranslational modifications of histones induce gene expression changes in pain states have been performed to elucidate epigenetic regulations relevant in nociception. Several investigators also employed multiple HDAC and HAT inhibitors to confirm that the observed effect can be reversed pharmacologically.

A 2011 study demonstrated how chronic pain can epigenetically suppress the transcription of Gad2, leading to impaired inhibitory function of GABAergic synapses in central pain-modulating neurons [81]. γ-Aminobutyric acid (GABA) is the predominant inhibitory neurotransmitter in the mammalian CNS. It is synthesized from glutamate by two isoforms of the rate-limiting enzyme glutamate decarboxylase, GAD67 and GAD65, which are encoded by the *Gad1* and *Gad2* genes, respectively. Epigenetic modification induced a decrease in *Gad2* transcription in persistent inflammatory and neuropathic pain in the rat brain stem nucleus raphe magnus, which is important for central mechanisms of chronic pain. This downregulation in transcription was due to decreased histone H3 acetylation in the transcription start site of the *Gad2* gene. The reduction in H3 acetylation was completely reversed by HDAC inhibitors in wild-type but not in *Gad2*-knockout mice, resulting in restoration of GABA synaptic function and histone hyperacetylation-attenuated pain [81].

Few studies have explored the role of HDAC inhibitors using rodent models of pain. Systemic administration of suberoylanilide hydroamic acid (SAHA) and MS-275 for 5 days significantly reduced the nociceptive response in the second phase of the formalin test in mice and upregulated the expression of mGlu2 receptors in DRG and spinal cord [82]. Upregulation of mGlu2 was mediated by increased acetylation of p65/RelA on lysine 310 and amplification of nuclear factor κB (NF-κB) transcriptional activity [82]. In another inflammatory pain model, preinjection of inhibitors targeting classes I and II including SAHA significantly delayed the thermal hyperalgesia induced by CFA in the hind paw [83]. Though MS-275 increased histone 3 acetylation in the spinal cord, similar to SAHA, it failed to alter the hyperalgesia [83]. Two different class I HDAC inhibitors (MS-275 and MGCD0103) significantly attenuated mechanical hypersensitivity in several different models of neuropathic pain and in a systemic model of antiretroviral drug-induced neuropathy [84]. However, drug treatment did not affect already established neuropathic pain, suggesting that histone acetylation may be involved in the emergence of hypersensitivity [84]. Increased histone acetylation in the spinal cord, but not in the DRG, suggests that these HDAC inhibitors exerted their effect centrally within the spinal cord [84].

In incisional model of pain, however, administration of SAHA significantly exacerbated incision-induced mechanical hypersensitivity. Conversely, anacardic acid, a HAT inhibitor derived from cashew nut shell, significantly attenuated incision-induced mechanical hypersensitivity and incision-induced hyperalgesic priming [85] (hyperalgesic priming refers to previous injury-induced vulnerability to exaggerated sensitization after subsequent injury). Acetylated histone H3 at lysine 9 (H3K9) was increased in spinal cord tissues after incision leading to increased association of H3K9 with the promoter regions of chemokine CC motif receptor 2 (CXCR2). CXCR2 is a proinflammatory receptor implicated in neuropathic and inflammatory pain. The expression of CXCR2 increased in spinal cord tissue after incision and SAHA treatment, suggesting that epigenetic regulation-induced CXCR2 signaling helps to control nociceptive sensitization after incision [85]. Our investigation of the effects of JNJ-26481585, a pan-HDAC inhibitor, showed that this compound, in clinical trials as a chemotherapeutic agent, enhanced mechanical sensitivity in mice. Transcriptional profiling of spinal cord from JNJ-26481585-treated mice indicates an overlap in mechanisms underlying neurotoxicity caused by other known chemotherapeutic agents [86]. Thus the effects of HDAC inhibitors may differ based on the properties of the compound studied.

Opioid-induced hyperalgesia, resulting from the long-term use of opioids for the treatment of pain, was prolonged for weeks if HDAC activity was inhibited during opioid treatment [87]. Chromatin immunoprecipitation (ChIP) assays demonstrated that promoter regions of *Pdyn* (prodynorphin) and *Bdnf* were associated with acetylated H3K9 after morphine and SAHA treatment. Morphine treatment caused an increase in spinal BDNF and dynorphin levels, and these levels were further increased in SAHA-treated mice [87].

The relationship between histone H3 modification and the upregulation of CCL2 and CCL3 in infiltrating macrophages was investigated in the injured peripheral nerve using a partial sciatic nerve ligation model of neuropathic pain in mice [88]. Using the ChIP assay it was demonstrated that the levels of acetylated H3K9 and lysine 4 trimethylated H3 in the promoter regions of the CCL2 and CCL3 chemokine genes were increased in the injured sciatic nerve. Furthermore, the upregulation of CCL2, CCL3, CCR2, CCR1, and CCR5 was suppressed by the HAT inhibitor anacardic acid. These studies thus demonstrated that histone modifications can mediate the upregulation of chemokines contributing to the development of neuropathic pain [88].

Two distinct but functionally related proteins, p300 and CREB-binding protein (CBP), that belong to the HAT family mediate gene expression in eukaryotes [75]. Curcumin is a p300/CBP inhibitor of HAT activity. The antinociceptive role of curcumin and its effect on the release of pronociceptive molecules BDNF and cyclooxygenase-2 (Cox-2) was investigated using a chronic constriction injury (CCI) model of neuropathic pain in rat spinal dorsal horn [89]. ChIP assay demonstrated that curcumin reduced the recruitment of p300/CBP and acetyl-histone H3/acetyl-histone H4 to the promoter of the BDNF and Cox-2 genes in a dose-dependent manner. BDNF and Cox-2 decreased in the spinal cord after curcumin treatment. Thus the therapeutic effect of curcumin in alleviating neuropathic pain could be mediated by the downregulation of p300/CBP HAT activity-induced gene expression changes in BDNF and Cox-2 [89].

Spinal cord injury results in loss of motor, sensory, and autonomic functions and often causes long-term deficits including chronic pain. Some researchers have proposed that enhancing the epigenetic plasticity of key cells participating in repair at different stages of recovery from spinal cord injury might harness the cells' developmental capacity to drive repair after injury [90]. A 2013 study established a correlation between diminished axon growth potential and histone 4 (H4) hypoacetylation. H4 acetylation is restored when neurons are triggered to grow and transcription of regeneration-associated genes is initiated [91]. Although these studies are in the early stages, it is conceivable that epigenetic alterations occurring over the course of the disease may offer a novel avenue for therapeutic interventions. Interestingly, although two different class I HDAC inhibitors (MS-275 and MGCD0103) significantly attenuated mechanical hypersensitivity after peripheral nerve injury by 40%–50% in rodents, pretreatment with HDAC inhibitor before the insult was necessary for the observed analgesic effect to occur [84]. When HDAC inhibitors were administered after L5 spinal nerve transection, MS-275 could no longer attenuate mechanical thresholds, suggesting that HDAC inhibitors, at least in this paradigm, hold little promise as therapeutic targets [84]. This effect could be specific to this class of compound, but additional studies are needed to explore the therapeutic potential of HDAC inhibitors for pain therapy.

Two studies investigated the role of sirtuin 1 (SIRT1) in pain states. SIRT1 is a class III NAD^+-dependent HDAC that plays an important role in cell survival, inflammation, energy metabolism, and aging [92]. Chronic morphine tolerance induced by repeated administration of morphine in rats resulted in the downregulation of SIRT1 and an increase in H3 acetylation in the dorsal horn of the spinal cord. Resveratrol treatment, which is a classic agonist of SIRT1, increased SIRT1 expression, suppressed global H3 acetylation compared to the group with morphine tolerance, and reversed morphine antinociceptive tolerance [93]. SIRT1 expression was also decreased in the spinal cord from a mouse model of CCI, again suggesting that the reduction in SIRT1 deacetylase activity may be a contributing factor in the development of neuropathic pain. Using a pharmacological approach, this study also showed that upregulating SIRT1 activity may be a useful therapeutic intervention strategy for treating neuropathic pain [94].

A beneficial role for HDAC inhibitors in attenuating nociception induced by prolonged exposure of the amygdala to corticosteroids was shown when attenuation of corticosteroids induced anxiety-like behavior, somatic allodynia, and visceral hyperalgesia in rats [95]. Coordinated action of the NAD^+-dependent protein deacetylase sirtuin 6 and NF-κB resulted in the sequestration of glucocorticoid receptors and thus the release of corticotropin-releasing factor. This was mediated by deacetylation of H3K9, indicating epigenetic modulation of chronic anxiety and pain by histone deacetylation [95].

C-fiber dysfunction underlies negative symptoms in neuropathic pain and is accompanied by long-lasting downregulation of the Nav1.8 sodium channel in DRG. There is now evidence for epigenetic repression of the Nav1.8 gene as a key factor in C-fiber hypoesthesia (injury-induced insensitivity) [96]. The injury-induced decrease in Nav1.8 expression and C-fiber hypoesthesia were reversed by HDAC inhibitors in a dose-dependent manner. There was an increase in histone acetylation at the regulatory sequence of Nav1.8. These authors also investigated a few other genes important for C-fiber nociception. Nerve-injury-induced repression of TRPA1 and TRPM8 was reversed by SAHA treatment, but not that of calcitonin gene-related peptide, indicating that epigenetic regulation can differ among genes [96].

HDACs lack intrinsic DNA-binding activity and are recruited to target genes via their direct association with transcriptional activators and repressors, as well as their incorporation into large multiprotein transcriptional complexes [79,97]. Thus, the specificity of HDACs depends on cell type, availability of interacting partner proteins, and the signaling milieu of the cell [79]. The utility of HDAC inhibitors will depend on the disease itself. Lack of specificity and side effects may be more acceptable when the goal is an increase in life span for diseases such as cancer, whereas the approval of a novel therapeutic agent for pain therapy requires a profile that is superior to existing options. As more selective HDAC inhibitors become available, these drugs will undoubtedly find broader applications in treating a variety of nononcological disorders, including inflammatory and neuropathic pain. Another aspect to reconcile is that, depending on the rodent model, route of administration, and doses used, both HDAC and HAT inhibitors were found to attenuate or increase hypersensitivity. Thus assigning analgesic or pronociceptive attributes to compounds capable of inducing global acetylation changes should be considered with caution. These compounds often also regulate the acetylation status of a variety of other nonhistone substrates [98]. Ultimately, it may depend on the specific epigenetic (see nucleosome positioning below) and transcriptional status of the cells and tissues being exposed to the drug to determine the region to be acetylated or deacetylated, triggering a downstream signaling cascade that we monitor as the behavioral end point, which is altered pain sensitivity. As more global epigenome maps and selective compounds become available, we will have a better understanding of the pathways through which these drugs mediate their effects.

8. NUCLEOSOME POSITIONING AND CHROMATIN REMODELERS

The fundamental repeating unit of eukaryotic chromatin is the nucleosome. Thus the nucleosome is the smallest packaging unit of the chromatin, consisting of 147 bp of DNA that is sharply bent and tightly wrapped in nearly two superhelical turns around the histone core comprising two copies each of the four canonical histone proteins (H3, H4, H2A, and H2B). A number of factors, ranging from DNA sequence to protein complexes, affect nucleosome packaging [99–102]. Nucleosomal DNA is much less accessible to many proteins, including transcription factors because of (1) the steric hindrance due to the proximity of the histone core and (2) the sharp bending and altered helical twist of the nucleosomal DNA, which differ from the conformations favored by DNA-binding proteins. However, DNA in the linker region between the nucleosomes is much more accessible. In general, promoters and other regulatory sequences are present in nucleosome-depleted regions, whereas transcribed DNA tends to be located in high-density nucleosomal arrays. Thus the nucleosome positioning pattern and its dynamic regulation contribute to the differences in transcriptional activity in different cell types. DNA sequences differ in their abilities to bend and change their helical conformation and this can lead to a wide range of sequence-dependent affinities for nucleosome formation. The position of a nucleosome relative to the underlying DNA sequence is typically identified using micrococcal nuclease, which preferentially digests linker DNA before nucleosome-protected DNA. Nucleosomes are not static; movement of a given histone octamer in reference to the underlying DNA is referred to as sliding, while exchange in and out of chromatin is called replacement [102]. Cells contain diverse sets of chromatin remodeling complexes (remodelers) that work in concert with other factors influencing nucleosome positioning to provide access to DNA in a regulated manner to implement gene transcription and DNA replication, repair, and recombination [103]. They have an ATPase domain and use the energy of ATP hydrolysis to slide nucleosomes laterally or eject them from DNA [99,103]. There are four different families of chromatin remodeling complexes [103]. The SWI/SNF family remodelers slide and eject nucleosomes at many loci and for diverse processes but lack roles in chromatin assembly; the ISWI family remodelers contain two to four subunits and they optimize nucleosome spacing to promote chromatin assembly and the repression of transcription. Some of the CHD family remodelers slide or eject nucleosomes to promote transcription but others have repressive roles. The INO80 family remodelers contain more than 10 subunits and have diverse functions, including promoting transcriptional activation and DNA repair. SWR1, although related to INO80, is unique in its ability to restructure the nucleosome by removing canonical H2A–H2B dimers and replacing them with H2A.Z–H2B dimers [103]. There is a paucity of studies investigating the roles of nucleosome position and remodelers in pain states. Studies investigating nucleosome positioning, chromatin remodelers, and nonhistone chromatin factors will enable the construction of genomic maps of nucleosomes under chronic pain states and can improve our understanding of how these epigenetic changes can bring about gene expression changes.

9. GENOMIC IMPRINTING

Genomic imprinting is the process that causes monoallelic expression (expression from one of the two parental chromosomes) of a subset of genes [104]. Both maternal and paternal genes can be imprinted or epigenetically marked. There are more than 100 imprinted genes and most of them are organized in clusters. These clusters are regulated in a coordinated manner by a single imprinting control region and contain at least one noncoding gene and multiple protein-coding genes [105]. The role of differential DNA methylation in genomic imprinting is well established but allele-specific histone modifications, large noncoding RNAs, and alterations in chromatin can also contribute to the epigenetic changes [105]. Imprinted expression can differ between tissues, developmental stages, and species.

There are a number of human disorders associated with aberrant genomic imprinting, one of which is Prader–Willi syndrome, a neurodevelopmental disorder with deficits in thermal and mechanosensory perception [106]. The necdin gene on chromosome 15q11–q12 is maternally imprinted, paternally transcribed, and not expressed in Prader–Willi syndrome. Necdin is highly expressed and important in the development of nerve growth factor (NGF)-dependent sensory neurons. Mice lacking the paternal necdin allele displayed significantly higher tolerance to thermal pain, which is often seen in individuals with Prader–Willi syndrome. This suggests that paternally expressed necdin facilitates TrkA signaling to promote the survival of NGF-dependent nociceptive neurons [107].

10. NONCODING RNAS

The actions of noncoding RNAs are often considered a third mode of epigenetic regulation. Protein-coding exons comprise less than 2% of the human genome sequence [108] and findings from high-throughput sequencing have led researchers to describe transcription of mammalian genomes as "pervasive" [109]. The GENCODE Consortium, within the framework of the ENCODE project, aims to identify all gene features in the human genome using a combination of computational analysis, manual annotation, and experimental validation [110]. These researchers have identified 20,078 messenger RNAs [110], and in the most complete human long noncoding RNA (lncRNA) annotation to date they have demonstrated 9277 manually annotated genes producing 14,880 transcripts [111]. Their analyses also showed that lncRNAs are generated through pathways similar to those of protein-coding genes and that they have similar histone modification profiles, splicing signals, and exon/intron lengths [111]. LncRNAs are longer than 200 nucleotides, a distinction that originated based on RNA purification protocols that exclude small RNAs. Noncoding RNAs are further classified based on their biogenesis, structure, and function, and in addition to lncRNAs they include transfer RNA, ribosomal RNA, microRNA (miRNA), PIWI-interacting RNA, small nucleolar RNA, and antisense RNA [112]. Much effort has been dedicated to elucidating the function of miRNAs that regulate gene expression by binding to mRNA and inducing translational repression or RNA degradation [10]. The role of these small, noncoding endogenous miRNAs as a new class of regulators of gene function [11,113] and as biomarkers [114] is now well established. DNA methylation and histone modification-induced epigenetic regulation leading to a modulation in expression are not limited to protein-encoding genes but affect miRNAs too. Thus DNA methylation and histone modifications can influence the levels of miRNAs transcribed from the genome, and aberrant DNA methylation and histone modifications are major causes of miRNA dysregulation [115]. The converse is also true, with miRNAs regulating DNMTs and histone deacetylases [116]. miRNAs can also be effectors of epigenetic modifier genes, such as *MeCP2*, leading to global changes in the epigenome [117]. This results in a complex network of interactions, the elucidation of which would be highly beneficial for better understanding of diseases including pain and might provide valuable insights into the molecular mechanisms underlying signaling under both normal physiological conditions and disease states.

11. MICRORNA AND PAIN

Earlier work in the pain field primarily focused on identification of dysregulated miRNAs in various rodent models of pain [118–123]. The first study to demonstrate a role for miRNAs in mediating pain used a conditional knockout model to delete the enzyme Dicer, an enzyme that is critical for miRNA biosynthesis, in a subset of sensory neurons that are Nav1.8 positive [124]. Contrary to expectations, these animals had a reduction in several nociceptor-associated transcripts and exhibited attenuated inflammatory pain [124]. Because it is well established that miRNAs play a significant role in pathophysiological stress signaling [11,113], various strategies for overexpression or downregulation of specific miRNAs are being pursued for therapeutic intervention for various diseases including pain. Studies have shown that miRNAs can reverse hyperalgesia in various rodent models of pain [125–129]. Cav1.2-comprising L-type calcium channel (Cav1.2-LTC) plays a crucial role in neuropathic pain. miR-103 simultaneously regulated the expression of all three of the subunits forming Cav1.2-LTC, and this effect was bidirectional given that knocking down or overexpressing miR-103 increased or decreased the level of Cav1.2-LTC translation, respectively. Knockdown of miR-103 in naïve rats resulted in hypersensitivity to pain, and its intrathecal delivery attenuated pain [125], indicating that miR-103 may be a therapeutic target for chronic neuropathic pain. miR-124a is involved in inflammatory nociception and MeCP2 is one of its targets [127]; an miRNA-124a mimic reversed formalin-induced inflammatory pain and downregulated MeCP2 [127]. miR-203, a skin- and keratinocyte-specific miRNA with a role in skin inflammation, was downregulated in paw keratinocytes, and both the mRNA and the protein levels of its target gene phospholipase A_2 activating protein were upregulated, in a postoperative incisional pain model [130].

A 2013 study demonstrated the potential for miRNA therapy in alleviating cancer pain [131]. In DRG from a mouse model of metastatic bone cancer pain, 57 miRNAs involved in the maintenance but not in the development phase of tumor-mediated hyperalgesia were deregulated. *Clcn3*, a gene encoding a chloride channel, was identified and validated as a key miRNA target in sensory neurons, contributing to tumor-induced nociceptive hypersensitivity in vivo [131]. Another study showed a role for miR-7a in alleviating the maintenance of neuropathic pain through regulation of neuronal excitability [132]. miR-7a targets the β2 subunit of the voltage-gated sodium channel, and reduction in miR-7a was associated with neuropathic pain and an increase in β2 subunit protein expression. This study also suggests that an overexpression of miRNA (here miR-7a) is a potential therapeutic strategy in pain therapy [132]. Although these studies highlight the potential for miRNA therapy, the challenges associated with miRNA delivery and design of effective synthetic RNAs remain. In the CFA model, miR-219 was downregulated in spinal cord, and CFA-induced thermal hyperalgesia, mechanical allodynia, and spinal neuronal sensitization could be prevented and reversed by the overexpression of miR-219 [133]. Downregulation of miR-219 induced thermal hyperalgesia and mechanical allodynia, suggesting that miR-219 plays a role in the modulation of chronic pain. This study linked two different epigenetic mechanisms, DNA methylation and noncoding RNAs, by demonstrating that chronic inflammatory pain can induce hypermethylation of CpG islands in the miR-219 promoter. A demethylation agent increased miR-219 and decreased CaMKII expression, resulting in alleviation of pain in the CFA model [133].

miRNAs are known to act as physiological ligands for Toll-like receptors (TLRs) [134], and TLR activation leads to production of inflammatory cytokines. A 2014 study reported a role for extracellular miRNA let-7b in the activation and excitation of nociceptor neurons [135]. Binding of let-7b to TLR7 at the plasma membrane of nociceptive neurons activated TRPA1 channels. Intraplantar injection of let-7b produced rapid spontaneous pain. Let-7b-evoked nocifensive pain was abolished in TLR7- and reduced in TRPA1-knockout mice and was abrogated by coadministration of a TRPA1 antagonist, demonstrating signaling via TLR7 and TRPA1. This study suggests that miRNAs enriched in DRG, such as let-7b, could be released to the circulation in an activity-dependent manner.

The presence of stable miRNAs in body fluids and the noninvasive nature of exploring biomarkers greatly enhanced interest in pursuing novel biomarkers for both physiological and pathological conditions. The purification and quantification of circulating miRNAs have improved over the years and quantitative RT-PCR has enabled detection of low-abundance miRNAs. Lack of reliable normalizers that can be used as universal standards is to be expected considering the dynamic nature of the composition of circulating RNAs. It is still necessary to standardize the methodologies, including sample collection, storage, normalization, and data and statistical analysis, as well as the clinical parameters, before an miRNA can be developed into a reliable biomarker that can be used routinely. A central database indicating the levels of miRNAs in both healthy male and female populations, stratified by ethnicity, age, and geography, would serve as a useful reference and would greatly reduce the challenges associated with data reproducibility. In one of the first studies investigating circulating miRNAs in whole blood, we profiled miRNAs in patients suffering from complex regional pain syndrome (CRPS). CRPS is a chronic neuropathic pain syndrome characterized by pain, inflammation, and aberrant sensory motor and trophic disturbances. The pathophysiology of CRPS is not completely understood, but its complex multifactor pathogenesis includes inflammatory, vascular, sympathetic nervous system, cortical, and spinal mechanisms. Diagnosis is based solely on clinical observations. We therefore explored the utility of circulating miRNAs in patient stratification. Of the three different groups that emerged from miRNA profiling, one group comprised 60% of CRPS patients; miRNA profiles from the remaining patients were interspersed among control samples from the other two groups. These findings suggest that clinically relevant patient stratification is possible on the basis of alterations in miRNA expression. Eighteen miRNAs were significantly different in

CRPS patients. Analysis of inflammatory markers showed that vascular endothelial growth factor, interleukin-1 receptor antagonist, and monocyte chemotactic protein-1 were significantly elevated in CRPS patients. Several other markers exhibited trends that did not reach significance. Analysis of the patients who were clustered according to their miRNA profile revealed correlations that were not significant in the total patient population, again indicating that miRNA profiles can be useful in grouping patients and in identifying additional biomarkers, the significance of which may be diminished when the patient population is considered overall. This is particularly important in CRPS, a syndrome with differing symptoms. Stratification based on molecular signature might provide insights into treatment options. Correlation analysis of miRNAs with clinical parameters identified miRNAs associated with comorbidities such as headache and use of narcotics and antiepileptic drugs [136]. Differentially expressed miRNAs can provide molecular insights into gene regulation and could lead to new therapeutic intervention strategies for CRPS.

Aberrant miRNA expression has been identified in both bodily fluids and tissue biopsy samples from patients with other painful conditions. In patients with bladder pain syndrome, miR-199a-5p has been suggested to play a role in the regulation of urothelial permeability [137], and expression of miR-449b and miR-500 in the bladder smooth muscle cells is increased [138]. Profiling of cerebrospinal fluid from 10 women with fibromyalgia and 8 healthy control subjects showed that the expression of nine miRNAs was significantly lower in patients than in controls [139]. miR-146a is upregulated in different cell types and tissues in patients with rheumatoid arthritis, a systemic autoimmune disorder characterized by the inflammation of synovial tissue [140], as well as in osteoarthritis [141].

We have currently undertaken another study to investigate whether intravenous ketamine treatment induced miRNA alterations in CRPS patients. We have identified significant differential expression of miRNAs distinguishing good and poor responders to treatment, both before and after treatment. Thus our preliminary data on 16 patients with CRPS indicate that miRNA signatures may be useful in predicting treatment response. The prospect of miRNAs serving as biomarkers that permit patient stratification and prediction of treatment response would be of tremendous value in the development of pain therapeutics. In addition to its utility in prognosis, miRNA-based patient stratification might serve as selection criteria for clinical trials and thereby increase the chances of a successful outcome by targeting the appropriate population. Experimental paradigms that are more predictive of trial outcomes are critically needed [142,143]. Patient stratification based on underlying molecular mechanisms will be extremely valuable in better understanding the disease and in choosing treatment options. The ability to predict treatment outcome by a simple qPCR-based blood test could revolutionize both treatment and clinical trial design. Efficacy trials could be conducted in mechanistically defined patient groups, guided by information obtained in preclinical and human volunteer models. Proof-of-concept studies evaluating the feasibility and utility of assessing miRNA signatures, performed using existing clinical trial samples, can thus help us determine the predictive validity of this approach. Specifically, assessing miRNA changes before and after treatment will provide insight into miRNA signatures and its alterations in good responders, nontreatment (placebo) responders, and poor responders. The high rate of failure of drugs targeted to treat neuropathic pain warrants urgent changes in the manner in which new analgesics are discovered and developed [142,143]. Studies targeting well-defined patient populations for clinical trials and delineating placebo response are crucial in developing drugs that may be efficacious in a subset of patients. Pharmaceutical companies can undertake retrospective investigations using stored plasma or serum samples from completed (unsuccessful or successful) clinical trials. It will be a worthwhile exercise to determine whether a patient's miRNA signature differed even prior to treatment initiation and whether that signature correlated with treatment outcome. From a mechanistic perspective, identifying the potential target genes for individual miRNAs could provide insights into the molecular bases of different types of pain.

Much of the focus in biomarker discovery in the pain field has been on secreted inflammatory mediators [144]. While these are enormously useful, combining cytokine and chemokine data with miRNAs will undoubtedly be beneficial in determining treatment options and better understanding the disease. Unlike cancer biology, in which researchers have access to tissue samples, CNS disorders including pain are dependent on biological markers that can be identified from bodily fluids, predominantly blood. Identifying several miRNAs as biomarkers, rather than relying on one specific molecule or parameter, may increase the chances of successful treatment in an extremely heterogeneous group of patients suffering from pain. Although pain is the common denominator in a variety of disorders, the altered processing of sensory stimuli leading to allodynia, hyperalgesia, and the underlying neurobiological mechanisms differ. Thus it is reasonable to expect overlap in miRNAs altered in different painful conditions, but every disorder, including different painful conditions, may have its own unique miRNA signature.

The identification of noncoding RNAs in the circulation within extracellular vesicles [145] has also generated enormous interest both from a biomarker perspective and as a novel avenue for delivery of therapeutic agents [146]. Intercellular communication can be mediated through direct cell–cell contact or through transfer of secreted molecules. A third mode of communication involving extracellular vesicles has now been recognized to play an important role in intercellular information transfer [145,147,148]. This method of transport of biomolecular cargo has gained significant attention because of the presence of RNAs, proteins, and lipids in these vesicles. It is now established that uptake of these vesicles can impart functional consequences in the recipient cells.

Exosomes are extracellular vesicles 30–100 nm in size and are derived from the endocytic compartment of the cell. Exosomes are generated by the inward budding of an endosomal membrane that results in the formation of multivesicular bodies (MVBs) [149,150]. MVBs can then follow either the secretory pathway or the lysosomal pathway. In the secretory pathway, MVBs fuse with the plasma membrane, releasing the exosomes either to the adjacent cell or into the circulation. In the lysosomal pathway, MVBs fuse with lysosomes and are thus degraded. Exosomes are released by various cell types, reticulocytes, and platelets and are present in most bodily fluids [151].

The protein and lipid composition of exosomes differs from that of other types of extracellular vesicles including apoptotic bodies. In fact, exosomes are enriched in certain proteins and lipids and they are used as markers to differentiate exosomes from other types of extracellular vesicles, organelles, or cellular debris [152,153]. Exosomal composition is reflective of the cell secreting them. Changes due to infection, activation, inflammation, or transformation (in tumor cells) will influence and alter the composition of the exosome and thus are the basis for the biological consequences upon its uptake as well as its biomarker utility. However, not everything that is present in the parent cell is incorporated into the exosomes, suggesting that this well-regulated process is dynamically altered by signaling cues. The precise molecular mechanisms and mediators responsible for determining the exosomal contents are still unknown. Manipulation of the parent cells can influence the packaging of biomolecules, and because exosomes can cross the blood–brain barrier, artificial manipulation of exosomal content, whether loading with nucleic acids or small molecules, is being explored for therapeutic intervention, including in CNS disorders [145].

Exosomes derived from antigen-presenting cells have been found to be protective in painful conditions. Dendritic cell-derived exosomes suppress the onset of collagen-induced arthritis in mice and reduce the severity of established arthritis [154]. Exosomes derived from dendritic cells expressing IL-10 and indoleamine 2,3-dioxygenase (IDO) increased the effectiveness of exosomal intervention [154,155]. Thus exosomes secreted by both naïve and IDO-expressing dendritic cells were immunosuppressive and anti-inflammatory and successfully reversed established arthritis. We have shown that a single intraplantar injection of exosomes derived from a mouse macrophage cell line into a CFA-induced mouse model of inflammatory pain significantly reduced paw swelling. Macrophage-derived exosomes also attenuated thermal hyperalgesia associated with CFA-induced inflammatory pain [156]. Thus exosomes derived from antigen-presenting cells such as dendritic cells and macrophages are capable of therapeutically relevant adaptive immune responses. Preclinical studies exploiting the immunogenicity of dendritic cell-derived exosomes for anticancer therapy are ongoing [157].

Release and uptake of exosomes between neurons have been demonstrated in vitro [158]. Secretion of exosomes in an activity-dependent manner could thus mediate the interneuronal transfer of information, allowing anterograde and retrograde signaling across synapses, necessary for plasticity [158]. Exosomes can carry pathogenic proteins linked to neurodegenerative disorders such as Parkinson, prion, and Alzheimer diseases, suggesting that exosomes might contribute to the spreading of the pathology throughout interconnected cortical areas [158].

A reciprocal cell communication in the nervous system has been identified with the signal-mediated transfer of exosomes from oligodendrocytes to neurons. The report demonstrated that exosome-mediated glia–neuron communication is bidirectional and is important for supporting metabolism and for maintaining neuronal integrity under conditions of cellular stress [159,160].

Another interesting aspect of exosomes is the high abundance of transcripts mediating epigenetic regulation. Our transcriptome analysis of exosomes derived from naïve or lipopolysaccharide (LPS)-stimulated RAW 264.7 cells by next-generation sequencing showed that the majority of the transcripts encode proteins, and various subunits of histone H1 were robustly represented in unstimulated exosomes [156]. "Reads" that mapped to miRNAs were more abundant after LPS stimulation. We hypothesize that mature miRNAs may fine-tune the regulation of inflammation by altering the mRNA levels of inflammatory proteins immediately and that the pre-miRNAs offer a second wave of regulation at a later time, enabling temporal epigenetic regulation in recipient cells [156]. Exosomes thus confer the advantage of delivering proteins and miRNAs that are primed to act directly and immediately along with mRNAs that can be readily translated, enabling regulation of the course of inflammatory gene expression.

Though the mechanisms that determine the loading of exosomes and the means by which they reach the recipient cells at distant locations though the circulation are not completely known, pharmacological interventions capable of selectively inducing the production of exosomes that are protective represent a novel strategy to be explored. Conversely, blocking the uptake of specific exosomes that may be pathogenic might be beneficial. Though much work lies ahead, exosome biology will undoubtedly open novel avenues in the understanding of normal physiology, disease mechanisms, and therapeutic intervention strategies.

12. LONG NONCODING RNA AND PAIN

Compared with miRNAs, the functions of most of the identified lncRNAs remain largely uncharacterized. There appear to be no unifying structural, biochemical, or functional characteristics that define a given transcript as a lncRNA [161]. In addition to binding to proteins, lncRNAs can bind DNA or RNA in a sequence-specific manner. As opposed to miRNAs, which negatively regulate gene expression, lncRNAs can influence gene expression either negatively or positively. Gene regulation may occur *in cis* (in proximity to the transcribed lncRNA) or *in trans* (at a distance from the transcription site). Many lncRNAs are processed to directly yield small RNAs or they can modulate how other RNAs are processed. In addition to direct transcriptional regulation, they are involved in recruitment of chromatin-modifying complexes to appropriate genomic loci [162,163]. A role for lncRNA in the induction and maintenance of neuropathic pain was reported in an elegant study [164]. This study identified a role for Kcna2 antisense RNA in the regulation of Kcna2 channel expression and in neuronal excitability. Expression of Kcna2 antisense RNA increased in ipsilateral DRG after nerve injury, and Kcna2 mRNA and protein decreased. Binding of the transcriptional activator myeloid zinc-finger protein 1 to the promoter of Kcna2 antisense RNA induced its activation. Furthermore, in vivo overexpression or blocking of Kcna2 antisense RNA in DRG led to alleviation of neuropathic pain in rats [164].

13. CONCLUSIONS AND FUTURE PERSPECTIVES

The main challenges in epigenetics research include access to diseased cells and the heterogeneity in tissue or cells. Thus human epigenomes pose a fundamental challenge, especially for studies in CNS disorders, including pain. Furthermore, in a single cell, the epigenome changes in response to the environment over time. A single organ (or tissue) consists of numerous cell types with different epigenomes, and disease can result from the dysfunction of a specific or multiple cell types. A variety of endogenous and exogenous etiological factors contribute to epigenetic changes leading to heterogeneity of disease processes. Human diseases are typically very complex processes, involving alterations in epigenomes, transcriptomes, proteomes, and metabolomes. The critical need for a complete environmental exposure assessment in epidemiological studies led to the concept of the exposome. The exposome is defined as the composite of every exposure to which an individual is subjected from conception to death [165]. Thus the unique genome of an individual and the distinct combination of exposome, epigenomes, transcriptomes, proteomes, and metabolomes in specific cell types makes disease processes in each human unique and provides the basis for the concept of the "unique disease principle" [166].

Many epigenetic studies rely on cells from blood to investigate alterations in other diseased cell types. Though this approach is not ideal, most chronic pain conditions have an inflammatory component that may be reflected in the epigenome of components from blood. For personalized medicine to be feasible and cost-effective, development of biomarkers that can be reliably assessed in bodily fluids is crucial. These markers may include nucleic acids, proteins, and metabolites or a combination that is reflective of genomic variations and epigenetic modifications. Noninvasive procurement of adequate quantities of sample specimens that can be stored for future use or retrospective studies will be ideal. Thus blood is emerging as the material of choice. It is encouraging that in one of the first studies investigating methylation changes in individuals with different heat pain thresholds [47], a significant positive correlation was observed between mean levels of methylation in blood and brain across the 100 top-ranked differentially methylated regions [47]. Similar methylation levels in the blood and in inaccessible tissues relevant to the disorder of interest will greatly enhance the validity of these epigenetic markers as diagnostics.

Compiling data from selected patient populations along with controls can serve as a starting reference and epigenetic alterations can be tracked over a period of time to capture the influence of chronic pain and/or lifestyle changes in these individuals. Rapid advances in technology may make feasible the detection of molecules that is not currently possible. Thus biobanking of samples will ensure that retrospective studies can be performed when needed. To realize the full potential of high-throughput data from "omics" research, collaboration is crucial among clinicians and basic researchers including biologists, statisticians and computational biologists, pharmaceutical and diagnostic companies, patients, and policymakers. In addition to finding the right patient population, finding the right molecules, targets, and doses will lead to better treatment outcomes (Figure 2). Identifying patients likely to experience pain relief without side effects and predicting abuse risk prior to initiating opioid therapy, though highly desirable, remain huge unmet needs. Data-based personalized prescribing of opioid analgesics as a means to achieve this goal has been discussed before, and several recommendations have been proposed [167]. Most acknowledge that personalized medicine is years away from being realized in practice. Leadership, guidance, and funding from organizations such as the National Institutes of Health and the International Association for the Study of Pain will undoubtedly be crucial in transforming personalized pain medicine. Ideally, preventive recommendations or guidelines tailored to individuals based on available biomolecular signature or risk or susceptibility factors, including lifestyle variations influencing epigenetic traits, will be of enormous value.

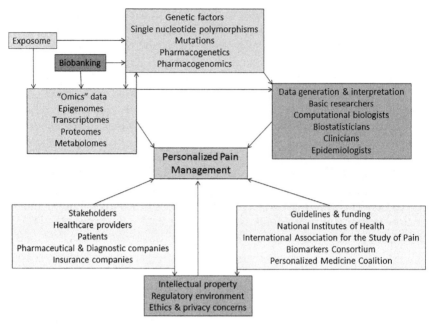

FIGURE 2   Various aspects of personalized medicine. The goal of personalized medicine is to establish patient-specific diagnosis, treatment strategies, or preventive measures based on the unique genetic and environmental profile of the individual. An integrated approach encompassing all aspects of disease mechanisms and partnerships among the various stakeholders will be essential for the successful realization of personalized medicine.

Accurate diagnostic tests that are specific and sensitive enough to identify patients who can benefit from targeted therapies will be crucial for personalized medicine to succeed. Clinical trials should be conducted on selected patient populations that have molecular markers associated with drug targets. Clinically beneficial new products and approaches successfully incorporated into clinical practice will be the measure of progress in making personalized medicine a reality [6]. Most chronic conditions such as pain are complex, and causes are multifactorial, genetic, and epigenetic. Alterations in genetic and epigenetic mechanisms may be responsible for both the development and the chronic maintenance of pain. A convergence of different epigenetic mechanisms underlying pain can be simultaneous or sequential. Changes in DNA methylation in response to pain or chronic drug exposure can lead to modulations in multiple gene-regulatory networks. Conversely, pain and treatment can lead to epigenetic changes that in turn can trigger additional biological responses that may help in alleviating or maintaining pain. Understanding the role of influences mediated by different epigenetic players will help us study a confluence of epigenetic factors pertaining to pain states. It has been hypothesized that an injury or disease might result in a type of molecular memory that could affect a person's risk of developing chronic pain later in life [168].

Tremendous genetic progress has led to the identification of targets for therapeutic intervention [3,5,169]. However, current treatment options are far from optimal. Several classes of drugs, though moderately effective, do not provide complete or near-complete relief. Very few resources have been directed toward developing an understanding of the role of epigenetics underlying pain. Increasing evidence points toward epigenetic cross talk involving DNA methylation and histone acetylation, though the hierarchical order of epigenetic events and dependencies resulting in gene silencing is not completely known [9]. The dynamic nature of posttranslational modifications of histone tails is crucial in maintaining accessibility of regulatory DNA. Comparative genome-wide gene expression analyses, coupled with miRNA expression and DNA methylation profiling, will enable us to link the epigenetic mechanisms to gene regulation in pain. It is highly likely that the relationship will not be a direct one-to-one regulation, but rather a complex network enabling checks and balances through a concurrent modulation of events reflecting the plasticity of the nervous system. Investigating how epigenetic alterations can induce chronic pain, influence risk factors, and maintain chronicity can provide guidance in the identification of novel pain targets or intervention strategies to complement existing therapies.

Acknowledgments

Studies in my lab are supported by grants from the Rita Allen Foundation, National Institutes of Health (1R21NS082991-01), PhRMA Foundation, Commonwealth Universal Research Enhancement, and Drexel University Clinical and Translational Research Institute.

References

[1] Schmader KE, Baron R, Haanpaa ML, Mayer J, O'Connor AB, Rice AS, et al. Treatment considerations for elderly and frail patients with neuropathic pain. Mayo Clin Proc 2010;85(3 Suppl.):S26–32.
[2] Dworkin RH, O'Connor AB, Audette J, Baron R, Gourlay GK, Haanpaa ML, et al. Recommendations for the pharmacological management of neuropathic pain: an overview and literature update. Mayo Clin Proc 2010;85(3 Suppl.):S3–14.
[3] Lacroix-Fralish ML, Mogil JS. Progress in genetic studies of pain and analgesia. Annu Rev Pharmacol Toxicol 2009;49:97–121.
[4] Lotsch J, Geisslinger G. A critical appraisal of human genotyping for pain therapy. Trends Pharmacol Sci 2010;31(7):312–7.
[5] LaCroix-Fralish ML, Austin JS, Zheng FY, Levitin DJ, Mogil JS. Patterns of pain: meta-analysis of microarray studies of pain. Pain 2011;152(8):1888–98.
[6] Hamburg MA, Collins FS. The path to personalized medicine. N. Engl J Med 2010;363(4):301–4.
[7] Jirtle RL, Skinner MK. Environmental epigenomics and disease susceptibility. Nat Rev Genet 2007;8(4):253–62.

[8] Holliday R. Epigenetics: a historical overview. Epigenetics 2006;1(2):76–80.
[9] Vaissiere T, Sawan C, Herceg Z. Epigenetic interplay between histone modifications and DNA methylation in gene silencing. Mutat Res 2008;659(1–2):40–8.
[10] Bartel DP. MicroRNAs: target recognition and regulatory functions. Cell 2009;136(2):215–33.
[11] Esteller M. Non-coding RNAs in human disease. Nat Rev Genet 2011;12(12):861–74.
[12] Bird A. DNA methylation patterns and epigenetic memory. Genes Dev 2002;16(1): 6–21.
[13] Lister R, Pelizzola M, Dowen RH, Hawkins RD, Hon G, Tonti-Filippini J, et al. Human DNA methylomes at base resolution show widespread epigenomic differences. Nature 2009;462(7271):315–22.
[14] Portela A, Esteller M. Epigenetic modifications and human disease. Nat Biotechnol 2010;28(10):1057–68.
[15] Laurent L, Wong E, Li G, Huynh T, Tsirigos A, Ong CT, et al. Dynamic changes in the human methylome during differentiation. Genome Res 2010;20(3):320–31.
[16] Jones PA, Liang G. Rethinking how DNA methylation patterns are maintained. Nat Rev Genet 2009;10(11):805–11.
[17] Illingworth R, Kerr A, DeSousa D, Jørgensen H, Ellis P, Stalker J, et al. A novel CpG island set Identifies tissue-specific methylation at developmental gene loci. PLoS Biol 2008;6(1):e22.
[18] Z-x C. Riggs AD. DNA methylation and demethylation in mammals. J Biol Chem 2011;286(21):18347–53.
[19] Meissner A, Mikkelsen TS, Gu H, Wernig M, Hanna J, Sivachenko A, et al. Genome-scale DNA methylation maps of pluripotent and differentiated cells. Nature 2008;454(7205):766–70.
[20] Portela A, Esteller M. Epigenetic modifications and human disease. Nat Biotech 2010;28(10):1057–68.
[21] Bird A. The methyl-CpG-binding protein MeCP2 and neurological disease. Biochem Soc Trans 2008;36(Pt 4):575–83.
[22] Jakovcevski M, Akbarian S. Epigenetic mechanisms in neurological disease. Nat Med 2012;18(8):1194–204.
[23] Tahiliani M, Koh KP, Shen Y, Pastor WA, Bandukwala H, Brudno Y, et al. Conversion of 5-methylcytosine to 5-hydroxymethylcytosine in mammalian DNA by MLL partner TET1. Science 2009;324(5929):930–5.
[24] Kriaucionis S, Heintz N. The nuclear DNA base 5-hydroxymethylcytosine is present in Purkinje neurons and the brain. Science 2009;324(5929):929–30.
[25] Kohli RM, Zhang Y. TET enzymes, TDG and the dynamics of DNA demethylation. Nature 2013;502(7472):472–9.
[26] Branco MR, Ficz G, Reik W. Uncovering the role of 5-hydroxymethylcytosine in the epigenome. Nat Rev Genet 2012;13(1):7–13.
[27] Mellen M, Ayata P, Dewell S, Kriaucionis S, Heintz N. MeCP2 binds to 5hmC enriched within active genes and accessible chromatin in the nervous system. Cell 2012;151(7):1417–30.
[28] Globisch D, Munzel M, Muller M, Michalakis S, Wagner M, Koch S, et al. Tissue distribution of 5-hydroxymethylcytosine and search for active demethylation intermediates. PLoS One 2010;5(12):e15367.
[29] Viet CT, Ye Y, Dang D, Lam DK, Achdjian S, Zhang J, et al. Re-expression of the methylated EDNRB gene in oral squamous cell carcinoma attenuates cancer-induced pain. Pain 2011;152(10):2323–32.
[30] Wang Y, Liu C, Guo Q-L, Yan J-Q, Zhu X-Y, Huang C-S, et al. Intrathecal 5-azacytidine inhibits global DNA methylation and methyl- CpG-binding protein 2 expression and alleviates neuropathic pain in rats following chronic constriction injury. Brain Res 2011;1418(0):64–9.

[31] Tajerian M, Alvarado S, Millecamps M, Dashwood T, Anderson KM, Haglund L, et al. DNA methylation of SPARC and chronic low back pain. Mol Pain 2011;7:65.

[32] Qi F, Zhou Y, Xiao Y, Tao J, Gu J, Jiang X, et al. Promoter demethylation of cystathionine-β-synthetase gene contributes to inflammatory pain in rats. Pain 2013;154(1):34–45.

[33] Doehring A, Geisslinger G, Lotsch J. Epigenetics in pain and analgesia: an imminent research field. Eur J Pain 2010;15(1):11–6.

[34] Geranton SM. Targeting epigenetic mechanisms for pain relief. Curr Opin Pharmacol 2012;12(1):35–41.

[35] Denk F, McMahon SB. Chronic pain: emerging evidence for the involvement of epigenetics. Neuron 2012;73(3):435–44.

[36] Crow M, Denk F, McMahon S. Genes and epigenetic processes as prospective pain targets. Genome Med 2013;5(2):12.

[37] Stone LS, Szyf M. The emerging field of pain epigenetics. Pain 2013;154(1):1–2.

[38] D'Addario C, Di Francesco A, Pucci M, Finazzi Agrò A, Maccarrone M. Epigenetic mechanisms and endocannabinoid signalling. FEBS J 2013;280(9):1905–17.

[39] Hwang CK, Song KY, Kim CS, Choi HS, Guo XH, Law PY, et al. Epigenetic programming of mu-opioid receptor gene in mouse brain is regulated by MeCP2 and Brg1 chromatin remodelling factor. J Cell Mol Med 2009;13(9B):3591–615.

[40] Hwang CK, Kim CS, Kim do K, Law PY, Wei LN, Loh HH. Up-regulation of the mu-opioid receptor gene is mediated through chromatin remodeling and transcriptional factors in differentiated neuronal cells. Mol Pharmacol 2010;78(1):58–68.

[41] Millecamps M, Tajerian M, Sage EH, Stone LS. Behavioral signs of chronic back pain in the SPARC-null mouse. Spine 2011;36(2):95–102.

[42] Bock C, Tomazou EM, Brinkman AB, Muller F, Simmer F, Gu H, et al. Quantitative comparison of genome-wide DNA methylation mapping technologies. Nat Biotechnol 2010;28(10):1106–14.

[43] Harris RA, Wang T, Coarfa C, Nagarajan RP, Hong C, Downey SL, et al. Comparison of sequencing-based methods to profile DNA methylation and identification of mono-allelic epigenetic modifications. Nat Biotechnol 2010;28(10):1097–105.

[44] Plongthongkum N, Diep DH, Zhang K. Advances in the profiling of DNA modifications: cytosine methylation and beyond. Nat Rev Genet 2014;15(10):647–61.

[45] Hernandez HG, Tse MY, Pang SC, Arboleda H, Forero DA. Optimizing methodologies for PCR-based DNA methylation analysis. BioTechniques 2013;55(4):181–97.

[46] Menzies V, Lyon DE, Archer KJ, Zhou Q, Brumelle J, Jones KH, et al. Epigenetic alterations and an increased frequency of micronuclei in women with fibromyalgia. Nurs Res Pract 2013;2013:795784.

[47] Bell JT, Loomis AK, Butcher LM, Gao F, Zhang B, Hyde CL, et al. Differential methylation of the TRPA1 promoter in pain sensitivity. Nat Commun 2014;5:2978.

[48] Julius D. TRP channels and pain. Annu Rev cell Dev Biol 2013;29:355–84.

[49] Bautista DM, Pellegrino M, Tsunozaki M. TRPA1: a gatekeeper for inflammation. Annu Rev Physiol 2013;75(1):181–200.

[50] Hartung T, Zhang L, Kanwar R, Khrebtukova I, Reinhardt M, Wang C, et al. Diametrically opposite methylome-transcriptome relationships in high- and low-CpG promoter genes in postmitotic neural rat tissue. Epigenetics 2012;7(5):421–8.

[51] Tajerian M, Alvarado S, Millecamps M, Vachon P, Crosby C, Bushnell MC, et al. Peripheral nerve injury is associated with chronic, reversible changes in global DNA methylation in the mouse prefrontal cortex. PLoS One 2013;8(1):e55259.

[52] Klein CJ, Botuyan M-V, Wu Y, Ward CJ, Nicholson GA, Hammans S, et al. Mutations in DNMT1 cause hereditary sensory neuropathy with dementia and hearing loss. Nat Genet 2011;43(6):595–600.

[53] Downs J, Geranton SM, Bebbington A, Jacoby P, Bahi-Buisson N, Ravine D, et al. Linking MECP2 and pain sensitivity: the example of Rett syndrome. Am J Med Genet A 2010;152A(5):1197–205.

VII. PERSONALIZED EPIGENETICS OF DISORDERS AND DISEASE MANAGEMENT

[54] Peters SU, Hundley RJ, Wilson AK, Warren Z, Vehorn A, Carvalho CMB, et al. The behavioral phenotype in MECP2 duplication syndrome: a comparison with idiopathic autism. Autism Res 2013;6(1):42–50.

[55] Calfa G, Percy AK, Pozzo-Miller L. Experimental models of Rett syndrome based on Mecp2 dysfunction. Exp Biol Med (Maywood) 2011;236(1):3–19.

[56] Gemelli T, Berton O, Nelson ED, Perrotti LI, Jaenisch R, Monteggia LM. Postnatal loss of methyl-CpG binding protein 2 in the forebrain is sufficient to mediate behavioral aspects of Rett syndrome in mice. Biol Psychiatry 2006;59(5):468–76.

[57] Samaco RC, Fryer JD, Ren J, Fyffe S, Chao H-T, Sun Y, et al. A partial loss of function allele of Methyl-CpG-binding protein 2 predicts a human neurodevelopmental syndrome. Hum Mol Genet 2008;17(12):1718–27.

[58] Samaco RC, McGraw CM, Ward CS, Sun Y, Neul JL, Zoghbi HY. Female Mecp2(+/-) mice display robust behavioral deficits on two different genetic backgrounds providing a framework for pre-clinical studies. Hum Mol Genet 2013;22(1):96–109.

[59] Geranton SM, Morenilla-Palao C, Hunt SP. A role for transcriptional repressor methyl-CpG-binding protein 2 and plasticity-related gene serum- and glucocorticoid-inducible kinase 1 in the induction of inflammatory pain states. J Neurosci 2007;27(23):6163–73.

[60] Geranton SM, Fratto V, Tochiki KK, Hunt SP. Descending serotonergic controls regulate inflammation-induced mechanical sensitivity and methyl-CpG-binding protein 2 phosphorylation in the rat superficial dorsal horn. Mol Pain 2008;4:35.

[61] Zhang Z, Tao W, Hou YY, Wang W, Kenny PJ, Pan ZZ. MeCP2 repression of G9a in regulation of pain and morphine reward. J Neurosci 2014;34(27):9076–87.

[62] Chahrour M, Jung SY, Shaw C, Zhou X, Wong ST, Qin J, et al. MeCP2, a key contributor to neurological disease, activates and represses transcription. Science 2008;320(5880):1224–9.

[63] Maze I, Covington HE, Dietz DM, LaPlant Q, Renthal W, Russo SJ, et al. Essential role of the histone methyltransferase G9a in cocaine-induced plasticity. Science 2010;327(5962):213–6.

[64] Heyn H, Esteller M. DNA methylation profiling in the clinic: applications and challenges. Nat Rev Genet 2012;13(10):679–92.

[65] Kelly TK, De Carvalho DD, Jones PA. Epigenetic modifications as therapeutic targets. Nat Biotechnol 2010;28(10):1069–78.

[66] Shir Y, Ratner A, Raja SN, Campbell JN, Seltzer Z. Neuropathic pain following partial nerve injury in rats is suppressed by dietary soy. Neurosci Lett 1998;240(2):73–6.

[67] Zachariah RM, Rastegar M. Linking epigenetics to human disease and Rett syndrome: the emerging novel and challenging concepts in MeCP2 research. Neural Plast 2012;2012:415825.

[68] Skene PJ, Illingworth RS, Webb S, Kerr ARW, James KD, Turner DJ, et al. Neuronal MeCP2 is expressed at near histone-octamer levels and globally alters the chromatin state. Mol Cell 2010;37(4):457–68.

[69] Doehring A, Oertel BG, Sittl R, Lötsch J. Chronic opioid use is associated with increased DNA methylation correlating with increased clinical pain. Pain 2013;154(1):15–23.

[70] Arand J, Spieler D, Karius T, Branco MR, Meilinger D, Meissner A, et al. In vivo control of CpG and non-CpG DNA methylation by DNA methyltransferases. PLoS Genet 2012;8(6):e1002750.

[71] Burns Kathleen H, Boeke Jef D. Human transposon tectonics. Cell 2012;149(4):740–52.

[72] Liang DY, Sun Y, Clark JD. Dietary methyl content regulates opioid responses in mice. J Pain Res 2013;6:281–7.

[73] Ivanov M, Kacevska M, Ingelman-Sundberg M. Epigenomics and interindividual differences in drug response. Clin Pharmacol Ther 2012;92(6):727–36.

[74] Tobi EW, Lumey LH, Talens RP, Kremer D, Putter H, Stein AD, et al. DNA methylation differences after exposure to prenatal famine are common and timing- and sex-specific. Hum Mol Genet 2009;18(21):4046–53.

[75] Kouzarides T. Chromatin modifications and their function. Cell 2007;128(4):693–705.

[76] Li B, Carey M, Workman JL. The role of chromatin during transcription. Cell 2007;128(4):707–19.

[77] Lee J-S, Smith E, Shilatifard A. The language of histone crosstalk. Cell September 3, 2010;142(5):682–5.

[78] Henikoff S, Shilatifard A. Histone modification: cause or cog? Trends Genet 2011;27(10):389–96.

[79] Haberland M, Montgomery RL, Olson EN. The many roles of histone deacetylases in development and physiology: implications for disease and therapy. Nat Rev Genet 2009;10(1):32–42.

[80] Morris MJ, Monteggia LM. Unique functional roles for class I and class II histone deacetylases in central nervous system development and function. Int J Dev Neurosci 2013;31(6):370–81.

[81] Zhang Z, Cai YQ, Zou F, Bie B, Pan ZZ. Epigenetic suppression of GAD65 expression mediates persistent pain. Nat Med 2011;17(11):1448–55.

[82] Chiechio S, Zammataro M, Morales ME, Busceti CL, Drago F, Gereau RWt, et al. Epigenetic modulation of mGlu2 receptors by histone deacetylase inhibitors in the treatment of inflammatory pain. Mol Pharmacol 2009;75(5):1014–20.

[83] Bai G, Wei D, Zou S, Ren K, Dubner R. Inhibition of class II histone deacetylases in the spinal cord attenuates inflammatory hyperalgesia. Mol Pain 2010;6:51.

[84] Denk F, Huang W, Sidders B, Bithell A, Crow M, Grist J, et al. HDAC inhibitors attenuate the development of hypersensitivity in models of neuropathic pain. PAIN 2013;154(9):1668–79.

[85] Sun Y, Sahbaie P, Liang DY, Li WW, Li XQ, Shi XY, et al. Epigenetic regulation of spinal CXCR2 signaling in incisional hypersensitivity in mice. Anesthesiology 2013;119(5):1198–208.

[86] Capasso K, Manners M, Quershi R, Tian Y, Gao R, Hu H, et al. Effect of histone deacetylase inhibitor JNJ-26481585 in pain. J Mol Neurosci 2014:1–9.

[87] Liang D-Y, Sun Y, Shi X-Y, Sahbaie P, Clark J. Epigenetic regulation of spinal cord gene expression controls opioid-induced hyperalgesia. Mol pain 2014;10(1):59.

[88] Kiguchi N, Kobayashi Y, Maeda T, Fukazawa Y, Tohya K, Kimura M, et al. Epigenetic augmentation of the macrophage inflammatory protein 2/C-X-C chemokine receptor type 2 axis through histone H3 acetylation in injured peripheral nerves elicits neuropathic pain. J Pharmacol Exp Ther 2012;340(3):577–87.

[89] Zhu X, Li Q, Chang R, Yang D, Song Z, Guo Q, et al. Curcumin alleviates neuropathic pain by inhibiting p300/CBP histone acetyltransferase activity-regulated expression of BDNF and Cox-2 in a rat model. PLoS ONE 2014;9(3):e91303.

[90] York E, Petit A, Roskams AJ. Epigenetics of neural repair following spinal cord injury. Neurotherapeutics 2013;10(4):757–70.

[91] Finelli MJ, Wong JK, Zou H. Epigenetic regulation of sensory axon regeneration after spinal cord injury. J Neurosci 2013;33(50):19664–76.

[92] Sinclair DA, Guarente L. Small-molecule allosteric activators of sirtuins. Annu Rev Pharmacol Toxicol 2014;54(1):363–80.

[93] He X, Ou P, Wu K, Huang C, Wang Y, Yu Z, et al. Resveratrol attenuates morphine antinociceptive tolerance via SIRT1 regulation in the rat spinal cord. Neurosci Lett 2014;566(0):55–60.

[94] Shao H, Xue Q, Zhang F, Luo Y, Zhu H, Zhang X, et al. Spinal SIRT1 activation attenuates neuropathic pain in mice. PLoS One 2014;9(6):e100938.

[95] Tran L, Schulkin J, Ligon CO, Greenwood-Van Meerveld B. Epigenetic modulation of chronic anxiety and pain by histone deacetylation. Mol Psychiatry 2014:1–13.

[96] Matsushita Y, Araki K, Omotuyi O, Mukae T, Ueda H. HDAC inhibitors restore C-fibre sensitivity in experimental neuropathic pain model. Br J Pharmacol 2013;170(5): 991–8.

[97] Shahbazian MD, Grunstein M. Functions of site-specific histone acetylation and deacetylation. Annu Rev Biochem 2007;76(1):75–100.

[98] Glozak MA, Sengupta N, Zhang X, Seto E. Acetylation and deacetylation of non-histone proteins. Gene 2005;363(0):15–23.

[99] Radman-Livaja M, Rando OJ. Nucleosome positioning: how is it established, and why does it matter? Dev Biol 2010;339(2):258–66.

[100] Segal E, Widom J. What controls nucleosome positions? Trends Genet 2009;25(8):335–43.

[101] Bai L, Morozov AV. Gene regulation by nucleosome positioning. Trends Genet 2010;26(11):476–83.

[102] Rando OJ, Ahmad K. Rules and regulation in the primary structure of chromatin. Curr Opin Cell Biol 2007;19(3):250–6.

[103] Clapier CR, Cairns BR. The biology of chromatin remodeling complexes. Annu Rev Biochem 2009;78(1):273–304.

[104] Ferguson-Smith AC. Genomic imprinting: the emergence of an epigenetic paradigm. Nat Rev Genet 2011;12(8):565–75.

[105] Adalsteinsson B, Ferguson-Smith A. Epigenetic control of the genome—lessons from genomic imprinting. Genes 2014;5(3):635–55.

[106] Tan PL, Katsanis N. Thermosensory and mechanosensory perception in human genetic disease. Hum Mol Genet 2009;18(R2):R146–55.

[107] Kuwako K-i, Hosokawa A, Nishimura I, Uetsuki T, Yamada M, Nada S, et al. Disruption of the paternal necdin gene diminishes TrkA signaling for sensory neuron survival. J Neurosci 2005;25(30):7090–9.

[108] Alexander RP, Fang G, Rozowsky J, Snyder M, Gerstein MB. Annotating non-coding regions of the genome. Nat Rev Genet 2010;11(8):559–71.

[109] Djebali S, Davis CA, Merkel A, Dobin A, Lassmann T, Mortazavi A, et al. Landscape of transcription in human cells. Nature 2012;489(7414):101–8.

[110] Harrow J, Frankish A, Gonzalez JM, Tapanari E, Diekhans M, Kokocinski F, et al. GEN-CODE: the reference human genome annotation for the ENCODE Project. Genome Res 2012;22(9):1760–74.

[111] Derrien T, Johnson R, Bussotti G, Tanzer A, Djebali S, Tilgner H, et al. The GENCODE v7 catalog of human long noncoding RNAs: analysis of their gene structure, evolution, and expression. Genome Res 2012;22(9):1775–89.

[112] Lee Tong I, Young Richard A. Transcriptional regulation and its misregulation in disease. Cell 2013;152(6):1237–51.

[113] Mendell JT, Olson EN. MicroRNAs in stress signaling and human disease. Cell 2012;148(6):1172–87.

[114] Rao P, Benito E, Fischer A. MicroRNAs as biomarkers for CNS disease. Front Mol Neurosci 2013;6.

[115] Dai E, Yu X, Zhang Y, Meng F, Wang S, Liu X, et al. EpimiR: a database of curated mutual regulation between miRNAs and epigenetic modifications. Database 2014;2014.

[116] Sato F, Tsuchiya S, Meltzer SJ, Shimizu K. MicroRNAs and epigenetics. FEBS J 2011;278(10):1598–609.

[117] Varela M, Roberts T, Wood MA. Epigenetics and ncRNAs in brain function and disease: mechanisms and prospects for therapy. Neurotherapeutics 2013;10(4):621–31.

[118] Bai G, Ambalavanar R, Wei D, Dessem D. Downregulation of selective microRNAs in trigeminal ganglion neurons following inflammatory muscle pain. Mol Pain 2007;3:15.

[119] Aldrich BT, Frakes EP, Kasuya J, Hammond DL, Kitamoto T. Changes in expression of sensory organ-specific microRNAs in rat dorsal root ganglia in association with mechanical hypersensitivity induced by spinal nerve ligation. Neuroscience 2009;164(2):711–23.

[120] Imai S, Saeki M, Yanase M, Horiuchi H, Abe M, Narita M, et al. Change in microRNAs associated with neuronal adaptive responses in the nucleus accumbens under neuropathic pain. J Neurosci : official J Soc Neurosci 2011;31(43):15294–9.

[121] Poh KW, Yeo JF, Ong WY. MicroRNA changes in the mouse prefrontal cortex after inflammatory pain. Eur J Pain 2011;15(8):801 e1–12.

[122] Kusuda R, Cadetti F, Ravanelli MI, Sousa TA, Zanon S, De Lucca FL, et al. Differential expression of microRNAs in mouse pain models. Mol Pain 2011;7:17.

[123] von Schack D, Agostino MJ, Murray BS, Li Y, Reddy PS, Chen J, et al. Dynamic changes in the microRNA expression profile reveal multiple regulatory mechanisms in the spinal nerve ligation model of neuropathic pain. PLoS One 2011;6(3):e17670.

[124] Zhao J, Lee MC, Momin A, Cendan CM, Shepherd ST, Baker MD, et al. Small RNAs control sodium channel expression, nociceptor excitability, and pain thresholds. J Neurosci 2010;30(32):10860–71.

[125] Favereaux A, Thoumine O, Bouali-Benazzouz R, Roques V, Papon MA, Salam SA, et al. Bidirectional integrative regulation of Cav1.2 calcium channel by microRNA miR-103: role in pain. EMBO J 2011;30(18):3830–41.

[126] Willemen H, Huo X-J, Mao-Ying Q-L, Zijlstra J, Heijnen C, Kavelaars A. MicroRNA-124 as a novel treatment for persistent hyperalgesia. J Neuroinflammation 2012;9(1):143.

[127] Kynast KL, Russe OQ, Möser CV, Geisslinger G, Niederberger E. Modulation of central nervous system–specific microRNA-124a alters the inflammatory response in the formalin test in mice. Pain 2013;154(3):368–76.

[128] Kress M, Hüttenhofer A, Landry M, Kuner R, Favereaux A, Greenberg DS, et al. microRNAs in nociceptive circuits as predictors of future clinical applications. Front Mol Neurosci 2013;6.

[129] Kynast KL, Russe OQ, Geisslinger G, Niederberger E. Novel findings in pain processing pathways: implications for miRNAs as future therapeutic targets. Expert Rev Neurother 2013;13(5):515–25.

[130] Sun Y, Li XQ, Sahbaie P, Shi XY, Li WW, Liang DY, et al. miR-203 regulates nociceptive sensitization after incision by controlling phospholipase A2 activating protein expression. Anesthesiology 2012;117(3):626–38.

[131] Bali KK, Selvaraj D, Satagopam VP, Lu J, Schneider R, Kuner R. Genome-wide identification and functional analyses of microRNA signatures associated with cancer pain. EMBO Mol Med 2013;5(11):1740–58.

[132] Sakai A, Saitow F, Miyake N, Miyake K, Shimada T, Suzuki H. miR-7a alleviates the maintenance of neuropathic pain through regulation of neuronal excitability. Brain 2013;136(Pt 9):2738–50.

[133] Pan Z, Zhu L-J, Li Y-Q, Hao L-Y, Yin C, Yang J-X, et al. Epigenetic modification of spinal miR-219 expression regulates chronic inflammation pain by targeting CaMKIIγ. J Neurosci 2014;34(29):9476–83.

[134] Fabbri M, Paone A, Calore F, Galli R, Croce CM. A new role for microRNAs, as ligands of toll-like receptors. RNA Biol 2013;10(2):169–74.

[135] Park C-K, Xu Z-Z, Berta T, Han Q, Chen G, Liu X-J, et al. Extracellular microRNAs activate nociceptor neurons to elicit pain via TLR7 and TRPA1. Neuron 2014;82(1):47–54.

[136] Orlova IA, Alexander GM, Qureshi RA, Sacan A, Graziano A, Barrett JE, et al. MicroRNA modulation in complex regional pain syndrome. J Transl Med 2011;9(1):195.

[137] Monastyrskaya K, Sánchez-Freire V, Hashemi Gheinani A, Klumpp DJ, Babiychuk EB, Draeger A, et al. miR-199a-5p regulates urothelial permeability and may play a role in bladder pain syndrome. Am J Pathol 2013;182(2):431–48.

[138] Sanchez Freire V, Burkhard FC, Kessler TM, Kuhn A, Draeger A, Monastyrskaya K. MicroRNAs may mediate the down-regulation of neurokinin-1 receptor in chronic bladder pain syndrome. Am J Pathol 2010;176(1):288–303.

[139] Bjersing JL, Lundborg C, Bokarewa MI, Mannerkorpi K. Profile of cerebrospinal microRNAs in fibromyalgia. PLoS One 2013;8(10):e78762.

[140] Ceribelli A, Nahid MA, Satoh M, Chan EKL. MicroRNAs in rheumatoid arthritis. FEBS Lett 2011;585(23):3667–74.

[141] Li X, Gibson G, Kim JS, Kroin J, Xu S, van Wijnen AJ, et al. MicroRNA-146a is linked to pain-related pathophysiology of osteoarthritis. Gene 2011;480(1–2):34–41.

[142] Kola I, Landis J. Can the pharmaceutical industry reduce attrition rates? Nat Rev Drug Discov 2004;3(8):711–5.

[143] Woodcock J, Witter J, Dionne RA. Stimulating the development of mechanism-based, individualized pain therapies. Nat Rev Drug Discov 2007;6(9):703–10.

[144] Marchi A, Vellucci R, Mameli S, Rita Piredda A, Finco G. Pain biomarkers. Clin Drug Investig 2009;29(Suppl. 1):41–6.

[145] El Andaloussi S, Mager I, Breakefield XO, Wood MJA. Extracellular vesicles: biology and emerging therapeutic opportunities. Nat Rev Drug Discov 2013;12(5):347–57.

[146] Alvarez-Erviti L, Seow Y, Yin HF, Betts C, Lakhal S, Wood MJA. Delivery of siRNA to the mouse brain by systemic injection of targeted exosomes. Nat Biotechnol 2011;29(4):341–5.

[147] Kowal J, Tkach M, Thery C. Biogenesis and secretion of exosomes. Curr Opin Cell Biol 2014;29C:116–25.

[148] Raposo G, Stoorvogel W. Extracellular vesicles: exosomes, microvesicles, and friends. J cell Biol 2013;200(4):373–83.

[149] Raiborg C, Rusten TE, Stenmark H. Protein sorting into multivesicular endosomes. Curr Opin Cell Biol 2003;15(4):446–55.

[150] Stoorvogel W, Kleijmeer MJ, Geuze HJ, Raposo G. The biogenesis and functions of exosomes. Traffic 2002;3(5):321–30.

[151] Kalra H, Simpson RJ, Ji H, Aikawa E, Altevogt P, Askenase P, et al. Vesiclepedia: a compendium for extracellular vesicles with continuous community annotation. PLoS Biol 2012;10(12):e1001450.

[152] Gyorgy B, Szabo TG, Pasztoi M, Pal Z, Misjak P, Aradi B, et al. Membrane vesicles, current state-of-the-art: emerging role of extracellular vesicles. Cell Mol Life Sci 2011;68(16):2667–88.

[153] Record M, Subra C, Silvente-Poirot S, Poirot M. Exosomes as intercellular signalosomes and pharmacological effectors. Biochem Pharmacol 2011;81(10):1171–82.

[154] Kim SH, Lechman ER, Bianco N, Menon R, Keravala A, Nash J, et al. Exosomes derived from IL-10-treated dendritic cells can suppress inflammation and collagen-induced arthritis. J Immunol 2005;174(10):6440–8.

[155] Bianco NR, Kim SH, Ruffner MA, Robbins PD. Therapeutic effect of exosomes from indoleamine 2,3-dioxygenase-positive dendritic cells in collagen-induced arthritis and delayed-type hypersensitivity disease models. Arthritis Rheum 2009;60(2):380–9.

[156] McDonald MK, Tian Y, Qureshi RA, Gormley M, Ertel A, Gao R, et al. Functional significance of macrophage-derived exosomes in inflammation and pain. Pain® 2014;155(8):1527–39.

[157] Viaud S, Théry C, Ploix S, Tursz T, Lapierre V, Lantz O, et al. Dendritic cell-derived exosomes for cancer immunotherapy: what's next? Cancer Res 2010;70(4):1281–5.

[158] Chivet M, Hemming F, Pernet-Gallay k, Fraboulet S, Sadoul R. Emerging role of neuronal exosomes in the central nervous system. Front Physiol 2012;3.

[159] Fruhbeis C, Frohlich D, Kuo WP, Amphornrat J, Thilemann S, Saab AS, et al. Neurotransmitter-triggered transfer of exosomes mediates oligodendrocyte-neuron communication. PLoS Biol 2013;11(7):e1001604.

[160] Frühbeis C, Fröhlich D, Kuo WP, Krämer-Albers E-M. Extracellular vesicles as mediators of neuron-glia communication. Front Cell Neurosci 2013;7.

[161] Sabin LR, Delas MJ, Hannon GJ. Dogma derailed: the many influences of RNA on the genome. Mol Cell 2013;49(5):783–94.

[162] Vucicevic D, Schrewe H, Andersson Ørom U. Molecular mechanisms of long ncRNAs in neurological disorders. Front Genet 2014;5.

[163] Yang L, Froberg JE, Lee JT. Long noncoding RNAs: fresh perspectives into the RNA world. Trends Biochem Sci 2014;39(1):35–43.

[164] Zhao X, Tang Z, Zhang H, Atianjoh FE, Zhao JY, Liang L, et al. A long noncoding RNA contributes to neuropathic pain by silencing Kcna2 in primary afferent neurons. Nat Neurosci 2013;16(8):1024–31.
[165] Wild CP. The exposome: from concept to utility. Int J Epidemiol 2012;41(1):24–32.
[166] Ogino S, Lochhead P, Chan AT, Nishihara R, Cho E, Wolpin BM, et al. Molecular pathological epidemiology of epigenetics: emerging integrative science to analyze environment, host, and disease. Mod Pathol 2013;26(4):465–84.
[167] Bruehl S, Apkarian AV, Ballantyne JC, Berger A, Borsook D, Chen WG, et al. Personalized medicine and opioid analgesic prescribing for chronic pain: opportunities and challenges. J Pain 2013;14(2):103–13.
[168] Denk F, McMahon SB, Tracey I. Pain vulnerability: a neurobiological perspective. Nat Neurosci 2014;17(2):192–200.
[169] Lötsch J, Doehring A, Mogil JS, Arndt T, Geisslinger G, Ultsch A. Functional genomics of pain in analgesic drug development and therapy. Pharmacol Ther 2013;139(1):60–70.

CHAPTER

15

Understanding Interindividual Epigenetic Variations in Obesity and Its Management

Sonal Patel, Arpankumar Choksi,
Samit Chattopadhyay

National Centre for Cell Science, Pune University Campus,
Ganeshkhind, Pune, India

OUTLINE

1. INTRODUCTION

Obesity is a highly predominant nutritional problem and one of the world's greatest health issues. According to the World Health Organization report on the global epidemic of obesity, an estimated 1 billion adults are considered to be overweight and more then 300 million people come under the category of clinically obese [1]. Moreover, obesity is not just limited to one group of people but it is prevalent in all age and ethnic groups. It increases the menace of type 2 diabetes (T2D), cardiovascular diseases, stroke, osteoarthritis, and mortality. According to medical convention, overweight and obesity are defined on the basis of body mass index (BMI), which is calculated by dividing weight (in kilograms) by height squared (in meters). The BMI of healthy, overweight, and obese individuals is defined as 18.5 to <25, 25–29.9, and ≥30 kg/m^2, respectively. However, individuals may have varying amounts of body fat at any given BMI [1]. The major cause of obesity is storage of exceptionally high amounts of triglycerides in adipose tissue and their release as free fatty acids from the adipose tissue, which in turn creates adverse effects [2].

Obesity is a complex disorder involving genetic and environmental factors. Research has identified more than 50 loci associated with obesity [3]. The major contributors to obesity are environmental factors; changes in lifestyle, including the increased availability of tempting, energy-rich foods; a reduction in physical activity; and sleeplessness [4,5].

Epigenetics is defined as the study of hereditary and reversible mitotic as well as meiotic changes, which affect gene expression without any modifications of DNA sequence [6]. Epigenetic modifications are tissue specific and include DNA methylation, histone modifications, and various noncoding RNAs. Epigenetic marks have been shown to play a pivotal role in the susceptibility to obesity. Moreover, there are many reports suggesting that epigenetic regulations respond to environmental signals such as dietary factors [7–13]. A number of genetic variants have been identified based on genome-wide association studies (GWAS) that partially explain interindividual variation in disease susceptibility [14]. Although the literature citing the involvement of epigenetic factors in the interplay between genes and environment is continually increasing, the molecular mechanisms underlying such interplay still need to be understood better.

2. IDENTIFICATION OF SUSCEPTIBILITY LOCI FOR OBESITY

Until 2006, the major approaches to tracking factors influencing obesity were (1) genome-wide linkage mapping in families and (2) association studies within candidate genes [15]. The genome-wide linkage mapping approach lacked any sensible susceptibility models, as linkage is best used to detect variants with high penetrance. The second approach was compromised by difficulty in selecting suitable candidates. Hence, these approaches led to the identification of very few genuine obesity-susceptibility variants. The "Human Obesity Gene Map" gives a better picture by enlisting 11 single-gene mutations, 50 loci related to Mendelian syndromes relevant to human obesity, 244 knockout or transgenic animal models, and 127 candidate genes [16]. During 2012–2014, with the advent of the human genome sequence, the International HapMap Consortium, and novel genotyping methods, a new hypothesis-free genome-wide association testing was used to generate significant association results. Other forms of obesity such as early onset, extreme obesity, and morbid adult obesity have been the subjects of focus in some GWAS. A few studies suggest that factors affecting BMI might also underlie some of the more severe forms of obesity [17]. More recently, confirmation of 14 of the identified obesity susceptibility loci came from a large-scale meta-analysis of 249,769 individuals, which also identified 18 novel loci associated with adiposity [18]. Such large-scale genome-wide association studies mark the beginning of the identification of new variants with a role to play in overall obesity and weight regulation.

3. HOW DO EPIGENETIC CHANGES CONTRIBUTE TO OBESITY?

Epigenetic modulation of gene expression can be mediated by three major molecular mechanisms: DNA methylation, histone modification, and noncoding RNA. These mechanisms bring about heritable changes without changing the underlying sequence and are discussed in detail below.

3.1 DNA Methylation

DNA methylation involves the covalent attachment of a methyl group to the $5'$ position of cytosine present in the DNA sequence in mammalian genomes by DNA methyltransferases (DNMTs). The methyl group bulges out into the major groove of the DNA double helix and impedes the binding of transcription factors, which in turn results in repression of a particular locus. In mammals, DNA methylation predominantly occurs at CpG dinucleotides (the p represents the intervening phosphate group). CpG islands are the huddled CpG dinucleotides at the 5′ ends of genes. The complementary DNA sequence of the CpG is the same and this palindromic sequence is utilized for the inheritance of DNA methylation during cellular divisions. During DNA replication, the methylated CpGs on the template strand are recognized by DNA methyltransferases, which add a methyl group to the newly synthesized complementary DNA strand. Because DNA methylation is transmitted through DNA replication and cell division, it is considered to be a more stable epigenetic mark [19]. Hypermethylation of CpG islands results in transcriptional repression, whereas hypomethylation results in transcriptional activation [20,21].

DNA methylation is one of the important mechanisms for gene imprinting [22], X-chromosome inactivation [23,24], and cellular differentiation as well as tissue-specific gene expression [19]. Evidence suggests that DNA methylation shows a relatively plastic nature particularly in the case of replication-independent methylation and demethylation [25].

The contribution of DNA methylation to obesity is not entirely known, but many instances in which environmental factors elicit gene expression changes via methylation of DNA are known. One such obesity-related gene is leptin, which codes for leptin hormone produced by fat cells, which regulates the amount of fat stored in the body. Promoter methylation of the leptin gene in adipose tissue is regulated by a high-fat diet [26]. Another important candidate gene in obesity is PGC1α (peroxisome proliferator-activated receptor γ coactivator 1α). It belongs to a family of transcription coactivators known to regulate cellular energy metabolism including both carbohydrate and lipid metabolism. Hence, it is a target molecule of the study of obesity and type 2 diabetes [27]. Promoter methylation of this PGC1α (or PPARGC1α) in human umbilical cord and human muscle has been respectively correlated to maternal pregestational BMI and to high-fat overfeeding [28,29]. Moreover, promoter methylation of the retinoid X receptor-α, endothelial nitric oxide synthase, superoxide dismutase-1 (SOD1), interleukin-8, and phosphoinositide 3-kinase, catalytic, δ-polypeptide genes at birth has been associated with adiposity in children [30]. On the basis of genome-wide methylation studies candidate regions with extreme interindividual variability in DNA methylation have been identified. These variably methylated regions (VMRs) can be divided into stable and dynamic groups. Dynamic VMRs have been identified as subject to environmental influence and their expression can be controlled by environmental factors via changes in DNA methylation [31].

Research suggests that environmental factors may have an effect on methylation–demethylation status of DNA, which ultimately affects gene expression and contributes to disease such as obesity [7,8,11,12]. Cultured myotubes, when exposed to palmitate or oleate, induce hypermethylation of the PGC1α promoter [32]. A study involving healthy sedentary men and women exposed to an acute bout of exercise showed a decrease in whole-genome methylation in the skeletal muscles as studied by methylated DNA capture, followed by qPCR and bisulfite sequencing [7]. Hypomethylation on the promoter of the PGC1α, PDK4, PPARδ, TFAM, and MEF2A genes was associated with exercise-induced dose-dependent expression [7]. Caffeine is known to elevate cytoplasmic calcium ion levels thereby mimicking the effect of exercise on the expression of genes. Compounds that inhibit calcium ion release from the sarcoplasmic reticulum, such as dantrolene, decreased promoter hypomethylation, leading to reduced gene expression [7]. In the context of adipose tissue, one of the genome-wide DNA methylation studies showed the presence of hypomethylated regions in tissue-specific regulatory elements [33]. Epigenetics thus can play a pivotal role in modifying gene expression based on environmental clues and predispose to obesity and other metabolic diseases.

3.2 Histone Modifications

Chromatin is made up of nucleosomes, which comprise 147 bp of DNA wrapped around a core histone octamer (composed of two copies each of histones H2A, H2B, H3, and H4). Histone H1 (linker histone) further compacts the DNA into a helical structure by occupying the exit and entry of the DNA into the nucleosome [34]. Histone proteins are made up of two domains—a globular domain and an N-terminal tail domain. Modifications such as acetylation, methylation, ubiquitination, sumoylation, and phosphorylation occur at the N-terminal amino acids of histones [35]. Histone modifications facilitate the entry of various transcription factors, which in turn activate or repress gene expression and ultimately result in a specific cellular response. These modifications on histones are collectively referred to as the histone code. Histone acetylation is code for the activation of gene expression, whereas the methylation of histones represents either an active or a repressive state of chromatin depending on the specific lysine involved. For example, methylation of histone H3 on lysine 9 represents transcriptional repression, while methylation of histone H3 on lysine 4 (H3K4me) is associated with transcriptional activation [36]. Writers of the histone code include histone acetyltransferases (HATs) and histone methyltransferases (HMTs), whereas the erasers of histone marks include the histone deacetylases (HDACs) and histone demethylases.

Cross talk between DNA methylation and histone modification regulates gene expression. Methyl CpG-binding protein-2 binds to methylated DNA, which in turn recruits both HDACs and HMTs [37]. However, research has identified that DNMT1 can be engaged by a number of histone-modifying enzymes such as HDACs and HMTs, which suggests that DNA methylation and histone modifications have a reciprocal relationship [38,39].

Research in the field of adipogenesis suggests that histone modifications are fundamentally involved in the epigenetic regulation of adipogenesis and play a pivotal role in obesity development. According to one study, modulation of five key regulatory genes of adipogenesis, Pref-1 (preadipocyte factor-1), CCAAT-enhancer-binding protein β (C/EBPβ), C/EBPα, PPARγ, and adipocyte protein 2 (aP2), was done by histone modifications during adipocyte differentiation [40]. In this study, changes in the distribution pattern of various histone marks at these genes during the differentiation of C3H 10T1/2 mouse mesenchymal stem cells (MSCs) and 3T3-L1 preadipocytes into adipocytes were identified [40]. Regulation of the transcription network by histone modifications at these genes is a key event in adipogenesis [40]. The roles of these genes with emphasis on obesity and their epigenetic regulation are discussed in detail in a later section.

Histone modifications such as demethylation have also been linked to metabolic disorders. Acetyl-CoA is a metabolite required for acetylation of histones by HATs. ATP-citrate lyase is the enzyme responsible for production of acetyl-CoA from citrate. Reports suggest that histone acetylation can be affected by the availability of glucose in an ATP-citrate lyase-dependent manner [41]. A nonfunctional histone demethylase, Jhdm2a, has been associated with obesity [42]. Also, nicotinamide adenine dinucleotide (NAD^+)-dependent sirtuins (class III HDACs) are known to epigenetically regulate metabolism by targeting both histone an nonhistone proteins [43]. SIRT1, one of the most studied members, is involved in various pathways such as adipogenesis, glucose utilization, and fatty acid oxidation. These reports strengthen the links between epigenetics and metabolism.

3.3 Noncoding RNAs

A wide variety of noncoding RNAs (ncRNAs) have been identified to play crucial roles in the epigenetic regulation of gene expression. There are two ways in which noncoding RNAs regulate genes. They can act either *in cis* or *in trans*. The *cis*-acting ncRNAs are generally the long ncRNAs (up to 100,000 nucleotides). For example, Xist plays an important role in X-chromosome inactivation. The *trans*-acting ncRNAs include the microRNAs, which mostly target the 3′ untranslated region of mRNAs and ultimately result in their degradation [44]. Studies suggest that miRNAs also have an important function in chromatin remodeling [45,46].

Studies suggest that miRNAs play vital roles in glucose homeostasis and lipid metabolism. Moreover, miRNAs have been identified to have important functions in adipocyte differentiation and obesity development. This indicates that miRNAs play a pivotal role in the focalization of type 2 diabetes and obesity [47]. It has been speculated that certain environmental agents can change the expression of genes through modulation of miRNAs in the cell [48]. Feedback between miRNA expression and epigenetic modifications has been evidenced [49]. DNA methylation and histone modifications can modulate the expression of various miRNAs. On the other hand some miRNAs can alter the expression of DNMT3A, DNMT3B, and polycomb genes [50,51]. Along with intracellular sources of miRNAs, there is some evidence suggesting a role for exogenous miRNAs in gene regulation and cellular metabolism. One such study suggests that exogenous miRNAs obtained from rice can pass through the mammalian gastrointestinal tract and reach target organs to play important regulatory roles. MiR-168a, an miRNA abundantly found in rice, binds to human and mouse low-density lipoprotein receptor adaptor protein 1 (LDLRAP1) mRNA and inhibits its expression in the liver, which in turn results in elevated plasma LDL-cholesterol levels [52]. This study provides a vital link between diet and epigenetic regulation of metabolism.

3.4 Obesity-Associated Genes and Their Epigenetic Regulation

There are many important genes that are involved in the development of obesity and related metabolic disorders, and the expression of these genes is under the influence of epigenetic regulation (Figure 1). Some of the genes involved in the development of obesity and which are regulated by epigenetic modulation are discussed here.

One such gene is ADIPOQ, which encodes adiponectin, the most abundant hormone that plays an important role in glucose regulation, inflammation, and fatty acid oxidation. Adiponectin is exclusively secreted by adipocytes in adults and also from the placenta during pregnancy. Moreover,

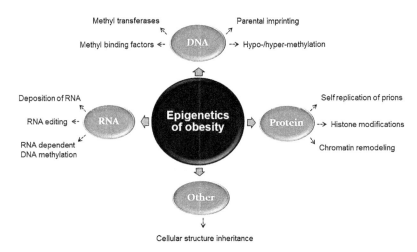

FIGURE 1   Various epigenetic factors influencing obesity through different modes.

an inverse correlation between serum adiponectin with insulin resistance in obese men with type 2 diabetes has been reported [53]. Regulation of adiponectin during pregnancy is very important for proper fetal growth. According to a few studies adiponectin expression is regulated by epigenetic mechanisms such as DNA methylation and histone acetylation [54,55]. According to one such study the DNA methylation status of the ADIPOQ gene was associated with maternal glucose status and adiponectin concentration. As adiponectin has insulin-sensitizing proprieties, these epigenetic adaptations have the potential to induce sustained glucose metabolism changes in the mother and offspring later in life. Epigenetic regulation has potential importance in the maintenance of glucose metabolism in the mother and in the adult life of the offspring [54]. According to another study methylation of histone H3 at lysine 9 was replaced by acetylation upon stimulation of adipocyte differentiation [55]. Moreover, upon TNF-α treatment, the adiponectin concentration was decreased owing to a decrease in acetylation of histone H3 at lysine 9 [55].

Adipocyte protein 2 is exclusively produced in adipocytes and macrophages. It is also known as fatty acid-binding protein 4. The major role of aP2 in these tissues is transport of fatty acids. It binds to both long-chain fatty acids and retinoic acid and delivers them to their cognate receptors in the nucleus of the adipocytes. aP2 has been considered as one of the markers of differentiated adipocytes. According to a study on C3H 10T1/2 mouse MSCs and 3T3-L1 preadipocytes, increased acetylation on H3 and H4 histone tails and in turn activation of aP2 was observed during adipogenesis [40]. This study sheds light on the epigenetic regulation of important regulatory genes of adipogenesis and their role in adipocyte differentiation [40].

The C/EBPs belong to a family of transcription factors that contains six members called C/EBPα to C/EBPζ. They activate gene expression through interaction with promoter regions of target genes. C/EBPα is an important protein for adipogenesis as well as for normal adipocyte function. It also plays a vital role in adipocyte differentiation and regulation of various obesity-related genes. It has been shown that C/EBPα expression is regulated by histone acetylation, and HDAC1 negatively regulates C/EBPα expression and adipogenesis [56]. Another important gene of this family is C/EBPβ, which also plays an important role in the expression of insulin-responsive genes that regulate glucose transport and metabolism in adipocytes [57]. C/EBPβ is also a critical factor for adipogenesis and generation of neurons [58,59]. Deletion of the C/EBPβ gene has been correlated with impaired carbohydrate metabolism [60]. Moreover, C/EBPβ is induced during early stages of adipogenesis, which activates other important factors of adipogenesis such as C/EBPα and PPARγ by cooperating with C/EBPδ to fully establish adipocyte differentiation [61,62]. According to an adipocyte differentiation study on 3T3-L1 preadipocytes,

C/EBPβ has increased activation marks on histones in its exonic and conserved 3′ UTR region including H3K4 trimethylation, K9/K14 acetylation, and H4K20 monomethylation during the early stages of adipogenesis [40]. In the same study, H3K27 trimethylation of the C/EBPα gene was found to be marked by H3K27 trimethylation in preadipocytes, and a decrease in these repressive marks was observed upon gene activation. In addition, trimethylation of the C/EBPα gene was also observed during adipogenesis, which is a histone code for active transcription. From these results it was inferred that the C/EBPα gene is under the control of bivalent histone modifications in preadipocytes during adipogenesis [40].

PPARα is one of the ligand-activated transcription factors belonging to the steroid hormone receptor superfamily. PPARα expression is predominantly high in tissue with high fatty acid catabolism, such as liver, heart, and muscle. A number of genes critical for lipid and lipoprotein metabolism are regulated by PPARα. One study suggests that PPARα is regulated by DNA methylation. According to this study the offsprings of pregnant rats fed a protein-restricted diet show hypomethylation of specific CpG dinucleotides in their hepatic PPARα promoter [63]. Another receptor of this family is PPARγ, which plays an important role in adipogenesis and obesity development. PPARγ has many common names such as the glitazone receptor and NR1C3 (nuclear receptor subfamily 1, group C, member 3). There are two isoforms of PPARγ in human and mouse: PPARγ1 (expressed in all tissues except muscle) and PPARγ2 (expressed exclusively in adipose tissue and the intestine). PPARγ has been reported to regulate fatty acid storage and glucose metabolism, and genes activated by PPARγ stimulate lipid uptake and adipocyte differentiation [64]. A PPARγ-knockout study in mice suggests that adipocyte generation was impaired, which further provides evidence for the importance of PPARγ in adipogenesis [65]. Moreover, a study on 10T1/2 mouse MSCs and 3T3-L1 preadipocytes showed that during adipogenesis H3K9 acetylation increases significantly at the PPARγ2 promoter, which proves that epigenetic modulation of PPARγ has an important function in adipocyte differentiation [40].

Pref-1 belongs to the Delta family of epidermal growth factor-like repeat-containing proteins. Pref-1 is also known as DLK1 (Delta, *Drosophila*, homolog-like 1) and FA1 (fetal antigen 1). Owing to exclusive expression of Pref-1 in preadipocytes, it has been utilized as an excellent marker for identification of preadipocytes [66]. From the study involving inhibition of adipogenesis by constitutive expression of Pref-1, it can be speculated that Pref-1 acts as an inhibitor of adipogenesis [67]. In the study involving 10T1/2 MSCs and 3T3-L1 preadipocytes, there was an increase in H3K27 trimethylation (a repressive histone mark) at the Pref-1 gene in 10T1/2 MSCs compared to 3T3-L1 preadipocytes. This identification suggests that Pref-1 is under epigenetic regulation by means of histone methylation during adipogenesis [40].

Another important gene playing a vital role in adipogenesis is uncoupling protein 1 (UCP1). UCP1, also known as thermogenin, is a mitochondrial transporter protein found in brown adipose tissue (BAT). UCP1 creates proton leaks across the inner mitochondrial membrane, which in turn results in uncoupling of oxidative phosphorylation from ATP synthesis and as a result energy is dissipated in the form of heat. UCP1 expression is BAT-specific and regulation of UCP1 is by DNA methylation in the enhancer region [68].

Leptin is a hormone produced exclusively by adipocytes in response to increased fat reserves and plays an important role in the regulation of the adipose-tissue mass and body weight. This regulation is via inhibition of food intake and stimulation of energy expenditure. Disruption in leptin function has been correlated with severe obesity, which was further proved by an obese phenotype in leptin-knockout mice [69]. According to a few studies the leptin gene is regulated by DNA methylation [70,71]. On the basis of a 2010 report the epigenetic profile of the leptin gene is affected by plasma glucose levels (only above a critical threshold). Changes in the leptin gene promoter methylation have been correlated with impaired glucose tolerance during pregnancy in obese women and the differential methylation pattern of the leptin gene has been associated with increased risk of developing obesity and type 2 diabetes during adult life of an affected child [71]. Moreover, a decrease in promoter methylation in the leptin gene was observed in women who had undergone successful dietary intervention, which suggests that the methylation status of the leptin gene can be utilized as an epigenetic biomarker to check response to a low-calorie diet in obese individuals [70].

The melanocortin 4 receptor (MC4R) is a membrane-bound receptor that interacts with adrenocorticotropic and melanocyte-stimulating hormones. Signal transduction by MC4R is mediated by G proteins, which further stimulate adenylate cyclase. The expression of MC4R is exclusive to the nervous system and it plays a crucial role in the regulation of energy homeostasis by regulation of food intake. Mutations in the MC4R gene have been associated with human obesity. According to a 2010 study, decreased methylation status in the MC4R gene was observed in response to a long-term high-fat diet in the obese Berlin fat mouse inbred line and the lean B6 mouse line [72].

Neuropeptide Y (NPY) is a peptide neurotransmitter that acts as a ligand for a family of five G-protein-coupled receptors (Y1–Y5) found in a number of tissues. NPY is associated with food uptake and obesity. NPY decreases lipolysis and promotes adipogenesis in, adipocytes which suggests that NPY has advantageous effects on lipid uptake and storage in fat [73,74]. Another important gene related to appetite regulation is pro-opiomelanocortin (POMC). It is a complex propeptide that acts as precursor of a number of hormones and neuropeptides released by tissue-specific proteolytic processing. POMC is further processed to yield peptides that play important roles in various biological processes such as skin pigmentation and control of adrenal growth and function. These peptides include adrenocorticotropin, melanocyte stimulating hormone, β-lipotropin, and β-endorphin. POMC expression is very high in the arcuate nucleus of the hypothalamus and POMC-expressing neurons are critical in appetite regulation and energy homeostasis [75]. A few studies on POMC suggest that mice and humans having a defect in the production of POMC-derived peptides show severe obesity [76]. According to a 2013 study, the methylation status of the NPY and POMC promoters in the leukocytes of obese men can be correlated with weight regain in these individuals [77]. From this study it can be speculated that NPY and POMC genes are differentially regulated by DNA methylation in the promoter region in cases of obesity, and the methylation status of these genes can be used as prognostic markers for obesity management [77].

The fat mass and obesity-associated (FTO) gene was discovered in a GWAS. In these studies, FTO was associated with obesity and type 2 diabetes [78,79]. FTO is mainly expressed in the hypothalamus and it encodes an enzyme called 2-oxoglutarate-dependent nucleic acid demethylase. It has been speculated that FTO might have an important role in the regulation of energy homeostasis, nucleic acid demethylation, and lipolysis [80,81]. A genome-wide survey in human peripheral white blood cells revealed predisposing diabetes type 2-related DNA methylation variation in the FTO gene. In this study, a small but significant amount of hypomethylation was observed on the CpG site present in the first intron of the FTO gene in T2D individuals compared to control [82].

GLUT4 belongs to the family of glucose transporters, which facilitates the rate-limiting step in glucose metabolism by enhancing uptake of glucose by skeletal muscles and adipose tissue in response to insulin signaling. Reduction in GLUT4 gene expression has been observed in the case of reduced glucose uptake due to insulin resistance. Furthermore, insulin resistance has been demonstrated in obesity and a reduction in GLUT4 has been correlated with obesity [83]. There is some evidence demonstrating regulation of GLUT4 expression by epigenetic modulation. In one such study, it was proposed that GLUT4 gene expression in 3T3-L1 cell differentiation from preadipocytes to adipocytes is regulated by changes in DNA methylation status of 5′ CpG sites present in the GLUT4 promoter region [84].

Insulin-like growth factor 2 (IGF-2) plays an important role in growth and development. IGF-2 promotes the growth and division of various kinds of cells. IGF-2 gene expression is specific for fetal tissue during development. In adults, expression and activity of IGF-2 are much less. According to various studies, only the paternal IGF-2 allele is transcribed in normal human tissues. This phenomenon of genomic imprinting of

IGF-2 is regulated by two differentially methylated regions upstream of the IGF-2 promoter [85]. According to one study on adipose tissue and skeletal muscle taken from lean and morbidly obese individuals with or without type 2 diabetes, tissue-specific DNA methylation of the IGF-2 gene was correlated with obesity and T2D [86]. In another such study, paternal obesity was associated with a decrease in DNA methylation of differentially methylated regions of the IGF-2 gene [87].

Insulin is a hormone exclusively secreted by pancreatic β cells in response to blood glucose level. Insulin acts as a master regulator of metabolic homeostasis. Insulin gene expression begins in the early embryonic development of the pancreas and remains under tight regulation throughout adult life. Disruption of insulin gene expression in pancreatic β cells leads to diabetes. It has been reported that regulation of insulin expression by demethylation of the CpG site present in the insulin promoter plays an important role in pancreatic β cell maturation and tissue-specific insulin expression [88].

Insulin receptor substrate 1 (IRS-1) is the primary substrate of the insulin receptor and it plays a central role in insulin signaling. It has been speculated that IRS-1 acts as a multisite docking protein that binds to various signal transduction proteins and links the insulin receptor kinase to a variety of biological functions in response to insulin signaling. Defect in IRS-1 plays a crucial role in insulin resistance in the case of T2D [89]. Differential gene expression and DNA methylation status of IRS-1 have been observed in adipose tissue of subjects with T2D compared with control subjects. This suggests that IRS-1 expression in adipose tissue is regulated by promoter DNA methylation [90].

Hypoxia-inducible factor 1α (HIF1α) is a transcription factor whose expression increases in response to hypoxia. Under normoxic conditions, HIF1α is hydroxylated on proline and asparagine residues, which leads to ubiquitination-mediated degradation of HIF1α. Under hypoxic conditions, HIF1α degradation is inhibited, which leads to increased activity of HIF1α in cells. During obesity the hypoxia response has been reported in adipose tissue in various studies [91]. Moreover, the epigenetic regulation of HIF1α by DNA methylation and histone modification has been reported in various studies [92]. From these studies it can be speculated that epigenetic regulation of HIF1α might be playing an important role in the development of obesity.

Inflammation in adipose tissue in the severe cases of obesity has been reported. Moreover, it has been speculated that obesity-related inflammation at the systemic level leads to insulin resistance and diabetes [93]. Inflammation of adipose tissue by macrophages plays an important role in the development of insulin resistance in obesity [93]. However, other lymphocyte subsets have also been associated with obesity, including T cells and natural killer (NK) cells [94,95]. Interferon-γ (IFN-γ) is a cytokine secreted by T cells and NK cells that regulates macrophage response and triggers the formation and release of reactive oxygen species at the site of infection or inflammation. IFN-γ has been associated with obesity-related inflammation and insulin resistance [95]. The other important mediator of obesity-related inflammation is TNF-α. It is in fact a multifunctional pro-inflammatory cytokine that plays important roles in the pathogenesis of obesity and obesity-associated insulin resistance [96]. Epigenetic regulation of IFN-γ and TNF-α expression has been reported in various studies. According to a study on peripheral blood DNA, IFN-γ, and TNF-α are regulated based on DNA methylation status of these genes [97]. Another example is the NR3C1 gene, which encodes nuclear receptor subfamily 3, group C, member 1, which belongs to the glucocorticoid receptor family. NR3C1 plays a vital role in cell proliferation and differentiation. Moreover, it has been reported that NR3C1 expression is involved in biological processes that influence birth weight [98]. Changes in methylation status of NR3C1 in newborns have been associated with maternal prenatal stress exposure, newborn birth weight, and adult-onset diseases [98].

Fatty acid synthase (FASN) is an important lipogenic enzyme that plays a key role in the synthesis of long-chain fatty acids from the acetyl-CoA precursors. Synthesis of long-chain fatty acids by fatty acid synthase is known as lipogenesis, which is under tight regulation by hormones and nutritional signaling. FASN expression is exclusive to lactating mammary gland, liver, and adipose tissue. Obesity-related overexpression of FASN is due to a specific region of the FAS promoter. Dysregulated expression of FASN has been correlated with pathological conditions such as obesity [99]. Additionally, a study of laboratory-induced overweight rats, owing to high fat intake, suggested that the methylation patterns of several CpG islands present in the promoter region of the FASN gene are associated with metabolic alterations related to obesity [100].

LDLRAP1 is a cytosolic protein that interacts with the cytoplasmic tail of the low-density lipoprotein (LDL) receptor. LDLRAP1 has a phosphotyrosine-binding domain, which interacts with the LDL receptor. The major function of LDLRAP1 is to remove excessive cholesterol from the bloodstream, and mutation in this gene leads to autosomal recessive hypercholesterolemia due to malfunction of LDL receptor signaling [101]. MicroRNA-mediated regulation of LDLRAP1 expression has been reported. According to this study, miR-168a binds to LDLRAP1 mRNA and regulates the expression of this gene in liver [52].

SOD2 and SOD3 belong to the superoxide dismutase enzyme family, which converts two superoxide radicals to hydrogen peroxide and diatomic oxygen. SOD2 is exclusively expressed in mitochondria, which neutralize superoxide by-products of oxidative phosphorylation. Moreover, it has been reported that oxidative damage associated with obesity and T2D is prevented by overexpression of SOD2 [102].

SOD3, also known as extracellular superoxide dismutase, is secreted into the extracellular space and forms a glycosylated homotetramer anchored to the extracellular matrix and cell surfaces. SOD3 has been speculated to protect the brain, lungs, and pulmonary arteries from oxidative stress. According to one study, increase in SOD3 gene expression protects against high-fat diet-induced obesity, fatty liver, and insulin resistance [103]. Both SOD2 and SOD3 have been reported to be regulated by epigenetic phenomena. Hypermethylation of the single CpG dinucleotide within the SOD2 gene in response to disrupted redox homeostasis and oxidative stress has been observed in the case of defective cardiorespiratory homeostasis [104]. A study on SOD3 suggests that cell-specific and IFN-γ-inducible expression of SOD3 in cells of the pulmonary artery is regulated by epigenetic mechanisms including DNA methylation and histone acetylation [105].

Methylenetetrahydrofolate reductase (MTHFR) is an enzyme that converts 5,10-methylenetetrahydrofolate to 5-methyltetrahydrofolate. This reaction is important in the synthesis of the important amino acid methionine and folic acid (vitamin B9). Methionine is an important amino acid of many proteins and it is important for the synthesis S-adenosylmethionine, which is a crucial mediator of DNA methylation. Moreover, a single-nucleotide polymorphism of the MTHFR gene has been correlated with obesity and related metabolic syndrome owing to a defect in folate and S-adenosylmethionine (SAM) metabolism [106]. DNA methylation status of the MTHFR gene promoter in peripheral blood cells has been correlated with end-stage renal disease [107]. Moreover, miR-22- and miR-29b-mediated downregulation of the MTHFR gene promotes liver carcinogenesis [108]. From these studies it can be speculated that epigenetic regulation of MTHFR might have an important function in the development of obesity and associated metabolic disorders.

The genes discussed above are summarized in Table 1.

4. NUTRITIONAL EPIGENETICS

Research indicates that certain dietary factors can alter gene expression via changes in DNA methylation and histone modification marks [109,110]. There are some reports suggesting that vitamin B deficiency may lead to changes in DNA methylation and histone modifications because of insufficient production of SAM, the universal methyl donor [111]. Chronic alcohol consumption can also alter epigenetic marks via reduction of methionine, choline, and vitamin B availability, which ultimately results in reduction in the availability of SAM [112]. The roles of various dietary factors in the alteration of epigenetic modifications are included in Table 2.

TABLE 1 Important Genes Involved in the Development of Obesity that Are Regulated by Epigenetic Mechanisms

Metabolic process	Gene symbol	Common gene name	Gene locus	Epigenetic mechanism	Reference
Adipogenesis	ADIPOQ	Adiponectin	3q27	Histone acetylation and methylation	[54,55]
	aP2	Adipocyte protein 2	8q21.13	Histone acetylation	[40]
	C/EBPα	CCAAT/enhancer-binding protein α	19q13.1	Histone acetylation and methylation	[40]
	C/EBPβ	CCAAT/enhancer-binding protein β	20q13.13	Histone methylation	[40]
	PPARα	Peroxisome proliferator-activated receptor α	22q12-q13.1	DNA methylation	[63]
	PPARγ	Peroxisome proliferator-activated receptor γ	3p25.2	Histone acetylation	[40]
	Pref-1	Preadipocyte factor 1	14q32.2	Histone methylation	[40]
	UCP1	Uncoupling protein 1 (mitochondrial, proton carrier)	4q28-q31	DNA methylation	[68]
Appetite regulation	LEP	Leptin	7q31	DNA methylation	[70]
	MC4R	Melanocortin 4 receptor	18q22	DNA methylation	[72]
	NPY	Neuropeptide Y	7p15.3	DNA methylation	[77]
	POMC	Pro-opiomelanocortin	2p23	DNA methylation, histone acetylation and methylation	[77]

Continued

TABLE 1 Important Genes Involved in the Development of Obesity that Are Regulated by Epigenetic Mechanisms—cont'd

Metabolic process	Gene symbol	Common gene name	Gene locus	Epigenetic mechanism	Reference
Body weight homeostasis	FTO	Fat mass and obesity associated	16q12.2	DNA methylation	[82]
Glucose homeostasis	GLUT4	Insulin-responsive glucose transporter 4	17p13	DNA methylation	[84]
	IGF-2	Insulin-like growth factor 2	11p15.5	DNA methylation and histone acetylation and methylation	[87]
	INS	Insulin	11p15.5	DNA methylation and histone acetylation	[88]
	IRS-1	Insulin receptor substrate 1	2q36	DNA methylation	[90]
Hypoxia	HIF1α	Hypoxia-inducible factor 1α	14q23.2	DNA methylation and histone acetylation and methylation	[92]
Inflammation	IFN-γ	Interferon γ	12q14	DNA methylation	[97]
	NR3C1	Nuclear receptor subfamily 3, group C, member 1 (glucocorticoid receptor)	5q31-q32	Histone acetylation	[98]
	TNF-α	Tumor necrosis factor α	6p21.3	DNA methylation	[97]
	FASN	Fatty acid synthase	17q25	DNA methylation	[100]
Lipid storage and metabolism	LDLRAP1	Low-density lipoprotein receptor adaptor protein 1	1p36-p35	MiRNA-mediated regulation	[52]
Oxidative stress	SOD2	Superoxide dismutase 2, mitochondrial	6q25	DNA methylation	[104]
	SOD3	Superoxide dismutase 3, extracellular	4p15.2	DNA methylation and histone acetylation	[105]
Vitamin metabolism	MTHFR	Methylenetetrahydrofolate reductase (NAD(P)H)	1p36.3	DNA methylation and miRNA-mediated regulation	[107,108]

Type of nutritional factor	Nutritional factor	Metabolic effects	Epigenetic mechanisms	Reference
Fatty acids	Eicosapentaenoic acid	Fatty acid metabolism	DNA methylation	[113]
	Docosahexaenoic acid	Fatty acid metabolism	DNA methylation	[114]
	Arachidonic acid	Fatty acid metabolism	DNA methylation	[115]
	Butyrate acetate propionate and other short chain fatty acids	Fatty acid metabolism, insulin resistance, inflammation	Histone acetylation and propionylation	[116]
Methyl donors	Betaine	Liver steatosis, insulin resistance	DNA and histone methylation	[117]
	Choline	Liver steatosis	DNA and histone methylation	[118]
	Folate	Adiposity, insulin resistance	DNA methylation	[119]
	Glycine, histidine, and serine	Amino acid metabolism	DNA and histone methylation	[120]
	Methionine	Insulin resistance, obesity	DNA and histone methylation	[119,120]
	Vitamin B12	Insulin resistance, obesity	DNA methylation	[119,120]
Phytochemicals (polyphenols and others)	Curcumin	Inflammation, obesity	DNA methylation, histone acetylation and microRNA	[121]
	Epigallocatechin 3-gallate	Obesity, insulin resistance, liver steatosis	DNA methylation and histone acetylation	[122]
	Genistein	Obesity	DNA methylation and histone acetylation	[123]
	Resveratrol	Obesity, liver steatosis	Histone acetylation	[124]
	Soy isoflavones	Body weight, insulin sensitivity	DNA methylation	[125]
	Sulforaphane	Adipocyte differentiation	Histone acetylation	[126]
Vitamins	Ascorbate (vitamin C)	Antioxidant processes	DNA methylation	[127]
	Retinol	Antioxidant processes	Histone acetylation	[128]
	Tocopherols	Antioxidant processes	Histone acetylation	[123]

5. WHEN DO DIETARY FACTORS INFLUENCE THE EPIGENOME?

Nutrition is known to induce epigenetic modifications, but the underlying reasons are still to be deciphered. Links between diet and epigenetic modifications can be categorized into the following: (1) during embryonic development (fetal and neonatal growth) and lactation and (2) during adult life.

5.1 Embryonic Development and Lactation

Epigenetic variations during early stages of development, when the majority of organ systems undergo rapid development, contribute to maintaining undifferentiated stem cells as well as organogenesis in embryo [129]. Thus, the establishment of the epigenetic pattern is a very important process during early embryogenesis in mammals. The genome-wide demethylation takes place during fertilization and implantation of the embryo into the uterus [130]. Remethylation of the genome sets epigenetic marks for different cell types soon after implantation. These early developmental periods, therefore, during which the epigenetic code is partially removed or reset, are vital. Once established, these epigenetic marks remain highly stable. Dysregulation of gene expression might occur owing to failure of cells to complete these epigenetic reprogramming mechanisms in time. Such regulatory period is susceptible to environmental influences, especially dietary factors that can disturb the precise establishment of the epigenetic code. This accounts for why dietary influence during the early developmental period might have such delayed effects on the metabolism of the individual [131].

Studies of the Dutch Hunger Winter, a scarcity of food that occurred in the Netherlands during the winter of 1944 for a period of 4 months, sheds light on the effect of maternal diet on subsequent disease susceptibility. According to those studies, the disease susceptibility of the offspring of women who were pregnant during the Dutch Famine was different depending on time of development. Individuals affected early in pregnancy with maternal malnutrition had normal birth weight, but in adult life they had cardiovascular abnormalities along with a proatherogenic lipid profile and reduced cognitive function [132]. Maternal malnutrition during midgestation was correlated with defective kidney and lung function [132]. Moreover, individuals who suffered maternal malnutrition at the end of gestation had low birth weight and at adult age they showed increased incidence of insulin resistance and hypertension [133]. The role of epigenetic modifications for this difference in disease susceptibility remains unknown. But from these studies it can be speculated that different tissue-dependent epigenetic alterations may be induced by various environmental factors depending on the time of development, the intensity, and the duration of exposure.

Changes in risk of colon cancer in the progeny are correlated with intake of vitamin B in the maternal diet. Moreover, colorectal cancer has been correlated with loss of imprinting of the IGF-2 gene in rodents as well as in humans [134,135]. Increase in methylation of the IGF-2 gene in the offspring has been correlated with folic acid supplementation during pregnancy [136]. Furthermore, maternal protein restriction has been correlated with reduced methylation and elevated expression of PPARα in rats [137]. Reports on pigs suggest that maternal diets with high and low protein contents effectively change global DNA methylation through changes in DNMT1, DNMT2, and DNMT3 levels in the liver and skeletal muscle of newborn offspring [138]. Moreover, many studies have reported the role of calorie restriction, total energy intake, and diet composition during development and lactation to have a major impact on epigenetic modulation of obesity development [139]. Furthermore, some evidence suggests that intake of a sufficient amount of leptin from breast milk prevents obesity in adult life [140]. These observations suggest that maternal diet has a large influence on progeny in terms of the development of various diseases such as cancer and obesity by epigenetic modulation of various important factors (Figure 2).

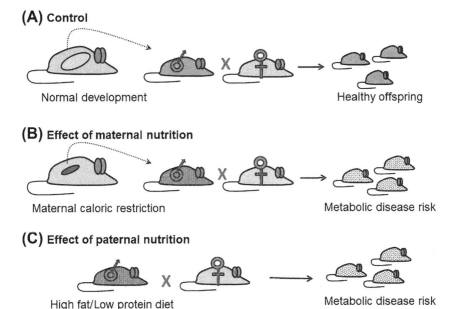

(A) Control

Normal development   Healthy offspring

(B) Effect of maternal nutrition

Maternal caloric restriction   Metabolic disease risk

(C) Effect of paternal nutrition

High fat/Low protein diet   Metabolic disease risk

FIGURE 2 **Effects of parental nutrition on susceptibility of offspring to obesity.** (A) Under control conditions, the embryo undergoes normal development, giving rise to a healthy offspring. (B) When the mother is subjected to caloric restriction, the embryo is not healthy, giving birth to a child with high metabolic disease risk. (C) When the father is fed with a diet rich in fat and deficit in protein, the offspring is again at higher risk of metabolic disease.

5.2 Nutritional Epigenetics in Adult Life

Epigenetic alterations are not limited only to early development but can occur throughout the life of the organism. Such epigenetic modifications over time accumulate and may finally lead to disease susceptibility. There are some reports that epigenetic changes can be influenced by the limitation or overconsumption of various dietary factors during adulthood. Intake of methyl donors has been correlated to DNA methylation and seems to play an important role in FASN methylation and ultimately affect normal liver function [141]. In rodents, intake of high fat and sugar has been correlated with alterations in DNA methylation patterns, which lead to dysregulation of genes important for energy homeostasis and obesity. These genes include LEP, NADH dehydrogenase (ubiquinone) 1β subcomplex subunit 6, and FASN [26,100].

Along with diet-induced epigenetic changes, there have been reports suggesting that exercise has also some role in epigenetic modulation of obesity- and T2D-associated genes [15,19]. Studies suggest that acute exercise increases the expression of PGC1α, mitochondrial pyruvate dehydrogenase lipoamide kinase isozyme 4 (PDK4), and PPARδ in skeletal muscle by decreasing promoter methylation of these genes with respect to intensity of exercise [7]. In this study, in response to acute exercise, a decrease in whole-genome methylation was observed in skeletal muscle biopsies obtained from healthy sedentary men and women [7]. Dose-dependent expression of PGC1α, PDK4, and PPARδ, along with significant hypomethylation of the promoters of their respective genes, was observed in response to exercise [7]. Additionally, a decrease in promoter methylation of PGC1α, PDK4, and PPAR-δ was observed in a study involving mouse muscles 45 min after ex vivo contraction [7]. These suggest that lifestyle factors such as exercise play an important role in regulation of body weight and energy homeostasis via epigenetic modulation of various important regulatory genes involved in obesity development [7].

6. GENOMIC IMPRINTING IN OBESITY

In mammals, both parents contribute two complete sets of chromosomes to the offspring and most of the autosomal genes are expressed from both maternally and paternally inherited alleles [142]. A few genes express characters from only one allele, their expression being determined by the parent during gametogenesis. These are dubbed as imprinted genes and represent only a small subgroup of mammalian genes. Imprinting is essentially a differential expression pattern of the two parental alleles (Figure 3). Such imprinted genes are targets for environmental agents to induce epigenetic changes without altering the underlying DNA coding region. Since

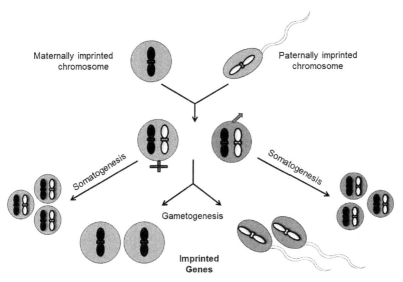

FIGURE 3   Mechanism of genomic imprinting. Somatic cells inherit genes from both parents. However, in all gametes, instead of random shuffling, the same set of genes is received from the parent.

the early 1980s genomic imprinting has been studied in humans and has been accounted for in several human disorders. Differential methylation patterns of CpG islands underlie the phenomenon of genomic imprinting [19]. Imprinted domains comprise clusters of imprinted genes on different chromosomes, which are under the control of an imprinting center. Imprinted genes show tissue- and stage-specific expression patterns [143]. One of the potential roles of imprinting is differentiation of tissue types by determining the transcription rate of genes with greater influence on cellular growth.

The first report in humans of genomic imprinting was Prader–Willi syndrome (PWS). PWS occurs because of a paternal deletion or uniparental disomy (both chromosomes from only one parent) of chromosome 15. Angelman syndrome is another reported example of a rare genetic disorder occurring owing to uniparental disomy [142]. PWS is a neurodevelopmental genetic complication with obesity as its major health problem [144]. The subject in early childhood develops insatiable appetite leading to weight gain and subsequent obesity with caloric restriction. Other peculiar characteristics of PWS are a particular facial appearance (narrow bifrontal diameter with almond-shaped eyes, small upturned nose, and downturned corners of the mouth), hypogonadism and hypogenitalism in both sexes, growth hormone deficiency, mild learning and behavioral problems (e.g., skin picking, temper tantrums), and hyperphagia leading to early childhood obesity [145]. If uncontrolled, PWS may turn out to be the most common genetic cause of morbid obesity in children. The decreased muscle tone and mass in adults with PWS contributes to a lower metabolic rate and physical inactivity subsequently causing obesity. The only management option would then be weight control by restricting dietary and caloric intake. Human recombinant growth hormone therapy will also decrease body weight and fat and increase muscle mass and is so used for managing PWS [146]. Approximately 70% of PWS cases have sporadic deletion of the 15q11–q13 region [144]. Either there is a defect in the imprinting center controlling the activity of imprinted genes or it is due to other chromosome 15 rearrangements [145]. Many genes, such as SNURF–SNRNP, small nucleolar RNAs, NDN, MAGEL2, MKRN3, etc., have been mapped to the 15q11–q13 region, the part that is interstitially deleted from chromosome 15 in PWS [147].

A study was carried out to detect obesity-related genetic loci that are potentially imprinted [148]. A genome-wide parent-of-origin linkage analyses using a European American sample of 1297 individuals under an allele-sharing model for discrete obesity-related traits was performed. A maternal effect in chromosome region 10p12 was found on trait analysis; however, the strongest evidence for a maternal effect was seen in region 12q24. A paternal effect was mapped to region 13q32 [148]. The results suggest that genomic imprinting may play a role in human obesity.

7. OBESITY MANAGEMENT

Being a global health burden, proper preventive and therapeutic interventions are necessary to manage obesity and obesity-related health problems. Preventive interventions for obesity are more focused on dietary habits, increased physical activity, and other lifestyle changes, whereas therapeutic interventions to treat obesity and obesity-related complications include more aggressive approaches such as pharmacologic treatments and surgery along with dietary habits, exercise, and other lifestyle factors. The type of treatment of obesity varies with the individual based on the severity of disease condition as well as the socioeconomic background of the patient. For individuals with low to moderate severity, dietary restrictions and exercise are recommended. In cases in which weight loss is not sufficient by dietary restrictions and exercise, pharmacological treatments are recommended, which include FDA-approved drugs such as orlistat and lorcaserin. Orlistat is a pancreatic lipase inhibitor, widely used for the treatment of obesity, which reduces the absorption of dietary fat. It is also effective for both short-term and long-term weight reduction when supplemented with proper dietary and exercise interventions [149]. Despite its great therapeutic application in obesity, orlistat is correlated with a higher incidence of gastrointestinal adverse effects, which limits its usage as an effective therapeutic option for obesity [149]. Another therapeutic drug used for the treatment of obesity is sibutramine, which is a serotonin–norepinephrine reuptake inhibitor that increases feeling of fullness after food intake and promotes energy expenditure. Clinical studies of this drug demonstrate constant effects on weight loss and obesity treatment. Moreover, the beneficial effects of sibutramine on triglycerides, high-density lipoprotein cholesterol, and glycemic control have also been documented [150]. It has been anticipated that sibutramine can cause increased blood pressure in obese individuals. Because of this adverse effect sibutramine is not recommended for use in obese patients with hypertension, which limits its therapeutic usefulness [150].

In most of cases surgery is reserved for severely high obesity with BMI of 35–40 kg/m^2, when no other nonsurgical treatments provide effective results. However, the use of surgery for the treatment of obesity is increasing irrespective of the above criteria in private health clinics in developed as well as developing countries. Because of a lack of sufficient clinical evidence the comparative safety and efficacy of different surgical procedures used for the treatment of obesity is still unclear [151].

Studies suggest that epigenetic regulation plays a key role in the onset of obesity and a correlation of obesity with epigenetic modulation may be utilized for the development of an important therapeutic approach for the treatment of obesity [152]. The inhibitors of important enzymes regulating epigenetic modifications (HATs, HDACs, and DNMTs) are being extensively screened for their role as anticancer agents. As epigenetic modulations play important roles in obesity development, the use of drugs affecting epigenetic modifications in the treatment of obesity can become a novel approach to managing obesity. For example, increased expression of endogenous GLUT4 mRNA was observed in preadipocytes when the levels of class II HDACs in the nucleus were reduced upon treatment with phenylephrine (an α-adrenergic receptor antagonist), leading to translocation of HDACs from the nucleus to the cytoplasm, and by small interfering RNA-mediated knockdown of class II HDACs [153].

8. CHALLENGES AND FUTURE PERSPECTIVES

Higher energy uptake compared to expenditure is the major cause of obesity. Moreover, research suggests that inherited genetic factors under the influence of environmental signals determine susceptibility of individuals to developing obesity and associated complications. According to emerging evidence, environmental factors such as diet and other lifestyle factors play pivotal roles in obesity development via epigenetic modulation of obesity-related genes. There have been some reports suggesting that many nutritional factors have the ability to modulate DNA methylation or histone modifications, some of which may be utilized in the treatment of obesity owing to their epigenetic mechanisms. Moreover, the time period of susceptibility of the epigenome for such environmental factors needs to be defined in a proper manner.

Future research focusing on the identification of more susceptibility loci for obesity along with understanding the mechanisms and networking between DNA methylation, histone modifications, and miRNAs can pave the way for effective therapeutic approaches to obesity management. Moreover, applications of drugs targeting epigenetic modulators such as HATs, HDACs, and DNMTs can be utilized as a novel therapy for obesity and related conditions.

LIST OF ACRONYMS AND ABBREVIATIONS

aP2	Adipocyte protein 2
BMI	Body mass index
C/EBP	CCAAT-enhancer-binding protein
DNMT	DNA methyltransferase
FASN	Fatty acid synthase
FDA	Food and Drug Administration
GLUT4	Glucose transporter type 4
GWAS	Genome-wide association studies
HAT	Histone acetyltransferase
HDAC	Histone deacetylase
HMT	Histone methyltransferase
IGF-2	Insulin-like growth factor 2
LDLRAP1	Low-density lipoprotein receptor adapter protein 1
LEP	Leptin
MAGEL2	MAGE-like 2
MEF2A	Myocyte-specific enhancer factor 2A
miRNA	MicroRNA
MKRN3	Makorin ring finger protein 3
ncRNA	Noncoding RNA
NDN	Necdin
PDK4	Mitochondrial pyruvate dehydrogenase lipoamide kinase isozyme 4
PGC1α/PPARGC1α	Peroxisome proliferator-activated receptor γ coactivator 1α

PPAR	Peroxisome proliferator-activated receptor
Pref-1	Preadipocyte factor-1
PWS	Prader–Willi syndrome
SAM	S-adenosylmethionine
SNRPN	Small nuclear ribonucleoprotein polypeptide N
SNURF	SNRPN upstream reading frame
SOD1	Superoxide dismutase-1
T2D	Type 2 diabetes
TFAM	Mitochondrial transcription factor A
VMR	Variably methylated region

References

[1] Obesity and overweight. World Health Organization 2011.
[2] Kelly T, Yang W, Chen CS, Reynolds K, He J. Global burden of obesity in 2005 and projections to 2030. Int J Obes (Lond) 2008;32(9):1431–7.
[3] Day FR, Loos RJ. Developments in obesity genetics in the era of genome-wide association studies. J Nutr Nutr 2011;4:222–38.
[4] McAllister EJ, Dhurandhar NV, Keith SW, Aronne LJ, Barger J, Baskin M, et al. Ten putative contributors to the obesity epidemic. Crit Rev Food Sci Nutr 2009;49:868–913.
[5] Mavanji V, Billington CJ, Kotz CM, Teske JA. Sleep and obesity: a focus on animal models. Neurosci Biobehav Rev 2012;36:1015–29.
[6] Bird A. Perceptions of epigenetics. Nature 2007;447(7143):396–8.
[7] Barres R, Yan J, Egan B, Treebak JT, Rasmussen M, Fritz T, et al. Acute exercise remodels promoter methylation in human skeletal muscle. Cell Metab 2012;15(3):405–11.
[8] Nitert MD, Dayeh T, Volkov P, Elgzyri T, Hall E, Nilsson E, et al. Impact of an exercise intervention on DNA methylation in skeletal muscle from first-degree relatives of patients with type 2 diabetes. Diabetes 2012;61(12):3322–32.
[9] Ling C, Groop L. Epigenetics: a molecular link between environmental factors and type 2 diabetes. Diabetes 2009;58(12):2718–25.
[10] Rönn T, Volkov P, Tornberg A, Elgzyri T, Hansson O, Eriksson KF, et al. Extensive changes in the transcriptional profile of human adipose tissue including genes involved in oxidative phosphorylation after a 6-month exercise intervention. Acta Physiol (Oxf) 2014;211(1):188–200.
[11] Rönn T, Ling C. Effect of exercise on DNA methylation and metabolism in human adipose tissue and skeletal muscle. Epigenomics 2013;5(6):603–5.
[12] Rönn T, Ling C. The impact of exercise on DNA methylation of genes associated with type 2 diabetes and obesity in human adipose tissue. US Endocrinol 2014;10(1):64–6.
[13] Jiménez-Chillarón JC, Díaz R, Martínez D, Pentinat T, Ramón-Krauel M, Ribó S, et al. The role of nutrition on epigenetic modifications and their implications on health. Biochimie 2012;94:2242–63.
[14] Brown PO, Hartwell L. Genomics and human disease–variations on variation. Nat Genet 1998;18(2):91–3.
[15] Herrera BM, Keildson S, Lindgren CM. Genetics and epigenetics of obesity. Maturitas 2011;69(1):41–9.

[16] Herbert A. The fat tail of obesity as told by the genome. Curr Opin Clin Nutr Metab Care 2008;11(4):366–70.

[17] Cotsapas C, Speliotes EK, Hatoum IJ, Greenawalt DM, Dobrin R, Lum PY, et al. Common body mass index-associated variants confer risk of extreme obesity. Hum Mol Genet 2009;18(18):3502–7.

[18] Speliotes EK, Willer CJ, Berndt SI, Monda KL, Thorleifsson G, Jackson AU, et al. Association analyses of 249,796 individuals reveal eighteen new loci associated with body mass index. Nat Genet 2011;42(11):937–48.

[19] Bird A. DNA methylation patterns and epigenetic memory. Genes Dev 2002;16(1):6–21.

[20] Bird AP. CpG-rich islands and the function of DNA methylation. Nature 1986; 321(6067):209–13.

[21] Reik W, Dean W. DNA methylation and mammalian epigenetics. Electrophoresis 2001;22(14):2838–43.

[22] Li E, Beard C, Jaenisch R. Role for DNA methylation in genomic imprinting. Nature 1993;366(6453):362–5.

[23] Walsh CP, Chaillet JR, Bestor TH. Transcription of IAP endogenous retroviruses is constrained by cytosine methylation. Nat Genet 1998;20(2):116–7.

[24] Waterland RA, Jirtle RL. Transposable elements: targets for early nutritional effects on epigenetic gene regulation. Mol Cell Biol 2003;23(15):5293–300.

[25] Yamagata Y, Szabo P, Szuts D, Bacquet C, Aranyi T, Paldi A. Rapid turnover of DNA methylation in human cells. Epigenetics 2012;7:141–5.

[26] Milagro FI, Campión J, García-Díaz DF, Goyenechea E, Paternain L, Martínez JA. High fat diet-induced obesity modifies the methylation pattern of leptin promoter in rats. J Physiol Biochem 2009;65:1–9.

[27] Liang H, Walter F. Ward PGC-1α: a key regulator of energy metabolism. Advan Physiol Edu 2006;30:145–51.

[28] Gemma C, Sookoian S, Alvarinas J, García SI, Quintana L, Kanevsky D, et al. Maternal pregestational BMI is associated with methylation of the PPARGC1A promoter in newborns. Obes (Silver Spring) 2009;17:1032–9.

[29] Brons C, Jacobsen S, Nilsson E, Ronn T, Jensen CB, Storgaard H, et al. Deoxyribonucleic acid methylation and gene expression of PPARGC1A in human muscle is influenced by high-fat overfeeding in a birth-weight-dependent manner. J Clin Endocrinol Metab 2010;95:3048–56.

[30] Godfrey KM, Sheppard A, Gluckman PD, Lillycrop KA, Burd - ge GC, McLean C, et al. Epigenetic gene promoter methylation at birth is associated with child's later adiposity. Diabetes 2011;60:1528–34.

[31] Feinberg A, Irizarry R, Fradin D, Aryee MJ, Murakami P, Aspelund T, et al. Personalized epigenomic signatures that are stable over time and covary with body mass index. Sci Transl Med 2010;2:1–16.

[32] Barrès R, Osler ME, Yan J, Rune A, Fritz T, Caidahl K, et al. Non-CpG methylation of the PGC-1alpha promoter through DNMT3B controls mitochondrial density. Cell Metab 2009;10:189–98.

[33] Grundberg E1, Meduri E, Sandling JK, Hedman AK, Keildson S, Buil A, et al. Global analysis of DNA methylation variation in adipose tissue from twins reveals links to disease-associated variants in distal regulatory elements. Am J Hum Genet 2013;93:876–90.

[34] Gonzalo S, Jaco I, Fraga MF, Chen T, Li E, Esteller M, et al. DNA methyltransferases control telomere length and telomere recombination in mammalian cells. Nat Cell Biol 2006;8(4):416–24.

[35] Turner BM. Histone acetylation and an epigenetic code. Bioessays 2000;22(9):836–45.

[36] Litt MD, Simpson M, Gaszner M, Allis CD, Felsenfeld G. Correlation between histone lysine methylation and developmental changes at the chicken betaglobin locus. Science 2001;293(5539):2453–5.

[37] Nakayama J, Rice JC, Strahl BD, Allis CD, Grewal SI. Role of histone H3 lysine 9 methylation in epigenetic control of heterochromatin assembly. Science 2001;292(5514):110–3.

[38] Rountree MR, Bachman KE, Baylin SB. DNMT1 binds HDAC2 and a new co-repressor, DMAP1, to form a complex at replication foci. Nat Genet 2000;25(3):269–77.

[39] Viré E, Brenner C, Deplus R, Blanchon L, Fraga M, Didelot C, et al. The polycomb group protein EZH2 directly controls DNA methylation. Nature 2006;439(7078):871–4.

[40] Zhang Q, Ramlee MK, Brunmeir R, Villanueva CJ, Halperin D, Xu F. Dynamic and distinct histone modifications modulate the expression of key adipogenesis regulatory genes. Cell Cycle 2012;11(23):4310–22.

[41] Wellen KE, Hatzivassiliou G, Sachdeva UM, Bui TV, Cross JR, Thompson CB. ATP-citrate lyase links cellular metabolism to histone acetylation. Science 2009;324:1076–80.

[42] Tateishi K, Okada Y, Kallin EM, Zhang Y. Role of Jhdm2a in regulating metabolic gene expression and obesity resistance. Nature 2009;458:757–61.

[43] Schwer B, Verdin E. Conserved metabolic regulatory functions of sirtuins. Cell Metab 2008;7:104–12.

[44] Siomi H, Siomi MC. On the road to reading the RNA-interference code. Nature 2009;457(7228):396–404.

[45] Kim DH, Sætrom P, Snøve Jr O, Rossi JJ. MicroRNA-directed transcriptional gene silencing in mammalian cells. Proc Natl Acad Sci USA 2008;105(42):16230–5.

[46] Bayne EH, Allshire RC. RNA-directed transcriptional gene silencing in mammals. Trends Genet 2005;21(7):370–3.

[47] Dehwah MA, Xu A, Huang Q. MicroRNAs and type 2 diabetes/obesity. J Genet Genomics 2012;39:11–8.

[48] Carthew RW, Sontheimer EJ. Origins and mechanisms of miRNAs and siRNAs. Cell 2009;136(4):642–55.

[49] Barski A, Jothi R, Cuddapah S, Cui K, Roh TY, Schones DE, et al. Chromatin poises miRNA- and protein-coding genes for expression. Genome Res 2009;19(10):1742–51.

[50] Fabbri M, Garzon R, Cimmino A, Liu Z, Zanesi N, Callegari E, et al. MicroRNA-29 family reverts aberrant methylation in lung cancer by targeting DNA methyltransferases 3A and 3B. Proc Natl Acad Sci USA 2007;104(40):15805–10.

[51] Juan AH, Kumar RM, Marx JG, Young RA, Sartorelli V. Mir-214-dependent regulation of the polycomb protein Ezh2 in skeletal muscle and embryonic stem cells. Mol Cell 2009;36(1):61–74.

[52] Zhang L, Hou D, Chen X, Li D, Zhu L, Zhang Y, et al. Exogenous plant MIR168a specifically targets mammalian LDLRAP1: evidence of cross-kingdom regulation by microRNA. Cell Res 2011;22(1):273–4.

[53] Izadi M, Goodarzi MT, Khalaj HS, Khorshidi D, Doali H. Serum adiponectin levels are inversely correlated with insulin resistance in obese men with type 2 diabetes. Int J Endocrinol Metab 2011;9(1):253–7.

[54] Bouchard L, Hivert MF, Guay SP, St-Pierre J, Perron P, Brisson D. Placental adiponectin gene DNA methylation levels are associated with mothers' blood glucose concentration. Diabetes 2012;61:1272–80.

[55] Sakurai N, Mochizuki K, Goda T. Modifications of histone H3 at lysine 9 on the adiponectin gene in 3T3-L1 adipocytes. J Nutr Sci Vitaminol (Tokyo) 2009;55(2):131–8.

[56] Kuzmochka C, Abdou HS, Haché RJ, Atlas E. Inactivation of histone deacetylase 1 (HDAC1) but not HDAC2 is required for the glucocorticoid-dependent CCAAT/enhancer binding proteinα (C/EBPA) expression and preadipocyte differentiation. Endocrinology 2014;155(12):4762–73. http://dx.doi.org/10.1210/en.2014-1565.

[57] Jain R, Police S, Phelps K, H Pekala P. Tumour necrosis factor-alpha regulates expression of the CCAAT-enhancer-binding proteins (C/EBPs) alpha and beta and determines the occupation of the C/EBP site in the promoter of the insulin-responsive glucose-transporter gene in 3T3-L1 adipocytes. Biochem J 1999;338(3):737–43.

[58] Tang QQ, Gronborg M, Huang H, Kim JW, Otto TC, Pandey A, et al. Sequential phosphorylation of CCAAT enhancer-binding protein beta by MAPK and glycogen synthase kinase 3beta is required for adipogenesis. Proc Natl Acad Sci USA 2005;102:9766–71.

[59] Paquin A, Barnabe-Heider F, Kageyama R, Miller FD. CCAAT/enhancer-binding protein phosphorylation biases cortical precursors to generate neurons rather than astrocytes in vivo. J Neurosci 2005;25:10747–58.

[60] Croniger CM, Millward C, Yang J, Kawai Y, Arinze IJ, Liu S, et al. Mice with a deletion in the gene for CCAAT/enhancer-binding protein beta have an attenuated response to cAMP and impaired carbohydrate metabolism. J Biol Chem 2001;276:629–38.

[61] Lane MD, Tang QQ, Jiang MS. Role of the CCAAT enhancer binding proteins (C/EBPs) in adipocyte differentiation. Biochem Biophys Res Commun 1999;266:677–83.

[62] Hamm JK, Park BH, Farmer SR. A role for C/EBPbeta in regulating peroxisome proliferator-activated receptor gamma activity during adipogenesis in 3T3-L1 preadipocytes. J Biol Chem 2001;276:18464–71.

[63] Karen Lillycrop KA, Phillips ES, Torrens C, Hanson MA, Jackson AA, Burdge GC. Feeding pregnant rats a protein-restricted diet persistently alters the methylation of specific cytosines in the hepatic PPARα promoter of the offspring. Br J Nutr 2008;100(2):278–82.

[64] Wahli W, Braissant O, Desvergne B. Peroxisome proliferator activated receptors: transcriptional regulators of adipogenesis, lipid metabolism and more. Chem Biol 1995;2(5):261–6.

[65] Jones JR, Barrick C, Kim KA, Lindner J, Blondeau B, Fujimoto Y, et al. Deletion of PPARγ in adipose tissues of mice protects against high fat diet-induced obesity and insulin resistance. Proc Natl Acad Sci USA 2005;102(17):6207–12.

[66] Shimomura I, Hammer RE, Richardson JA, Ikemoto S, Bashmakov Y, Goldstein JL, et al. Insulin resistance and diabetes mellitus in transgenic mice expressing nuclear SREBP-1c in adipose tissue: model for congenital generalized lipodystrophy. Genes Dev 1998;12:3182–94.

[67] Wang Y, Zhao L, Smas C, Sul HS. Pref-1 interacts with fibronectin to inhibit adipocyte differentiation. Mol Cell Biol 2010;30(14):3480–92.

[68] Shore A, Karamitri A, Kemp P, Speakman JR, Lomax MA. Role of Ucp1 enhancer methylation and chromatin remodelling in the control of Ucp1 expression in murine adipose tissue. Diabetologia 2010;53:1164–73.

[69] Drel VR, Mashtalir N, Ilnytska O, Shin J, Li F, Lyzogubov VV, et al. The leptin-deficient (ob/ob) mouse: a new animal model of peripheral neuropathy of type 2 diabetes and obesity. Diabetes 2006;55(12):3335–43.

[70] Cordero P, Campion J, Milagro FI, Goyenechea E, Steemburgo T, Javierre BM, et al. Leptin and TNF-alpha promoter methylation levels measured by MSP could predict the response to a low-calorie diet. J Physiol Biochem 2011;67(3):463–70.

[71] Bouchard L, Thibault S, Guay SP, Santure M, Monpetit A, St-Pierre J, et al. Leptin gene epigenetic adaptation to impaired glucose metabolism during pregnancy. Diabetes Care 2010;33(11):2436–41.

[72] Widiker S, Karst S, Wagener A, Brockmann GA. High-fat diet leads to a decreased methylation of the Mc4r gene in the obese BFMI and the lean B6 mouse lines. J Appl Genet 2010;51(2):193–7.

[73] Kos K, Baker AR, Jernas M, Harte AL, Clapham JC, O'Hare JP, et al. DPP-IV inhibition enhances the antilipolytic action of NPY in human adipose tissue. Diabetes Obes Metab 2009;11(4):285–92.

[74] Yang K, Guan H, Arany E, Hill DJ, Cao X. Neuropeptide Y is produced in visceral adipose tissue and promotes proliferation of adipocyte precursor cells via the Y1 receptor. FASEB J 2008;22:2452–64.

[75] Morton GJ, Cummings DE, Baskin DG, Barsh GS, Schwartz MW. Central nervous system control of food intake and body weight. Nature 2006;443(7109):289–95.

[76] Coll AP, Farooqi IS, Challis BG, Yeo GS, O'Rahilly S. Proopiomelanocortin and energy balance: insights from human and murine genetics. J Clin Endocrinol Metab 2004;89(6):2557–62.

[77] Crujeiras AB, Campion J, Díaz-Lagares A, Milagro FI, Goyenechea E, Abete I, et al. Association of weight regain with specific methylation levels in the NPY and POMC promoters in leukocytes of obese men: a translational study. Regul Pept 2013;10(186):1–6.

[78] Dina C, Meyre D, Gallina S, Durand E, Korner A, Jacobson P, et al. Variation in FTO contributes to childhood obesity and severe adult obesity. Nat Genet 2007;39:724–6.

[79] Scott LJ, Mohlke KL, Bonnycastle LL, Willer CJ, Li Y, Duren WL, et al. A genome-wide association study of type 2 diabetes in Finns detects multiple susceptibility variants. Science 2007;316:1341–5.

[80] Olszewski PK, Fredriksson R, Olszewska AM, Stephansson O, Alsiö J, Radomska KJ, et al. Hypothalamic FTO is associated with the regulation of energy intake not feeding reward. BMC Neurosci 2009;10:129–40.

[81] Wahlen K, Sjolin E, Hoffstedt J. The common rs9939609 gene variant of the fat mass and obesity-associated gene FTO is related to fat cell lipolysis. J Lipid Res 2008;49:607–11.

[82] Toperoff G, Aran D, Kark JD, Rosenberg M, Dubnikov T, Nissan B, et al. Genome-wide survey reveals predisposing diabetes type 2-related DNA methylation variations in human peripheral blood. Hum Mol Genet 2012;21:371–83.

[83] Garvey WT, Maianu L, Huecksteadt TP, Birnbaum MJ, Molina JM, Ciaraldi TP. Pre-translational suppression of a glucose transporter protein causes insulin resistance in adipocytes from patients with non-insulin-dependent diabetes mellitus and obesity. J Clin Invest 1991;87(3):1072–81.

[84] Yokomori N, Tawata M, Onaya T. DNA demethylation during the differentiation of 3T3–L1 cells affects the expression of the mouse GLUT4 gene. Diabetes 1999;48:685–90.

[85] Boissonnas CC, Abdalaoui HE, Haelewyn V, Fauque P, Dupont JM, Gut I, et al. Specific epigenetic alterations of IGF2-H19 locus in spermatozoa from infertile men. Eur J Hum Genet 2010;18:73–80.

[86] Chen M, Macpherson A, Owens J, Wittert G, Heilbronn L. Obesity alone or with type 2 diabetes is associated with tissue specific alterations in DNA methylation and gene expression of PPARGC1A and IGF2. J Diabetes Res Clin Metabolism 2012;1:1–8.

[87] Soubry A, Schildkraut JM, Murtha A, Wang F, Huang Z, Berna A, et al. Paternal obesity is associated with IGF2 hypomethylation in newborns: results from a Newborn Epigenetics Study (NEST) cohort. BMC Med 2013;11:29–38.

[88] Kuroda A, Rauch TA, Todorov I, Ku HT, Al-Abdullah IH, Kandeel F, et al. Insulin gene expression is regulated by DNA methylation. PLoS One 2009;4(9):e6953.

[89] Schmitz-Peiffer C. Signalling aspects of insulin resistance in skeletal muscle: mechanisms induced by lipid oversupply. Cell Signal 2000;12:583–94.

[90] Nilsson E, Jansson PA, Perfilyev A, Volkov P, Pedersen M, Svensson MK, et al. Altered DNA methylation and differential expression of genes influencing metabolism and inflammation in adipose tissue from subjects with type 2 diabetes. Diabetes 2014;63:2962–76.

[91] Ye J. Emerging role of adipose tissue hypoxia in obesity and insulin resistance. Int J Obes (Lond) 2009;33:54–66.

[92] Brigati C, Banelli B, di Vinci A, Casciano I, Allemanni G, Forlani A, et al. Inflammation, HIF-1, and the epigenetics that follows. Mediat Inflamm 2010;2010:5. Article ID 263914.

[93] Lumeng CN, Bodzin JL, Saltiel AR. Obesity induces a phenotypic switch in adipose tissue macrophage polarization. J Clin Invest 2007;117(1):175–84.

[94] Feuerer M, Herrero L, Cipolletta D, Atia Naaz A, Wong J, Nayer A, et al. Lean, but not obese, fat is enriched for a unique population of regulatory T cells that affect metabolic parameters. Nat Med 2009;15:930–9.

[95] O'Rourke RW, Metcalf MD, White AE, Madala A, Winters BR, Maizlin II, et al. Depot-specific differences in inflammatory mediators and a role for NK cells and IFN-gamma in inflammation in human adipose tissue. Int J Obes (Lond) 2009;33:978–90.

[96] Hotamisligil GS. Inflammatory pathways and insulin action. Int J Obes Relat Metab Disord 2003;27(3):53–5.

[97] Bollati V, Baccarelli A, Sartori S, Tarantini L, Motta V, Rota F, et al. Epigenetic effects of shiftwork on blood DNA methylation. Chronobiol Int 2010;27(5):1093–104.

[98] Mulligan CJ, D'Errico NC, Stees J, Hughes DA. Methylation changes at NR3C1 in newborns associate with maternal prenatal stress exposure and newborn birth weight. Epigenetics 2012;7(8):1–5.

VII. PERSONALIZED EPIGENETICS OF DISORDERS AND DISEASE MANAGEMENT

[99] Schleinitz D, Klöting N, Körner A, Berndt J, Reichenbächer M, Tönjes A, et al. Effect of genetic variation in the human fatty acid synthase gene (FASN) on obesity and fat depot-specific mRNA expression. Obes (Silver Spring) 2010;18(6):1218–25.

[100] Lomba A, Martinez JA, Garcia-Diaz DF, Paternain L, Marti A, Campion J, et al. Weight gain induced by an isocaloric pair-fed high fat diet: a nutriepigenetic study on FAS and NDUFB6 gene promoters. Mol Genet Metab 2010;101:273–8.

[101] Garcia CK, Wilund K, Arca M, Zuliani G, Fellin R, Maioli M, et al. Autosomal recessive hypercholesterolemia caused by mutations in a putative LDL receptor adaptor protein. Science 2001;292(5520):1394–8.

[102] Aguer C, Pasqua M, Thrush AB, Moffat C, McBurney M, Jardine K, et al. Increased proton leak and SOD2 expression in myotubes from obese non-diabetic subjects with a family history of type 2 diabetes. Biochim Biophys Acta 2013;1832(10):1624–33.

[103] Cui R, Gao M, Qu S, Liu D. Overexpression of superoxide dismutase 3 gene blocks high-fat diet-induced obesity, fatty liver and insulin resistance. Gene Ther 2014;21:840–8.

[104] Nanduri J, Makarenko V, Reddy VD, Yuan G, Pawar A, Wang N, et al. Epigenetic regulation of hypoxic sensing disrupts cardiorespiratory homeostasis. Proc Natl Acad Sci USA 2012;109(7):2515–20.

[105] Zelko IN, Stepp MW, Vorst AL, Folz RJ. Histone acetylation regulates the cell-specific and interferon-gamma-inducible expression of extracellular superoxide dismutase in human pulmonary arteries. Am J Respir Cell Mol Biol 2011;45:953–61.

[106] Di Renzo L, Bigioni M, Bottini FG, Del Gobbo V, Premrov MG, Cianci R, et al. Normal weight obese syndrome: role of single nucleotide polymorphism of IL-15Rα and MTHFR 677C→T genes in the relationship between body composition and resting metabolic rate. Eur Rev Med Pharmacol Sci 2006;10:235–45.

[107] Ghattas M, El-Shaarawy F, Mesbah N, Abo-Elmatty D. DNA methylation status of the methylenetetrahydrofolate reductase gene promoter in peripheral blood of end-stage renal disease patients. Mol Biol Rep 2014;41(2):683–8.

[108] Koturbash I, Melnyk S, James SJ, Beland FA, Pogribny IP. Role of epigenetic and miR-22 and miR-29b alterations in the downregulation of Mat1a and Mthfr genes in early preneoplastic livers in rats induced by 2-acetylaminofluorene. Mol Carcinog 2013;52(4):318–27.

[109] Hardy TM, Tollefsbol TO. Epigenetic diet: impact on the epigenome and cancer. Epigenomics 2011;3(4):503–18.

[110] Park L, Friso S, Choi S. Nutritional influences on epigenetics and age-related disease. Proc Nutr Soc 2012;71(1):75–83.

[111] Niculescu MD, Zeisel SH. Diet, methyl donors and DNA methylation: interactions between dietary folate, methionine and choline. J Nutr 2002;132(8):2333S–5S.

[112] Mason JB, Choi SW. Effects of alcohol on folate metabolism: implications for carcinogenesis. Alcohol 2005;35(3):235–41.

[113] Ceccarelli V, Racanicchi S, Martelli MP, Nocentini G, Fettucciari K, Riccardi C, et al. Eicosapentaenoic acid demethylates a single CpG that mediates expression of tumor suppressor CCAAT/enhancer-binding protein delta in U937 leukemia cells. J Biol Chem 2011;286:27092–102.

[114] Kulkarni A, Dangat K, Kale A, Sable P, Chavan-Gautam P, Joshi S. Effects of altered maternal folic acid, vitamin B12 and docosahexaenoic acid on placental global DNA methylation patterns in Wistar rats. PLoS One 2011;6:e17706.

[115] Kiec-Wilk B, Sliwa A, Mikolajczyk M, Malecki MT, Mathers JC. The CpG island methylation regulated expression of endothelial proangiogenic genes in response to b-carotene and arachidonic acid. Nutr Cancer 2011;63:1053–63.

[116] Gao Z, Yin J, Zhang J, Ward RE, Martin RJ, Lefevre M, et al. Butyrate improves insulin sensitivity and increases energy expenditure in mice. Diabetes 2009;58:1509–17.

[117] Wang Z, Yao T, Pini M, Zhou Z, Fantuzzi G, Song Z. Betaine improved adipose tissue function in mice fed a high-fat diet: a mechanism for hepatoprotective effect of betaine in nonalcoholic fatty liver disease. Am J Physiol Gastrointest Liver Physiol 2010;298:G634–42.

[118] Guerrerio AL, Colvin RM, Schwartz AK, Molleston JP, Murray KF, Diehl A, et al. Choline intake in a large cohort of patients with nonalcoholic fatty liver disease. Am J Clin Nutr 2012;95:892–900.

[119] Sinclair KD, Allegrucci C, Singh R, Gardner DS, Sebastian S, Bispham J, et al. DNA methylation, insulin resistance, and blood pressure in offspring determined by maternal periconceptional B vitamin and methionine status. Proc Natl Acad Sci USA 2007;104:19351–6.

[120] Wang J, Wu Z, Li D, Li N, Dindot SV, Satterfield MC, et al. Nutrition, epigenetics, and metabolic syndrome. Antioxid Redox Signal 2012;17:282–301.

[121] Shao W, Yu Z, Chiang Y, Yang Y, Chai T, Foltz W, et al. Curcumin prevents high fat diet induced insulin resistance and obesity via attenuating lipogenesis in liver and inflammatory pathway in adipocytes. PLoS One 2012;7:e28784.

[122] Yun JM, Jialal I, Devaraj S. Effects of epigallocatechin gallate on regulatory T cell number and function in obese v. lean volunteers. Br J Nutr 2010;103:1771–7.

[123] Dolinoy DC, Weidman JR, Waterland RA, Jirtle RL. Maternal genistein alters coat color and protects Avy mouse offspring from obesity by modifying the fetal epigenome. Environ Health Perspect 2006;114:567–72.

[124] Bujanda L, Hijona E, Larzabal M, Beraza M, Aldazabal P, García-Urkia N, et al. Resveratrol inhibits nonalcoholic fatty liver disease in rats. BMC Gastroenterol 2008;8:40.

[125] Howard TD, Ho SM, Zhang L, Chen J, Cui W, Slager R, et al. Epigenetic changes with dietary soy in cynomolgus monkeys. PLoS One 2011;6:e26791.

[126] Nian H, Delage B, Ho E, Dashwood RH. Modulation of histone deacetylase activity by dietary isothiocyanates and allyl sulfides: studies with sulforaphane and garlic organosulfur compounds. Environ Mol Mutagen 2009;50:213–21.

[127] Chung TL, Brena RM, Kolle G, Grimmond SM, Berman BP, Laird PW, et al. Vitamin C promotes widespread yet specific DNA demethylation of the epigenome in human embryonic stem cells. Stem Cells 2010;28:1848–55.

[128] Moreira JC, Dal-Pizzol F, Rocha AB, Klamt F, Ribeiro NC, Ferreira CJ, et al. Retinol-induced changes in the phosphorylation levels of histones and high mobility group proteins from Sertoli cells. Braz J Med Biol Res 2000;33:287–93.

[129] Reik W. Stability and flexibility of epigenetic gene regulation in mammalian development. Nature 2007;447:425–32.

[130] Santos F, Hendrich B, Reik W, Dean W. Dynamic reprogramming of DNA methylation in the early mouse embryo. Dev Biol 2002;241:172–82.

[131] Faulk C, Dolinoy DC. Timing is everything: the when and how of environmentally induced changes in the epigenome of animals. Epigenetics 2011;6:791–7.

[132] Roseboom T, de Rooij S, Painter R. The Dutch famine and its long-term consequences for adult health. Early Hum Dev 2006;82:485–91.

[133] Ravelli AC, van der Meulen JH, Michels RP, Osmond C, Barker DJ, Hales CN, et al. Glucose tolerance in adults after prenatal exposure to famine. Lancet 1998;351:173–7.

[134] Cui H, Cruz-Correa M, Giardiello FM, Hutcheon DF, Kafonek DR, Brandenburg S, et al. Loss of IGF2 imprinting: a potential marker of colorectal cancer risk. Science 2003;299(5613):1753–5.

[135] Sakatani T, Kaneda A, Iacobuzio-Donahue CA, Carter MG, de Boom Witzel S, Okano H, et al. Loss of imprinting of Igf2 alters intestinal maturation and tumorigenesis in mice. Science 2005;307(5717):1976–8.

[136] Steegers-Theunissen RP, Obermann-Borst SA, Kremer D, Lindemans J, Siebel C, Steegers EA, et al. Periconceptional maternal folic acid use of 400 lg per day is related to increased methylation of the IGF2 gene in the very young child. PLoS ONE 2009;4(11):e7845.

[137] Lillycrop KA, Phillips ES, Jackson AA, Hanson MA, Burdge GC. Dietary protein restriction of pregnant rats induces and folic acid supplementation prevents epigenetic modification of hepatic gene expression in the offspring. J Nutr 2005;135(6):1382–6.

[138] Altmann S, Murani E, Schwerin M, Metges CC, Wimmers K, Ponsuksili S. Maternal dietary protein restriction and excess affects offspring gene expression and methylation of non-SMC subunits of condensin I in liver and skeletal muscle. Epigenetics 2012;7(3):239–52.

[139] Pico C, Palou M, Priego T, Sánchez J, Palou A. Metabolic programming of obesity by energy restriction during the perinatal period: different outcomes depending on gender and period, type and severity of restriction. Front Physiol 2012;3:436.

[140] Pico C, Oliver P, Sánchez J, Miralles O, Caimari A, Priego T, et al. The intake of physiological doses of leptin during lactation in rats prevents obesity in later life. Int J Obes (Lond) 2007;31:1199–209.

[141] Cordero P, Gomez-Uriz AM, Campion J, Milagro FI, Martinez JA. Dietary supplementation with methyl donors reduces fatty liver and modifies the fatty acid synthase DNA methylation profile in rats fed an obesogenic diet. Genes Nutr 2013;8:105–13.

[142] Butler MG. Genomic imprinting disorders in humans: a mini-review. J Assist Reprod Genet 2009;26:477–86.

[143] Butler MG. Prader-Willi syndrome: obesity due to genomic imprinting. Curr Genomics 2011;12:204–15.

[144] Elena G, Bruna C, Benedetta M, Stefania DC, Giuseppe C. Prader-Willi syndrome: clinical aspects. J Obes 2012;2012:473941.

[145] Medeiros CB, Bordallo AP, Souza FM, Collett-Solberg PF. Endocrine management of children with Prader–Willi syndrome. Pediatr Health, Med Ther 2013;4:117–25.

[146] Murthy SK, al-Nassar KE, Verghese L. Sporadic occurrence of nondeletion Prader-Willi syndrome in two cases: a female with maternal uniparental disomy and a male with complex chromosomal rearrangement. Nutrition 1995;11:650–2.

[147] Chamberlain SJ. RNAs of the human chromsome 15q11-q13 imprinted region. Wiley Interdiscip Rev RNA 2013;4(2):155–66.

[148] Dong C, Li WD, Geller F, Lei L, Li D, Gorlova OY, Hebebrand J, Amos CI, Nicholls RD, Price RA, et al. Possible genomic imprinting of three human obesity-related genetic loci. Am J Hum Genet 2005;76:427–37.

[149] O'Meara S, Riemsma R, Shirran L, Mather L, ter Reit G. A systematic review of the clinical effectiveness of orlistat used for the management of obesity. Obes Rev 2004;5:51–68.

[150] Arterburn DE, Crane PK, Veenstra DL. The efficacy and safety of sibutramine for weight loss: a systematic review. Arch Intern Med 2004;164:994–1003.

[151] Colquitt J, Clegg A, Loveman E, Royle P, Sidhu M, Colquitt J. Surgery for morbid obesity. Cochrane Database Syst Rev 2005;4:CD003641.

[152] Chatterjee TK, Idelman G, Blanco V, Blomkalns AL, Piegore Jr MG, Weintraub DS, et al. Histone deacetylase 9 is a negative regulator of adipogenic differentiation. J Biol Chem 2011;286:27836–47.

[153] Weems J, Olson AL. Class II histone deacetylases limit GLUT4 gene expression during adipocyte differentiation. J Biol Chem 2011;286:460–8.

CHAPTER

16

Epigenetic Modifications of miRNAs in Cancer

Ammad A. Farooqi[1], Muhammad Z. Qureshi[2], Muhammad Ismail[3]

[1]Laboratory for Translational Oncology and Personalized Medicine, Rashid Latif Medical College, Lahore, Pakistan; [2]Department of Chemistry, GCU, Lahore, Pakistan; [3]IBGE, Islamabad, Pakistan

OUTLINE

1. INTRODUCTION

Gene activation and inactivation are complicated, multistep, and tightly controlled molecular mechanisms. Increasingly it is being recognized that the promoter CpG island of a silenced gene is occupied by a polycomb group complex that modulates chromatin remodeling. These changes include the easily reversible modification H3K27me3, which is catalyzed by Polycomb Repressive Complex 2 (PRC2) that expectedly methylates flanking nucleosomes. Methylated regions are recognized by PRC1. It is noteworthy that covalent modifications of histones and methylation of cytosine in CpG dinucleotides are widely and extensively studied mechanisms underlying transcriptional regulation. Histones are positively charged proteins that can bind with negatively charged DNA very tightly. Active genes are characteristically different, as evidenced by H3K4me3 marks at the promoter, the presence of H2A.Z (variant histone), acetylation of key histone H3 and H4 lysines, and H3K36me3 to facilitate the intricate process of transcription. Moreover, nucleosomes are not positioned over the transcription start region.

2. DNA METHYLATION

Research over the years has considerably expanded the epigenetic landscape and DNA methylation is now the most widely studied mechanism in mammals. Laboratory methodologies have helped in developing a molecular network that operates during this biological phenomenon and it is now known that de novo methylation is mediated by DNA methyltransferases DNMT3A and DNMT3B and later maintained by DNMT1. With tremendously growing information, it is now more understandable that methylated regions are invaded by methyl-CpG-binding proteins and multiprotein nano-machinery consisting of MeCP2, histone deacetylase

(HDAC), and Sin3a to induce histone deacetylation. Histone methyltransferases are shipped toward deacetylated histones to methylate the lysine 9 residue on histone H3.

3. DNA METHYLTRANSFERASES

DNA methyltransferases (DNMTs) are involved in transfer of methyl groups to DNA. Mechanistically it has been revealed that DNA methyltransferases utilize a methyl donor, which is S-adenosylmethionine. DNMTs interact with histone deacetylases to exert inhibitory effects on transcription. In mammals DNMT1, DNMT3A, and DNMT3B are widely studied. Functionally they are divided into two subcategories including de novo methyltransferases and maintenance methyltransferases. Overexpression of DNMTs is frequently reported in various cancers.

4. microRNA: GENERAL OVERVIEW

microRNAs (miRNAs) are small noncoding RNAs and doubtlessly the most extensively and deeply investigated regulators of a wide range of cellular activities. Research over the years has gradually and sequentially brought miRNA biology into the limelight as one of the most widely studied molecular mechanisms in oncology. There is an overwhelmingly increasing interest in understanding the quantitative regulation of genes. miRNAs have emerged as master regulators of the gene network and it is now known that miRNA modulates mRNA quantity by different approaches, for example, perfect complementarity of binding leads to mRNA degradation of the target gene. Imperfect pairing inhibits translation of mRNA to protein.

It has been reported in giant cell tumors of bone that miRNAs within the Dlk1-Dio3 region at human chromosome 14.32 undergo epigenetic silencing [1]. It has been convincingly revealed that miR-148a is silenced by CpG island hypermethylation in glioma cells. Cells reconstructed with miR-148a displayed remarkably reduced tumorigenic properties [2].

In the upcoming sections we briefly provide an overview of advances in emerging roles of DNMTs in cancer progression and miRNA regulation of DNMTs (Figure 1).

5. DNA METHYLTRANSFERASES 1

DNMT1-mediated inhibition of the Runt-related transcription factor 3 (RUNX3) gene expression is frequently noted (shown in Figure 2). miR-148a negatively regulates DNMT1. Aza-2'-deoxycytidine-treated gastric cancer

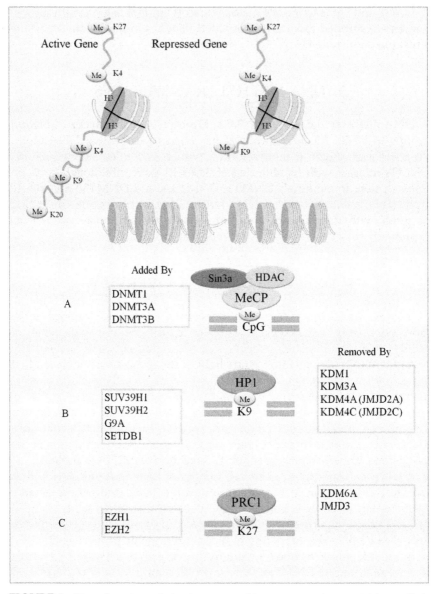

FIGURE 1   Shows how transcription is suppressed in promoter regions containing methylated CpGs that are bound by. This leads to chromatin condensation owing to histone deacetylation, which results in a limited accessibility of the transcriptional machinery to promoter regions. Methylated CpG islands indicate condensed, closed chromatin structure (heterochromatin) and transcriptional silencing because of assembly of MeCP2 (methyl-CpG-binding protein 2), SIN3A (a transcriptional corepressor), and HDAC at methylated DNA. Unmethylated CpG islands within the promoter region present euchromatin (an open chromatin structure).

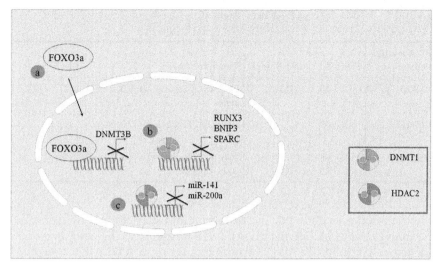

FIGURE 2   Shows role of different regulators in epigenetic silencing. (A) FOXO3A accumulates in nucleus and inhibits expression of DNMT3B. (B) DNMT1 represses expression of target genes. (C) HDAC2 downregulates expression of miR-1414 and miR-2000a.

AGS and BGC-823 cells show a considerable increase in the nonmethylated form of the RUNX3 promoter. Similar results were obtained upon enforced expression of miR-148a in AGS and BGC-823 cancer cells, thus substantiating miRNA-mediated decrease in expression of DNMT1 and decline in methylation levels of tumor suppressor genes [3]. The migratory and invasive potential of miR-148a-overexpressing A549 and H1299 lung cancer cells was reduced notably [4]. DNMT1 expression was downregulated in miR-148b- and miR-152-overexpressing pancreatic cancer MIA Paca-2 and AsPC-1 cells. More importantly, proliferation potential and DNMT1-mediated methylation of tumor suppressor genes including BNIP3 and SPARC were reduced significantly in AsPC-1 and MIA Paca-2 cancer cells (shown in Figure 2) [5]. Gastrokine 1-transfected MKN28, AGS, and MKN1 cancer cells had a higher expression of miR-185. Moreover, DNMT1 was notably reduced, thus showing that miR-185 quantitatively controlled DNMT1 [6]. DNMT1 is quantitatively controlled by miR-152 and miR-185. Intraperitoneal injection of miR-152 mimic-transfected SKOV3/DDP cells in nude mice notably improved cisplatin sensitivity [7]. Targets of DNMT1 and miRNA regulation of DNMT1 are shown in Table 1.

6. DNA METHYLTRANSFERASES 3A

It has been shown that bioactive ingredients, specifically epigallocatechin gallate (EGCG), induce a decrease in mRNA levels of DNMT3A and degradation of DNMT3A in HCT 116 human colon cancer cells

TABLE 1   DNMT-Induced Epigenetic Silencing of Tumor Suppressor Genes that Consequently Results in Cancer Development and Progression

DNA methyl transferase	miRNA	Target gene	Cancer/cell line	Reference
DNMT1	miR-148a	RUNX3	Gastric cancer AGS and BGC-823 cells	[3]
DNMT1	miR-148a		A549 and H1299 lung cancer cells	[4]
DNMT1	miR-148b miR-152	BNIP3 and SPARC	Pancreatic cancer MIA PaCa-2 and AsPC-1 cells	[5]

(methylation-sensitive) [8]. Trichostatin A and decitabine synergistically inhibited DNMT3A/3B and HDAC1/2 expression in the ovarian cancer ascites cell line SKOV3 [9]. Intriguingly, treating AML1/ETO-positive cells with a low concentration of DNMT inhibitors (azacitidine and decitabine) facilitated H3K27me3 loss and gain of acetylated histone H4 at the interleukin-3 (IL-3) promoter [10].

7. DNMT3A REGULATION BY microRNA

DNMT3A is quantitatively controlled by miR-143 in breast cancer cells. Cancer cells reconstituted with miR-143 displayed a marked decrease in proliferation potential [11].

8. DNA METHYLTRANSFERASES 3B

DNMT3B expression is negatively regulated by Forkhead O transcription factor 3a (FOXO3a) in lung cancer cells. Inhibiting MDM2 E3-ligase-mediated degradation of FOXO3a revealed greater nuclear accumulation of FOXO3a and enhanced DNMT3B promoter bound FOXO3a. Gene silencing of FOXO3a relieved expression of DNMT3B (shown in Figure 2) [12]. It is now known that downregulation of miR-148b, miR-29c, and miR-26b resulted in an overexpression of DNMT3B [13]. DNMT3B overexpression in older primary acute myeloid leukemia patients was associated with fewer complete remissions and shorter overall and disease-free survival [14]. Cervical cancer cell lines treated with trichostatin A, an HDAC inhibitor, indicated a decrease in expression of DNMT3B [15]. Interfering with DNMT1 and DNMT3b in pancreatic ductal adenocarcinoma cells induced apoptosis [16]. Also, it has been shown that DNMT3B overexpression correlated considerably with the hypermethylation of miR-124a-3 in breast cancer [17].

9. DNMT3B REGULATION BY microRNA

miR-200b- and miR-200c-overexpressing gastric cancer cells had a notably reduced expression of DNMT3A and DNMT3B [18]. Transfecting miR-29b into MDA-MB-231 cells dramatically inhibited DNMT3A and DNMT3B expression. Moreover, miR-29b-transfected MCF-7 cells revealed a remarkable decrease in methylation levels of hypermethylated tumor suppressor genes [19].

10. HISTONE MODIFICATIONS

Histone alteration takes place in the N-terminal region. In active genes, lysine 4 of histone H3 (H3K4) is methylated and in inactive genes lysines 9 and 27 of histone H3 are methylated. H3 hypoacetylation, methylation of H3K9, and phosphorylation are the main transcriptional changes. These are responsible for changes in chromatin structure and silencing of genes. It has been experimentally verified that the mono- and dimethylation of histone H3 at lysine 9 (H3K9me1 and H3K9me2) inhibit transcription and modulated by G9a histone methyltransferase [20].

11. HISTONE DEACETYLASES AND microRNA

PELP1 (proline-, glutamic acid-, and leucine-rich protein 1), a nuclear receptor coregulator, is frequently overexpressed in breast cancer and negatively regulates expression of miR-141 and miR-200a. Mechanistically it was shown that PELP1 downregulated miR-141 and miR-200a expression by increasing the positioning of HDAC2 at promoter regions of these miRNAs (shown in Figure 2). PELP1-silenced cells displayed a marked decline in miR-200a targets ZEB2 and ZEB1. PELP1-competent cells had notable tumor-forming and metastasizing potential; however, reintroducing miR-141 and miR-200a mimetics into PELP1-overexpressing cancer cells dramatically reduced tumor growth and metastasis in xenografted mice [21]. CG-1521, hydroxamic acid-based HDAC inhibitor, reportedly modulated the expression of 35 miRNAs in inflammatory breast cancer SUM190PT cells and 63 miRNAs in SUM149PT cells [22]. *Cryptosporidium parvum*, a microorganism, is reported to upregulate CX3CL1 in infected cells by promoting installment of HDACs and nuclear factor κB (NF-κB) p50 at the promoter region of the miR-424-503 gene [23]. Entinostat, a class I HDAC inhibitor, considerably enhanced miR-205, miR-125a, and miR-125b in breast cancer cells and synchronous functionality of these miRNAs significantly reduced erbB2/erbB3 levels [24].

12. LONG NONCODING RNAs

Data obtained through high-throughput technologies have started to shed light on the fact that various long noncoding RNAs facilitate recruitment of PRC2 complexes to specific target genes to repress their expression. In line with this approach, enhancer of zeste homolog 2 (EZH2) has emerged as a deeply studied modulator reported to mediate H3 lysine 27 trimethylation (H3K27me3) to the targeted genes.

The long noncoding RNA HOTAIR has been shown to promote the loading of two histone modification complexes, PRC2 and LSD1. HOTAIR-mediated reprogramming is mediated through histone H3K27 methylation and H3K4 demethylation to promote cancer metastasis. miR-7 is indirectly inhibited by HOTAIR in cancer. It has been experimentally verified that SETDB1 increased STAT3 expression by binding to its promoter. However, miR-7-mediated negative regulation of SETDB1 exerted inhibitory effects on SETDB1-mediated upregulation of STAT3 in MDA-MB-231 cells as well as in xenografted mice [25]. It has been convincingly revealed that ANRIL, a long noncoding RNA, extensively enhanced expression of miR-99a/miR-449a in both SGC-7901 and BGC-823 cell lines. Mechanistically it has been shown that ANRIL directed recruitment of EZH2 to promoters of miR-99a/miR-449a and consequent H3K27 trimethylation in cells. Knockdown of ANRIL resulted in an increase in expression of miR-99a/miR-449a [26]. Esophageal squamous cell carcinoma cells overexpress long intergenic nonprotein coding (linc) RNAs encoded by a gene located next to POU3F3 (linc-POU3F3). Detailed analysis revealed that linc-POU3F3-overexpressing cancer cells had hypermethylated CpG islands in the POU3F3 gene. Inhibition of linc-POU3F3 or EZH2 remarkably decreased DNMT1, DNMT3A, and DNMT3B bound to the promoter of POU3F3 [2].

13. DNA HYPOMETHYLATION-DEPENDENT ACTIVATION OF microRNAs

Certain hints have emerged suggesting that often oncogenic miRNAs harboring cancer-germline transcripts undergo DNA hypomethylation-dependent activation in various tumors. In accordance with this notion, it has been shown that a novel cancer-germline transcript (CT-GABRA3) is activated in various tumors in a DNA hypomethylation-dependent manner. Intriguingly, CT-GABRA3 harbored cancer-promoting miRNAs including miR-105 and miR-767. miR-767-mediated negative regulation of the ten–eleven translocation family of tumor suppressor genes is associated with cancer progression [28].

14. miR-370 AND miR-373 AND miR-375

miR-370 is hypermethylated in IGROV1 and TOV112D endometrioid ovarian cancer cells, and cells reconstituted with miR-370 indicated enhanced response to chemotherapeutic drugs [29]. miR-373 is epigenetically silenced in nonsmall-cell lung cancer. HDAC inhibitor-treated cancer cells displayed significantly enhanced expression of miR-373 and it negatively regulated IRAK2 and LAMP1. Transfecting miR-373 into A549 and Calu-6 cells resulted in a decrease in cell proliferation, migration, and invasion [30]. Increasingly it is being realized that epigenetic inhibition of miR-375 induced expression of IGF-1R in trastuzumab-resistant HER2-positive breast cancer cells. Cancer cells reconstructed with miR-375 displayed a marked increase in sensitivity to trastuzumab. Similar results were obtained upon blockade of DNA methylation and histone deacetylation in trastuzumab-resistant cells [31]. Targets of various miRNAs are summarized in Table 2.

15. microRNA REGULATION OF FANCA, KIF14, AND KLF6

Nonsmall-cell lung cancer cells, upon treatment with 5-aza-2′-deoxycytidine (demethylating agent), displayed an increase in miR-503 expression, and Fanconi anemia complementation group A protein (FANCA) was downregulated [27].

KIF14 (kinesin family member 14) is an oncogene frequently overexpressed in cancer. Sp1 and YY1 triggered expression of KIF14 in the OvCa cell line as evidenced by chromatin immunoprecipitation results. Expression of miR-382 was significantly lower in KIF14^{HIGH} primary OvCa tumors and a miR-382 mimic significantly reduced KIF14 expression [32].

Krüppel like factor 6 (KLF6) undergoes alternative mRNA splicing to generate a tumor suppressor full-length KLF6 (KLF6-FL) and oncogenic KLF6 splice variant 1. Surprisingly, a 2014 report emphasized the fact that KLF6-FL was negatively regulated by miR-210 and miR-1301. Cancer cells

TABLE 2   Targets of miRNAs in Various Cancers

miRNA	Target	Cancer/cell line	Reference
miR-373	IRAK2 and LAMP1	A549 and Calu-6 cells	[30]
miR-375	IGF-1R	Trastuzumab-resistant HER2-positive breast cancer cells	[31]
miR-503	FANCA	Nonsmall-cell lung cancer	[27]
miR-382	KIF14	Ovarian cancer	[32]

stably expressing KLF6-FL showed a decrease in cell proliferation, migration, and angiogenesis [33].

16. INTERPLAY OF NF-κB AND microRNAs

Let-7c is considerably reduced in arsenite-treated keratinocyte HaCaT cells. It was revealed that arsenite exerted its inhibitory effects by enhancing hypermethylation of the let-7c promoter. In contrast, HaCaT cells treated with 5-aza-2′-deoxycytidine showed an increase in let-7c expression. Moreover, the Ras/NF-κB-induced signaling axis was also inhibited by let-7c [34]. It has been convincingly revealed that NF-κB subunit p65 was associated with the transcription start site of misrepresented miRNAs in Epstein–Barr virus (EBV)-infected B cells. Intriguingly, H3K27me3 and H3K4me3 were also noted to be modified in EBV-infected cells [35]. Detailed mechanistic insights provided persuasive evidence that STAT3-triggered miR-146b expression was impaired in cancer cells with a methylated miR-146b promoter. Moreover, it was shown that miR-146b inhibited NF-κB-dependent IL-6 production that signaled through receptors to activate STAT3-mediated intracellular signaling to mediate migration and invasion in breast cancer cells [36].

17. CARCINOGENS

7,12-Dimethylbenz(a)anthracene (DMBA), ultraviolet B irradiation, and 12-O-tetradecanoyl phorbol-1,3-acetate have been shown to induce skin cancer in mouse models. DMBA-induced cancer models had significantly enhanced HDAC levels and DNMT-induced promoter methylation of the tumor suppressor miR-203 [37]. By combining gene expression data and whole-genome DNA methylation with the expected contributory roles of various transcription factors, a molecular network was assembled that focused on transcription factor-target gene pairs, and results revealed that arsenic mediated genes that function as transcription factors, thus controlling a network of target genes [38]. Methylnitrosamino-1-(3-pyridyl)-1-butanone, a tobacco smoke carcinogen, has been shown to induce DNMT1-mediated hypermethylation of the tumor suppressor gene retinoic acid receptor β [39].

18. IN VIVO

miR-941 is frequently hypermethylated in hepatocellular carcinoma (HCC) cell lines. KDM6B, a demethylase, is a target of miR-941, and HCC cells overexpressing miR-941 displayed a marked decrease in cellular proliferation, invasion, and migration in xenografted mice [40]. miR-615-5p is frequently hypermethylated in pancreatic ductal adenocarcinoma and tumor growth was notably reduced in mice xenografted with miR-615-5p-overexpressing cancer cells [41].

Promoter methylation of miR-31 is noted in nasopharyngeal carcinoma C666-1 cells. Subcutaneously injecting stably miR-31-expressing C666-1 cells into nude mice considerably reduced tumor growth [42]. miR-199a is methylated in prostate cancer. The miR-199a-3p agomir negatively regulated aurora kinase A and tumor growth was considerably inhibited in xenografted mice [43]. miR-203 expression is significantly reduced in cancer cells via promoter methylation. Injecting SUM159 cells expressing miR-203 into mice did not result in the development of metastases in the lung even after 15 weeks [44]. Epigenetically silenced miR-338-3p was noted in gastric cancer. It was shown that the tumor volume in nude mice xenografted with miR-338-3p-expressing gastric cancer SGC-7901 cells was significantly less [45]. DNA hypermethylation is frequently noted in the miR-886-3p promoter. Xenografting miR-886-3p-overexpressing NCI-H446 cells into nude mice substantially inhibited tumor growth, invasion, and lung metastasis [46].

19. CONCLUSION

There has been a tremendous expansion in our rapidly developing view of the roles of miRNAs in various cancers and how inactivation of tumor suppressor miRNAs and overexpression of oncogenic miRNAs underlie cancer development, progression, and metastasis. It is also intriguing to note how the methylation machinery is functionally active in cancer cells to epigenetically silence tumor suppressor miRNAs. Wide-ranging approaches have been used to restore the expression of tumor suppressor miRNAs using natural and synthetic agents. However, it still needs to be seen how miRNA subsets cooperatively induce resistance against chemotherapeutic drugs. We still have an incomplete picture of how the epigenetic machinery works mechanistically and how it can be targeted to improve the efficacy of therapeutics. Another major stumbling block is to differentially target the epigenetic machinery of cancer cells while leaving normal cells undisturbed. A better understanding of the nature of cancer and the modulators of epigenetic mechanisms will be helpful to get a step closer to personalized medicine.

References

[1] Lehner B, Kunz P, Saehr H, Fellenberg J. Epigenetic silencing of genes and micrornas within the imprinted Dlk1-Dio3 region at human chromosome 14.32 in giant cell tumor of bone. BMC Cancer July 9, 2014;14:495.

[2] Li S, Chowdhury R, Liu F, Chou AP, Li T, Mody RR, et al. Tumor suppressive miR-148a is silenced by CpG island hypermethylation in IDH1 mutant gliomas. Clin Cancer Res September 15, 2014. pii: clincanres.0234.2014.

[3] Zuo J, Xia J, Ju F, Yan J, Zhu A, Jin S, et al. MicroRNA-148a can regulate runt-related transcription factor 3 gene expression via modulation of DNA methyltransferase 1 in gastric cancer. Mol Cells April 2013;35(4):313–9.

[4] Chen Y, Min L, Zhang X, Hu S, Wang B, Liu W, et al. Decreased miRNA-148a is associated with lymph node metastasis and poor clinical outcomes and functions as a suppressor of tumor metastasis in non-small cell lung cancer. Oncol Rep October 2013;30(4):1832–40.

[5] Azizi M, Teimoori-Toolabi L, Arzanani MK, Azadmanesh K, Fard-Esfahani P, Zeinali S. MicroRNA-148b and microRNA-152 reactivate tumor suppressor genes through suppression of DNA methyltransferase-1 gene in pancreatic cancer cell lines. Cancer Biol Ther April 2014;15(4):419–27.

[6] Yoon JH, Choi YJ, Choi WS, Ashktorab H, Smoot DT, Nam SW, et al. GKN1-miR-185-DNMT1 axis suppresses gastric carcinogenesis through regulation of epigenetic alteration and cell cycle. Clin Cancer Res September 1, 2013;19(17):4599–610.

[7] Xiang M, Birkbak NJ, Vafaizadeh V, Walker SR, Yeh JE, Liu S, et al. STAT3 induction of miR-146b forms a feedback loop to inhibit the NF-κB to IL-6 signaling axis and STAT3-driven cancer phenotypes. Sci Signal January 28, 2014;7(310). http://dx.doi.org/10.1126/scisignal.2004497. ra11.

[8] Moseley VR, Morris J, Knackstedt RW, Wargovich MJ. Green tea polyphenol epigallocatechin 3-gallate, contributes to the degradation of DNMT3A and HDAC3 in HCT 116 human colon cancer cells. Anticancer Res December 2013;33(12):5325–33.

[9] Meng F, Sun G, Zhong M, Yu Y, Brewer MA. Inhibition of DNA methyltransferases, histone deacetylases and lysine-specific demethylase-1 suppresses the tumorigenicity of the ovarian cancer ascites cell line SKOV3. Int J Oncol August 2013;43(2):495–502.

[10] Buchi F, Masala E, Rossi A, Valencia A, Spinelli E, Sanna A, et al. Redistribution of H3K27me3 and acetylated histone H4 upon exposure to azacitidine and decitabine results in de-repression of the AML1/ETO target gene IL3. Epigenetics March 2014;9(3):387–95.

[11] Ng EK, Li R, Shin VY, Siu JM, Ma ES, Kwong A. MicroRNA-143 is downregulated in breast cancer and regulates DNA methyltransferases 3A in breast cancer cells. Tumour Biol March 2014;35(3):2591–8. http://dx.doi.org/10.1007/s13277-013-1341-7.

[12] Yang YC, Tang YA, Shieh JM, Lin RK, Hsu HS, Wang YC. DNMT3B overexpression by deregulation of FOXO3a-mediated transcription repression and MDM2 overexpression in lung Cancer. J Thorac Oncol September 2014;9(9):1305–15.

[13] Sandhu R, Rivenbark AG, Coleman WB. Loss of post-transcriptional regulation of DNMT3b by microRNAs: a possible molecular mechanism for the hypermethylation defect observed in a subset of breast cancer cell lines. Int J Oncol August 2012;41(2):721–32.

[14] Niederwieser C, Kohlschmidt J, Volinia S, Whitman SP, Metzeler KH, Eisfeld AK, et al. Prognostic and biologic significance of DNMT3B expression in older patients with cytogenetically normal primary acute myeloid leukemia. Leukemia 2015;29(3):567–75.

[15] Liu N, Zhao LJ, Li XP, Wang JL, Chai GL, Wei LH. Histone deacetylase inhibitors inducing human cervical cancer cell apoptosis by decreasing DNA-methyltransferase 3B. Chin Med J Engl September 2012;125(18):3273–8.

[16] Gao J, Wang L, Xu J, Zheng J, Man X, Wu H, et al. Aberrant DNA methyltransferase expression in pancreatic ductal adenocarcinoma development and progression. J Exp Clin Cancer Res November 5, 2013;32:86.

[17] Ben Gacem R, Ben Abdelkrim O, Ziadi S, Ben Dhiab M, Trimeche M. Methylation of miR-124a-1, miR-124a-2, and miR-124a-3 genes correlates with aggressive and advanced breast cancer disease. Tumour Biol May 2014;35(5):4047–56.

[18] Tang H, Deng M, Tang Y, Xie X, Guo J, Kong Y, et al. miR-200b and miR-200c as prognostic factors and mediators of gastric cancer cell progression. Clin Cancer Res October 15, 2013;19(20):5602–12.

[19] Starlard-Davenport A, Kutanzi K, Tryndyak V, Word B, Lyn-Cook B. Restoration of the methylation status of hypermethylated gene promoters by microRNA-29b in human breast cancer: a novel epigenetic therapeutic approach. J Carcinog July 26, 2013;12:15.

[20] Tachibana M, Sugimoto K, Nozaki M, Ueda J, Ohta T, Ohki M, et al. G9a histone methyltransferase plays a dominant role in euchromatic histone H3 lysine 9 methylation and is essential for early embryogenesis. Genes Dev 2002;16:1779–91.

[21] Roy SS, Gonugunta VK, Bandyopadhyay A, Rao MK, Goodall GJ, Sun LZ, et al. Significance of PELP1/HDAC2/miR-200 regulatory network in EMT and metastasis of breast cancer. Oncogene July 10, 2014;33(28):3707–16.

[22] Chatterjee N, Wang WL, Conklin T, Chittur S, Tenniswood M. Histone deacetylase inhibitors modulate miRNA and mRNA expression, block metaphase, and induce apoptosis in inflammatory breast cancer cells. Cancer Biol Ther July 2013;14(7):658–71.

[23] Zhou R, Gong AY, Chen D, Miller RE, Eischeid AN, Chen XM. Histone deacetylases and NF-kB signaling coordinate expression of CX3CL1 in epithelial cells in response to microbial challenge by suppressing miR-424 and miR-503. PLoS One May 28, 2013;8(5):e65153.

[24] Wang S, Huang J, Lyu H, Lee CK, Tan J, Wang J, et al. Functional cooperation of miR-125a, miR-125b, and miR-205 in entinostat-induced downregulation of erbB2/erbB3 and apoptosis in breast cancer cells. Cell Death Dis March 21, 2013;4:e556.

[25] Zhang EB, Kong R, Yin DD, You LH, Sun M, Han L, et al. Long noncoding RNA ANRIL indicates a poor prognosis of gastric cancer and promotes tumor growth by epigenetically silencing of miR-99a/miR-449a. Oncotarget April 30, 2014;5(8):2276–92.

[26] Zhang H, Cai K, Wang J, Wang X, Cheng K, Shi F, et al. MiR-7, inhibited indirectly by LincRNA HOTAIR, directly inhibits SETDB1 and reverses the EMT of breast cancer stem cells by downregulating the STAT3 pathway. Stem Cells 2014;32(11):2858–68.

[27] Li W, Zheng J, Deng J, You Y, Wu H, Li N, et al. Increased levels of the long intergenic non-protein coding RNA POU3F3 promote DNA methylation in esophageal squamous cell carcinoma cells. Gastroenterology June 2014;146(7):1714–26. e5.

[28] Loriot A, Van Tongelen A, Blanco J, Klaessens S, Cannuyer J, van Baren N, et al. A novel cancer-germline transcript carrying pro-metastatic miR-105 and TET-targeting miR-767 induced by DNA hypomethylation in tumors. Epigenetics August 1, 2014;9(8):1163–71.

[29] Chen XP, Chen YG, Lan JY, Shen ZJ. MicroRNA-370 suppresses proliferation and promotes endometrioid ovarian cancer chemosensitivity to cDDP by negatively regulating ENG. Cancer Lett October 28, 2014;353(2):201–10.

[30] Seol HS, Akiyama Y, Shimada S, Lee HJ, Kim TI, Chun SM, et al. Epigenetic silencing of microRNA-373 to epithelial-mesenchymal transition in non-small cell lung cancer through IRAK2 and LAMP1 axes. Cancer Lett October 28, 2014;353(2):232–41.

[31] Ye XM, Zhu HY, Bai WD, Wang T, Wang L, Chen Y, et al. Epigenetic silencing of miR-375 induces trastuzumab resistance in HER2-positive breast cancer by targeting IGF1R. BMC Cancer February 26, 2014;14:134.

[32] Thériault BL, Basavarajappa HD, Lim H, Pajovic S, Gallie BL, Corson TW. Transcriptional and epigenetic regulation of KIF14 overexpression in ovarian cancer. PLoS One March 13, 2014;9(3):e91540.

[33] Liang WC, Wang Y, Xiao LJ, Wang YB, Fu WM, Wang WM, et al. Identification of miRNAs that specifically target tumor suppressive KLF6-FL rather than oncogenic KLF6-SV1 isoform. RNA Biol June 12, 2014;11(7).

[34] Jiang R, Li Y, Zhang A, Wang B, Xu Y, Xu W, et al. The acquisition of cancer stem cell-like properties and neoplastic transformation of human keratinocytes induced by arsenite involves epigenetic silencing of let-7c via Ras/NF-κB. Toxicol Lett June 5, 2014;227(2):91–8.

[35] Vento-Tormo R, Rodríguez-Ubreva J, Di Lisio L, Islam AB, Urquiza JM, Hernando H, et al. NF-κB directly mediates epigenetic deregulation of common microRNAs in Epstein-Barr virus-mediated transformation of B-cells and in lymphomas. Nucleic Acids Res 2014;42(17):11025–39.

[36] Xiang Y, Ma N, Wang D, Zhang Y, Zhou J, Wu G, et al. Mir-152 and miR-185 co-contribute to ovarian cancer cells cisplatin sensitivity by targeting DNMT1 directly: a novel epigenetic therapy independent of decitabine. Oncogene January 16, 2014;33(3):378–86.

[37] Tiwari P, Gupta KP. Modulation of miR-203 and its regulators as a function of time during the development of 7, 12 dimethylbenz [a] anthracene induced mouse skin tumors in presence or absence of the antitumor agents. Toxicol Appl Pharmacol July 15, 2014;278(2):148–58.

[38] van Breda SG, Claessen SM, Lo K, van Herwijnen M, Brauers KJ, Lisanti S, et al. Epigenetic mechanisms underlying arsenic-associated lung carcinogenesis. Arch Toxicol September 9, 2014. http://dx.doi.org/10.1007/s00204-014-1351-2.

[39] Wang J, Zhao SL, Li Y, Meng M, Qin CY. 4-(Methylnitrosamino)-1-(3-pyridyl)-1-butanone induces retinoic acid receptor β hypermethylation through DNA methyltransferase 1 accumulation in esophageal squamous epithelial cells. Asian Pac J Cancer Prev 2012;13(5):2207–12.

[40] Zhang PP, Wang XL, Zhao W, Qi B, Yang Q, Wan HY, et al. DNA methylation-mediated repression of miR-941 Enhances lysine (K)-specific demethylase 6B expression in Hepatoma cells. J Biol Chem August 29, 2014;289(35):24724–35.

[41] Gao W, Gu Y, Li Z, Cai H, Peng Q, Tu M, et al. miR-615-5p is epigenetically inactivated and functions as a tumor suppressor in pancreatic ductal adenocarcinoma. Oncogene April 28, 2014;0. http://dx.doi.org/10.1038/onc.2014.101.

[42] Cheung CC, Chung GT, Lun SW, To KF, Choy KW, Lau KM, et al. miR-31 is consistently inactivated in EBV-associated nasopharyngeal carcinoma and contributes to its tumorigenesis. Mol Cancer August 7, 2014;13:184.

[43] Qu Y, Huang X, Li Z, Liu J, Wu J, Chen D, et al. miR-199a-3p inhibits aurora kinase A and attenuates prostate cancer growth: new avenue for prostate cancer treatment. Am J Pathol May 2014;184(5):1541–9.

[44] Taube JH, Malouf GG, Lu E, Sphyris N, Vijay V, Ramachandran PP, et al. Epigenetic silencing of microRNA-203 is required for EMT and cancer stem cell properties. Sci Rep 2013;3:2687. http://dx.doi.org/10.1038/srep02687.

[45] Li P, Chen X, Su L, Li C, Zhi Q, Yu B, et al. Epigenetic silencing of miR-338-3p contributes to tumorigenicity in gastric cancer by targeting SSX2IP. PLoS One June 24, 2013;8(6):e66782.

[46] Cao J, Song Y, Bi N, Shen J, Liu W, Fan J, et al. DNA methylation-mediated repression of miR-886-3p predicts poor outcome of human small cell lung cancer. Cancer Res June 1, 2013;73(11):3326–35.

17

Managing Autoimmune Disorders through Personalized Epigenetic Approaches

Christopher Chang

Division of Rheumatology, Allergy and Clinical Immunology,
University of California at Davis, CA, USA

OUTLINE

1. INTRODUCTION

The immune system is finely balanced to protect the organism from invasion by dangerous predators such as microbial pathogens and inorganic threats, while at the same time allowing the organism to tolerate nonharmful contacts with the environment. When the immune system fails to protect, immunodeficiency results and the organism becomes susceptible to attack by pathogens or other dangerous entities. On the other hand, autoimmune diseases result from a derangement in the homeostasis of the immune system, leading to the immune system turning on itself and causing a variety of life-threatening diseases—in essence, a breach of tolerance [1]. The specific mechanisms of how the various autoimmune diseases occur are not entirely clear, but probably involve many aspects of the immune system. There are clearly commonalities among the various disease states, but at the same time, there are differences, and the extent of the role that each of the pathogenic mechanisms plays in the development of autoimmune diseases may vary from disease to disease.

Out of this complexity, the challenge to characterize the pathogenic mechanism of diseases such as systemic lupus erythematosus (SLE), rheumatoid arthritis (RA), psoriasis, polymyositis, myasthenia gravis, autoimmune thyroid disease, autoimmune hepatitis, and others becomes daunting but necessary. Understanding of the pathogenesis of each of these diseases will allow for the development of specific treatments based on the genetics and epigenetics of each patient. Not all patients will respond with equal efficacy and safety to any particular medication or treatment. This is the premise of personalized medicine. And the goal is to develop knowledge of which medication to use to treat each individual patient with maximum efficacy and safety.

Autoimmune diseases are common, affecting more than 23.5 million people in the United States. Globally, it is estimated that autoimmune diseases may affect up to 5% of the world's population. Autoimmune diseases affect both adults and children, and as an example, the incidence of SLE in children is about 0.36–2.5 per 100,000 individuals per year [2]. With most autoimmune diseases, there is a female predominance, and clearly, there are ethnic differences as well [3]. The X-chromosome inactivation skew theory provides a potential explanation for the predominance of autoimmune diseases in women, but other factors may also play a part. Moreover, the specific environmental triggers that tilt the immune balance and lead to the development of autoimmune diseases are many, and as a species, we are constantly seeking out healthy lifestyles to prolong life and ameliorate the risk of disease. With autoimmune diseases, each patient is different and each disease is different. There are often multiple phenotypes for each disease. Personalized medicine becomes very difficult and complicated. Identification of biomarkers for prognosis and predictive efficacy and safety of various medications is an ongoing area of research to help identify the best and most customized approach to the treatment of these patients. Epigenetics is now recognized to be a phenomenon that is widespread in normal physiology and disease and has the potential to help delineate genotypes and endotypes to help accomplish the goals of personalized and genomic medicine (Figure 1).

2. EVOLUTION OF THE TREATMENT OF AUTOIMMUNITY

Autoimmune diseases have been recognized since early recorded medical history, although it was only in the nineteenth century that these diseases were attributed to defects in the immune system. Descriptions of patients with deformed hands and joints as well as damaged bones have been reported in ancient Egyptian and Chinese literature. As our understanding of the cellular and humoral elements of the immune system increased, along with the elucidation of the function of the various elements, the recognition that many of these diseases may stem from abnormalities in immune function led to the development of a series of medications. Early treatment of autoimmune disease utilized herbal medication preparations, including aspirin, derived from the bark of the willow tree, and licorice, colchicine, or ginseng, which all possess anti-inflammatory properties [4–6].

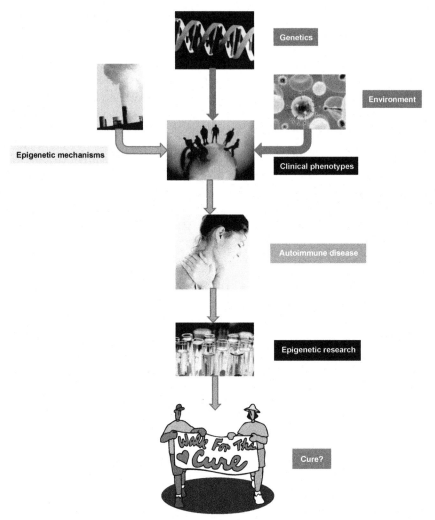

FIGURE 1   How epigenetic changes affect disease and why this may present an opportunity for the development of customizable or personalized medicine treatments.

It is ironic that even after the concept of autoimmunity was proposed as a potential mechanism for many of the diseases described in early history, many scientists and physicians, including the proponent of autoimmunity himself, Paul Ehrlich, could not come to terms with the fact that the body could turn on itself. The concept of "horror autotoxicus" in the early twentieth century had a significant dampening effect on the development of the specialty of autoimmunity, and it was not until the 1940s to the 1960s that much advancement was made in the understanding of the pathophysiology of autoimmunity [7,8]. Out of this research came a new group of medications initially known as compounds A, B, C, D, E, and F.

The exciting revelation of the anti-inflammatory effects of compound E, the original name for cortisone, opened the door for the extensive use of corticosteroids in not just autoimmune disease, but any disease in which inflammation may be playing a role. Later developments brought us the disease-modifying antirheumatic drugs or DMARDs, and this alleviated some of the side effects of corticosteroids that became apparent after chronic use in autoimmune diseases. While chronologically, the DMARDs postdate corticosteroids, several DMARDs, such as chloroquine and cyclophosphamide, actually have a long history of use in autoimmune diseases. The development of molecular biology techniques allowed for the production of large amounts of antibodies, leading to an entire new industry encompassing a completely novel class of drugs, the biological modulators. Mouse and human monoclonal antibodies were mass produced, in particular, those directed against the inflammatory response, such as the antitumor necrosis factor antibodies as well as anti-B cell monoclonals such as rituximab. These molecules had fewer long-term side effects than chronic corticosteroid use and have revolutionized the treatment of autoimmune diseases. However, not all patients respond equally to each of the drugs, and the challenge is to determine which patients carrying which phenotypes are suitable for treatment with which particular drug.

It is interesting to note that the mechanisms of epigenetics, which are discussed later, are potential targets not just for new drugs, but also for drugs that have been used for decades. Many of these drugs were developed years ago with little understanding of the mechanism, and it is only now, after many years of use, that they have been found to have an impact on DNA and protein changes related to epigenetics. For example, 5-azacytidine is an anticancer drug that has DNA-demethylating activity [9]. Procainamide is also a DNA-demethylating agent and both 5-azacytidine and procainamide have been associated with drug-induced lupus [9–11].

3. EPIGENETIC BASIS FOR AUTOIMMUNITY

The pathophysiology of autoimmunity may involve changes at the cellular, humoral, and even DNA level. Multiple factors probably contribute to the development of autoimmunity, including but not limited to the presence of autoantibodies, apoptosis, activation or deactivation of regulatory cells, and activation and suppression of genes based on their DNA and histone structure. It is the last that is influenced by epigenetic changes. Epigenetic changes are those that do not involve changes in DNA sequence and therefore are not mutation-based changes. Instead, it is the secondary or tertiary structure of DNA or proteins that changes the way genes are expressed. There are two major mechanisms for epigenetic mod-ification—DNA methylation and histone acetylation. Other changes that have an impact on DNA and histone structure would also be considered epigenetic changes. A third mechanism involving the action of microR-NAs is controversial with regard to whether these are actually epigenetic modifications. But many feel that microRNAs should be characterized as an epigenetic phenomenon.

In autoimmune diseases, studies involving discordant twin sets have shown that there is more to genetic and environmental influences in the pathogenesis of disease [12–14]. In SLE, the highest concordance rate among identical twins is less than 60%. It has also been observed that the degree of methylation of DNA in patients with SLE is far less than that of normal controls.

Hypomethylation generally leads to activation of DNA, although the gene involved may be an activating or a suppressive gene, so the effect may vary. When the methylation occurs at the 5′ position of cytosine in CpG dinucleotides that are located in promoter regions, this can lead to a repression of transcription of the index gene. DNA-methylating agents play a significant role in the development of disease. It has been shown that DNA methylation patterns in CD4+ cells correlate with disease phe-notype in autoimmune diseases such as SLE [15].

It should be noted that methylation is not the only structural alteration implicated in epigenetics. DNA hydroxymethylation has been extensively studied in a number of diseases, including neurologic disorders such as Alzheimer disease [16]. Evidence that epigenetic modifications may play a role in these common diseases was shown in a study in which global reduc-tions in DNA methylation and hydroxymethylation were found in autopsy specimens of the hippocampus of patients with Alzheimer disease [17].

In addition to effects of methylation of DNA, histone modification also has the ability to open or close DNA, thereby either facilitating or inhibiting expression of genes, respectively. Chromatin consists of DNA and protein in a complex, and within that complex are multiple proteins. Histones form a complex around DNA, and these proteins can be modified by acetylation, methylation, or some other chemical reaction. The most common way to influence epigenetic changes is by acetylation or deacetylation of histones. This is accomplished by histone acetyltransferases (HATs) or histone deacet-ylases (HDACs), respectively. There are approximately 18 HDACs, which are assigned to one of four classes of HDACs (Table 1) [18]. The action of HDACs on a variety of cells and signaling pathways leads to regulation of chromatin architecture and downstream immunological function [19]. The class I, II, and IV HDACs are dependent on zinc ion, while class III includes the sirtuins [20]. The class I HDACs are the most widely studied and they are usually located in the nucleus and regulate production of cyto-kines and chemokines. Class II HDACs are important in the regulation of

TABLE 1 The Histone Deacetylases (HDAC) Enzymes

Class I	Class II		Class III	Class IV
	IIA	IIB		
HDAC1	HDAC4	HDAC6	SIRT1	HDAC11
HDAC2	HDAC5	HDAC10	SIRT2	
HDAC3	HDAC7		SIRT3	
HDAC8	HDAC9		SIRT4	
			SIRT5	
			SIRT6	
			SIRT7	

lymphocyte activation and differentiation [19]. This group of HDACs starts in the periphery and translocates to the nucleus to exert their action.

MicroRNAs constitute the third mechanism of regulation of gene expression and while some contend that microRNAs should not be considered an epigenetic phenomenon, there is clearly a relationship between DNA methylation and microRNA activity. Several studies have indicated that DNA methylation and microRNAs may act in a concerted fashion to regulate gene expression. Methylation may affect the activity of certain subsets of microRNAs and it has been also shown that the reverse may be true, that microRNAs may affect de novo DNA methylation through regulation of transcriptional repression in a mouse embryonic stem cell line [21]. Two microRNAs, microRNA-21 and microRNA-148, are associated with a hypomethylated DNA state in CD4+ T cells in animal models and humans with SLE. Inhibition of microRNA-148 and microRNA-21 resulted in a reversal of this hypomethylation. In these patients, elevated cellular markers of inflammation CD70 and Lymphocyte function associated antigen-1 (LFA-1) were detected in conjunction with levels of the two microRNAs in CD4+ cells [22,23]. This illustrates the interactions between different facets of epigenetics (Figure 2).

4. LESSONS FROM ONCOLOGY

Both HDAC inhibitors and DNA-demethylating agents have been used in the treatment of cancers well before their use in autoimmune diseases. Therefore, the experience gained in terms of efficacy and safety can provide important information regarding their use in autoimmune diseases. HDACs in particular have been found to have a proinflammatory effect by virtue of their action on inhibitors of NF-κB, such as inhibitor of nuclear factor κB kinase subunit β.

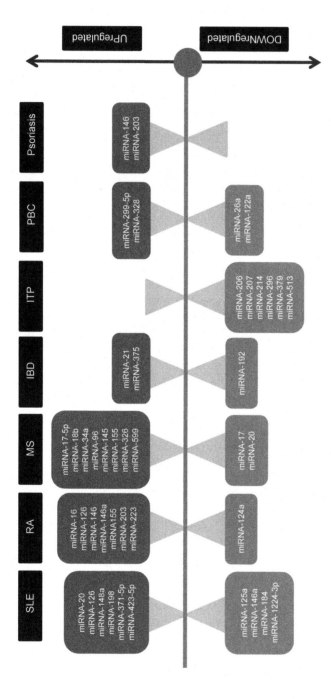

DM = Dermatomyositis; IBD = Inflammatory bowel disease; ITP = Idiopathic thrombocytopenia; MS = Multiple sclerosis; PBC = Primary biliary cirrhosis; PM = Polymyositis; RA = Rheumatoid arthritis; SLE = Systemic lupus erythematosus

FIGURE 2 MicroRNAs that are upregulated or downregulated in autoimmune diseases. IBD=inflammatory bowel disease; ITP=idiopathic thrombocytopenia; MS=multiple sclerosis; PBC=primary biliary cirrhosis; RA=rheumatoid arthritis; SLE=systemic lupus erythematosus.

HDACs can also affect type 1 interferon production and can play a role in T regulatory (Treg) cell homeostasis by virtue of the action on Forkhead box P3 (Foxp3) transcription factor [24,25]. A complex consisting of HDACs 7 and 9, as well as tip60 and Foxp3, regulates the stability and function of Treg cells. In general, the effect of HDACs on Treg cell function is inhibitory, and cytokine production is also suppressed by the activity of HDACs on activated T cells as well. Deletion of HDAC1 in a mouse allergy model can lead to exacerbation of asthma in the form of increased airway inflammation, hypersecretion of mucus, parenchymal lung inflammation, and increased airway resistance. In highly active cancer cells, the use of DNA methylation and HDAC enzymes as a treatment methodology can potentially induce cell cycle arrest, cell differentiation, and apoptotic death of "transformed cells." An analogy can be made between autoimmunity and cancer. Both involve cells in which normal regulation has been compromised. So both diseases should be amenable to reconstitution of such control, whether by epigenetic mechanisms or some other strategy. The immune system is infinitely complex, and whether the end result of manipulation by HDAC inhibitors is pro- or anti-inflammatory may depend on the balance of multiple molecules or pathways that are targeted by this process. Again, the challenge is to identify which pathway to suppress and how to render the suppression specific so as to limit any untoward side effects.

4.1 Polycomb-Mediated Gene and Polycomb-Response Elements

Polycomb-mediated gene silencing (PcG) proteins are transcriptional repressors that are thought to play a role in cancer pathogenesis through their effects on the development and differentiation of stem cells and, thus, oncogenesis [26–30]. These proteins may act via DNA methylation, rendering certain genes more prone to methylating enzymes and thereby differentially predisposing certain patients to certain cancers. It has been estimated that in some cases, certain genes can be up to 12 times as likely to be aberrantly methylated as a result of PcGs [30]. It should be noted that PcGs can also act independent of DNA methylation [31]. Polycomb-response elements (PREs) have been described in *Drosophila*, but not yet in humans, as the targets of PcGs that are found in genomic regulatory loci. How PcGs are recruited to these PREs is not yet clear, but it has been proposed that long noncoding RNA sequences or sequence-specific transcription factors may be facilitators of this process [32]. The normal physiologic processes of aging and the abnormal chronic inflammation seen in autoimmune and autoinflammatory diseases have been suggested to be triggers of abnormal methylation [33,34], thus raising the question as to whether epigenetic changes such as DNA methylation may be involved in a positive feedback loop that can perpetuate disease and prevent normal healing. Viruses and other environmental factors can also modulate DNA methylation through the action of PcGs [35,36].

5. PERSONALIZED EPIGENETIC THERAPY IN AUTOIMMUNE DISEASES

The targets of epigenetic therapy in autoimmune diseases are dependent upon the pathophysiology of autoimmune diseases. Several mechanisms are in play, which all lead to a disease state whereby there is a breakdown of tolerance and the host immune system turns on itself. The pathways in our innate and adaptive immune systems are numerous and each is highly complex, and therefore one or more defects are likely to be present at any point and at any given time. In many cases, the defect is clinically insignificant, owing to the redundancy of the immune system, and it is likely that the advent of disease is preceded by an alignment of derangements similar to the Swiss cheese model of disaster analysis. The targets for epigenetic therapy are essentially anything that can be regulated by gene expression and can include cells such as lymphocytes or dendritic cells, molecules such as cytokines and chemokines, signaling pathways, and gene clusters. Epigenetic modifications can be specific to a single gene or can be more global. In most cases, the derangement is specific, but the challenge is that any epigenetic treatment may have an impact on a wide range of genes. This is where personalized epigenetics may be of benefit, but it requires a better understanding of pathophysiology.

It is now known that autoimmune diseases can result from a breach of tolerance, which can result from multiple immune dysfunctions. One pathway that has been extensively studied is the T helper cell 17 (Th17) subset [37]. Both Th1 and Th17 cells have been shown to play a role in various autoimmune diseases, including psoriasis, central nervous system autoimmunity, and SLE [38]. In addition, Treg cells and their associated cytokines tumor growth factor-β (TGF-β) and interleukin-10 (IL-10) have been found to be compromised in patients with autoimmune diseases.

6. SPECIFIC EPIGENETIC AGENTS IN AUTOIMMUNITY

6.1 HDAC Inhibitors

6.1.1 Butyrates, Benzamides, and Cyclic Peptides

Earlier compounds with HDAC activity include butyrates such as phenylbutyrate, benzamides, and cyclic peptides such as depsipeptides and apicidin [39]. The HDAC inhibitors phenylbutyrate and trichostatin

A (TSA) have been studied in a rat model of RA. The rationale for the use of HDAC inhibitors in RA was related to the role of regulation of the cell cycle inhibitor genes p16^{INK4} and p21^{CIP1} by epigenetic manipulation. Previous work in cancer had demonstrated that HDAC inhibitors could regulate proinflammatory cytokines including IL-1, IL-6, IL-8, and tumor necrosis factor-α (TNF-α). Because synovial inflammation and tumorigenesis could both represent a derangement in immune regulation, it was believed that HDAC inhibitors may play a parallel role in the treatment of autoimmune diseases such as RA. However, from a theoretical standpoint, depending on the gene targets of HDAC inhibitors, a paradoxical effect can be generated whereby there may be an increase in the proinflammatory state.

Chung et al. conducted a study of HDAC inhibitors that demonstrated an upregulation of p16^{INK4} and p21^{CIP1} [40]. This was associated with an increase in the levels of acetylated histones H3 and H4 when cultured synovial fibroblasts were incubated with phenylbutyrate or TSA. The effect of HDAC inhibitors on upregulation of p16^{INK4} in particular was sustained in the synovial fibroblasts of rats with adjuvant arthritis. The application of topical phenylbutyrate also led to increased expression of these two genes only in the synovium of rats with adjuvant arthritis, and not in normal synovial cells. Clinical significance was also demonstrated, as joint swelling was reduced. Topical 10% phenylbutyrate or 1% trichostatin ointment was also able to suppress paw swelling and pannus formation and also to promote wound healing. These clinical changes were associated with a suppression of TNF-α in the affected tissues of arthritic rats.

6.1.2 Trichostatin A

Trichostatin A is an HDAC inhibitor that targets the class I and class II HDAC inhibitors [41]. Patients with SLE demonstrated an altered level of gene expression, which can be associated with the histone code. The histone code describes posttranslational modification and the positioning of histone proteins alongside genetic material and has the ability to regulate DNA expression. Histones may also control the methylation of DNA, an independent mechanism for epigenetics, and the histone code is therefore considered to be a "master" regulator of DNA expression [42,43]. In an MRL-lpr/lpr mouse model of SLE, trichostatin A was found to be able to affect acetylation of histones H3 and H4, and this is accompanied by improvements in clinical status [44]. Other effects of trichostatin A include its effects on proinflammatory molecules such as cytokines and chemokines. It has been demonstrated that trichostatin A is able to upregulate the expression of cyclooxygenase-2 and CXCL12 in mouse macrophages, while at the same time downregulating proinflammatory genes encoding

TNF-α, IL-12p40, IL-6, endothelin 1, and the chemokines CCL2/monocyte chemotactic protein-1 (MCP-1) and CCL17 [45,46].

Trichostatin may also play a role in the treatment of multiple sclerosis (MS), a disease in which a number of immune derangements may be present. Using the animal model of experimental autoimmune encephalomyelitis (EAE), it has been demonstrated that an imbalance in the secretion of interferon-γ may play a role in the pathogenesis of MS. The downstream effect of this is a skewing of the T helper cell paradigm to a predominantly Th1 environment. Th1 cells secrete IL-2, a master activator of T cells, creating a positive feedback loop leading to inflammation. HDAC inhibitors have been found to be able to inhibit the expression of IL-2 and increase the expression of the regulatory cytokine IL-10 in the EAE animal model [47,48].

6.1.3 *Suberoylanilide Hydroxamic Acid*

The successful use of the HDAC inhibitor suberoylanilide hydroxamic acid (SAHA) in the treatment of various cancers including cutaneous T cell lymphoma, Sezary syndrome, gliomas, and other solid tumors led to the attempts to utilize this drug to treat lung fibrosis. The rationale was that the effect of SAHA would be to regulate the transcription of genes encoding proinflammatory cytokines. SAHA and the class I-selective HDAC inhibitor MS275 have been shown to possess antirheumatic properties in human RA E11 synovial fibroblast cells. These agents appear to act via suppression of several transcription pathways including the NF-κB pathway and the p38 mitogen-activated protein kinase (MAPK) signaling pathway. Initially, both agents were found to be able to induce growth arrest and suppress lipopolysaccharide-induced NF-κB nuclear accumulation [49], but subsequently, both agents were also found to be able to suppress expression of p38 MAPK and increase expression of MAPK phosphatase-1 [50], suggesting that HDACs indeed act on multiple critical signaling pathways. In the latter study, other effects were noted as well, including suppression of granulocyte chemotactic protein-2, macrophage migration-inhibitory factor, and MCP-2, leading to reduced migration of THP-1 and U937 monocytes, suggesting that the HDAC class I inhibitors play a significant role in the clinical development of autoimmune diseases via their impact on immunomodulatory pathways [50].

6.2 DNA-Methylation Agents

DNA-methylating agents that have been used in the treatment of cancers include 5-azacytidine, procainamide, decitabine, zebularine, procain, and several others. Inhibition of gene expression generally follows a hypomethylation pattern, and the DNA methyltransferases (DNMTs) include DNMT1, DNMT2, DNMT3A, and DNMT3B, of which DNMT1 is the most well studied.

6.2.1 Procainamide

Procainamide is the prototype drug for drug-induced lupus. It is a drug that is seldom used now, but acts as a competitive inhibitor of DNA methyltransferases. This leads to reduced methylation and increased transcription of genes encoding proinflammatory cytokines, including LFA-1 and CD70. Thus, procainamide can lead to drug-induced lupus [51–53]. Hydralazine is another drug that is associated with drug-induced lupus and acts by inhibiting the extracellular signal-regulated kinase (ERK) pathway signaling and DNMT activity, also leading to a hypomethylated state and clinical manifestations of drug-induced lupus [54].

6.2.2 5-Azacytidine

5-Azacytidine has been widely used in the treatment of various cancers, primarily myelodysplastic syndromes. It acts by inhibiting DNA methyltransferases, thus leading to a hypomethylated state, as well as by a direct cytotoxic effect on neoplastic hematologic cells. It has been shown that 5-azacytidine increases the expression of the chemokine CXCL12 through hypomethylation of the promoter region of CXCL12 in RA synovial fibroblasts. CXCL12 is also known as stromal-derived factor-1 and is involved in the trafficking of progenitor cells in response to tissue damage [55–57]. Other effects of CXCL12 include chemotaxis of lymphocytes and monocytes into affected joints in RA patients, as well as increased expression of matrix metalloproteinases, collagenase activity, and joint destruction [58]. The effects mediated by DNA-demethylating agents through the regulation of CXCL12 and other cytokines or chemokines such as CD11a may be a mechanism by which these agents lead to the generation of a proinflammatory state in animal models of SLE [59]. As in the case of procainamide, this is an example of an agent that can be used to treat cancer, but may contrarily induce an autoimmune disease.

Other examples of possible biotherapies in the treatment of autoimmune disease stem from research identifying target genes that play a role in pathogenesis. Human endogenous retrovirus (HERV) elements are known to exist in large amounts in primate genomes. It has been postulated that overexpression of these genes is increased in autoimmunity and that these genes are a potential marker for disease. HERVs play a role by affecting the production of cross-reactive autoantibodies or by interfering with other neighboring immune-related genes. Fali et al. demonstrated that B cells in SLE patients were unable to methylate the promoter for HRES-1/p28, leading to an increase in expression of HRES-1/p28. The dysregulation of HERV expression can therefore be a potential epigenetic target for therapeutic studies of SLE [60].

An interesting study on patients with Sjogren syndrome illustrates the cell-specific nature of epigenetic modifications. It was found that global DNA methylation was reduced in salivary gland epithelial cells (SGECs) but not peripheral T or B cells from Sjogren syndrome patients compared to controls. This led to a sevenfold decrease in DNMT1 expression and a one-to twofold increase in expression of Gadd45-α. The discovery of aberrant DNA methylation in traditional nonimmune cells such as SGECs may open the door for new therapeutic perspectives in Sjogren syndrome [61].

6.3 MicroRNAs

MicroRNAs are short 21- to 23-nt noncoding RNAs that act as master regulators of gene expression at the posttranscriptional level. This group of compounds was discovered in the early 1990s and has been credited with controlling more than half the genes in the mammalian genome. Over the years, it has been recognized that many genes may be under the control of one microRNA. MicroRNAs act by binding to partially complementary nucleotides in the 3′ untranslated region of messenger RNA. In doing so, they trigger one of two effects—inhibition of translation or degradation of RNA. Downstream effects include loading of "mature microRNAs" onto microRNA-induced silencing complexes (miRISCs) [62]. The miRISC then seeks out a messenger RNA sequence that is complementary to the seed sequence on the mature microRNA. The effects of microRNAs have been extensively studied since their discovery and abnormalities of expression of microRNAs have been found in many disease states, including autoimmune diseases.

6.3.1 MicroRNAs that Play a Role in Autoimmunity

There are currently approaching 1000 known microRNAs, and because they bind to a wide variety of genes, their effects can be extremely complex. Since each microRNA generally has more than one target, the effects may be different depending on the messenger RNA to which the microRNA binds. It is perhaps too simplistic to attempt to delineate the effect of a microRNA without considering the entire genome and the balance of upregulation and downregulation of various genes affected by a particular microRNA. Nevertheless, the study of microRNAs generally involves searching for a difference in the expression of a given microRNA in a particular disease being studied, e.g., SLE. And the results of these studies are discussed below for particular microRNAs known to affect the mechanisms of autoimmune and autoinflammatory diseases.

In a study by Zhao et al., the association of microRNA-126, extent of DNA methylation, and immune activity was investigated in SLE patients [63]. The authors found that microRNA-126 is highly expressed in CD4+ T cells of patients with SLE, and this correlated inversely with DNMT1 levels. This appeared to lead to the activation of several immune-related genes, including LFA-1 and CD20. In vitro knockdown of microRNA-126 in the CD4+ cells from patients with SLE resulted in a decrease in auto-immune activity, as well as reduced immunoglobulin production by B cells.

Multiple microRNAs have been identified as having a potential role in the pathogenesis of SLE and RA. One microRNA that is important in SLE is microRNA-146a. MicroRNA-146a happens to be downregu-lated in SLE, and miRNA-146a is known to affect the expression of TNF receptor-associated factor 6 and IL-1 receptor-associated kinase 1. Both of these factors play a role in the activation of NF-κB, a transcrip-tion factor that when activated leads to the stimulation of many pro-inflammatory cytokines. Studies have shown that reduced expression of miRNA-146a is associated with high interferon expression, which potentially represents a biomarker for increased disease activity and, in particular, renal involvement [64]. In addition to the above-mentioned activities, microRNA-146a appears to have many functions in the regu-lation of inflammation, and it has been shown to have a regulatory role in T cell apoptosis and secretion of proinflammatory cytokines such as IL-2 [65].

Other microRNAs involved in RA and SLE include microRNA-125a, which regulates the expression of KLF13 and CCL5 and whose downreg-ulation has been implicated in increased inflammatory activity in SLE in the kidneys and joints [66], as well as microRNA-155, which is involved in activation of IL-6, IL-17, and IL-22 in a K/BxN mouse serum trans-fer arthritis model [67,68]. In this case, the expression of microRNA-155 appears to have a stimulatory role in the development of arthritis, rather than a regulatory role. Other microRNAs that have demon-strated an upregulation in autoimmune diseases include microRNA-223 and microRNA-346 in RA and microRNA-21, microRNA-148a, and microRNA-198 in SLE.

Impaired apoptosis of fibroblasts may be a contributing mechanism in RA. It has been found that microRNA-34a plays a role in apopto-sis. Demethylation of the microRNA-34a promoter leads to enhanced expression of microRNA-34a, which in turn leads to increased levels of FasL-mediated apoptosis in RA synovial fibroblasts [69]. In addition, microRNA-34a levels correlated with the level of X-linked inhibitor of apoptosis protein, which plays a role in the pathogenesis of RA.

In other autoimmune diseases, microRNAs have also been associated with clinical disease. In addition to its implication in SLE, microRNA-146a is used as a biomarker in Sjogren syndrome, because it has been found that it is overexpressed in this disease compared with healthy controls. Functional assays have shown that microRNA-146a is asso-ciated with upregulation of phagocytic activity in human monocytic

THP-1 cells. This particular microRNA plays a regulatory role for the expression of proinflammatory cytokines TNF-α, IP-10, IL-7, IL-1β, and MIP1α [70]. One of the best-studied diseases with respect to microR-NAs is multiple sclerosis. Because of the existence of a well-established animal model in the EAE mouse, several microRNAs have been associated with the disease. For example, microRNA-326 is upregulated in patients with disease, and in the animal model, it was associated with an increase in Th17 cell numbers and clinical severity [71]. In addition, microRNA-34a and microRNA-155, which was previously discussed in regard to RA, are also upregulated in active multiple sclerosis lesions. It has been suggested that the mechanism of action is through the regulation of macrophage signaling through CD47 induction [72].

In primary biliary cirrhosis, microRNAs that are downregulated include microRNA-122a and microRNA-26a, while those that are upregulated include microRNA-328 and microRNA-299-5p [73]. The target pathways for regulation by these microRNAs involve a range of pathways related to apoptosis, inflammation, and inflammatory cell proliferation.

It is important to state that often an association is found between the expression of a microRNA and disease activity without really understanding the specific gene or genes to which the particular microRNA is targeting. Thus, while we may be able to apply this clinically in the development of a new treatment targeting that particular microRNA, we run the risk of not completely understanding the dangers of using this agent. This has been the case in the development of biological modulators as well, and at any stage of development or even after marketing the drug, new safety concerns may arise.

6.3.2 *The Use of Antagomirs as Epigenetic Therapeutics*

MicroRNAs are attractive targets for the development of new therapeutics because of their direct involvement in gene regulation. Attempts to create antagonists, known as antagomirs, began as early as 2005. Antagomirs are synthetic analogs of microRNAs that function as silencing agents. Antagomirs are being studied in animal trials to reduce levels of microRNAs that are upregulated in disease states. An antagomir of microRNA-21 has been studied in the treatment of heart and lung fibrosis, whereas microRNA inhibition has also been studied in hepatitis C virus [74] and hypercholesterolemia [75]. Interestingly, a study of microRNA in dust mite allergy involved an antagomir directed against microRNA-145, leading to a reduction in airway hyperresponsiveness, mucus secretion, Th2 cytokine release, and eosinophilic inflammation [76], and one against microRNA-126, which also reduced Th2 cell activity in a mouse model for dust mite allergy [77].

Antagomirs to microRNA-33a/b have also been studied in high-density lipoproteinemia [78,79], and antagomirs against microRNA-103/107 have been associated with effects on insulin signaling and sensitivity [80].

7. ENVIRONMENTAL AGENTS INVOLVED IN AUTOIMMUNITY THAT INFLUENCE EPIGENETICS

While many environmental exposures can play a role in clinical epigenetics, some of the more important or better accepted entities range from infectious pathogens such as viruses, to cigarette smoke and, more recently, vitamins and minerals such as vitamin D. Some examples are given below.

7.1 Viruses

7.1.1 Epstein–Barr Viruses

Epstein–Barr virus (EBV) is a common infection. Most people infected by EBV experience a subclinical infection during childhood or are affected by a self-limiting disease called infectious mononucleosis. Following infection, EBV exists in the memory B cells in a latent form. In this form, there is not expected to be any sequelae or ongoing chronic infection—normally. On the other hand, EBV is associated with a number of cancers, including lymphoma, or autoimmune diseases, including SLE or RA. Latent viral oncoproteins or other untranslated RNAs may be responsible for the occurrence of cancers and autoimmune diseases, potentially as a result of molecular mimicry. The control of production of latent viral proteins or the reactivation of EBV is under the control of epigenetic mechanisms including DNA methylation or histone deacetylation.

It has been demonstrated that DNA methylation is associated with silencing of Wp and Cp, which are alternative promoters for transcripts coding for nuclear antigens EBNA 1–6, as well as LMP1p, LMP2Ap, and LMP2Bp, which are promoters for transcripts encoding transmembrane proteins [81,82]. In addition, histone protein changes resulting from HDAC and HAT activity may play a role in the regulation of latent EBV promoters. These promoters are located on so-called "acetylation islands," which contain a high level of diacetylated H3 and tetra-acetylated H4 [83]. These epigenetic phenomena may play a role in the reactivation of EBV toward malignant transformation of various cell types, leading to potential neoplasms or autoimmune diseases.

7.2 Pollutants

7.2.1 Cigarette Smoke and Other Pollutants

It is well known that pollutants including cigarette smoke and diesel exhaust represent significant risk factors for many diseases, including cardiovascular, pulmonary, hematologic, oncologic, and autoimmune diseases. Environmental exposures play an important role in molding epigenetic changes. The mechanism of action of these pollutants has been attributed to the production of reactive oxygen species, dysfunctional apoptosis, inflammatory cell recruitment, and other physiologic changes. But epigenetics also plays a role. For example, ozone exposure has been associated with altered microRNA expression within the bronchial airways. A study of sputum samples after ozone exposure revealed that the expression of 10 microRNAs was upregulated. These included microRNAs miR-132, miR-143, miR-145, miR-199a*, miR-199b-5p, miR-222, miR-223, miR-25, miR-424, and miR-582-5p. Several of these have been associated with regulation of inflammatory pathways [84].

Cigarette smoke has been a known cause of lung cancer and cardiovascular disease for many years. It is now recognized that cigarette smoke significantly alters the methylation status both globally and gene-specifically [85]. In fact, it has been noted that even prenatal smoking will affect global and gene-specific methylation patterns [86]. Cigarette smoke can affect DNA methylation through four different mechanisms, including (1) recruitment of DNMT1 to damaged DNA sites during DNA repair [87]; (2) alteration of DNA methylation through the expression and activation of DNA binding factors, such as Sp1, which binds to CG motifs and inhibits de novo methylation during embryogenesis [88,89]; (3) nicotine-induced downregulation of DNMT1 mRNA and protein expression [90]; and (4) hypoxia-induced HIF1α-dependent increase in methionine adenosyltransferase 2A activity [91], which may account for the damaging effects of prenatal cigarette smoke exposure [85].

Cigarette smoking also affects epigenetics through changes in histone acetylation. In a study by Ito et al., cigarette smoking was found to be associated with a decrease in histone deacetylase expression, which was thought to lead to increased expression of inflammatory cytokines and an inhibition of glucocorticoid activity in alveolar macrophages [92].

While lung cancer and chronic obstructive pulmonary disease have garnered the most attention as adverse health effects of cigarette smoke, one should not discount the other potential risks, including RA and lupus [93,94]. The pathogenic mechanisms outlined above may be in play in the increase in autoimmune diseases observed in cigarette smokers. The oxidative stress associated with cigarette smoking can potentially lead to modifications in DNA methylation, and these methylation changes have been associated with the development of lupus in a mouse model. Li et al.

have demonstrated that T cells from lupus patients with active disease display decreased ERK pathway signaling in conjunction with decreased DNA methylation and overexpress genes that are normally suppressed by DNA methylation in T cells [95].

Another study looked at ambient polycyclic aromatic hydrocarbons and their effect on the methylation status of Foxp3. In addition to an increase in Foxp3 methylation, Treg cell function was also impaired. This correlated with an increased expression of interferon-γ and a decreased expression of IL-10 and was linked to high IgE levels. The authors suggest that epigenetic modifications involving Treg cell biomarkers are associated with atopic diseases including asthma and allergic rhinitis [96].

Further evidence that air pollution can have an impact on health through epigenetic modifications can be found in the Normative Aging Study carried out between 1999 and 2009, in which investigators evaluated 777 patients and reported the relationship between air pollution and the expression of key immunologic molecules. The authors of the study found that air pollution, in particular, ozone and particulate matter, was associated with hypomethylation of F3, ICAM-1, and TLR-2 and hypermethylation of IL-6 and interferon-γ. The authors were also able in some cases to pin down an effect of carbon fibers on F3 expression and effects of sulfate and ozone on ICAM-1 expression [97].

Ultraviolet (UV) radiation has also been associated with SLE. It has been demonstrated that elevated Gadd45A mRNA expression is associated with global demethylation of CD4$^+$ T cells in SLE patients. UV irradiation was associated with increased Gadd45A expression, which was accompanied by increases in CD11a and CD70, both of which play a role in the development of autoreactive CD4$^+$ T cells and excessive B cell stimulation. These effects of DNA methylation are thought to play a role in the development of SLE [98].

7.3 Vitamins and Minerals

7.3.1 Vitamin D

Vitamin D is considered a hormone that has been linked to various regulatory processes in the immune system. Vitamin D appears to play a key role in aging, immune dysfunction, and cancer. The epigenetic mechanisms involved in vitamin D metabolism involve the activating enzymes CYP27A1 and CYP27A2, as well as the inhibiting enzyme CYP24. The vitamin D receptor (VDR) binds to specific genomic sequences known as vitamin D response elements (VDREs). The numerous immunologic effects of vitamin D that have been discovered are so varied and complex that a discussion of this topic is beyond the scope of this chapter. Suffice it to say that vitamin D has been found to have regulatory activity on Th1, Th2, Th17, and Treg cells; dendritic cells; mucosal immunity; and immune tolerance [99].

But these activities of vitamin D can be modulated by epigenetic mechanisms. For example, histone acetylation by HATs and histone deacetylation by HDACs both contribute to the normalization of vitamin D metabolism. The effects of vitamin D are modulated through binding to the VDR, which then binds to other receptors to form a dimer, such as the VDR/retinoid X receptor (RXR) dimer. These dimers interact with HATs, which tend to activate transcription [100]. Conversely, when the VDR/RXR complex interacts with negative VDREs, which recruit transcription corepressors such as SMRT or NCOR1, then this promotes histone deacetylation and suppression of transcription. The interaction of VDREs with corepressors is important in the regulation of vitamin D activity, and it has been shown that deletion of corepressors will enhance VDR activity [101].

8. BIG DATA AND EPIGENETICS

The explosion of data that has already accumulated in the study of DNA methylation patterns and histone deacetylation patterns, not to mention the hundreds of microRNAs that may influence each of the autoimmune diseases, has made it evident that interpretation of results of studies on biomarkers and therapeutics is to be no easy task. No longer are the days in which one may find one single test that can indicate disease, such as a blood sugar or a blood pressure measurement. We have become much more sophisticated in our efforts to identify patients with different natural histories or prognoses of their disease, or even to predict who might develop disease. A prime example of how this may affect treatment of diseases is the choice that some women with a family history of breast cancer make to undergo a mastectomy before the disease can rear its ugly head. Defining treatment programs promises to be just as complex, and early studies using computer-generated models involving biological processes are already under way.

8.1 Big Data

Big data are used to describe collections of data sets that are larger and more complex than those traditionally generated. Big data generally results from our ability to now collect and store huge amounts of data in servers with unlimited capacity. Big data can range from patient demographic and health records, to financial data, to engineering test data. These data sets cannot be handled by traditional data processing tools or applications [102]. Challenges in handling big data sets include collection, storage, sharing, searching, transfer, analysis, interpretation, and intellectual property issues.

In our context, epigenetics, the availability of microarray testing and the ability to run traditional biotesting at a nanomolecular level, thus requiring very little space per test, have led to an explosion in available results to analyze. Lab-on-a-chip technology has revolutionized the way we do experiments, and instead of collecting information on one assay, we are bombarded with information on hundreds or thousands or more tests at the same time. Thus, the results are often expressed not as a single numerical value, but as a pattern that can only be visualized using a computer. This has brought forth the age of genomics, proteomics, transcriptomics, metabolomics, and other related fields.

8.2 Transcriptome Analysis

The transcriptome can be thought of as the precursor to proteomics. The transcriptome is a term used to describe the relationships between the expression of multiple mRNAs that are being simultaneously transcribed in a given cell population. Predictors of diseases may in the future be based on the relative expression of multiple genetic and epigenetic elements, which is affected by identifiable or nonidentifiable genetic triggers. A gene expression heat map is a graphical representation of the pattern of alterations in gene expression that may be characteristic of certain disease states. Figure 3 shows an example of a gene expression heat map, but these colorful results can be generated for DNA methylation patterns in epigenetics as well.

Another challenge is how one might evolve from a "shotgun" approach of looking at the entire genome to a more directed approach with greater relevance to the disease. An example of this application in autoimmunity is the use of modular transcriptional analysis to evaluate interferon signatures in SLE [103]. Newer techniques that are based on more efficient systems of sequencing DNA or RNA will help to usher in a more targeted approach to identifying predictive factors and biomarkers of autoimmune diseases [104].

A detailed discussion of bioinformatics with regard to transcriptome and DNA methylation patterns is beyond the scope of this chapter, but this new science addresses all of the challenges of big data, including data processing, normalization of the data, filtering of nonrelevant genes, replicability, clustering, statistical analysis, gene ontology, pathway analysis, and clinical relevance.

9. DISCUSSION

As we understand more about the roles of epigenetics, genetics, and environment in the pathogenesis of diseases including autoimmune diseases, the challenge is how to translate this knowledge into disease

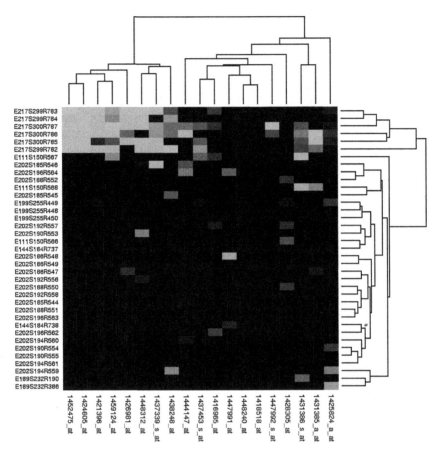

FIGURE 3   An example of a gene expression heat map.

management tools that are clinically useful. It is clear that people do not present with diseases in the same way, nor do they respond to medications or treatment plans in the same manner. Diseases are known to occur as a result of genetic predisposition and are triggered by the environment, most likely through epigenetic mechanisms. These epigenetic changes, by definition, are reversible, if further triggers change the way genes are expressed, and heritable, which may contribute to the changes in prevalence of various types of diseases, such as autoimmune diseases or allergic disease that have been observed through the years.

The variability in host genetics plays only a partial role in the difference in disease phenotypes found in autoimmune and inflammatory disease. Environment further modulates this, rendering disease states heterogeneous and therefore making it difficult to find optimal treatment regimens for all patients alike. Identifying the primary pathogenic mechanisms may help determine which type of treatment to administer to which patients, but as our experience from the promise of gene therapy in cystic fibrosis has taught us, it is not simply a matter of switching off or turning on one single gene. An understanding of the precise epigenetic changes that occur in a patient with an autoimmune disease may at least partially guide us in the knowledge of which gene may need to be regulated and by which epigenetic mechanisms. This is where epigenetics may show promise in personalized medicine (Table 2).

As of this writing, epigenetic drugs are used in the treatment of cancers, but there are none that are approved for the treatment of autoimmune diseases. This will probably change, as animal and human studies begin to establish the efficacy and safety of these agents. The fact that epigenetic changes are reversible means that there is the possibility of identifying the epigenetic mechanism involved in a particular patient with a particular disease and undo the changes leading to the disease. The fact that epigenetic changes are heritable means that we may have the ability to change disease trends by inhibiting vertical transmission of the disease or disease propensity to offspring of the affected patient.

The concept of epigenetic drift attempts to explain why, as we age, we become more susceptible to certain diseases, including autoimmune diseases. Aging is accompanied by immunosenescence, which has been linked to an increase in susceptibility to infections and a greater risk of illness or death as a result. But in addition to immunosenescence, epigenetic drift describes the divergence in the epigenome as a result of stochastic changes in DNA methylation. While epigenetics affects disease, epigenetic drift appears to be under the influence of genetics and the environment [105]. Epigenetic drift may be either tissue-specific or non-tissue-specific, the latter affecting stem cell differentiation and possibly explaining the decline in stem cell function with age [106,107]. However, the value of studying how epigenetic signatures vary with age is in the prediction of disease risk. As epigenetic signatures change with relation to age, this may influence disease risk, in particular, cancer or autoimmune diseases.

Of all the challenges in developing drugs that play on the epigenetic control of gene expression, two factors perhaps most limit their effectiveness. The first is that alterations in epigenetics is cell-specific and may vary from one autoimmune disease to another. Therefore, drugs need to be designed individually for each type of autoimmune dysfunction. But perhaps the second consideration is even more important, which is that we have yet to determine how to utilize epigenetic modifications to selectively turn on or off certain culprit genes. Development of epigenetic drugs therefore often leads to global effects, which may lead to side effects that may outweigh any benefits seen with the use of these agents.

TABLE 2   Targets for Personalized Epigenetic Therapy for Autoimmune Diseases

CELLS
T cells
Th1
Th17
Th2
Treg
B cells
Dendritic cells

MOLECULES
Cytokines
CD70
IL-1
IL-1β
IL-2
IL-6
IL-10
IL-12/23
IL-17
IL-18
IL-23
Interferon-γ
LFA-1
TGF-β
Cyclin-dependent kinase inhibitor p21
Cytokine receptors
CXCL9
CXCL10
CXCL11
CXCR12
Signaling molecules
CTLA-4
PCD-1

TABLE 2   Targets for Personalized Epigenetic Therapy for
Autoimmune Diseases—cont'd

MOLECULES

CD40 ligand

Matrix metalloproteinases

Pathways

Apoptosis

BCR signaling

Complement system

Death receptor signaling

Glucocorticoid receptor activation

Innate immunity

   Toll-like receptors

NF-κB

TCR signaling

Gene clusters

MHC

FCγRs

TCR ε-chain gene

10. CONCLUSIONS

The prospects of personalized epigenetics in the treatment of auto-immune disease are in its infancy and requires improved knowledge of genotypes, endotypes, phenotypes, and pathophysiology of the various autoimmune diseases, including how all of these factors interact with one another. The challenges are to find biomarkers of pathophysiology, prognostication, and response to medications. Modulation of gene expression using epigenetic tools such as DNA methylation or histone deacetylation will be a difficult challenge to overcome, and specificity will most certainly be a concern for safety to patients. The treatment should not be worse than the disease. MicroRNAs may actually be a more promising avenue for the development of cutting-edge molecules capable of reversing the breach of tolerance seen in autoimmunity.

References

[1] Wang L, Wang FS, Chang C, Gershwin ME. Breach of tolerance: primary biliary cirrhosis. Semin Liver Dis 2014;34(3):297–317.

[2] Pineles D, Valente A, Warren B, Peterson MG, Lehman TJ, Moorthy LN. Worldwide incidence and prevalence of pediatric onset systemic lupus erythematosus. Lupus 2011;20(11):1187–92.

[3] Morey C, Avner P. Genetics and epigenetics of the X chromosome. Ann NY Acad Sci 2010;1214:E18–33.

[4] Mackay IR. Travels and travails of autoimmunity: a historical journey from discovery to rediscovery. Autoimmun Rev 2010;9(5):A251–8.

[5] Chang C. Autoimmunity: from black water fever to regulatory function. J Autoimmun 2014;48–49:1–9.

[6] Chang C. Unmet needs in the treatment of autoimmunity: from aspirin to stem cells. Autoimmun Rev 2014;13(4–5):331–46.

[7] Silverstein AM. Autoimmunity versus horror autotoxicus: the struggle for recognition. Nat Immunol 2001;2(4):279–81.

[8] Steinman L. Escape from "horror autotoxicus": pathogenesis and treatment of autoimmune disease. Cell 1995;80(1):7–10.

[9] Cheishvili D, Boureau L, Szyf M. DNA demethylation and invasive cancer: implications for therapeutics. Br J Pharmacol August 18, 2014; http://dx.doi.org/10.1111/bph.12885.

[10] Singh V, Sharma P, Capalash N. DNA methyltransferase-1 inhibitors as epigenetic therapy for cancer. Curr Cancer Drug Targets 2013;13(4):379–99.

[11] Xiao X, Chang C. Diagnosis and classification of drug-induced autoimmunity (DIA). J Autoimmun 2014;48–49:66–72.

[12] Crow MK, Kirou KA, Wohlgemuth J. Microarray analysis of interferon-regulated genes in SLE. Autoimmunity 2003;36(8):481–90.

[13] Crow MK, Wohlgemuth J. Microarray analysis of gene expression in lupus. Arthritis Res Ther 2003;5(6):279–87.

[14] Baechler EC, Batliwalla FM, Karypis G, Gaffney PM, Ortmann WA, Espe KJ, et al. Interferon-inducible gene expression signature in peripheral blood cells of patients with severe lupus. Proc Natl Acad Sci USA 2003;100(5):2610–5.

[15] Zhao M, Liu S, Luo S, Wu H, Tang M, Cheng W, et al. DNA methylation and mRNA and microRNA expression of SLE CD4+ T cells correlate with disease phenotype. J Autoimmun November 2014;54:127–36.

[16] Kristensen LS, Treppendahl MB, Gronbaek K. Analysis of epigenetic modifications of DNA in human cells. In: Haines JL, et al., editors. Current Protocols in Human Genetics/Editorial Board. 2013. [Chapter 20:Unit20 2].

[17] Chouliaras L, Mastroeni D, Delvaux E, Grover A, Kenis G, Hof PR, et al. Consistent decrease in global DNA methylation and hydroxymethylation in the hippocampus of Alzheimer's disease patients. Neurobiol Aging 2013;34(9):2091–9.

[18] de Ruijter AJ, van Gennip AH, Caron HN, Kemp S, van Kuilenburg AB. Histone deacetylases (HDACs): characterization of the classical HDAC family. Biochem J 2003;370(Pt 3):737–49.

[19] Wang L, Tao R, Hancock WW. Using histone deacetylase inhibitors to enhance Foxp3(+) regulatory T-cell function and induce allograft tolerance. Immunol Cell Biol 2009;87(3):195–202.

[20] Stunkel W, Campbell RM. Sirtuin 1 (SIRT1): the misunderstood HDAC. J Biomol Screen 2011;16(10):1153–69.

[21] Sinkkonen L, Hugenschmidt T, Berninger P, Gaidatzis D, Mohn F, Artus-Revel CG, et al. MicroRNAs control de novo DNA methylation through regulation of transcriptional repressors in mouse embryonic stem cells. Nat Struct Mol Biol 2008;15(3):259–67.

[22] Ceribelli A, Yao B, Dominguez-Gutierrez PR, Chan EK. Lupus T cells switched on by DNA hypomethylation via microRNA? Arthritis Rheum 2011;63(5):1177–81.

[23] Pan W, Zhu S, Yuan M, Cui H, Wang L, Luo X, et al. MicroRNA-21 and microRNA-148a contribute to DNA hypomethylation in lupus CD4+ T cells by directly and indirectly targeting DNA methyltransferase 1. J Immunol 2010;184(12):6773–81.

[24] Beier UH, Akimova T, Liu Y, Wang L, Hancock WW. Histone/protein deacetylases control Foxp3 expression and the heat shock response of T-regulatory cells. Curr Opin Immunol 2011;23(5):670–8.

[25] Beier UH, Wang L, Bhatti TR, Liu Y, Han R, Ge G, et al. Sirtuin-1 targeting promotes Foxp3+ T-regulatory cell function and prolongs allograft survival. Mol Cell Biol 2011;31(5):1022–9.

[26] Bracken AP, Helin K. Polycomb group proteins: navigators of lineage pathways led astray in cancer. Nat Rev Cancer 2009;9(11):773–84.

[27] Simon JA, Kingston RE. Mechanisms of polycomb gene silencing: knowns and unknowns. Nat Rev Mol Cell Biol 2009;10(10):697–708.

[28] Francis NJ, Follmer NE, Simon MD, Aghia G, Butler JD. Polycomb proteins remain bound to chromatin and DNA during DNA replication in vitro. Cell 2009;137(1):110–22.

[29] Khalil AM, Guttman M, Huarte M, Garber M, Raj A, Rivea Morales D, et al. Many human large intergenic noncoding RNAs associate with chromatin-modifying complexes and affect gene expression. Proc Natl Acad Sci USA 2009;106(28):11667–72.

[30] Widschwendter M, Fiegl H, Egle D, Mueller-Holzner E, Spizzo G, Marth C, et al. Epigenetic stem cell signature in cancer. Nat Genet 2007;39(2):157–8.

[31] Kondo Y, Shen L, Cheng AS, Ahmed S, Boumber Y, Charo C, et al. Gene silencing in cancer by histone H3 lysine 27 trimethylation independent of promoter DNA methylation. Nat Genet 2008;40(6):741–50.

[32] Gieni RS, Hendzel MJ. Polycomb group protein gene silencing, non-coding RNA, stem cells, and cancer. Biochem Cell Biol 2009;87(5):711–46.

[33] Hahn MA, Hahn T, Lee DH, Esworthy RS, Kim BW, Riggs AD, et al. Methylation of polycomb target genes in intestinal cancer is mediated by inflammation. Cancer Res 2008;68(24):10280–9.

[34] Teschendorff AE, Menon U, Gentry-Maharaj A, Ramus SJ, Weisenberger DJ, Shen H, et al. Age-dependent DNA methylation of genes that are suppressed in stem cells is a hallmark of cancer. Genome Res 2010;20(4):440–6.

[35] Kwiatkowski DL, Thompson HW, Bloom DC. The polycomb group protein Bmi1 binds to the herpes simplex virus 1 latent genome and maintains repressive histone marks during latency. J Virol 2009;83(16):8173–81.

[36] Violot S, Hong SS, Rakotobe D, Petit C, Gay B, Moreau K, et al. The human polycomb group EED protein interacts with the integrase of human immunodeficiency virus type 1. J Virol 2003;77(23):12507–22.

[37] Singh RP, Hasan S, Sharma S, Nagra S, Yamaguchi DT, Wong D, et al. Th17 cells in inflammation and autoimmunity. Autoimmun Rev December 2014;13(12):1174–81. http://dx.doi.org/10.1016/j.autrev.2014.08.019. Epub 2014 August 23.

[38] Chong WP, Horai R, Mattapallil MJ, Silver PB, Chen J, Zhou R, et al. IL-27p28 inhibits central nervous system autoimmunity by concurrently antagonizing Th1 and Th17 responses. J Autoimmun 2014;50:12–22.

[39] Halili MA, Andrews MR, Sweet MJ, Fairlie DP. Histone deacetylase inhibitors in inflammatory disease. Curr Top Med Chem 2009;9(3):309–19.

[40] Chung YL, Lee MY, Wang AJ, Yao LF. A therapeutic strategy uses histone deacetylase inhibitors to modulate the expression of genes involved in the pathogenesis of rheumatoid arthritis. Mol Ther 2003;8(5):707–17.

[41] Drummond DC, Noble CO, Kirpotin DB, Guo Z, Scott GK, Benz CC. Clinical development of histone deacetylase inhibitors as anticancer agents. Annu Rev Pharmacol Toxicol 2005;45:495–528.

[42] Morley M, Molony CM, Weber TM, Devlin JL, Ewens KG, Spielman RS, et al. Genetic analysis of genome-wide variation in human gene expression. Nature 2004;430(7001):743–7.

[43] Sims 3rd RJ, Reinberg D. From chromatin to cancer: a new histone lysine methyltransferase enters the mix. Nat Cell Biol 2004;6(8):685–7.

[44] Garcia BA, Busby SA, Shabanowitz J, Hunt DF, Mishra N. Resetting the epigenetic histone code in the MRL-lpr/lpr mouse model of lupus by histone deacetylase inhibition. J Proteome Res 2005;4(6):2032–42.

[45] Aung HT, Schroder K, Himes SR, Brion K, van Zuylen W, Trieu A, et al. LPS regulates proinflammatory gene expression in macrophages by altering histone deacetylase expression. FASEB J 2006;20(9):1315–27.

[46] Bode KA, Schroder K, Hume DA, Ravasi T, Heeg K, Sweet MJ, et al. Histone deacetylase inhibitors decrease Toll-like receptor-mediated activation of proinflammatory gene expression by impairing transcription factor recruitment. Immunology 2007;122(4):596–606.

[47] Takahashi I, Miyaji H, Yoshida T, Sato S, Mizukami T. Selective inhibition of IL-2 gene expression by trichostatin A, a potent inhibitor of mammalian histone deacetylase. J Antibiot (Tokyo) 1996;49(5):453–7.

[48] Camelo S, Iglesias AH, Hwang D, Due B, Ryu H, Smith K, et al. Transcriptional therapy with the histone deacetylase inhibitor trichostatin A ameliorates experimental autoimmune encephalomyelitis. J Neuroimmunol 2005;164(1–2):10–21.

[49] Choo QY, Ho PC, Tanaka Y, Lin HS. Histone deacetylase inhibitors MS-275 and SAHA induced growth arrest and suppressed lipopolysaccharide-stimulated NF-kappaB p65 nuclear accumulation in human rheumatoid arthritis synovial fibroblastic E11 cells. Rheumatology (Oxford) 2010;49(8):1447–60.

[50] Choo QY, Ho PC, Tanaka Y, Lin HS. The histone deacetylase inhibitors MS-275 and SAHA suppress the p38 mitogen-activated protein kinase signaling pathway and chemotaxis in rheumatoid arthritic synovial fibroblastic E11 cells. Molecules 2013;18(11):14085–95.

[51] Chang C, Gershwin ME. Drug-induced lupus erythematosus: incidence, management and prevention. Drug Saf 2011;34(5):357–74.

[52] Yung R, Powers D, Johnson K, Amento E, Carr D, Laing T, et al. Mechanisms of drug-induced lupus. II. T cells overexpressing lymphocyte function-associated antigen 1 become autoreactive and cause a lupuslike disease in syngeneic mice. J Clin Invest 1996;97(12):2866–71.

[53] Yung RL, Quddus J, Chrisp CE, Johnson KJ, Richardson BC. Mechanism of drug-induced lupus. I. Cloned Th2 cells modified with DNA methylation inhibitors in vitro cause autoimmunity in vivo. J Immunol 1995;154(6):3025–35.

[54] Chang C, Gershwin ME. Drugs and autoimmunity–a contemporary review and mechanistic approach. J Autoimmun 2010;34(3):J266–75.

[55] Nanki T, Hayashida K, El-Gabalawy HS, Suson S, Shi K, Girschick HJ, et al. Stromal cell-derived factor-1-CXC chemokine receptor 4 interactions play a central role in CD4+ T cell accumulation in rheumatoid arthritis synovium. J Immunol 2000;165(11):6590–8.

[56] Nanki T, Lipsky PE. Cytokine, activation marker, and chemokine receptor expression by individual CD4(+) memory T cells in rheumatoid arthritis synovium. Arthritis Res 2000;2(5):415–23.

[57] Blades MC, Ingegnoli F, Wheller SK, Manzo A, Wahid S, Panayi GS, et al. Stromal cell-derived factor 1 (CXCL12) induces monocyte migration into human synovium transplanted onto SCID mice. Arthritis Rheum 2002;46(3):824–36.

[58] Konttinen YT, Ainola M, Valleala H, Ma J, Ida H, Mandelin J, et al. Analysis of 16 different matrix metalloproteinases (MMP-1 to MMP-20) in the synovial membrane: different profiles in trauma and rheumatoid arthritis. Ann Rheum Dis 1999;58(11):691–7.

[59] Richardson BC, Strahler JR, Pivirotto TS, Quddus J, Bayliss GE, Gross LA, et al. Phenotypic and functional similarities between 5-azacytidine-treated T cells and a T cell subset in patients with active systemic lupus erythematosus. Arthritis Rheum 1992;35(6):647–62.

[60] Fali T, Le Dantec C, Thabet Y, Jousse S, Hanrotel C, Youinou P, et al. DNA methylation modulates HRES1/p28 expression in B cells from patients with Lupus. Autoimmunity 2014;47(4):265–71.

[61] Thabet Y, Le Dantec C, Ghedira I, Devauchelle V, Cornec D, Pers JO, et al. Epigenetic dysregulation in salivary glands from patients with primary Sjogren's syndrome may be ascribed to infiltrating B cells. J Autoimmun 2013;41:175–81.

[62] Fabian MR, Sonenberg N. The mechanics of miRNA-mediated gene silencing: a look under the hood of miRISC. Nat Struct Mol Biol 2012;19(6):586–93.

[63] Zhao S, Wang Y, Liang Y, Zhao M, Long H, Ding S, et al. MicroRNA-126 regulates DNA methylation in CD4+ T cells and contributes to systemic lupus erythematosus by targeting DNA methyltransferase 1. Arthritis Rheum 2011;63(5):1376–86.

[64] Tang Y, Luo X, Cui H, Ni X, Yuan M, Guo Y, et al. MicroRNA-146A contributes to abnormal activation of the type I interferon pathway in human lupus by targeting the key signaling proteins. Arthritis Rheum 2009;60(4):1065–75.

[65] Curtale G, Citarella F, Carissimi C, Goldoni M, Carucci N, Fulci V, et al. An emerging player in the adaptive immune response: microRNA-146a is a modulator of IL-2 expression and activation-induced cell death in T lymphocytes. Blood 2010;115(2):265–73.

[66] Zhao X, Tang Y, Qu B, Cui H, Wang S, Wang L, et al. MicroRNA-125a contributes to elevated inflammatory chemokine RANTES levels via targeting KLF13 in systemic lupus erythematosus. Arthritis Rheum 2010;62(11):3425–35.

[67] Stanczyk J, Pedrioli DM, Brentano F, Sanchez-Pernaute O, Kolling C, Gay RE, et al. Altered expression of MicroRNA in synovial fibroblasts and synovial tissue in rheumatoid arthritis. Arthritis Rheum 2008;58(4):1001–9.

[68] Leng RX, Pan HF, Qin WZ, Chen GM, Ye DQ. Role of microRNA-155 in autoimmunity. Cytokine Growth Factor Rev 2011;22(3):141–7.

[69] Niederer F, Trenkmann M, Ospelt C, Karouzakis E, Neidhart M, Stanczyk J, et al. Down-regulation of microRNA-34a* in rheumatoid arthritis synovial fibroblasts promotes apoptosis resistance. Arthritis Rheum 2012;64(6):1771–9.

[70] Pauley KM, Stewart CM, Gauna AE, Dupre LC, Kuklani R, Chan AL, et al. Altered miR-146a expression in Sjogren's syndrome and its functional role in innate immunity. Eur J Immunol 2011;41(7):2029–39.

[71] Du C, Liu C, Kang J, Zhao G, Ye Z, Huang S, et al. MicroRNA miR-326 regulates TH-17 differentiation and is associated with the pathogenesis of multiple sclerosis. Nat Immunol 2009;10(12):1252–9.

[72] Junker A, Krumbholz M, Eisele S, Mohan H, Augstein F, Bittner R, et al. MicroRNA profiling of multiple sclerosis lesions identifies modulators of the regulatory protein CD47. Brain 2009;132(Pt 12):3342–52.

[73] Padgett KA, Lan RY, Leung PC, Lleo A, Dawson K, Pfeiff J, et al. Primary biliary cirrhosis is associated with altered hepatic microRNA expression. J Autoimmun 2009;32(3–4):246–53.

[74] Shrivastava S, Petrone J, Steele R, Lauer GM, Di Bisceglie AM, Ray RB. Up-regulation of circulating miR-20a is correlated with hepatitis C virus-mediated liver disease progression. Hepatology 2013;58(3):863–71.

[75] Rotllan N, Fernandez-Hernando C. MicroRNA regulation of cholesterol metabolism. Cholesterol 2012;2012:847849.

[76] Collison A, Mattes J, Plank M, Foster PS. Inhibition of house dust mite-induced allergic airways disease by antagonism of microRNA-145 is comparable to glucocorticoid treatment. J Allergy Clin Immunol 2011;128(1):160–7. e4.

VII. PERSONALIZED EPIGENETICS OF DISORDERS AND DISEASE MANAGEMENT

[77] Mattes J, Collison A, Plank M, Phipps S, Foster PS. Antagonism of microrna-126 suppresses the effector function of TH2 cells and the development of allergic airways disease. Proc Natl Acad Sci USA 2009;106(44):18704–9.

[78] Najafi-Shoushtari SH. MicroRNAs in cardiometabolic disease. Curr Atheroscler Rep 2011;13(3):202–7.

[79] Fernandez-Hernando C, Suarez Y, Rayner KJ, Moore KJ. MicroRNAs in lipid metabolism. Curr Opin Lipidol 2011;22(2):86–92.

[80] Trajkovski M, Hausser J, Soutschek J, Bhat B, Akin A, Zavolan M, et al. MicroRNAs 103 and 107 regulate insulin sensitivity. Nature 2011;474(7353):649–53.

[81] Niller HH, Wolf H, Ay E, Minarovits J. Epigenetic dysregulation of epstein-barr virus latency and development of autoimmune disease. Adv Exp Med Biol 2011;711:82–102.

[82] Niller HH, Wolf H, Minarovits J. Viral hit and run-oncogenesis: genetic and epigenetic scenarios. Cancer Lett 2011;305(2):200–17.

[83] Takacs M, Banati F, Koroknai A, Segesdi J, Salamon D, Wolf H, et al. Epigenetic regulation of latent Epstein-Barr virus promoters. Biochim Biophys Acta 2010;1799(3–4):228–35.

[84] Fry RC, Rager JE, Bauer R, Sebastian E, Peden DB, Jaspers I, et al. Air toxics and epigenetic effects: ozone altered microRNAs in the sputum of human subjects. Am J Physiol Lung Cell Mol Physiol 2014;306(12):L1129–37.

[85] Lee KW, Pausova Z. Cigarette smoking and DNA methylation. Front Genet 2013;4:132.

[86] Breton CV, Byun HM, Wenten M, Pan F, Yang A, Gilliland FD. Prenatal tobacco smoke exposure affects global and gene-specific DNA methylation. Am J Respir Crit Care Med 2009;180(5):462–7.

[87] Mortusewicz O, Schermelleh L, Walter J, Cardoso MC, Leonhardt H. Recruitment of DNA methyltransferase I to DNA repair sites. Proc Natl Acad Sci USA 2005;102(25):8905–9.

[88] Mercer BA, Wallace AM, Brinckerhoff CE, D'Armiento JM. Identification of a cigarette smoke-responsive region in the distal MMP-1 promoter. Am J Respir Cell Mol Biol 2009;40(1):4–12.

[89] Di YP, Zhao J, Harper R. Cigarette smoke induces MUC5AC protein expression through the activation of Sp1. J Biol Chem 2012;287(33):27948–58.

[90] Satta R, Maloku E, Zhubi A, Pibiri F, Hajos M, Costa E, et al. Nicotine decreases DNA methyltransferase 1 expression and glutamic acid decarboxylase 67 promoter methylation in GABAergic interneurons. Proc Natl Acad Sci USA 2008;105(42):16356–61.

[91] Liu Q, Liu L, Zhao Y, Zhang J, Wang D, Chen J, et al. Hypoxia induces genomic DNA demethylation through the activation of HIF-1alpha and transcriptional upregulation of MAT2A in hepatoma cells. Mol Cancer Ther 2011;10(6):1113–23.

[92] Ito K, Lim S, Caramori G, Chung KF, Barnes PJ, Adcock IM. Cigarette smoking reduces histone deacetylase 2 expression, enhances cytokine expression, and inhibits glucocorticoid actions in alveolar macrophages. FASEB J 2001;15(6):1110–2.

[93] Miller FW, Alfredsson L, Costenbader KH, Kamen DL, Nelson LM, Norris JM, et al. Epidemiology of environmental exposures and human autoimmune diseases: findings from a National Institute of Environmental Health Sciences Expert Panel Workshop. J Autoimmun 2012;39(4):259–71.

[94] Costenbader KH, Gay S, Alarcon-Riquelme ME, Iaccarino L, Doria A. Genes, epigenetic regulation and environmental factors: which is the most relevant in developing autoimmune diseases? Autoimmun Rev 2012;11(8):604–9.

[95] Li Y, Gorelik G, Strickland FM, Richardson BC. Oxidative stress, T cell DNA methylation, and lupus. Arthritis Rheumatol 2014;66(6):1574–82.

[96] Hew KM, Walker AI, Kohli A, Garcia M, Syed A, McDonald-Hyman C, et al. Childhood exposure to ambient polycyclic aromatic hydrocarbons is linked to epigenetic modifications and impaired systemic immunity in T cells. Clin Exp Allergy January 2015;45(1):238–48.

[97] Bind MA, Lepeule J, Zanobetti A, Gasparrini A, Baccarelli A, Coull BA, et al. Air pollution and gene-specific methylation in the Normative Aging Study: association, effect modification, and mediation analysis. Epigenetics 2014;9(3):448–58.

[98] Li Y, Zhao M, Yin H, Gao F, Wu X, Luo Y, et al. Overexpression of the growth arrest and DNA damage-induced 45alpha gene contributes to autoimmunity by promoting DNA demethylation in lupus T cells. Arthritis Rheum 2010;62(5):1438–47.

[99] Yang CY, Leung PS, Adamopoulos IE, Gershwin ME. The implication of vitamin D and autoimmunity: a comprehensive review. Clin Rev Allergy Immunol 2013;45(2):217–26.

[100] Fujiki R, Kim MS, Sasaki Y, Yoshimura K, Kitagawa H, Kato S. Ligand-induced transrepression by VDR through association of WSTF with acetylated histones. EMBO J 2005;24(22):3881–94.

[101] Christian M, White R, Parker MG. Metabolic regulation by the nuclear receptor corepressor RIP140. Trends Endocrinol Metab 2006;17(6):243–50.

[102] Sejdic E. Medicine: adapt current tools for handling big data. Nature 2014;507(7492):306.

[103] Chiche L, Jourde-Chiche N, Whalen E, Presnell S, Gersuk V, Dang K, et al. Modular transcriptional repertoire analyses of adults with systemic lupus erythematosus reveal distinct type I and type II interferon signatures. Arthritis Rheumatol 2014;66(6):1583–95.

[104] Morozova O, Hirst M, Marra MA. Applications of new sequencing technologies for transcriptome analysis. Annu Rev Genomics Hum Genet 2009;10:135–51.

[105] Shah S, McRae AF, Marioni RE, Harris SE, Gibson J, Henders AK, et al. Genetic and environmental exposures constrain epigenetic drift over the human life course. Genome Res November 2014;24(11):1725–33.

[106] West J, Widschwendter M, Teschendorff AE. Distinctive topology of age-associated epigenetic drift in the human interactome. Proc Natl Acad Sci USA 2013;110(35):14138–43.

[107] Teschendorff AE, West J, Beck S. Age-associated epigenetic drift: implications, and a case of epigenetic thrift? Hum Mol Genet 2013;22(R1):R7–15.

Cardiovascular Diseases and Personalized Epigenetics

Adam M. Zawada, Gunnar H. Heine

Department of Internal Medicine IV, Nephrology and Hypertension,
Saarland University Medical Center, Homburg, Germany

OUTLINE

1. INTRODUCTION—CARDIOVASCULAR DISEASE AND EPIGENETICS

Cardiovascular disease (CVD) is a major cause of morbidity and mortality worldwide [1]. Despite significant therapeutic improvements, which comprise both pharmacological interventions, such as the use of statins, inhibitors of the renin–angiotensin system, and β-blockers, as well as innovations in the field of percutaneous and surgical vascular interventions, the high prevalence of CVD in industrialized nations and developing countries poses an enormous medical and economic burden for our health care systems. To allow the effective use of resources and vasculoprotective treatment strategies and to subsequently reduce disease burden and the substantial economic sequelae, early identification of individuals at high risk for CVD and a better understanding of the pathogenesis of CVD are of specific interest.

Pathophysiologically, atherosclerosis—as the central underlying mechanism of CVD—is characterized as chronic inflammation of large or intermediate arteries [2]. Within this pathophysiological process, endothelial dysfunction and subintimal infiltration of immune cells, mainly monocytes, are considered the first steps.

Endothelial dysfunction itself is induced by the effects of traditional risk factors—particularly arterial hypertension, hyperglycemia, smoking, and dyslipidemia with elevated plasma levels of oxidized low-density lipoprotein (oxLDL)—and other proinflammatory mediators—e.g., angiotensin II—on endothelial cells. Activated by these risk factors and mediators, endothelial cells begin to secrete chemokines and to express selectins and integrins on their surface, which induce circulating immune cells— particularly monocytes—to roll along and adhere to the endothelial layer (Figure 1).

Following their firm attachment to the endothelium, monocytes and other immune cells pass the endothelial cell layer by diapedesis into the intimal layer of the vascular wall (intima). Here, monocytes differentiate into tissue-resident macrophages, which engulf oxLDL and other lipids. These metabolites stimulate macrophages to produce proinflammatory cytokines and growth factors. Thus, the inflammatory reaction is amplified, leading to further attraction of immune cells into the intima and of smooth muscle cells from the media into the intima layer of the vasculature. This continuous recruitment and proliferation of cells within the intima leads to the formation of atherosclerotic plaques, which are characterized by a necrotic core and a cap consisting of smooth muscle cells and collagen-rich matrix. This fibrous cap may then finally be destabilized by macrophage-derived matrix metalloproteinases, resulting in plaque rupture and exposure of prothrombotic material. The subsequent coagulation cascade induces thrombus formation and may lead to vascular occlusion [3].

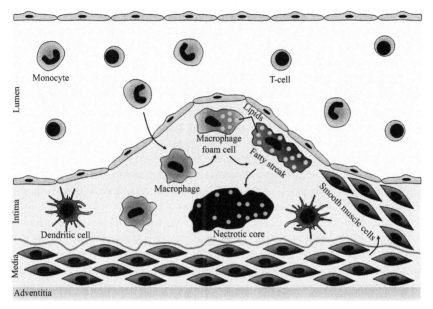

FIGURE 1 **The role of monocytes in atherogenesis.** After activation, endothelial cells facilitate the attachment and transmigration of monocytes into the subendothelial space. Here, monocytes differentiate into macrophages, which engulf lipoprotein particles and secrete proinflammatory cytokines as well as growth factors. The migration of smooth muscle cells from the tunica media into the tunica intima further promotes this atherogenic process. Finally, an advanced atherosclerotic plaque develops, which is composed of a lipid- and macrophage-rich necrotic core and an overlying fibrous cap. Rupture of this fibrous cap may lead to thrombus formation and vascular occlusion. *Reproduced with permission from Ref. [6].*

The clinical sequelae of atherosclerosis depends on the affected vessel segment and may lead to diverse pathologies such as myocardial infarction and stroke.

In contrast to these acute, potentially life-threatening events, which occur within a few minutes, the pathogenesis of atherosclerosis is a condition that develops over several decades and may begin very early in life. Thus, the development of diagnostic tools for early identification of atherosclerosis may allow us to introduce personalized prevention programs to individuals at a much younger age than in conventional cardiovascular medicine, in which preventive strategies are mostly offered to individuals who already have advanced atherosclerotic vascular changes, given that early changes are not recognized by current diagnostic tools [4]. These prevention programs currently comprise conventional pharmacological interventions as well as lifestyle changes.

Against this background, the implications of introducing epigenetic research into cardiovascular disease may be twofold: First, characterization of epigenetic features in CVD may allow a much earlier detection of atherosclerotic vascular changes in apparently healthy subjects compared to classical approaches that are currently applied. This may pave the way for the application of novel, personalized strategies for individual prevention in cardiovascular medicine. In atherosclerosis, many cell types undergo cellular transformation, which requires reprogramming of gene expression. While epigenetic mechanisms control cellular homeostasis and thus maintain cellular identity during subsequent cell divisions, epigenetic dysregulation may induce cellular transformations in atherogenesis. Importantly, these changes in epigenetic gene regulation may occur very early in atherogenesis, before the development of advanced vascular lesions that are detectable by conventional imaging modalities. Such epigenetic changes thus possess the potential to become a valuable diagnostic tool for early atherosclerotic disease [5].

Second, beyond such potential diagnostic applications of epigenetic research, the modifiable nature of epigenetic marks may allow the development of specific drugs for novel targeted therapeutic intervention in cardiovascular medicine. Thereby, epigenetic modifying drugs bear the potential to reduce cardiovascular morbidity and mortality in the future.

The field of epigenetics has attracted substantial interest in cardiovascular medicine. All components of epigenetic regulation, i.e., DNA methylation, histone modifications, and RNA interference, have been characterized in the context of CVD, and pioneering studies point toward a central contribution of epigenetic mechanisms in the development and progression of CVD. Notably, epigenetic dysregulation may affect both atherosclerosis-protective and atherosclerosis-susceptible genes in various cell types involved in the pathogenesis of atherosclerosis.

Among numerous environmental stimuli that affect epigenetic regulation, nutrition is one central factor directly involved in cardiovascular health and disease.

Specifically, intake of folate and vitamins B_6 and B_{12} may have a direct impact on the methylation status of an individual, owing to the central role of these vitamins in the generation of the universal methyl group donor S-adenosylmethionine in C1 (one-carbon) metabolism [6]. In line with this, vitamin supplementation allows the amelioration of epigenetic dysregulation that is seen in patients with a disturbed C1 metabolism, such as patients with chronic kidney disease [7].

Numerous other nutritional compounds may modify the epigenome. In their review article, Szarc vel Szic et al. [8] suggested food as "epigenetic medicine" and listed 58 bioactive dietary phytochemicals that have an impact on our epigenome. Most of them were investigated in the context of cancer; however, some of them link epigenetics and CVD, such as epicatechin (apples, cocoa, green and black tea), epigallocatechin 3-gallate (green tea), quercetin (citrus, capers, apples, berries, tea, wine), and resveratrol (grapes, blueberries, peanuts, red wine).

In this context, it is of interest that nutrition may influence the epigenome even before an individual's birth, as maternal nutrition influences the epigenetic regulation of the fetus with sequelae on cardiovascular health that may extend into adult age. Thus, atherosclerotic disease that becomes manifest in adulthood may partly be explained by exposure to environmental conditions in utero. As a classical example, individuals prenatally exposed to famine during the Dutch Hunger Winter of 1944–1945 were five decades later shown to have an altered DNA methylation pattern and a higher prevalence of coronary heart disease and obesity than those adults born before or conceived after that period [9,10]. These results are in line with animal studies that analyzed the pathophysiology of arterial hypertension. Offspring of mothers fed a low protein diet during pregnancy displayed an altered DNA methylation pattern of genes linked to blood pressure control [11].

Such fetal reprogramming occurs also beyond the context of nutrition, as other environmental factors during pregnancy may affect the development of CVD in adult age. For example, prenatal exposure to tobacco smoke induces persistent changes in DNA methylation patterns [12]. Moreover, prenatal exposure to cocaine induces a decrease in protein kinase Cε (PKCε) gene expression via DNA methylation [13]. PKCε plays a central role in cardioprotection during cardiac ischemia and reperfusion injury, and prenatal cocaine abuse might thereby modify the vulnerability of ischemic injury in the offspring's heart via epigenetic mechanisms.

Since the field of epigenetics in CVD is a relatively novel discipline, substantial efforts are needed to shed more light on the impact of distinct environmental factors in the context of epigenetics and CVD. In the following, we aim to give a general overview on topical knowledge in the field of epigenetic regulation in CVD. As a general limitation to most studies discussed, studies on epigenetics—unlike genetic studies—deal with the problem of cell specificity and reversibility. Therefore, both the specific cell type analyzed and confounding factors such as age, gender, comedication, and lifestyle habits are crucial for the interpretation of results from epigenetic studies in the field of CVD. The second general limitation is the lack of standardization of analytical techniques between studies, which complicates the interpretation and comparability of their findings. It is, however, of most importance that epigenetic research in the context of CVD takes account of patients' characteristics—comprising age, gender, and comedication—and critically discusses limitations of the applied analytical methods.

In this chapter we discuss the three major features of epigenetic gene regulation—DNA methylation, histone modifications, and microRNAs—in the context of CVD. Moreover, one section describes the effect of a disturbed C1 metabolism on epigenetic dysregulation in CVD, as a balanced

C1 metabolism is mandatory for physiological epigenetic gene regulation. This follows current research strategies, as most studies published hitherto have focused on single features of epigenetic gene regulation. It should, however, be kept in mind that these different epigenetic gene-regulatory mechanisms are closely interlinked and act together to regulate gene transcription to form a distinct phenotype, which has elegantly been shown in the context of CVD [14,15].

The discussion focuses on the issue of how far a CVD-specific pattern of epigenetic modifications will allow an earlier prediction of individual cardiovascular risk and thus support personalized vasculoprotective medicine in the future. Finally, as epigenetic marks are potentially reversible, we discuss the option of novel therapy strategies in CVD.

We generally center our chapter on atherosclerotic cardiovascular disease, which represents the single most important disease entity within the spectrum of human CVD. Other cardiovascular diseases are touched upon wherever deemed necessary.

2. DNA METHYLATION AND CARDIOVASCULAR DISEASE

Studies on DNA methylation in the context of atherosclerotic CVD can be broadly categorized into reports on global DNA methylation and analyses of gene-specific DNA methylation. Table 1 provides an overview of the results of these studies.

2.1 Assessment of Global DNA Methylation in Cardiovascular Disease

The main cell types involved in atherogenesis are peripheral blood cells and vascular tissue (atherosclerotic lesions in most cases); both cell types were analyzed in the context of atherosclerotic CVD.

A small cohort study recruited 17 male patients who had suffered from either stroke or myocardial infarction and found significantly lower DNA methylation in peripheral blood cells (measured by cytosine extension assay) compared to 15 male healthy controls [16]. An association between global DNA hypomethylation and prevalent cerebrovascular or coronary artery disease was confirmed among 712 elderly men from the Normative Aging Study, when DNA methylation of the LINE1 (long interspersed nucleotide element-1) repetitive elements in peripheral blood mononuclear cells was analyzed as a marker for global DNA methylation [17]. Moreover, lower LINE1 methylation at study initiation was a predictor of cardiovascular mortality.

TABLE 1 Studies on Global and Gene-Specific DNA Methylation in Human Atherosclerotic CVD

Disease	References	Source	Method	Gene	Regulation
Myocardial infarction/stroke	Castro et al. [16]	Peripheral blood cells	Cytosine extension assay	Global DNA methylation	Hypomethylation
Myocardial infarction/stroke	Kim et al. [19]	Peripheral blood cells	MethyLight assay	Global DNA methylation (Alu and satellite 2 repetitive elements)	Hypermethylation
Ischemic heart disease/stroke	Baccarelli et al. [17]	Peripheral blood cells	Pyrosequencing of bisulfite-treated DNA	Global DNA methylation (LINE1)	Hypomethylation
Coronary artery disease	Sharma et al. [18]	Peripheral blood cells	Cytosine extension assay	Global DNA methylation	Hypermethylation
Ischemic cardiopathy	Castillo-Diaz et al. [20]	Atherosclerotic artery samples	Microarray analysis	Genome-wide analysis	Mainly hypomethylation (e.g., NOTCH1, FOXP1)
Atherosclerotic vascular disease	Hiltunen et al. [21]	Atherosclerotic artery samples	High-performance liquid chromatography	Global DNA methylation analysis	Hypomethylation
Coronary artery disease	Post et al. [22]	Atherosclerotic artery samples	Southern blot analysis	ERα	Hypermethylation
Coronary artery/cerebrovascular/peripheral artery disease	Kim et al. [23]	Atherosclerotic artery samples	Methylation-specific PCR and combined bisulfite restriction analysis	ERβ	Hypermethylation

Continued

TABLE 1   Studies on Global and Gene-Specific DNA Methylation in Human Atherosclerotic CVD—cont'd

Disease	References	Source	Method	Gene	Regulation
Cerebrovascular disease	Zawadzki et al. [25]	Atherosclerotic artery samples	Methylation-specific PCR and pyrosequencing	TFPI-2	Hypermethylation
Coronary heart disease	Jiang et al. [26]	Peripheral blood cells	Pyrosequencing	PLA2G7	Hypermethylation
Coronary artery disease	Zhuang et al. [27]	Peripheral blood cells	MethyLight assay	p15INK4b	Hypermethylation
Myocardial infarction	Talens et al. [28]	Peripheral blood cells	Mass spectrometry	INS, GNASAS	Hypermethylation
Atherosclerotic vascular disease	Hiltunen et al. [21]	Atherosclerotic artery samples	Sequencing of bisulfite-modified DNA	15-Lipoxygenase	Hypomethylation
Coronary artery disease	Breitling et al. [29]	Whole blood	Sequenom mass spectrometer system	F2RL3	Hypomethylation

In contrast, other clinical studies point to global DNA hypermethylation rather than hypomethylation in atherosclerotic CVD. In one study, 137 Indian CAD (coronary artery disease) patients were reported to have significantly higher peripheral blood cell DNA methylation levels (measured by cytosine extension assay) compared to 150 controls [18]. Moreover, among 286 Singapore Chinese Health Study participants, subjects with prevalent myocardial infarction and/or stroke had higher leukocyte DNA methylation of repetitive elements—as a marker for global DNA methylation—than controls [19]. In contrast to the Normative Aging Study, male patients with incident cardiovascular events during follow-up had higher global DNA methylation than subjects without cardiovascular event.

Global DNA methylation analyses in vascular tissue rather point toward DNA hypo- rather than hypermethylation in atherosclerotic lesions. Using a microarray-based approach, Castillo-Díaz et al. [20] found a near-complete demethylation of normally hypermethylated CpG islands in 45 atherosclerotic coronary artery samples from 45 patients undergoing coronary revascularization compared to 16 control aorta samples from patients undergoing aortic valve replacement. These results are in accordance with data from Hiltunen et al. [21], who found reduced global DNA methylation in advanced atherosclerotic lesion compared to normal arteries when 55 human arteries were analyzed via high-performance liquid chromatography.

Furthermore, animal studies also support the role of DNA hypomethylation in atherosclerotic CVD. In atherosclerotic lesions from ApoE-knockout mice on a high-fat diet, and in smooth muscle cells during their transformation from a contractile to a synthetic phenotype, global DNA hypomethylation was detected [21]. Importantly, in the ApoE mouse model it was found that changes in DNA methylation pattern—both hypo- and hypermethylation—occur very early in atherogenesis before any visual signs of vascular lesions and thus have the potential to become a valuable diagnostic tool for atherosclerotic CVD [5].

In conclusion, most data on global DNA methylation in atherosclerotic CVD support the concept of global DNA hypomethylation, although some studies with contrary results exist. Most of these data have come from small cross-sectional observations without standardization of technical and clinical parameters across studies, which makes it somewhat difficult to compare their partly contradictory results. Importantly, the use of different tissues, storage conditions, and methods for the analysis of global DNA methylation may account for the partly contradictory study results. Moreover, differences in study size and design (comprising differences in, e.g., gender, age, and medication in the study population) as well as nonstandardized definition of atherosclerotic vascular disease are potential reasons for these discrepant findings. Therefore, there is an urgent need for well-powered longitudinal studies that are adequately designed to address the question of how far global changes in DNA methylation affect the individual cardiovascular outcome. As epigenetic marks are susceptible to a wide range of confounding factors, such as age, gender, and medication, the exact characterization of these factors is a must for these future study reports.

The major limitation of global DNA methylation analysis is that it does not provide any information on the dysregulation of distinct genes. In atherosclerosis, both DNA hypomethylation and DNA hypermethylation may occur at the same time: when hypomethylation of atherosclerosis-susceptible genes coincides with hypermethylation of atherosclerosis-protective genes, analysis of global DNA methylation may suggest global hypomethylation, global hypermethylation, or neutral results.

In summary, there are strong arguments that analysis of gene-specific DNA methylation patterns in atherosclerotic CVD may allow a better understanding of atherogenesis and the underlying pathways. Thus, the analysis of gene-specific DNA methylation may have a much higher potential to become a novel diagnostic tool in atherosclerotic CVD than the analysis of global DNA methylation.

2.2 Assessment of Gene-Specific DNA Methylation in Cardiovascular Disease

Most gene-specific DNA methylation studies in the context of atherosclerotic CVD selected a priori candidate genes and measured their methylation status under different atherosclerotic CVD conditions. Additionally, first genome-wide studies added valuable information about gene regulation in atherosclerotic CVD. These studies again used either histological samples of atherosclerotic lesions or peripheral blood cells.

DNA methylation of estrogen receptors (*ERα* and *ERβ*) was repeatedly selected as a target for gene-specific analyses, as broad effects of estrogens have been discussed in cardiovascular health and disease. The promoter regions of *ERα* and *ERβ* were shown to be hypermethylated in coronary plaques and in transforming smooth muscle cells [22,23].

Other genes analyzed in this context are monocarboxylate transporter 3 (*MCT3*) and tissue factor pathway inhibitor-2 (*TFPI-2*): transforming smooth muscle cells were found to display hypermethylation of *MCT3* [24], and hypermethylation of *TFPI-2* was reported in human atherosclerotic lesions [25].

Among studies that reported gene-specific methylation in peripheral blood cells in the context of human atherosclerotic CVD, a cross-sectional study with 36 coronary heart disease patients and 36 controls identified peripheral blood cell DNA hypermethylation of *PLA2G7*, the gene product of which is the phospholipase A2, group VII, to be associated with coronary heart disease in females, but, interestingly, not in males [26].

In peripheral blood cells from 95 Chinese patients with coronary artery disease, DNA methylation of a locus on chromosome 9p21 ($p15^{INK4b}$)— a region associated with coronary artery disease—was higher than in peripheral blood cells from 110 sex- and age-matched participants without prevalent coronary artery disease [27].

In the placebo group of the PROSPER trial, which tested pravastatin for reduction of coronary and cerebrovascular events in elderly subjects with either prevalent atherosclerotic CVD or a high risk for future atherosclerotic CVD, Talens et al. [28] compared individuals who developed myocardial infarction during a follow-up period of 3 years ($n = 122$) with an event-free control group ($n = 126$). They found higher peripheral blood cell DNA methylation of *INS* and *GNASAS*, two loci sensitive to prenatal conditions, in women who experienced myocardial infarction compared to women without event. No such difference was found in men.

Several studies point toward DNA hypomethylation of other distinct genes in the context of atherosclerotic CVD: in a genome-wide approach Castillo-Díaz et al. [20] identified 142 hypomethylated and 17 hypermethylated CpG islands in atherosclerotic samples compared to nonatherosclerotic aortic samples. A high number of those CpG islands were linked to genes coding for signaling and transcription factors, such as *PROX1*, *NOTCH1*, or *FOXP1*, and to genes involved in smooth muscle cell modulation and inflammation. Similarly, hypomethylation of the promoter region of the 15-lipoxygenase gene was found in human atherosclerotic lesions [21].

Among those studies that focused on gene-specific methylation in peripheral blood cells, the large-scale prospective KAROLA study measured whole-blood DNA methylation of *F2RL3*, which codes for the blood-clotting-associated protease-activated receptor-4. Among 1206 subjects undergoing cardiovascular inpatient rehabilitation, who were followed for the occurrence of future cardiovascular events, for cause-specific mortality, and for all-cause mortality, a lower methylation level of *F2RL3* was highly associated with all-cause mortality [29].

These pioneering studies suggest that DNA methylation—global or gene-specific—may contribute to the development and progression of atherosclerotic CVD. Moreover, they have identified various dysregulated candidate genes in atherosclerotic CVD, thus providing insights into the potential pathophysiological pathways. However, again most experimental data were not confirmed in subsequent reports from independent research groups. Therefore, future studies will have to validate these results in different cohorts of atherosclerotic CVD patients. They should furthermore analyze whether changes in the DNA methylation pattern in one specific cell type are mirrored by changes in other cell types. Of note, although the analysis of DNA methylation changes in atherosclerotic vascular tissue is of special interest for the characterization of the underlying pathophysiological processes in atherogenesis, such athero-sclerotic tissue is not easily accessible for routine diagnostic approaches. Instead, the analysis of DNA methylation in peripheral blood cells has a much higher potential to become a valuable diagnostic tool in the future. Unfortunately, this approach also has major limitations, as changes in the composition of peripheral blood cells occur in various human diseases, which inevitably leads to changes in DNA methylation when using a whole-blood approach. Instead, the isolation of specific blood cells, i.e., monocytes, for DNA methylation analyses would be time consuming and thus unfeasible for a diagnostic approach offered to a broad population. In conclusion, there is an unmet need for the identification of robust markers that are not restricted to a specific cell type.

3. C1 METABOLISM AND CARDIOVASCULAR DISEASE

DNA methylation is closely interlinked with the metabolism of homo-cysteine and its derivatives, referred to as "one-carbon metabolism." S-adenosylmethionine (SAM), a central component of the C1 metabolism, is the universal methyl group donor involved in DNA methylation (along with numerous other cellular methylation reactions) (Figure 2) [6].

Within the C1 metabolism, most scientific work in the past focused on homocysteine, which has been hypothesized as a nontraditional cardio-vascular risk factor based upon a bulk of experimental and epidemiologi-cal data [6]. Homocysteine is in equilibrium with S-adenosylhomocysteine (SAH), which is a powerful competitive inhibitor of SAM-dependent methyltransferases. Thus, removal of intracellular homocysteine is man-datory to allow regular DNA methylation; this removal is ensured by three different pathways: first via transsulfuration to cystathionine by the cystathionine-β-synthase in a vitamin B_6-dependent pathway, second via remethylation to methionine by the methionine synthase in a folate/vitamin B_{12}-dependent reaction, and third via betaine-homocysteine methyltrans-ferase using betaine as the methyl group donor. Any disturbances in these reactions—which may be due to low availability of the cofactors folate, vitamin B_6, and/or vitamin B_{12}—may increase intracellular homocysteine and subsequently SAH levels, leading to subsequent inhibition of trans-methylation reactions.

Notably, a well-balanced C1 metabolism is crucial, not only for avoid-ing accumulation of the methylation inhibitor SAH, but also for allowing adequate generation of SAM as the universal methyl group donor. SAM is generated from the amino acid methionine, which itself either derives from remethylation of homocysteine or is taken up with food. Methyl-transferases, e.g., DNA methyltransferases, transfer the methyl group of SAM to its target, e.g., DNA; by losing its methyl group, SAM is converted

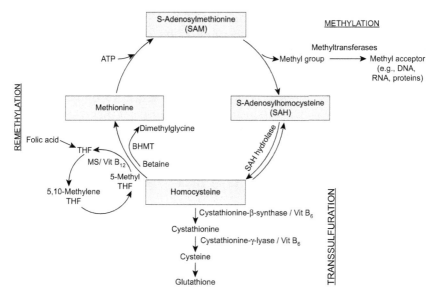

FIGURE 2   **C1 metabolism.** In C1 metabolism, homocysteine is metabolized via either the transsulfuration or the remethylation pathway. Within the remethylation pathway homocysteine is metabolized to methionine and subsequently to S-adenosylmethionine (SAM), the universal methyl group donor. By transferring its methyl group, SAM is converted to S-adenosylhomocysteine (SAH), a powerful inhibitor of SAM-dependent methyltransferases. Thus, in order not to interfere with cellular methylation reactions, SAH has to be hydrolyzed to homocysteine. Since the latter is a reversible reaction, any accumulation of homocysteine will directly increase SAH levels. Vit, vitamin; THF, tetrahydrofolate; ATP, adenosine triphosphate; MS, methionine synthase; BHMT, betaine homocysteine methyltransferase. *Reproduced with permission from Ref. [6].*

into SAH. SAH itself can be hydrolyzed to homocysteine and adenosine via SAH hydrolases. Importantly, the equilibrium of this reaction favors SAH formation from homocysteine rather than the hydrolysis of SAH. This illustrates that, to allow hydrolysis of SAH, rapid removal of intracellular homocysteine and adenosine is necessary.

Various states of disease in human medicine are characterized by hyperhomocysteinemia. Specifically, patients with chronic kidney disease (CKD) [7] display dramatically increased plasma levels of both homocysteine and—as a direct consequence of its equilibrium with homocysteine—SAH. In addition, kidneys play a direct role in the elimination of SAH [30]. As SAH inhibits methylation reactions, elevated SAH levels in CKD patients are accompanied by global DNA hypomethylation [7,31].

Hyperhomocysteinemia has been characterized as an independent cardiovascular risk factor in several prospective epidemiological studies [32–34], and addition of plasma homocysteine levels to the Framingham Risk Score improves risk prediction considerably [35]. Moreover, in experimental studies homocysteine was found to induce major proatherogenic pathways, including the production of reactive oxygen species (ROS) [36], proliferation of smooth muscle cells [37], recruitment of immune cells [38], and induction of a prothrombotic state [39].

These clinical and experimental findings stimulated many research groups to perform intervention studies, which tested the hypothesis that homocysteine lowering is beneficial for cardiovascular outcome [40–51].

As summarized in Table 2, homocysteine lowering in these studies was achieved via supplementation with folate, vitamin B_6, and/or vitamin B_{12}. However, despite their significant reductions in homocysteine, most of these studies failed to show a beneficial cardiovascular outcome. One single study reported beneficial effects on a combined end point; this effect was, however, driven by a reduction in coronary revascularization, whereas the arguably more relevant components of the combined study end point, namely myocardial infarctions or total mortality, were not significantly reduced. Therefore, the causal role of homocysteine in CVD remains at best questionable.

In past years the acknowledgment of a potential contribution of other C1 metabolites in CVD pointed toward an attractive explanation of these seemingly controversial findings.

Specifically, experimental studies analyzed the pathophysiological effect of SAH and found, in the ApoE mouse model, in which either homocysteine or SAH was increased by a specific diet, that only those animals with elevated SAH levels developed atherosclerotic lesions; moreover, SAH concentrations and aortic sinus lesions were highly correlated [52]. Further animal and in vitro studies found SAH to induce smooth muscle cell proliferation, apoptosis of endothelial cells, and aortic ROS production by activation of the extracellular signal-regulated kinases 1/2 pathway [53,54]. Pioneering clinical studies also support the detrimental role of SAH in the cardiovascular system. In a small epidemiological study, in which 30 patients with atherosclerotic CVD and 29 matched controls were recruited, SAH was more strongly increased than homocysteine [55]. These results were confirmed in a case–control study with 32 individuals [16] and 124 patients with chronic kidney disease [56]. In 420 apparently healthy subjects from the I Like HOMe study, traditional cardiovascular risk factors and subclinical atherosclerosis—measured as intima media thickness—were associated with SAH, but not with homocysteine. Moreover, plasma SAH was strongly associated with kidney function [57]. Finally, in dialysis patients, plasma SAH could be linked to a disturbed regulation of atherosclerosis-related genes [31].

The first prospective large-scale study on the role of SAH in atherosclerotic CVD supports these findings: among 1003 patients undergoing coronary angiography, plasma SAH was identified as independent predictor of cardiovascular events [58].

TABLE 2 Randomized Controlled Trials on the Effects of Lowering Homocysteine

Study	Participants (n)	Inclusion criterion	Homocysteine level (baseline)	Intervention	Primary outcome	Results
VISP, Toole et al. [40]	3680	Nondisabling cerebral infarction	A: 13.4 μmol/L B: 13.4 μmol/L	A: 2.5 mg folic acid 25 mg Vit B$_6$ 0.4 mg Vit B$_{12}$ B: 20 μg folic acid 200 μg Vit B$_6$ 6 μg Vit B$_{12}$	Recurrent cerebral infarction	No benefit
HOPE-2, Lonn et al. [41]	5522	Vascular disease or diabetes	A: 12.2 μmol/L B: 12.2 μmol/L	A: 2.5 mg folic acid 50 mg Vit B$_6$ 1 mg Vit B$_{12}$ B: Placebo	Myocardial infarction, stroke, CVD mortality	No benefit
NORVIT, Bonna et al. [42]	3749	Acute myocardial infarction within 7 days before recruitment	A: 13.1 μmol/L B: 12.9 μmol/L	A: 0.8 mg folic acid 40 mg Vit B$_6$ 0.4 mg Vit B$_{12}$ B: 0.8 mg folic acid 0.4 mg Vit B$_{12}$	Recurrent myocardial infarction, stroke, and sudden cardiac death	No benefit

Continued

TABLE 2   Randomized Controlled Trials on the Effects of Lowering Homocysteine—cont'd

Study	Participants (n)	Inclusion criterion	Homocysteine level (baseline)	Intervention	Primary outcome	Results
WENBIT, Ebbing et al. [43]	3096	Patients undergoing coronary angiography	A: 9.8 μmol/L B: 9.9 μmol/L C: 9.2 μmol/L D: 9.9 μmol/L	A: 0.8 mg folic acid 40 mg Vit B_6 0.4 mg Vit B_{12} B: 0.8 mg folic acid 0.4 mg Vit B_{12} C: 40 mg Vit B_6 D: Placebo	All-cause death, nonfatal acute myocardial infarction, acute hospitalization for unstable angina pectoris, and nonfatal thromboembolic stroke	No benefit
WAFACS, Albert et al. [44]	5442	Women aged ≥42 years with a history of CVD or three or more CVD risk factors	A: 12.1 μmol/L B: 12.5 μmol/L	A: 2.5 mg folic acid 50 mg Vit B_6 1 mg Vit B_{12} B: Placebo	Myocardial infarction, stroke, coronary revascularization, or CVD mortality	No benefit

Study	N	Patient population	Homocysteine	Treatment	Endpoint	Result
SEARCH, Armitage et al. [45]	12064	Survivors of myocardial infarction	A+B: 13.5 μmol/L	A: 2mg folic acid 1mg Vit B_{12}; B: Placebo	Major coronary event (coronary death, myocardial infarction, or coronary revascularization), fatal or nonfatal stroke, or noncoronary revascularization	No benefit
VITATOPS, Hankey et al. [46]	8164	Recent stroke or transient ischemic attack	A: 14.4 μmol/L; B: 14.2 μmol/L	A: 2mg folic acid 25mg Vit B_6 0.5mg Vit B_{12}; B: Placebo	Stroke, myocardial infarction, or vascular death	No benefit
Swiss Heart Study, Schnyder et al. [47]	553	Patients after successful angioplasty of at least one significant coronary stenosis (≥50%)	A: 11.4 μmol/L; B: 11.1 μmol/L	A: 1mg folic acid 10mg Vit B_6 400 μg Vit B_{12}; B: Placebo	Death, nonfatal myocardial infarction, and need for repeat revascularization	Benefit
ASFAST, Zoungas et al. [48]	315	Advanced chronic kidney disease	A: 27.0 μmol/L; B: 27.0 μmol/L	A: 15mg folic acid; B: Placebo	Rate of progression of mean maximum carotid intima-media thickness, myocardial infarction, stroke, and cardiovascular death	No benefit

Continued

TABLE 2   Randomized Controlled Trials on the Effects of Lowering Homocysteine—cont'd

Study	Participants (n)	Inclusion criterion	Homocysteine level (baseline)	Intervention	Primary outcome	Results
HOST, Jamison et al. [49]	2056	Advanced chronic kidney disease or end-stage renal disease and high homocysteine levels ($\geq$15 µmol/L)	A: 24.0 µmol/L B: 24.2 µmol/L	A: 40 mg folic acid 100 mg Vit B$_6$ 2 mg Vit B$_{12}$ B: Placebo	All-cause mortality	No benefit
Renal HOPE-2, Mann et al. [50]	619	$\geq$55 years with CKD and high cardiovascular risk	A: 15.9 µmol/L B: 15.7 µmol/L	A: 2.5 mg folic acid 50 mg Vit B$_6$ 1 mg Vit B$_{12}$ B: Placebo	Death from cardiovascular causes, myocardial infarction, and stroke	No benefit
Heinz et al. [51]	650	End-stage renal disease	A: 30.0 µmol/L B: 28.2 µmol/L	A: 5 mg folic acid 20 mg Vit B$_6$ 50 µg Vit B$_{12}$ B (placebo): 0.2 mg folic acid 1.0 mg Vit B$_6$ 4 µg Vit B$_{12}$	All-cause mortality	No benefit

These study results support the notion that SAH, as a central epigenetic regulator, may contribute to the initiation and progression of atherosclerotic CVD [52,55,58]. Importantly, supplementation with folate, vitamin B_6, and vitamin B_{12}, which effectively lowers plasma homocysteine levels, does not at all affect plasma SAH levels [59,60].

Interventional studies that test the impact of SAH lowering on cardiovascular outcome are thus eagerly awaited; they will ultimately solve the question whether SAH rather than homocysteine should be considered as a nontraditional cardiovascular risk factor implicated in the pathogenesis of atherosclerotic CVD. Unfortunately, no interventional strategy for lowering of SAH has been identified yet.

4. HISTONE MODIFICATIONS IN CARDIOVASCULAR DISEASE

DNA methylation and histone modifications are tightly connected in epigenetic gene regulation. Thus, dysregulated histone modifications can be expected to be associated with the development and progression of atherosclerotic CVD. Moreover, beyond their role in regulation of DNA methylation, C1 metabolites—particularly SAH and SAM—play a central role in the regulation of histone methylation.

Although some studies point toward dysregulation of distinct post-translational modifications of histones, i.e., methylation, acetylation, phosphorylation, ADP ribosylation, sumoylation, and biotinylation, the research on histone modifications in the context of atherosclerotic CVD is in its very beginning and more efforts are needed to better understand the complex interaction between histone modifications and the development of atherosclerotic CVD. Given the paucity of data on the implication of histone modifications in atherosclerotic CVD, this section aims to provide a short overview on the contribution of such histone modifications in a broader spectrum of CVD, comprising heart failure and cardiac hypertrophy.

Similar to DNA methylation, the regulation of a limited number of predefined atherosclerosis-related genes has been the focus of the few scientific studies published in this field until now. In the context of histone modifications, one of the best characterized genes is the endothelial *NOS3* gene, which codes for endothelial nitric oxide synthase (eNOS). eNOS is crucial for the production of the vasoprotective factor NO and its dysregulation is associated with increased atherosclerotic CVD risk [61]. *NOS3* expression is sensitive to environmental stimuli such as hypoxia, which causes *NOS3* silencing. This silencing of *NOS3* in hypoxia may be induced by decreased histone acetylation (H3ac and H4ac) and H3K4 dimethylation at the *NOS3* proximal promoter [62]. Moreover, other proatherogenic conditions, such as shear stress at the vascular endothelium, may also modulate histone modifications at the *NOS3* promoter [63]. Therefore, modulation of these epigenetic marks at the *NOS3* gene is discussed as a central mechanism that underlies endothelial dysfunction occurring under distinct proatherogenic conditions.

Similarly, histone acetylation has gained substantial interest in the context of smooth muscle cell differentiation and proliferation: in smooth muscle cell differentiation, serum response factor (SRF) is essential for the expression of specific genes, as SRF binds to specific regulatory sequences of these genes. Interestingly, this binding is inhibited by deacetylation of histone H4 in response to vascular injury [64]. Moreover, treatment of smooth muscle cells with the histone deacetylase (HDAC) inhibitor Scriptaid results in G1-phase cell cycle arrest and reduced proliferation; in vivo, HDAC inhibition decreased neointima formation in a murine model of vascular injury [65]. However, conflicting results have been published by other groups, who found HDAC inhibition (with trichostatin A) to enhance rather than to decrease proliferation and migration of smooth muscle cells [66].

In line with a proatherogenic effect of HDAC inhibition are experimental results from Ldlr$^{-/-}$ (Ldlr: low-density lipoprotein receptor) mice that were fed a high-fat diet. In this model, HDAC inhibition (with trichostatin A) exacerbated atherosclerosis and increased the uptake of oxLDL by macrophages in vitro [67].

Importantly, studies on the role of HDAC inhibition in atherogenesis applied different inhibitors and concentrations, which could partly account for these conflicting findings from different laboratory groups. To reconcile these conflicting results, it has been acknowledged that—depending on concentration and HDAC class specificity—HDAC inhibitors are able either to induce or to repress inflammation-related genes. Moreover, different inflammation-related genes may show different response to the same HDAC inhibition. For example, trichostatin A and suberoylanilide hydroxamic acid were found to suppress the lipopolysaccharide-induced expression of the proinflammatory-associated genes *Edn-1*, *Ccl-7*, and *Il-12p40*, but to induce expression of the proinflammatory-associated genes *Cox-2* and *Serpine1* in primary mouse bone marrow-derived macrophages [68]. Future studies are needed, which should analyze the inflammatory response to the inhibition of specific subtypes of HDACs in the context of atherosclerosis.

Similarly, several lines of evidence argue for a specific contribution of individual HDACs in atherosclerosis. As one single example, under experimental shear-stress conditions, HDAC5 regulates the expression of the shear-stress-responsive genes *NOS3* (eNOS) and *KLF2* [69], whereas HDAC3 is essential for endothelial cell survival during shear stress, in which its knockdown induced accelerated atherosclerosis [70].

In addition to histone acetylation, some data point toward a contribution of histone methylation in atherogenesis. In smooth muscle cells from type 2 diabetic db/db mice, H3K9 trimethylation, a key repressive chromatin mark, was shown to be involved in the increased expression of proinflammatory-associated genes. This dysregulation was associated with reduced protein levels of Suv39h1, a methyltransferase, which catalyzes H3K9 trimethylation. In line with this, Suv39h1 overexpression reversed this proinflammatory phenotype. Based on these results, dysregulation of histone methylation is discussed as one underlying mechanism for the proinflammatory phenotype of these cells [71].

Beyond atherosclerotic vascular disease, some work on histone modifications has been done in the context of cardiac hypertrophy and heart failure. H3K4 trimethylation was found to be essential for physiological functions in murine cardiomyocytes. Loss of H3K4 trimethylation induced dysregulation of cardiac-specific genes, which was associated with augmented intracellular calcium content and increased contractility in vitro as well as with systolic function in vivo [72].

In addition, methylation of H3K79, which is catalyzed by DOT1L, may be of importance for cardiomyocyte function. Cardiac-specific knockout of *Dot1L* in murine models leads to phenotypes similar to those observed in patients with dilated cardiomyopathy (DCM). In line with this, it was confirmed that DOT1L is downregulated in idiopathic DCM patient samples compared to control samples [73].

In left-ventricular myocardium from patients with terminal heart failure, upregulation of atrial natriuretic peptide (ANP) and brain natriuretic peptide (BNP)—which are secreted from cardiac cells in various types of heart diseases, and the plasma levels of which are biomarkers of the severity of heart failure—was associated with reduced H3K9 di- and trimethylation in the promoter regions of the respective genes. In a sophisticated approach with isolated working murine hearts, which were exposed to an acute increase in cardiac preload, the central role of H3K9 demethylation for activation of the *ANP* gene was confirmed [74].

Finally, two genome-wide studies aimed to characterize the dysregulation of histone modifications in heart failure. First, both in animal models and in human left-ventricular tissue, genome-wide analysis of H3K4 and H3K9 trimethylation identified differential histone methylation in the vicinity of genes encoding cardiac-relevant proteins during the development of heart failure [75]. Second, H3K36 trimethylation was analyzed in patients with cardiomyopathy and end-stage heart failure: compared to nondiseased human hearts, differential H3K36 trimethylation was found in numerous coding regions of the genome, such as the *DUX4* locus, which is linked to facioscapulohumeral muscular dystrophy [15].

In summary, a major contribution of dysregulation of histone modifications to atherosclerotic and nonatherosclerotic CVD has been suggested.

However, as of this writing, the vast majority of scientific evidence derives from experimental (mostly murine) models of CVD. In the future, more efforts will be needed to analyze histone modification in the context of human atherosclerotic and nonatherosclerotic CVD.

5. MICRORNAS IN CARDIOVASCULAR DISEASE

miRNAs are central regulators of cellular homeostasis and inflammatory processes [76]. As one single miRNA is able to control a multitude of target genes within specific pathways and signaling cascades such as inflammatory responses in endothelial cells and macrophages [77], dysregulation of miRNAs may directly lead to the development of diverse pathological conditions such as atherosclerosis [78,79].

In line with this, distinct miRNAs have been identified as biomarkers for diagnosis and prognosis in atherosclerotic and nonatherosclerotic CVD [80–111]. Importantly, miRNAs are very stable; they can be detected and analyzed not only intracellularly, but also in a variety of body fluids, such as plasma and urine. Table 3 lists cellular and circulating miRNAs that were analyzed in the context of human atherosclerotic and nonatherosclerotic CVD. The next sections separately summarize studies on cellular or circulating miRNAs in atherosclerotic and nonatherosclerotic CVD.

5.1 Cellular MicroRNAs in Cardiovascular Disease

In the context of atherosclerotic CVD, intracellular miRNA expression was analyzed in atherosclerotic plaques and in circulating leukocytes.

In human atherosclerotic lesions, a distinct miRNA signature has been found. Comparing the miRNA expression profiles of nonatherosclerotic left internal thoracic arteries to aortic, carotid, and femoral atherosclerotic plaques, the expression of miR-21, -34a, -146a, -146b, and -210 was upregulated in atherosclerotic lesions [83]. In line with this, the miRNA expression profile of advanced atherosclerotic plaques obtained during carotid endarterectomy differs from the miRNA expression profile of nonatherosclerotic internal thoracic arteries obtained during elective coronary artery bypass surgery, and the vast majority of miRNAs that are differentially expressed between atherosclerotic plaques and nonatherosclerotic arteries are upregulated in atherosclerotic plaques [84] (compare Table 3). Finally, when the miRNA expression profile of atherosclerotic intima obtained from patients suffering from peripheral artery disease was compared with the miRNA expression profile of normal intima obtained from the same patients, levels of miR-21, miR-130a, miR-27b, let-7f, and miR-210 were significantly increased in

TABLE 3 Overview of Data on miRNA Expression in Human Atherosclerotic and Nonatherosclerotic CVD

Disease	References	Source	Method	miRNA	Regulation
CAD	Fichtlscherer et al. [80]	Plasma	Array/real-time PCR	miR-126	↓
				miR-17	↓
				miR-92a	↓
				miR-155	↓
				miR-145	↓
CAD	Taurino et al. [81]	Whole blood	Array/real-time PCR	miR-140-3p	↑
				miR-182	↑
CAD	Hoekstra et al. [82]	PBMCs	Real-time PCR	miR-135a	↑
				miR-147	↓
CAD	Raitoharju et al. [83]	Atherosclerotic plaque	Array/real-time PCR	miR-21	↑
				miR-34a	↑
				miR-146a	↑
				miR-146b-5p	↑
				miR-210	↑
CAD	Bidzhekov et al. [84]	Atherosclerotic plaque	Array	miR-520b	↓
				miR-105	↓
				miR-10b	↑

Continued

TABLE 3 Overview of Data on miRNA Expression in Human Atherosclerotic and Nonatherosclerotic CVD—cont'd

Disease	References	Source	Method	miRNA	Regulation
				miR-218	↑
				miR-17-3p	↑
				ambi-miR-7039	↑
				miR-186	↑
				miR-489	↑
				miR-143	↑
				miR-524-5p	↑
				miR-220	↑
				miR-147	↑
				miR-422a	↑
				miR-15b	↑
				miR-185	↑
				miR-181a	↑
				miR-98	↑
				miR-152	↑
				miR-422b	↑
				miR-30e-5p	↑
				miR-26b	↑
				miR-125a	↑

CAD	Balderman et al. [85]	Calcified carotid artery	In situ hybridization	miR-30b	↓
CAD	Gao et al. [86]	Plasma	Real-time PCR	miR-122	↑
				miR-370	↑
CAD	Hulsmans et al. [87]	CD14+ Monocytes	Array/real-time PCR	miR-181a	↓
CAD	Satoh et al. [88]	Monocytes	Real-time PCR	let-7i	↓
CAD	Sondermeijer et al. [89]	Platelets	Array/real-time PCR	miR-340*	↑
				miR-624*	↑
CAD	Takahashi et al. [90]	PBMCs	Real-time PCR	miR-146a/b	↑
CAD	Weber et al. [91]	Whole blood	Array/real-time PCR	miR-19a	↓
				miR-584	↓
				miR-155	↓
				miR-222	↓
				miR-145	↓
				miR-29a	↓
				miR-378	↓
				miR-342	↓
				miR-181d	↓
				miR-150	↓
				miR-30e-5p	↓

Continued

TABLE 3 Overview of Data on miRNA Expression in Human Atherosclerotic and Nonatherosclerotic CVD—cont'd

Disease	References	Source	Method	miRNA	Regulation
CAD	Minami et al. [92]	EPC	Real-time PCR	miR-221	↑
				miR-222	↑
PAD	Li et al. [93]	Serum	Real-time PCR	miR-130a	↑
				miR-27b	↑
				miR-210	↑
PAD	Li et al. [93]	Atherosclerotic plaque	Real-time PCR	miR-21	↑
				miR-130a	↑
				miR-27b	↑
				let-7f	↑
				miR-210	↑
				miR-221	↓
				miR-222	↓
AMI	Wang et al. [94]	Plasma	Real-time PCR	miR-1	↑
				miR-133a	↑
				miR-499	↑
				miR-208a	↑
AMI	D'Alessandra et al. [95]	Plasma	Array/real-time PCR	miR-1	↑
				miR-133a	↑

Continued

Disease	Reference	Sample	Method	miRNA	Change
AMI	Corsten et al. [96]	Plasma	Real-time PCR	miR-133b	↑
				miR-499-5p	↑
				miR-122	↓
				miR-375	↓
AMI	Adachi et al. [97]	Plasma	Real-time PCR	miR-208b	↑
AMI	Ai et al. [98]	Plasma	Real-time PCR	miR-499	↑
AMI	Kuwabara et al. [99]	Serum	Real-time PCR	miR-133a	↑
				miR-223	↓
AMI	Cheng et al. [100]	Serum	Real-time PCR	miR-499	↑
AMI	Gidlof et al. [101]	Plasma	Real-time PCR	miR-1	↑
				miR-1	↑
				miR-133a	↑
AMI	Long et al. [102]	Plasma	Real-time PCR	miR-1	↑
				miR-133a	↑
				miR-208b	↑
				miR-499-5p	↑
AMI	Devaux et al. [103]	Plasma	Real-time PCR	miR-1	↑
				miR-126	↓
				miR-208b	↑
				miR-499	↑

TABLE 3   Overview of Data on miRNA Expression in Human Atherosclerotic and Nonatherosclerotic CVD—cont'd

Disease	References	Source	Method	miRNA	Regulation
AMI	Olivieri et al. [104]	Plasma	Real-time PCR	miR-1	↓
				miR-21	↑
				miR-133a	↑
				miR-423-5p	↑
				miR-499-5p	↑
AMI	Meder et al. [105]	Whole blood	Array/real-time PCR	miR-1291	↓
				miR-663b	↓
				miR-145	↑
				miR-30c	↑
ACS	Oerlemans et al. [106]	Serum	Real-time PCR	miR-1	↑
				miR-499	↑
				miR-208a	↑
				miR-21	↑
				miR-146a	↑
HF	Corsten et al. [96]	Plasma	Real-time PCR	miR-499	↑
				miR-122	↑
HF	Tijsen et al. [107]	Plasma	Array/real-time PCR	miR-423-5p	↑
				miR-129-5p	↑

Continued

Reference	Condition	Sample	Method	miRNA	
Fukushima et al. [108]	HF	Plasma	Real-time PCR	miR-675	↑
				miR-18b*	↑
				HS_202.1	↑
				miR-1254	↑
				miR-622	↑
Voellenkle et al. [109]	HF	PBMCs	Real-time PCR	miR-126	↓
				miR-107	↓
				miR-139	↓
				miR-142-5p	↓
Vogel et al. [111]	HF	Whole blood	Array	miR-520d-5p	↑
				miR-122*	↑
				miR-643	↑
				miR-548i	↑
				miR-718	↑
				miR-935	↑
				let-7e*	↑
				miR-376a	↑
				miR-1225-5p	↑
				miR-675*	↑
				miR-622	↑
				miR-582-3p	↑

TABLE 3 Overview of Data on miRNA Expression in Human Atherosclerotic and Nonatherosclerotic CVD—cont'd

Disease	References	Source	Method	miRNA	Regulation
				miR-551b*	↑
				miR-224	↑
				miR-670	↑
				miR-331-5p	↑
				miR-369-3p	↑
				miR-944	↑
				miR-200b*	↑
				miR-519e*	↑
				miR-558	↓
				miR-1302	↓
				miR-146b-3p	↓
				miR-345	↓
				miR-760	↓
				miR-218	↓
				miR-1301	↓
				miR-604	↓
				miR-370	↓
				miR-144	↓

Reference	Disease	Sample	Method	miRNA	Regulation
Goren et al. [110]	HF	Serum	Real-time PCR	miR-574-5p	↓
				miR-566	↓
				miR-1321	↓
				miR-143*	↓
				miR-551b	↓
				miR-20b*	↓
				miR-1914	↓
				miR-597	↓
				miR-623	↓
				miR-421	↓
Olivieri et al. [104]	HF	Plasma	Real-time PCR	miR-423-5p	↑
				miR-320a	↑
				miR-22	↑
				miR-92b	↑
				miR-21	↑
				miR-133a	↑
				miR-423-5p	↑
				miR-499-5p	↑

↑ upregulation; ↓ downregulation; CAD, coronary artery disease; PAD, peripheral artery disease; AMI, acute myocardial infarction; ACS, acute coronary syndrome; HF, heart failure.

atherosclerotic plaques, while only the levels of miR-221 and miR-222 were significantly decreased [93].

Moreover, several studies point toward a dysregulation of miRNA expression in circulating blood cells from patients with prevalent atherosclerotic CVD. In peripheral blood mononuclear cells (PBMCs) from patients with coronary artery disease, an increased expression of miR-135a and a reduced expression of miR-147 were found in comparison to healthy controls [82]. In a second study, analysis of the miRNA profile of whole blood from patients with CAD revealed upregulation of miR-140-3p and miR-182 compared to healthy controls [81].

In the context of nonatherosclerotic CVD, Vogel et al. [111] analyzed miRNA expression profiles in whole blood from 53 patients with nonischemic heart failure and 39 controls, both free of relevant CAD. Screening identified several miRNAs that are differentially expressed in patients with heart failure, among which miR-558, miR-122*, and miR-520d-5p best discriminated patients and controls. Combining eight miRNAs (miR-520d-5p, miR-558, miR-122*, miR-200b*, miR-622, miR-519e*, miR-1231, and miR-1228*) further improved discrimination performance. Moreover, the authors found that miR-622, miR-520d-5p, miR-519e*, miR-558, and miR-200b* had a similar or even slightly better diagnostic performance than NT-proBNP as an established biomarker for heart failure. Finally, in a retrospective outcome analysis, miR-519e* predicted the combined end point of hospitalization for heart failure, heart transplantation, stroke, and cardiovascular death.

Further studies are needed that should analyze the miRNA signature of distinct cell types, such as monocytes, from patients with prevalent atherosclerotic and nonatherosclerotic CVD. This is of particular interest as differences in miRNA profiles reported in PBMCs or whole blood analyses may merely reflect shifts in the relative distribution of leukocyte subpopulations in peripheral blood, which are frequently observed in patients suffering from cardiovascular disease. Indeed, first studies point toward a dysregulation of miR-181a and let-7i in monocytes from patients with prevalent CVD [87,88].

5.2 Circulating MicroRNAs in Cardiovascular Disease

In addition to their intracellular function as posttranscriptional gene regulators, miRNAs can be released by cells and circulate in the blood. Circulating miRNAs were found to be remarkably stable and protected against degradation, as they can be packaged in vesicles or associated with protein or lipid complexes [112]. As these circulating miRNAs can be taken up by distant cells and modulate their gene expression, a role of miRNAs in cell-to-cell communication has been postulated.

Given their high stability and their accessibility, circulating miRNAs entered the field as valuable biomarkers in cardiovascular medicine.

Emerging evidence has associated particular entities of cardiovascular disease with disease-specific patterns of distinct miRNAs.

Profiles of circulating miRNAs in patients with stable coronary artery disease identified levels of miR-126, miR-17, miR-92a, miR-155, and miR-145 to be significantly reduced in patients compared to healthy controls [80]. In patients with hyperlipidemia high circulating levels of miR-122 and miR-370 were associated with plasma triglycerides and total and LDL-cholesterol, as well as with the presence of coronary artery disease [86].

Next, the role of circulating miRNAs has widely been studied in patients suffering from acute myocardial infarction. The sudden necrosis of cardiomyocytes in acute myocardial infarction not only leads to the release of cell-specific proteins such as cardiac troponin, which are currently the hallmark of diagnosis, but also to the release of distinct miRNAs into the circulation. Several cardiac miRNAs were repeatedly analyzed and consistently found to be elevated after the onset of myocardial infarction, namely miR-208a/b, miR-499, miR-1, and miR-133a/b (compare Table 3). As miR-208 is encoded by the α-myosin heavy chain gene, this miRNA is exclusively expressed in the heart, while the others are also expressed in the skeletal muscle [112]. The levels of circulating miR-208 are not detectable in the plasma from healthy people, but they rapidly and dramatically increase in patients with acute myocardial infarction [94,96]. This explains why miR-208 bears the potential to outperform cardiac troponin I and troponin T in early diagnosis of acute myocardial infarction in future years [94].

Beyond the acute setting of myocardial infarction, a first large-scale prospective study aimed at identifying circulating miRNAs as predictors of future cardiovascular events among apparently healthy individuals [113]: in 820 participants from the Bruneck study, circulating levels of miR-126 were positively associated with risk for myocardial infarction, whereas miR-223 and miR-197 were inversely associated with cardiovascular outcome.

Cardiomyocyte-derived miRNAs were also investigated in the context of heart failure. Patients with acute heart failure were shown to have significantly elevated levels of circulating miR-499 compared with control subjects, whereas levels of miR-208b, miR-1, and miR-133a only tended to be higher [96]. Their moderate increase in heart failure compared to their dramatic increase in myocardial infarction may be a result of the rather low cardiac myocyte injury compared to myocardial infarction. In line with this, another study that performed miRNA arrays on plasma from patients with heart failure and controls found no significant increases in miR-1, miR-208a, miR-208b, and miR-499 [107]. Instead, seven other miRNAs were identified (compare Table 3), among which miR-423 was most strongly related to the clinical diagnosis of heart failure. miR-423 may be derived from the myocardium; but other cell types may contribute to the release of miR-423 into the circulation, given that myocardiac-specific miRNAs were not increased in this study. Future studies will have to address the question of how far elevated miRNAs in heart failure merely reflect their release from cell death or whether secretory pathways exist in cardiomyocytes.

In summary, the crucial role of miRNAs in the cardiovascular system has widely been acknowledged, and distinct miRNAs have been suggested as promising diagnostic markers in human atherosclerotic and nonatherosclerotic CVD.

Additionally, miRNA expression may be measured in a relatively rapid and cost-effective way. Nonetheless, the technology for miRNA analysis awaits further substantial improvements before its introduction into the clinical routine. For instance, until now, no consistent housekeeping miRNA has been established that may allow robust normalization of measurements, which is mandatory for comparison of measurements between laboratories.

6. CONCLUSIONS AND FURTHER PERSPECTIVES

Despite significant improvements in health care systems, cardiovascular disease remains one of the leading causes of death worldwide. All three major features of epigenetic regulation, i.e., DNA methylation, histone modifications, and miRNAs, are centrally involved in the pathophysiology of atherosclerotic and nonatherosclerotic CVD. Analysis of these three epigenetic features has been introduced into the field of cardiovascular medicine, which may in the near future allow us to define novel biomarkers for an earlier identification of individuals at high CVD risk. It is hoped that this will help us to apply personalized vasculoprotective prevention strategies to younger individuals, long before these affected individuals will develop clinically overt CVD. Compared to today's approach of late initiation of preventive measurements in cardiovascular medicine, such early interventions may allow a more potent reduction of the CVD burden and its economic sequelae for health care systems.

Beyond the application of epigenetic marks as biomarkers, research on epigenetic modulators for therapeutic intervention in cardiovascular medicine is arising. Of note, routinely used drugs in cardiovascular medicine—particularly statins—may have effects on epigenetic regulation as part of the pleiotropic effects of these drugs [114]. Beyond these epigenetic side effects of drugs that are currently applied to patients because of their conventional (nonepigenetic) modes of action, drugs that directly target epigenetic features—e.g., DNA methylation and histone modifications by DNA methyltransferase inhibitors and histone deacetylase inhibitors—have been designed and introduced into clinical practice [115].

Until now, four epigenetic drugs have been approved in oncology: azacitidine (Vidaza) and decitabine (Dacogen), two DNA methylation inhibitors, and vorinostat (Zolinza) and romidepsin (Istodax), two HDAC inhibitors [116]. As of this writing, several clinical trials are testing further agents, some of which for indications beyond the field of oncology. Importantly, such innovative epigenetic drugs are also being tested in the context of CVD: RVX-208 is in phase II trials for atherosclerosis and diabetes, since it significantly increases ApoA-I and HDL-cholesterol and enhances cholesterol efflux. RVX-208 targets bromodomain-containing proteins, a family of proteins that bind to acetylated residues of histones [117].

Because they modulate gene expression nonspecifically, however, DNA methyltransferase inhibitors and drugs that target histone modifications will inevitably cause undesirable systemic side effects, and gene-specific approaches for modulation of DNA methylation and histone modification should not be expected to enter cardiovascular medicine in the next decade. Instead, modulation of miRNA expression by synthetic antimirs, which directly inhibit distinct miRNAs, may allow a more directed intervention. Modulation of distinct miRNAs for vasculoprotection has already been tested in first animal studies, which found that therapeutic silencing of distinct miRNAs (miR-33, miR-145, miR-342) inhibited the development of atherosclerosis in ApoE$^{-/-}$ and Ldlr$^{-/-}$ mice [118–120]. The potential of miRNAs as targets for clinical intervention has spurred intense interest from pharmaceutical companies [121]. As of this writing, the first miRNA-modifying drug is in clinical trials for hepatitis C virus (HCV) infection: miravirsen is an antimir with a high affinity for miR-122, which is a liver-specific miRNA that can modulate HCV replication. In a phase II study, miravirsen strongly decreased serum HCV RNA in patients with chronic HCV infection, with no serious adverse effects [122]. Thus, miravirsen may become the first miRNA-modifying drug on the market. In the context of cancer, the first miRNA-mimicking drug entered a clinical trial: MRX34, which is a miR-34 mimic, was found in preclinical studies to inhibit tumor growth [123] and is now being tested in phase I in patients with primary liver cancer or metastatic cancer that has spread to the liver [121]. Many other miRNA-modifying drugs are being tested for different indications in preclinical studies; within the context of cardiovascular medicine, potential targets are miR-208 (for treatment of heart failure and cardiometabolic disease) as well as miR-195 (for treatment of postmyocardial infarction remodeling) [121]. Thus, miRNA-modifying drugs may become available for prevention and therapeutic intervention of human CVD in the next couple of years.

LIST OF ACRONYMS AND ABBREVIATIONS

ACS	Acute coronary syndrome
AMI	Acute myocardial infarction
ANP	Atrial natriuretic peptide
BNP	Brain natriuretic peptide
C1	One-carbon
CAD	Coronary artery disease
Ccl-7	Chemokine (C–C motif) ligand 7
CKD	Chronic kidney disease
Cox-2	Cytochrome c oxidase subunit II
CVD	Cardiovascular disease
DCM	Dilated cardiomyopathy
DM	Diabetes mellitus
DOT1L	DOT1-like histone H3K79 methyltransferase
DUX4	Double homeobox 4
Edn-1	Endothelin 1
eNOS/NOS3	Endothelial nitric oxide synthase
EPC	Endothelial progenitor cell
ERα	Estrogen receptor α
ERβ	Estrogen receptor β
ERK 1/2	Extracellular signal-regulated kinase 1/2
F2RL3	Coagulation factor II (thrombin) receptor-like 3
FOXP1	Forkhead box P1
GNASAS	Guanine nucleotide-binding protein, α stimulating complex locus antisense RNA
HDAC	Histone deacetylase
HF	Heart failure
Il-12p40	Interleukin 12b
I Like HOMe Study	Inflammation, Lipoprotein Metabolism, and Kidney Damage in Early Atherogenesis—the Homburg Evaluation Study
INS	Insulin
KLF2	Krüppel-like factor 2

LDL	Low density lipoprotein
LINE1	Long interspersed nucleotide element-1
MCT3	Monocarboxylate transporter 3
NOS3	Nitric oxide synthase 3 (endothelial cell)
NOTCH1	Notch 1
$p15^{INK4b}$	Cyclin-dependent kinase inhibitor 2B
PBMCs	Peripheral blood mononuclear cells
PKCε	Protein kinase Cε
PLA2G7	Phospholipase A2, group VII
PROX1	Prospero homeobox 1
oxLDL	Oxidized low-density lipoprotein
ROS	Reactive oxygen species
SAHA	Suberoylanilide hydroxamic acid
SAH	S-adenosylhomocysteine
SAM	S-adenosylmethionine
Serpine1	Serine (or cysteine) peptidase inhibitor, clade E, member 1
SRF	Serum response factor
TFPI-2	Tissue factor pathway inhibitor-2

References

[1] Murray CJ, Lopez AD. Mortality by cause for eight regions of the world: global Burden of disease Study. Lancet 1997;349:1269–76.
[2] Ross R. Atherosclerosis–an inflammatory disease. N Engl J Med 1999;340:115–26.
[3] Zawada AM, Rogacev KS, Schirmer SH, Sester M, Bohm M, Fliser D, et al. Monocyte heterogeneity in human cardiovascular disease. Immunobiology 2012;217:1273–84.
[4] Robinson JG, Gidding SS. Curing atherosclerosis should be the next major cardiovascular prevention goal. J Am Coll Cardiol 2014;63:2779–85.
[5] Lund G, Andersson L, Lauria M, Lindholm M, Fraga MF, Villar-Garea A, et al. DNA methylation polymorphisms precede any histological sign of atherosclerosis in mice lacking apolipoprotein E. J Biol Chem 2004;279:29147–54.
[6] Zawada AM, Rogacev KS, Heine GH. Clinical relevance of epigenetic dysregulation in chronic kidney disease-associated cardiovascular disease. Nephrol Dial Transpl 2013;28:1663–71.
[7] Ingrosso D, Cimmino A, Perna AF, Masella L, De Santo NG, De Bonis ML, et al. Folate treatment and unbalanced methylation and changes of allelic expression induced by hyperhomocysteinaemia in patients with uraemia. Lancet 2003;361:1693–9.
[8] Szarc vel Szic K, Ndlovu MN, Haegeman G, Vanden Berghe W. Nature or nurture: let food be your epigenetic medicine in chronic inflammatory disorders. Biochem Pharmacol 2010;80:1816–32.

[9] Painter RC, Roseboom TJ, Bleker OP. Prenatal exposure to the Dutch famine and disease in later life: an overview. Reprod Toxicol 2005;20:345–52.

[10] Heijmans BT, Tobi EW, Stein AD, Putter H, Blauw GJ, Susser ES, et al. Persistent epigenetic differences associated with prenatal exposure to famine in humans. Proc Natl Acad Sci U S A 2008;105:17046–9.

[11] Bogdarina I, Welham S, King PJ, Burns SP, Clark AJ. Epigenetic modification of the renin-angiotensin system in the fetal programming of hypertension. Circ Res 2007;100:520–6.

[12] Breton CV, Byun HM, Wenten M, Pan F, Yang A, Gilliland FD. Prenatal tobacco smoke exposure affects global and gene-specific DNA methylation. Am J Respir Crit Care Med 2009;180:462–7.

[13] Zhang H, Darwanto A, Linkhart TA, Sowers LC, Zhang L. Maternal cocaine administration causes an epigenetic modification of protein kinase Cepsilon gene expression in fetal rat heart. Mol Pharmacol 2007;71:1319–28.

[14] Haas J, Frese KS, Park YJ, Keller A, Vogel B, Lindroth AM, et al. Alterations in cardiac DNA methylation in human dilated cardiomyopathy. EMBO Mol Med 2013;5:413–29.

[15] Movassagh M, Choy MK, Knowles DA, Cordeddu L, Haider S, Down T, et al. Distinct epigenomic features in end-stage failing human hearts. Circulation 2011;124:2411–22.

[16] Castro R, Rivera I, Struys EA, Jansen EE, Ravasco P, Camilo ME, et al. Increased homocysteine and S-adenosylhomocysteine concentrations and DNA hypomethylation in vascular disease. Clin Chem 2003;49:1292–6.

[17] Baccarelli A, Wright R, Bollati V, Litonjua A, Zanobetti A, Tarantini L, et al. Ischemic heart disease and stroke in relation to blood DNA methylation. Epidemiology 2010;21:819–28.

[18] Sharma P, Kumar J, Garg G, Kumar A, Patowary A, Karthikeyan G, et al. Detection of altered global DNA methylation in coronary artery disease patients. DNA Cell Biol 2008;27:357–65.

[19] Kim M, Long TI, Arakawa K, Wang R, Yu MC, Laird PW. DNA methylation as a biomarker for cardiovascular disease risk. PLoS One 2010;5:e9692.

[20] Castillo-Diaz SA, Garay-Sevilla ME, Hernandez-Gonzalez MA, Solis-Martinez MO, Zaina S. Extensive demethylation of normally hypermethylated CpG islands occurs in human atherosclerotic arteries. Int J Mol Med 2010;26:691–700.

[21] Hiltunen MO, Turunen MP, Hakkinen TP, Rutanen J, Hedman M, Makinen K, et al. DNA hypomethylation and methyltransferase expression in atherosclerotic lesions. Vasc Med 2002;7:5–11.

[22] Post WS, Goldschmidt-Clermont PJ, Wilhide CC, Heldman AW, Sussman MS, Ouyang P, et al. Methylation of the estrogen receptor gene is associated with aging and atherosclerosis in the cardiovascular system. Cardiovasc Res 1999;43:985–91.

[23] Kim J, Kim JY, Song KS, Lee YH, Seo JS, Jelinek J, et al. Epigenetic changes in estrogen receptor beta gene in atherosclerotic cardiovascular tissues and in-vitro vascular senescence. Biochim Biophys Acta 2007;1772:72–80.

[24] Zhu S, Goldschmidt-Clermont PJ, Dong C. Inactivation of monocarboxylate transporter MCT3 by DNA methylation in atherosclerosis. Circulation 2005;112:1353–61.

[25] Zawadzki C, Chatelain N, Delestre M, Susen S, Quesnel B, Juthier F, et al. Tissue factor pathway inhibitor-2 gene methylation is associated with low expression in carotid atherosclerotic plaques. Atherosclerosis 2009;204:e4–14.

[26] Jiang D, Zheng D, Wang L, Huang Y, Liu H, Xu L, et al. Elevated PLA2G7 gene promoter methylation as a gender-specific marker of aging increases the risk of coronary heart disease in females. PLoS One 2013;8:e59752.

[27] Zhuang J, Peng W, Li H, Wang W, Wei Y, Li W, et al. Methylation of p15INK4b and expression of ANRIL on chromosome 9p21 are associated with coronary artery disease. PLoS One 2012;7:e47193.

[28] Talens RP, Jukema JW, Trompet S, Kremer D, Westendorp RG, Lumey LH, et al. Hypermethylation at loci sensitive to the prenatal environment is associated with increased incidence of myocardial infarction. Int J Epidemiol 2012;41:106–15.

[29] Breitling LP, Salzmann K, Rothenbacher D, Burwinkel B, Brenner H. Smoking, F2RL3 methylation, and prognosis in stable coronary heart disease. Eur Heart J 2012;33:2841–8.

[30] Garibotto G, Valli A, Anderstam B, Eriksson M, Suliman ME, Balbi M, et al. The kidney is the major site of S-adenosylhomocysteine disposal in humans. Kidney Int 2009;76:293–6.

[31] Zawada AM, Rogacev KS, Hummel B, Grun OS, Friedrich A, Rotter B, et al. SuperTAG methylation-specific digital karyotyping reveals uremia-induced epigenetic dysregulation of atherosclerosis-related genes. Circ Cardiovasc Genet 2012;5:611–20.

[32] Anderson JL, Muhlestein JB, Horne BD, Carlquist JF, Bair TL, Madsen TE, et al. Plasma homocysteine predicts mortality independently of traditional risk factors and C-reactive protein in patients with angiographically defined coronary artery disease. Circulation 2000;102:1227–32.

[33] Mallamaci F, Zoccali C, Tripepi G, Fermo I, Benedetto FA, Cataliotti A, et al. Hyperhomocysteinemia predicts cardiovascular outcomes in hemodialysis patients. Kidney Int 2002;61:609–14.

[34] Nygard O, Nordrehaug JE, Refsum H, Ueland PM, Farstad M, Vollset SE. Plasma homocysteine levels and mortality in patients with coronary artery disease. N Engl J Med 1997;337:230–6.

[35] Veeranna V, Zalawadiya SK, Niraj A, Pradhan J, Ference B, Burack RC, et al. Homocysteine and reclassification of cardiovascular disease risk. J Am Coll Cardiol 2011;58:1025–33.

[36] Au-Yeung KK, Woo CW, Sung FL, Yip JC, Siow YL, Karmin O. Hyperhomocysteinemia activates nuclear factor-kappaB in endothelial cells via oxidative stress. Circ Res 2004;94:28–36.

[37] Majors A, Ehrhart LA, Pezacka EH. Homocysteine as a risk factor for vascular disease. Enhanced collagen production and accumulation by smooth muscle cells. Arterioscler Thromb Vasc Biol 1997;17:2074–81.

[38] Poddar R, Sivasubramanian N, DiBello PM, Robinson K, Jacobsen DW. Homocysteine induces expression and secretion of monocyte chemoattractant protein-1 and interleukin-8 in human aortic endothelial cells: implications for vascular disease. Circulation 2001;103:2717–23.

[39] Bienvenu T, Ankri A, Chadefaux B, Montalescot G, Kamoun P. Elevated total plasma homocysteine, a risk factor for thrombosis. Relation to coagulation and fibrinolytic parameters. Thromb Res 1993;70:123–9.

[40] Toole JF, Malinow MR, Chambless LE, Spence JD, Pettigrew LC, Howard VJ, et al. Lowering homocysteine in patients with ischemic stroke to prevent recurrent stroke, myocardial infarction, and death: the Vitamin Intervention for Stroke Prevention (VISP) randomized controlled trial. JAMA 2004;291:565–75.

[41] Lonn E, Yusuf S, Arnold MJ, Sheridan P, Pogue J, Micks M, et al. Homocysteine lowering with folic acid and B vitamins in vascular disease. N Engl J Med 2006;354:1567–77.

[42] Bonaa KH, Njolstad I, Ueland PM, Schirmer H, Tverdal A, Steigen T, et al. Homocysteine lowering and cardiovascular events after acute myocardial infarction. N Engl J Med 2006;354:1578–88.

[43] Ebbing M, Bleie O, Ueland PM, Nordrehaug JE, Nilsen DW, Vollset SE, et al. Mortality and cardiovascular events in patients treated with homocysteine-lowering B vitamins after coronary angiography: a randomized controlled trial. JAMA 2008;300:795–804.

[44] Albert CM, Cook NR, Gaziano JM, Zaharris E, MacFadyen J, Danielson E, et al. Effect of folic acid and B vitamins on risk of cardiovascular events and total mortality among women at high risk for cardiovascular disease: a randomized trial. JAMA 2008;299:2027–36.

[45] Armitage JM, Bowman L, Clarke RJ, Wallendszus K, Bulbulia R, Rahimi K, et al. Effects of homocysteine-lowering with folic acid plus vitamin B_{12} vs placebo on mortality and major morbidity in myocardial infarction survivors: a randomized trial. JAMA 2010;303:2486–94.

[46] VITATOPS Trial Study Group. B vitamins in patients with recent transient ischaemic attack or stroke in the VITAmins TO Prevent Stroke (VITATOPS) trial: a randomised, double-blind, parallel, placebo-controlled trial. Lancet Neurol 2010;9:855–65.

[47] Schnyder G, Roffi M, Flammer Y, Pin R, Hess OM. Effect of homocysteine-lowering therapy with folic acid, vitamin B_{12}, and vitamin B_6 on clinical outcome after percutaneous coronary intervention: the Swiss Heart study: a randomized controlled trial. JAMA 2002;288:973–9.

[48] Zoungas S, McGrath BP, Branley P, Kerr PG, Muske C, Wolfe R, et al. Cardiovascular morbidity and mortality in the Atherosclerosis and Folic Acid Supplementation Trial (ASFAST) in chronic renal failure: a multicenter, randomized, controlled trial. J Am Coll Cardiol 2006;47:1108–16.

[49] Jamison RL, Hartigan P, Kaufman JS, Goldfarb DS, Warren SR, Guarino PD, et al. Effect of homocysteine lowering on mortality and vascular disease in advanced chronic kidney disease and end-stage renal disease: a randomized controlled trial. JAMA 2007;298:1163–70.

[50] Mann JF, Sheridan P, McQueen MJ, Held C, Arnold JM, Fodor G, et al. Homocysteine lowering with folic acid and B vitamins in people with chronic kidney disease–results of the renal Hope-2 study. Nephrol Dial Transpl 2008;23:645–53.

[51] Heinz J, Kropf S, Domrose U, Westphal S, Borucki K, Luley C, et al. B vitamins and the risk of total mortality and cardiovascular disease in end-stage renal disease: results of a randomized controlled trial. Circulation 2010;121:1432–8.

[52] Liu C, Wang Q, Guo H, Xia M, Yuan Q, Hu Y, et al. Plasma S-adenosylhomocysteine is a better biomarker of atherosclerosis than homocysteine in apolipoprotein E-deficient mice fed high dietary methionine. J Nutr 2008;138:311–5.

[53] Luo X, Xiao Y, Song F, Yang Y, Xia M, Ling W. Increased plasma S-adenosylhomocysteine levels induce the proliferation and migration of VSMCs through an oxidative stress-ERK1/2 pathway in apoE(-/-) mice. Cardiovasc Res 2012;95:241–50.

[54] Sipkens JA, Hahn NE, Blom HJ, Lougheed SM, Stehouwer CD, Rauwerda JA, et al. S-Adenosylhomocysteine induces apoptosis and phosphatidylserine exposure in endothelial cells independent of homocysteine. Atherosclerosis 2012;221:48–54.

[55] Kerins DM, Koury MJ, Capdevila A, Rana S, Wagner C. Plasma S-adenosylhomocysteine is a more sensitive indicator of cardiovascular disease than plasma homocysteine. Am J Clin Nutr 2001;74:723–9.

[56] Valli A, Carrero JJ, Qureshi AR, Garibotto G, Barany P, Axelsson J, et al. Elevated serum levels of S-adenosylhomocysteine, but not homocysteine, are associated with cardiovascular disease in stage 5 chronic kidney disease patients. Clin Chim Acta 2008;395:106–10.

[57] Zawada AM, Rogacev KS, Hummel B, Berg JT, Friedrich A, Roth HJ, et al. S-adenosylhomocysteine is associated with subclinical atherosclerosis and renal function in a cardiovascular low-risk population. Atherosclerosis 2014;234:17–22.

[58] Xiao Y, Zhang Y, Wang M, Li X, Su D, Qiu J, et al. Plasma S-adenosylhomocysteine is associated with the risk of cardiovascular events in patients undergoing coronary angiography: a cohort study. Am J Clin Nutr 2013;98:1162–9.

[59] Hubner U, Geisel J, Kirsch SH, Kruse V, Bodis M, Klein C, et al. Effect of 1 year B and D vitamin supplementation on LINE-1 repetitive element methylation in older subjects. Clin Chem Lab Med 2013:1–7.

[60] Green TJ, Skeaff CM, McMahon JA, Venn BJ, Williams SM, Devlin AM, et al. Homocysteine-lowering vitamins do not lower plasma S-adenosylhomocysteine in older people with elevated homocysteine concentrations. Br J Nutr 2010;103:1629–34.

[61] Forstermann U, Munzel T. Endothelial nitric oxide synthase in vascular disease: from marvel to menace. Circulation 2006;113:1708–14.

[62] Fish JE, Yan MS, Matouk CC, St Bernard R, Ho JJ, Gavryushova A, et al. Hypoxic repression of endothelial nitric-oxide synthase transcription is coupled with eviction of promoter histones. J Biol Chem 2010;285:810–26.

[63] Chen W, Bacanamwo M, Harrison DG. Activation of p300 histone acetyltransferase activity is an early endothelial response to laminar shear stress and is essential for stimulation of endothelial nitric-oxide synthase mRNA transcription. J Biol Chem 2008;283:16293–8.

[64] McDonald OG, Wamhoff BR, Hoofnagle MH, Owens GK. Control of SRF binding to CArG box chromatin regulates smooth muscle gene expression in vivo. J Clin Invest 2006;116:36–48.

[65] Findeisen HM, Gizard F, Zhao Y, Qing H, Heywood EB, Jones KL, et al. Epigenetic regulation of vascular smooth muscle cell proliferation and neointima formation by histone deacetylase inhibition. Arterioscler Thromb Vasc Biol 2011;31:851–60.

[66] Song S, Kang SW, Choi C. Trichostatin A enhances proliferation and migration of vascular smooth muscle cells by downregulating thioredoxin 1. Cardiovasc Res 2010;85:241–9.

[67] Choi JH, Nam KH, Kim J, Baek MW, Park JE, Park HY, et al. Trichostatin A exacerbates atherosclerosis in low density lipoprotein receptor-deficient mice. Arterioscler Thromb Vasc Biol 2005;25:2404–9.

[68] Halili MA, Andrews MR, Labzin LI, Schroder K, Matthias G, Cao C, et al. Differential effects of selective HDAC inhibitors on macrophage inflammatory responses to the Toll-like receptor 4 agonist LPS. J Leukoc Biol 2010;87:1103–14.

[69] Wang W, Ha CH, Jhun BS, Wong C, Jain MK, Jin ZG. Fluid shear stress stimulates phosphorylation-dependent nuclear export of HDAC5 and mediates expression of KLF2 and eNOS. Blood 2010;115:2971–9.

[70] Zampetaki A, Zeng L, Margariti A, Xiao Q, Li H, Zhang Z, et al. Histone deacetylase 3 is critical in endothelial survival and atherosclerosis development in response to disturbed flow. Circulation 2010;121:132–42.

[71] Villeneuve LM, Reddy MA, Lanting LL, Wang M, Meng L, Natarajan R. Epigenetic histone H3 lysine 9 methylation in metabolic memory and inflammatory phenotype of vascular smooth muscle cells in diabetes. Proc Natl Acad Sci U S A 2008;105:9047–52.

[72] Stein AB, Jones TA, Herron TJ, Patel SR, Day SM, Noujaim SF, et al. Loss of H3K4 methylation destabilizes gene expression patterns and physiological functions in adult murine cardiomyocytes. J Clin Invest 2011;121:2641–50.

[73] Nguyen AT, Xiao B, Neppl RL, Kallin EM, Li J, Chen T, et al. DOT1L regulates dystrophin expression and is critical for cardiac function. Genes Dev 2011;25:263–74.

[74] Hohl M, Wagner M, Reil JC, Muller SA, Tauchnitz M, Zimmer AM, et al. HDAC4 controls histone methylation in response to elevated cardiac load. J Clin Invest 2013;123:1359–70.

[75] Kaneda R, Takada S, Yamashita Y, Choi YL, Nonaka-Sarukawa M, Soda M, et al. Genome-wide histone methylation profile for heart failure. Genes Cells 2009;14:69–77.

[76] Liu G, Abraham E. MicroRNAs in immune response and macrophage polarization. Arterioscler Thromb Vasc Biol 2013;33:170–7.

[77] Nazari-Jahantigh M, Wei Y, Noels H, Akhtar S, Zhou Z, Koenen RR, et al. MicroRNA-155 promotes atherosclerosis by repressing Bcl6 in macrophages. J Clin Invest 2012;122:4190–202.

[78] Small EM, Olson EN. Pervasive roles of microRNAs in cardiovascular biology. Nature 2011;469:336–42.

[79] Wei Y, Nazari-Jahantigh M, Neth P, Weber C, Schober A. MicroRNA-126, -145, and -155: a therapeutic triad in atherosclerosis? Arterioscler Thromb Vasc Biol 2013;33:449–54.

[80] Fichtlscherer S, De Rosa S, Fox H, Schwietz T, Fischer A, Liebetrau C, et al. Circulating microRNAs in patients with coronary artery disease. Circ Res 2010;107:677–84.

[81] Taurino C, Miller WH, McBride MW, McClure JD, Khanin R, Moreno MU, et al. Gene expression profiling in whole blood of patients with coronary artery disease. Clin Sci (Lond) 2010;119:335–43.

[82] Hoekstra M, van der Lans CA, Halvorsen B, Gullestad L, Kuiper J, Aukrust P, et al. The peripheral blood mononuclear cell microRNA signature of coronary artery disease. Biochem Biophys Res Commun 2010;394:792–7.

[83] Raitoharju E, Lyytikainen LP, Levula M, Oksala N, Mennander A, Tarkka M, et al. miR-21, miR-210, miR-34a, and miR-146a/b are up-regulated in human atherosclerotic plaques in the Tampere Vascular Study. Atherosclerosis 2011;219:211–7.

[84] Bidzhekov K, Gan L, Denecke B, Rostalsky A, Hristov M, Koeppel TA, et al. microRNA expression signatures and parallels between monocyte subsets and atherosclerotic plaque in humans. Thromb Haemost 2012;107:619–25.

[85] Balderman JA, Lee HY, Mahoney CE, Handy DE, White K, Annis S, et al. Bone morphogenetic protein-2 decreases microRNA-30b and microRNA-30c to promote vascular smooth muscle cell calcification. J Am Heart Assoc 2012;1:e003905.

[86] Gao W, He HW, Wang ZM, Zhao H, Lian XQ, Wang YS, et al. Plasma levels of lipometabolism-related miR-122 and miR-370 are increased in patients with hyperlipidemia and associated with coronary artery disease. Lipids Health Dis 2012;11:55.

[87] Hulsmans M, Sinnaeve P, Van der Schueren B, Mathieu C, Janssens S, Holvoet P. Decreased miR-181a expression in monocytes of obese patients is associated with the occurrence of metabolic syndrome and coronary artery disease. J Clin Endocrinol Metab 2012;97:E1213–8.

[88] Satoh M, Tabuchi T, Minami Y, Takahashi Y, Itoh T, Nakamura M. Expression of let-7i is associated with Toll-like receptor 4 signal in coronary artery disease: effect of statins on let-7i and Toll-like receptor 4 signal. Immunobiology 2012;217:533–9.

[89] Sondermeijer BM, Bakker A, Halliani A, de Ronde MW, Marquart AA, Tijsen AJ, et al. Platelets in patients with premature coronary artery disease exhibit upregulation of miRNA340* and miRNA624*. PLoS One 2011;6:e25946.

[90] Takahashi Y, Satoh M, Minami Y, Tabuchi T, Itoh T, Nakamura M. Expression of miR-146a/b is associated with the Toll-like receptor 4 signal in coronary artery disease: effect of renin-angiotensin system blockade and statins on miRNA-146a/b and Toll-like receptor 4 levels. Clin Sci (Lond) 2010;119:395–405.

[91] Weber M, Baker MB, Patel RS, Quyyumi AA, Bao G, Searles CD. MicroRNA expression profile in CAD patients and the impact of ACEI/ARB. Cardiol Res Pract 2011;2011:532915.

[92] Minami Y, Satoh M, Maesawa C, Takahashi Y, Tabuchi T, Itoh T, et al. Effect of atorvastatin on microRNA 221/222 expression in endothelial progenitor cells obtained from patients with coronary artery disease. Eur J Clin Invest 2009;39:359–67.

[93] Li T, Cao H, Zhuang J, Wan J, Guan M, Yu B, et al. Identification of miR-130a, miR-27b and miR-210 as serum biomarkers for atherosclerosis obliterans. Clin Chim Acta 2011;412:66–70.

[94] Wang GK, Zhu JQ, Zhang JT, Li Q, Li Y, He J, et al. Circulating microRNA: a novel potential biomarker for early diagnosis of acute myocardial infarction in humans. Eur Heart J 2010;31:659–66.

[95] D'Alessandra Y, Devanna P, Limana F, Straino S, Di Carlo A, Brambilla PG, et al. Circulating microRNAs are new and sensitive biomarkers of myocardial infarction. Eur Heart J 2010;31:2765–73.

[96] Corsten MF, Dennert R, Jochems S, Kuznetsova T, Devaux Y, Hofstra L, et al. Circulating MicroRNA-208b and MicroRNA-499 reflect myocardial damage in cardiovascular disease. Circ Cardiovasc Genet 2010;3:499–506.

[97] Adachi T, Nakanishi M, Otsuka Y, Nishimura K, Hirokawa G, Goto Y, et al. Plasma microRNA 499 as a biomarker of acute myocardial infarction. Clin Chem 2010;56:1183–5.

[98] Ai J, Zhang R, Li Y, Pu J, Lu Y, Jiao J, et al. Circulating microRNA-1 as a potential novel biomarker for acute myocardial infarction. Biochem Biophys Res Commun 2010;391:73–7.

[99] Kuwabara Y, Ono K, Horie T, Nishi H, Nagao K, Kinoshita M, et al. Increased microRNA-1 and microRNA-133a levels in serum of patients with cardiovascular disease indicate myocardial damage. Circ Cardiovasc Genet 2011;4:446–54.

[100] Cheng Y, Tan N, Yang J, Liu X, Cao X, He P, et al. A translational study of circulating cell-free microRNA-1 in acute myocardial infarction. Clin Sci (Lond) 2010;119:87–95.

[101] Gidlof O, Andersson P, van der Pals J, Gotberg M, Erlinge D. Cardiospecific microRNA plasma levels correlate with troponin and cardiac function in patients with ST elevation myocardial infarction, are selectively dependent on renal elimination, and can be detected in urine samples. Cardiology 2011;118:217–26.

[102] Long G, Wang F, Duan Q, Chen F, Yang S, Gong W, et al. Human circulating microRNA-1 and microRNA-126 as potential novel indicators for acute myocardial infarction. Int J Biol Sci 2012;8:811–8.

[103] Devaux Y, Vausort M, Goretti E, Nazarov PV, Azuaje F, Gilson G, et al. Use of circulating microRNAs to diagnose acute myocardial infarction. Clin Chem 2012;58:559–67.

[104] Olivieri F, Antonicelli R, Lorenzi M, D'Alessandra Y, Lazzarini R, Santini G, et al. Diagnostic potential of circulating miR-499-5p in elderly patients with acute non ST-elevation myocardial infarction. Int J Cardiol 2013;167:531–6.

[105] Meder B, Keller A, Vogel B, Haas J, Sedaghat-Hamedani F, Kayvanpour E, et al. MicroRNA signatures in total peripheral blood as novel biomarkers for acute myocardial infarction. Basic Res Cardiol 2011;106:13–23.

[106] Oerlemans MI, Mosterd A, Dekker MS, de Vrey EA, van Mil A, Pasterkamp G, et al. Early assessment of acute coronary syndromes in the emergency department: the potential diagnostic value of circulating microRNAs. EMBO Mol Med 2012;4:1176–85.

[107] Tijsen AJ, Creemers EE, Moerland PD, de Windt LJ, van der Wal AC, Kok WE, et al. MiR423-5p as a circulating biomarker for heart failure. Circ Res 2010;106:1035–9.

[108] Fukushima Y, Nakanishi M, Nonogi H, Goto Y, Iwai N. Assessment of plasma miR-NAs in congestive heart failure. Circ J 2011;75:336–40.

[109] Voellenkle C, van Rooij J, Cappuzzello C, Greco S, Arcelli D, Di Vito L, et al. MicroRNA signatures in peripheral blood mononuclear cells of chronic heart failure patients. Physiol Genomics 2010;42:420–6.

[110] Goren Y, Kushnir M, Zafrir B, Tabak S, Lewis BS, Amir O. Serum levels of microRNAs in patients with heart failure. Eur J Heart Fail 2012;14:147–54.

[111] Vogel B, Keller A, Frese KS, Leidinger P, Sedaghat-Hamedani F, Kayvanpour E, et al. Multivariate miRNA signatures as biomarkers for non-ischaemic systolic heart failure. Eur Heart J 2013;34:2812–22.

[112] Creemers EE, Tijsen AJ, Pinto YM. Circulating microRNAs: novel biomarkers and extracellular communicators in cardiovascular disease? Circ Res 2012;110:483–95.

[113] Zampetaki A, Willeit P, Tilling L, Drozdov I, Prokopi M, Renard JM, et al. Prospective study on circulating MicroRNAs and risk of myocardial infarction. J Am Coll Cardiol 2012;60:290–9.

[114] Napoli C, Crudele V, Soricelli A, Al-Omran M, Vitale N, Infante T, et al. Primary prevention of atherosclerosis: a clinical challenge for the reversal of epigenetic mechanisms? Circulation 2012;125:2363–73.

[115] Yoo CB, Jones PA. Epigenetic therapy of cancer: past, present and future. Nat Rev Drug Discov 2006;5:37–50.

[116] DeWoskin VA, Million RP. The epigenetics pipeline. Nat Rev Drug Discov 2013;12:661–2.

[117] Bailey D, Jahagirdar R, Gordon A, Hafiane A, Campbell S, Chatur S, et al. RVX-208: a small molecule that increases apolipoprotein A-I and high-density lipoprotein cholesterol in vitro and in vivo. J Am Coll Cardiol 2010;55:2580–9.

[118] Lovren F, Pan Y, Quan A, Singh KK, Shukla PC, Gupta N, et al. MicroRNA-145 targeted therapy reduces atherosclerosis. Circulation 2012;126:S81–90.

[119] Wei Y, Nazari-Jahantigh M, Chan L, Zhu M, Heyll K, Corbalan-Campos J, et al. The microRNA-342-5p fosters inflammatory macrophage activation through an Akt1- and microRNA-155-dependent pathway during atherosclerosis. Circulation 2013;127:1609–19.

[120] Rotllan N, Ramirez CM, Aryal B, Esau CC, Fernandez-Hernando C. Therapeutic silencing of microRNA-33 inhibits the progression of atherosclerosis in ldlr-/- mice. Arterioscler Thromb Vasc Biol 2013;33:1973–7.

[121] Li Z, Rana TM. Therapeutic targeting of microRNAs: current status and future challenges. Nat Rev Drug Discov 2014;13:622–38.

[122] Janssen HL, Reesink HW, Lawitz EJ, Zeuzem S, Rodriguez-Torres M, Patel K, et al. Treatment of HCV infection by targeting microRNA. N Engl J Med 2013;368:1685–94.

[123] Bouchie A. First microRNA mimic enters clinic. Nat Biotechnol 2013;31:577.

CHALLENGES AND FUTURE DIRECTIONS

19

Future Challenges and Prospects for Personalized Epigenetics

Peng Zhang[1], Ying Liu[1], Qianjin Lu[1], Christopher Chang[2]

[1]Department of Dermatology, Hunan Key Laboratory of Medical Epigenomics, Second Xiangya Hospital, Central South University, Hunan, China; [2]Division of Rheumatology, Allergy and Clinical Immunology, University of California at Davis, Davis, CA, USA

OUTLINE

1. INTRODUCTION

The field of epigenetics, which is defined as the study of stable and heritable changes in gene expression without alterations in the underlying DNA sequence itself [1], has exploded since the late 1990s. Epigenetic marks are influenced by a mix of genetic and environmental variations. Epigenetic mechanisms include methylation of DNA at the cytosine residue of cytosine–phosphate–guanine (CpG) dinucleotides, covalent modifications of amino acid residues within histone proteins, and small noncoding RNAs (microRNAs or miRNAs). Epigenetic regulation in gene expression is a dynamic process, playing vital roles in both physiological and pathological events. With the advent of epigenetics, great advances have been made in exploring and understanding many diseases such as cancers and autoimmune diseases (AIDs). More and more biomarkers and therapies associated with epigenetics have been developed and have shown a promising future in the field of personalized medicine. The ultimate target of personalized medicine is to treat every patient based on the knowledge of the patient's concrete background information, including genetic mutations related to the disease and epigenetic aberrations contributing to the disease, to offer a personalized holistic therapeutic approach.

Since diseases can develop as a result of genetic or epigenetic modifications, any combination of these can play a role in the generation of the multitude of disease phenotypes and endotypes. As we identify genetic differences between those with or without a specific disease, or those with varying degrees of severity, it is imperative that we also consider the role of epigenetics and how these changes may influence the genetic susceptibility to disease. One must also not discount the fact that epigenetic changes may also occur as a function of environmental influences.

Here, we aim to describe the future challenges and prospects of personalized epigenetic therapy, involving the direction of future research in personalized epigenetics, promises and pitfalls of incorporating epigenetic data into personalized medicine, and the interaction of epigenetics with genetic and environmental conditions.

2. PERSONALIZED EPIGENETICS

An important challenge for every doctor in the everyday practice of medicine is to identify the right diagnosis and offer appropriate treatment. Currently, most patients with a specific type disease always receive the same therapy. However, it has become clear to clinical physicians that there is an individual variance in the response to treatment, resulting in an increasing demand for personalized diagnosis and treatment.

Personalized medicine or individualized medicine, first coined in the context of genetics, refers to a type of medical model characterized by the customization of health care using molecular technology during the process of medical practice specific to the individual patient. Currently, personalized medicine is defined by using molecular analysis to tailor the most appropriate medical treatment depending on the characteristics of each patient. Traditionally, the usage of the term personalized medicine has often been limited to the identification of the optimal drug and the optimal dosage for a subgroup of patients; however, the current concept of personalized medicine is far more broad and includes, but is not limited to, diagnostics, preventive interventions, and targeted treatment for individual patients. Diagnostic testing is useful for making the right diagnosis and selecting the appropriate therapies according to the patient's genetic and epigenetic profile. In the past, genetic information has played a major role in certain aspects of personalized medicine, but now it has broadened to all variations of personalized measures, including epigenetics (Figure 1).

Since the term epigenetics was first coined by Conrad Waddington in 1942, its definition has evolved [2]. The current definition of epigenetics, adopted in this chapter, refers to the study of mitotically heritable changes in gene expression that are not encoded in the original DNA sequence itself [1]. More concepts of epigenetics include Feinberg's epigenetic progenitor origin model, which states that cancers occur as a result of a polyclonal epigenetic disruption of progenitor cells, via the action of tumor

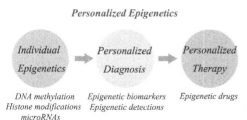

Personalized Epigenetics

FIGURE 1   **Process of personalized epigenetics.** Personalized epigenetics is based on the epigenetic information relative to the individual patient, with which the diagnosis is made using the epigenetic biomarkers, and the patient is treated with epigenetics-related therapies.

progenitor genes. Moreover, the heterogeneity of tumor cells results from epigenetic progression in progenitor cells, and tumor progression results from later epigenetic alterations that substitute for genetic variation. The model suggests that cancer risk assessment should be done before the development of cancer by the study of nonneoplastic progenitor or stem cells that may be epigenetically disrupted [3]. Baylin and Jones have reviewed the biological and translational implications of research on epigenetics in cancer since the turn of the century, describing how epigenetic alterations comprise one of the most promising areas for development of biomarkers for diagnosis, disease monitoring, prognosis, and treatment of cancers and other diseases [4].

3. FUTURE DIRECTION OF PERSONALIZED MEDICINE IN EPIGENETICS

Making the right diagnosis and offering the best therapeutic treatments are always the most important and challenging tasks for every clinical physician. There exists an unmet need to find new biomarkers and improve the accuracy of diagnosis, predict the prognosis of diseases such as cancers and autoimmune diseases, and provide customized or personalized treatment strategies. One of the ways to achieve this is to conduct translational research on epigenetic biomarkers and therapeutics and thereby bring the science of epigenetics from the bench to the bedside. The increasing demand for personalized diagnosis and treatment is derived from the individual differences in clinical presentation and response to medications. Although great progress has been made in the discovery of epigenetic mechanisms and epigenetic biomarkers, there is still a long way to go toward making the dream of personalized medicine come true.

To realize the goal of individualized medicine, we need not only the availability of accurate and convenient individual epigenetics-based detection methods to differentiate the geno- and phenotypes of the patients, but also the identification of useful epigenetic biomarkers to monitor the progress of the diseases. In addition, it would obviously be optimal if we could tailor specific medication according to the demographic and clinical background of the patient, the stage of his or her disease, and the response to current drugs and treatments.

3.1 Epigenetic Biomarkers in Personalized Medicine

The study of epigenetics has witnessed a rapid and multidisciplinary explosion of research productivity. More and more evidence indicates that many factors, such as lifestyle, diet, stress, drugs, and mental status, have a great influence on the epigenetic code via alterations of the DNA methylome, histone modifications, and miRNA expression. Searching for new biomarkers to improve the accuracy of diagnosis is one of the most meaningful objectives for clinical researchers to strive for. The three primary epigenetic mechanisms involved in gene expression include DNA methylation, histone posttranslational modifications, and miRNA expression.

A biomarker is defined as "a measurement, including but not limited to a genetic, biological, biochemical, molecular, or imaging event whose alterations correlate with the pathogenesis and/or manifestations of a disease and can be evaluated qualitatively and/or quantitatively in laboratories" [5]. Better biomarkers are needed to improve diagnosis, guide molecularly targeted therapy, and monitor therapeutic response. Epigenetic characteristics have surfaced as a rapidly increasing series of biomarkers in the diagnosis and prognosis of diseases [6].

3.1.1 DNA Methylation as Biomarkers

As of this writing, the vast majority of replicated and candidate clinical DNA methylation biomarkers come from cancer research [4]. Epigenetic biomarkers for cancers have generated a great deal of interest because of their potential usefulness in early detection and diagnosis, prediction of response to therapy, and assessment of prognosis. Epigenetic alterations often occur before the onset of the phenotype and can be detected in various types of samples (urine, saliva, feces, plasma, and sputum samples), providing an optimal solution for non- or minimally invasive diagnostic detection for cancer. Therefore, they are promising indicators for early tumor detection. Once proven to be more effective and accurate, epigenetic biomarkers will provide a complementary analysis for conventional methodologies, such as histopathology and immunohistochemistry, that are currently used for early diagnosis.

Compared to other types of biomarkers, biomarkers related to DNA methylation have several advantages. Changes in DNA methylation are found throughout the genome and are not restricted to CpG islands and can be measured by a group of sensitive and cost-efficient techniques [7]. In addition, DNA methylation is a stable biomarker not easily altered, which is important for monitoring the progression of diseases [8]. Methods of DNA methylation detection can be simplistically divided into two types: global and gene-specific methylation analysis. Global methylation analysis measures the overall level of methyl cytosines in a genome, and with the development of microarrays and next-generation sequencing platforms, more and more approaches are available now, such as DNA microarrays (Me-DIP-chip) and methylated DNA immunoprecipitation sequencing (Me-DIP-seq) [9]. For gene-specific methylation analysis, a great number of techniques have been developed. Among these, bisulfite genomic sequencing is regarded as a gold-standard technology for detection of DNA methylation because it provides an efficient way to identify 5-methylcytosine at single-base-pair resolution [10]. More than 30 years ago, DNA methylation as a marker for diagnosis was first proven to be associated with cancer with accompanying alterations in normal gene regulation [11]. The altered DNA methylation patterns in gene promoters provide possibilities for diagnosis and clinical monitoring in many different types of tumors [12], such as colorectal cancer [13], breast cancer [14], lung cancer [15], and urological cancer [16].

DNA methylation of promoter genes as a biomarker has been found not only in cancer, but also in other human disorders, such as metabolic diseases [17], autoimmune diseases [5], and cardiovascular diseases [18]. For example, hypomethylation in T cells and DNA hypomethylation of *TNFSF7* (CD70) are important epigenetic hallmarks of systemic lupus erythematosus (SLE) [19,20]. DNA demethylation of the gene coding for insulin is elevated in patients with new-onset type 1 diabetes [21]. A number of cardiac dysfunctions have been linked to abnormal DNA methylation patterns. Hypermethylation of the *HSD11B2* gene, playing roles in catalyzing conversion between cortisone and cortisol, has been correlated with hypertension [22]. Also, decreased methylation of long interspersed nuclear element-1 is a useful biomarker for ischemic heart disease and stroke [23]. Hypermethylation of GNASAS, IL-10, MEG3, and ABCA1 and hypomethylation of INSIGF and IGF2 have been associated with lipid metabolism [24]. It is important to understand that DNA methylation changes are found throughout the genome and are not restricted to CpG islands. However, the clinically and functionally relevant ones might be located at specific loci.

3.1.2 Histone Modifications as Biomarkers

The concept of the histone code was proposed for the first time in 2001 [25]. Subsequently, evidence of alterations in posttranslational modifications has emerged in a wide range of diseases such as cancers [26], autoimmune diseases [27], and neurological syndromes [28]. In 2005, Kurdistani and colleagues demonstrated that lower global/cellular levels of dimethylation of lysine 4 of histone H3 (H3K4me2) and acetylation of lysine 18 of H3 (H3K18ac) can predict a higher risk of prostate cancer recurrence [29]. Subsequently, they also observed that lower levels of H3K4me2 and H3K18ac are predictors for poorer survival probabilities in both lung and kidney cancers. In addition, lower cellular levels of H3K9me2 also predict poorer outcome in patients with prostate or kidney cancers [30].

3.1.3 miRNAs as Biomarkers

Unlike mRNA, miRNAs not only reside in the cells but also are abundant in circulating fluids and have strong stability in various types of biological samples, i.e., formalin-fixed, paraffin-embedded clinical tissue; fresh snap-frozen material; plasma or serum; and saliva, suggesting that they may be potentially useful as diagnostic, prognostic, and predictive biomarkers for various types of diseases [31]. MiRNA can be measured by conventional techniques such as Northern blotting and qRT-PCR and advanced techniques such as deep sequencing and genome-wide miRNA profiling techniques [32,33]. MiR-15 and miR-16 were the first two miRNAs detected in cancer and were implicated in the pathogenesis of chronic lymphocytic leukemia [25]. MiR423-5p is a promising potential candidate biomarker in blood and cardiomyocytes of patients with heart failure [34]. Circulating miR145, -155, -92a, -17, and -126 are predictive biomarkers for coronary artery disease [35]. With the advent of large-scale "miRNAome" analysis, we have reason to believe that more and more miRNAs will be identified to be associated with the diagnosis and prognosis of many diseases. In addition, miRNAs appear to be good candidates for noninvasive biomarkers as demonstrated by the diagnostic value of miRNAs found in body fluids in cancer patients [36].

3.2 Epigenetic Therapy in Personalized Medicine

Epigenetic alterations have now been associated with a great number of human disorders and the related treatment has been termed "epigenetic therapy" accordingly. Therefore, epigenetic therapies are defined as the usage of drugs or other epigenetics-associated methods to deal with medical conditions, offering potential alternatives in the treatment of various ailments.

3.2.1 Cancers

The field of the most intensive study of epigenetic mechanisms is cancer. Evidence related to epigenetic therapy in cancer is based on two key findings: first, epigenetic regulation is used by various cancers to deactivate cellular antitumor systems and second, most human cancers can activate oncogenes via epigenetic mechanisms [37,38].

As of this writing, there are four U.S. Food and Drug Administration (FDA)-approved epigenetic drugs: the DNA methyltransferase inhibitors 5-azacytidine (Vidaza) and decitabine (2′-deoxy-5-azacytidine, Dacogen) and the histone deacetylase (HDAC) inhibitors suberoylanilide hydroxamic acid (SAHA; Zolinza) and romidepsin (Istodax) [39]. The DNA methyltransferase (DNMT) inhibitors 5-azacytidine and 5-aza-20-deoxycytidine, with their ability to reactivate cellular antitumor systems, have been approved by the FDA for the treatment of many cancers [40]. Zebularine is also used as a prototype of epigenetic therapy for cancer chemoprevention [41], by inhibiting DNA methylation and tumor growth both in vitro and in vivo. SAHA is a chemically synthesized hydroxamate, inducing growth inhibition and differentiation in several types of neoplastic cells [42,43]. However, because of their wide effects throughout the cells, side effects are inevitable. Survival rates are improved significantly when these drugs are used for the treatment of various cancers (Table 1).

TABLE 1　Epigenetic Agents Currently Approved or Under Investigation

Drug	Available (FDA approved)	Indications	Mechanism of action	Comments/other names
ACY1216	Phase II	Multiple myeloma	HDAC6 inhibitor	
5-Azacytidine	Yes	Refractory anemia (RA) or RA with ringed sideroblasts (RARS) (if accompanied by neutropenia or thrombocytopenia or requiring transfusions), RA with excess blasts (RAEB), RA with excess blasts in transformation (RAEB-T), and chronic myelomonocytic leukemia (CMMoL)	DNA demethylation	Vidaza
Abexinostat	Phase II clinical trials	B cell lymphoma	HDAC inhibitor	PCI-24781
Belinostat	Approved	Peripheral T cell lymphoma (PTCL)	HDAC inhibitor	Beleodaq (also in trials for ovarian cancer)
Decitabine	Yes	Myelodysplastic syndromes (MDS) including previously treated and untreated, de novo and secondary MDS of all French-American-British subtypes (RA, RARS, RAEB, RAEB-T, and CMMoL) and intermediate 1, intermediate 2, and high-risk International Prognostic Scoring System groups	DNA demethylation	2'-Deoxy-5-azacytidine, Dacogen
Entinostat	Phase II clinical trials	Hodgkin lymphoma, breast cancer, metastatic lung cancer	Benzamide HDAC inhibitor	SBDX-275, MS-275
EPZ-5676	Phase I	Mixed-lineage rearranged leukemia	Histone methyltransferase inhibitor	
EVP-0334	Phase I	Neurodegenerative disorders	Class 1 HDAC inhibitor	

Givinostat	Phase II	Granted orphan drug status for polycythemia vera and juvenile idiopathic arthritis, phase II trials for multiple cancers	Inhibits class 1 and 2 histones	Gavinostat, ITF2357
GSK525762	Phase I	Carcinoma	Bromodomain-containing protein inhibitor	
Mocetinostat	Phase II clinical trials	Follicular lymphoma, Hodgkin lymphoma, and acute myelogenous leukemia	HDAC inhibitor	MGCD0103
OTX015	Class I	Hematologic malignancies	Bromodomain-containing protein inhibitor	
Pabinostat	Phase III clinical trials	Hodgkin lymphoma, cutaneous T cell lymphoma	Nonselective HDAC inhibitor	Faridak
Pracinostat	Phase II clinical trials	Acute myeloid leukemia (orphan drug status), metastatic or recurrent prostate cancer	HDAC inhibitor	SB-939, granted orphan drug status by FDA
Resminostat	Phase II clinical trials	Hepatocellular carcinoma (granted orphan drug status), Hodgkin lymphoma	HDAC inhibitor	RAS2410
Romidepsin	Yes	Cutaneous T cell lymphoma (CTCL) in patients who have received at least one prior systemic therapy, PTCL in patients who have received at least one prior therapy	HDAC inhibitor	Istodax
RVX-208	Phase II	Atherosclerosis, type 2 diabetes	Bromodomain-containing protein inhibitor	

TABLE 1 Epigenetic Agents Currently Approved or under Investigation—cont'd

Drug	Available (FDA approved)	Indications	Mechanism of action	Comments/other names
SAHA	Yes	Cutaneous manifestations in patients with CTCL who have progressive, persistent, or recurrent disease on or following two systemic therapies	HDAC inhibitor	Zolinza, Vorinostat, suberoylanilide hydroxamic acid
SGI110	Phase II	Myelodysplastic syndromes, acute myeloid leukemia, ovarian cancer	DNA methylation inhibitor	
SRT2104	Phase II	Plaque psoriasis, type 2 diabetes	Sirtuin 1 stimulant	
Valproic acid	Phase III clinical trials (already approved for other clinical uses)	Cervical cancer, ovarian cancer	HDAC inhibitor	Depakote, Epilim, Valparin, Valpro, Vilapro, and Stavzor (different brand names for different clinical uses, on the WHO essential medication list)

3.2.2 Autoimmune Diseases

Owing to the core pathology of AIDs, which involves the loss of tolerance to self-antigens and the formation of autoantibodies, therapeutic strategies against AIDs are developed to recover immune tolerance and inhibit inflammation. Inhibitors of DNMTs and HDACs have been applied to treat AIDs in animal models, resulting in remission of symptoms and decreased inflammatory cytokines. SLE is a prototypic autoimmune disease with unknown etiology in which genetic predispositions together with environmental factors trigger the disease [44]. The HDAC inhibitor trichostatin A has been found to improve lupus-like disease in NZB/W F1 mice via regulatory T cells [45].

4. EPIGENETIC DATA VERSUS GENETIC DATA IN PERSONALIZED MEDICINE

According to their definitions, the critical difference between epigenetics and genetics is whether the original DNA sequence is altered. Genetic modifications are limited by the total number of genes in human beings; therefore, epigenetic modifications may be more variable and abundant than genetic variability. Epigenetic changes are reversible and heritable and therefore may generate many more possibilities for adapting to the changing environment. As for personalized medicine, it is reasonable to expect to obtain more data from epigenetic studies than from genetic studies, providing the opportunity to discover more reliable and accurate biomarkers or therapeutic targets for epigenetics-associated disorders. However, the massive amounts of data that can be generated mean that the study and analysis of epigenetic data are very labor intensive and often require high-throughput computing systems to accomplish, utilizing analytical methods designed for the evaluation of "big data."

5. COMPUTATIONAL EPIGENETICS

Owing to the amazing explosion of epigenetic data sets, computational analysis plays an important role in all aspects of epigenetic research. Computational epigenetics consists of the development and application of bioinformatic analysis methods for solving epigenetic questions, computational data analysis, and theoretical modeling in the context of epigenetics. The great demand in the field of computational epigenetics was first initiated by the rapid development of increasingly large volumes of epigenetic data, which required appropriate bioinformatics for general and specialist databases, basic bioinformatics tools, and sophisticated algorithms for complicated analysis, modeling, and predication of DNA–protein interactions.

VIII. CHALLENGES AND FUTURE DIRECTIONS

A great number of computational, mathematical, and statistical methods have been developed to address questions in data mining, sequence analysis, and molecular interactions. ClustalW, BLAST, BLAT (BLAST-Like Alignment Tool), and MEGA (Molecular Evolutionary Genetics Analysis) are traditional sequence analysis tools, used for searching homologies of ortholog candidates for the KEGG/GENES database, predicting the secondary structures of histone deacetylases, homology modeling of DNA methyltransferases, and optimizing the activities of histone deacetylase inhibitors [46]. Table 2 shows the most common resources of databases for epigenetic bioinformatics analysis.

The most accurate and probably the most widely used protocol for analyzing DNA methylation is bisulfite sequencing, which consists of a selective conversion of unmethylated cytosines to uracils by bisulfite treatment and subsequent amplification, cloning, sequencing, and comparison to the genomic sequence [47,48]. Manual analysis of DNA methylation data from bisulfite sequencing is a complex and error-prone

TABLE 2   The Most Common Databases for Epigenetic Bioinformatics Analysis

Epigenetics databases	Summary
MethDB	19,905 DNA methylation contents data and 5382 methylation patterns for 48 species, 1511 individuals, 198 tissues and cell lines, and 79 phenotypes
PubMeth	Over 5000 records on methylated genes in various cancer types
REBASE	Over 22,000 DNA methyltransferase genes derived from GenBank
MeInfoText	Gene methylation information across 205 human cancer types
MethPrimerDB	259 primer sets from human, mouse, and rat for DNA methylation analysis
The Histone Database	254 sequences from histone H1, 383 from histone H2, 311 from histone H2B, 1043 from histone H3, and 198 from histone H4, altogether representing at least 857 species
ChromDB	9341 chromatin-associated proteins, including RNAi-associated proteins, for a broad range of organisms
CREMOFAC	1725 redundant and 720 nonredundant chromatin-remodeling factor sequences in eukaryotes
MethyLogiX DNA methylation database	DNA methylation data of human chromosomes 21 and 22, male germ cells, and late-onset Alzheimer disease
The Krembil Family Epigenetics Laboratory	DNA methylation data of human chromosomes 21 and 22, male germ cells, and DNA methylation profiles in monozygotic and dizygotic twins

process. A basic Microsoft Excel template invented by Anbazhagan can be used for the calculation of average methylation and similar statistics once methylation data have been generated [49]. In addition, an interactive software tool, BiQ Analyzer, provides start-to-end support for analysis of DNA methylation data from bisulfite sequencing [50]. The software has a friendly interactive interface, making it easier to import sequence files directly from the sequencer without any manual alteration, and provides help with all steps of alignment and quality control, performs basic statistics, and generates publication-quality diagrams. The program is available for free for noncommercial users and can be downloaded from http://biq-analyzer.bioinf.mpi-inf.mpg.de/.

Since the first miRNA Web resource, the miRBase database, was launched, a great number of new resources have been released. Through these databases and Web tools, a deeper insight into all aspects of miRNA biology and function can be achieved. Several online miRNA databases are listed in Table 3. However, it is necessary to be aware of

TABLE 3 Representative Online Tools of miRNA Databases

Resource	Description	URL
miRBase	The miRNA sequence and annotation database	www.mirbase.org
miRecords	Database of validated animal miRNA target interactions (also predicted targets from 11 algorithms)	mirecords.biolead.org
miRTarBase	Database of experimentally validated miRNA target interactions	mirtarbase.mbc.nctu.edu.tw/
miRvar	Database for genomic variations in miRNAs	genome.igib.res.in/mirlovd
TransmiR	Manually curated database of TF–miRNA regulations	202.38.126.151/hmdd/mirna/tf/
CID-miRNA	Web service for the prediction of miRNA precursors	mirna.jnu.ac.in/cidmirna/
dbDEMC	Database of differentially expressed miRNAs in human cancers	159.226.118.44/ dbDEMC/index.html
HMDD	Manually curated database of human miRNA–disease associations	202.38.126.151/hmdd/mirna/md
MapMi	Web service for mapping miRNA sequences to genomic loci across many species	www.ebi.ac.uk/enright-srv/ MapMi

Continued

VIII. CHALLENGES AND FUTURE DIRECTIONS

TABLE 3   Representative Online Tools of miRNA Databases—cont'd

Resource	Description	URL
microRNA.org	A resource for predicted miRNA targets (miRanda) and miRNA expression profiles	www.microrna.org
miR2Disease	Literature-curated database of human miRNA–disease relationships	www.mir2disease.org
miRdSNP	Database of disease-associated SNPs and their distance from miRNA target sites on the 3′ UTRs of human genes	mirdsnp.ccr.buffalo.edu/
TargetRank	Target prediction database with an integrated ranking of conserved and nonconserved miRNA targets	genes.mit.edu/targetrank/
TargetScan	Predicted targets of miRNA regulation based on the TargetScan algorithm	www.targetscan.org
CoGemiR	Collection of information on miRNA genomic location, conservation, and expression data	cogemir.tigem.it/
miRGen	Provides an overview of the genomic context of miRNAs (including clusters), target predictions from six algorithms, and experimentally supported targets	www.diana.pcbi.upenn.edu/miRGen.html
miRNAMap	Resource that collects information about known miRNAs and their targets, expression profiles, predicted targets, and tissue specificity	mirnamap.mbc.nctu.edu.tw
miRWalk	Database that integrates miRNA target predictions from various resources and information on validated targets	mirwalk.uni-hd.de/

some risks with the use of these resources and tools. First, the cross-platform comparisons of high-throughput miRNA expression experiments reveal high correlations, but the miRNA expression data have the risk of being unreliable as these databases were often generated under different experimental conditions with different setups [51]. One should also be aware of the release date or the date of the latest update and the changes compared to the previously released version. Second, to

VIII. CHALLENGES AND FUTURE DIRECTIONS

avoid false-positive predictions and a bias toward patterns in the target prediction algorithms, it is recommended to retrieve results of several algorithms and consider only commonly predicted targets for further investigation [52]. Entries with obsolete names/identifiers in miRNA resources may change or be deleted from the primary miRNA sequence registry (miRBase).

6. CHALLENGES OF PERSONALIZED EPIGENETICS

As the development and widespread application of personalized medicine increase, a number of challenges arise, including intellectual property rights, reimbursement policies, and issues related to patient privacy.

6.1 Intellectual Property Rights

When it comes to any innovation in personalized medicine, it is inevitable that one must address the investments and interests that are influenced by intellectual property rights. It is urgent and necessary to define the boundaries of patent protection. In June 2013, the U.S. Supreme Court passed a bill that no patent can be applied for natural occurring genes, while edited or artificially DNA can still be patented. Those who object to these patents state that this bill is an obstacle for further medical research, while those who favor patents believe that protecting the financial interests of researchers and their associated institutions is beneficial for the development of the industry in the long term [53]. It is clear that there are conflicts of interests of scientific, social, economic, and political nature, and it is up to society and the evolution of our values and morals to decide what is governed by the rule of law and applicable to the next generation of epigenetic research.

6.2 Reimbursement Policies

When personalized medicine is integrated into the health care system, a redefined reimbursement policy is needed. Efficacy of genetic tests, cost-effectiveness and the management of benefits and risks, and the adoption of new methods of therapy are considerations in defining fair and standard reimbursements to physicians and health care organizations. In most societies, reimbursement policies are defined by the government or by private industry. Given the sordid history of medical reimbursement in countries with or without a national health care system, this presents one of the major challenges to the fair distribution of new forms of therapy such as epigenetic therapy [5].

6.3 Patient Privacy

No one can deny the importance of patient privacy protection, although this goal seems to become more and more intangible in our high-tech world. Just as every coin has two sides, this is no exception when it comes to personalized medicine. Offering the possibilities and convenience for doctors to make the right diagnosis and decide the appropriate therapy, personalized medicine may, at the same time, give rise to the fear of patients who may be identified and targeted as predisposed to certain diseases or predicted to be insensitive to certain therapies, thereby compromising their access to fair and equitable health coverage. As the saying goes, "There is no free lunch!"

These concerns extend to patients' families, as relatives who do not have symptoms may also need to be managed, bringing into play a whole host of privacy issues. The fact that epigenetic changes are heritable and reversible provides daunting challenges in the management of families. No longer is it personalized medicine, but personalized "familial" medicine. For example, is it acceptable to share all or partial information with the patients and their family members? Will the circumstances affect how data are managed and transmitted? How will we deal with the ramifications and potential consequences once extensive data on predisposition are available? These are all important ethical and philosophical questions that will manifest themselves as significant challenges as we learn more about the relationship between genetics, epigenetics, environment, and disease.

7. CONCLUSIONS

Although many biomarkers and therapies associated with personalized epigenetics have been studied during the past decades, large-scale validation and multicenter studies are still necessary to ensure their applicability and practicability in medicine. Because of the heterogeneous nature of complex diseases, it is somehow reasonable to have conflicting results of some biomarkers in different studies. The present development of biomarkers primarily focuses on making a precise diagnosis or the ability to monitor disease progression. The next step is to discover biomarkers that may predict the onset of disease in predisposed individuals and the occurrence of flares in established patients, providing valuable information for doctors and patients to achieve "real-time" monitoring of the diseases.

For some complex diseases, no single biomarker will be sufficient to diagnose, monitor, and stratify all the patients with the disease. Therefore, it will be necessary to discover and validate biomarker "panels" of specific disorders. Moreover, developing a new series of pharmacodynamic biomarkers is needed to identify the patients who might respond favorably to a particular drug, select the specific type and dose of drug, and evaluate the therapeutic effect. Realizing the goals of personalized epigenetics will require collaborative efforts from many disciplines and hopefully revolutionize the clinical management of patients.

LIST OF ACRONYMS AND ABBREVIATIONS

AID	Autoimmune disease
CMMol	Chronic myelomonocytic leukemia
CTCL	Cutaneous T-cell lymphoma
DNMT	DNA methyltransferase
HDAC	Histone deacetylase
IM	Individualized medicine
MDS	Myelodysplastic syndrome
MiRNA	MicroRNA
PM	Personalized medicine
PTCL	Peripheral T cell lymphoma
PTMs	Posttranslational modifications
RA	Refractory anemia
RAEB	Refractory anemia with excess blasts
RAEB-T	Refractory anemia with excess blasts in transformation
RARS	Refractory anemia with ringed sideroblasts
SLE	Systemic lupus erythematosus

References

[1] Bird A. Perceptions of epigenetics. Nature 2007;447:396-8.
[2] Waddington CH. The epigenotype. 1942. Int J Epidemiol 2012;41:10–3.
[3] Feinberg AP, Ohlsson R, Henikoff S. The epigenetic progenitor origin of human cancer. Nat Rev Genet 2006;7:21–33.
[4] Baylin SB, Jones PA. A decade of exploring the cancer epigenome–biological and translational implications. Nat Rev Cancer 2011;11:726–34.
[5] Liu CC, Kao AH, Manzi S, Ahearn JM. Biomarkers in systemic lupus erythematosus: challenges and prospects for the future. Ther Adv Musculoskelet Dis 2013;5:210–33.
[6] Sandoval J, Peiro-Chova L, Pallardo FV, Garcia-Gimenez JL. Epigenetic biomarkers in laboratory diagnostics: emerging approaches and opportunities. Expert Rev Mol Diagn 2013;13:457–71.
[7] Esteller M, Corn PG, Baylin SB, Herman JG. A gene hypermethylation profile of human cancer. Cancer Res 2001;61:3225–9.

[8] Carmona FJ, Esteller M. DNA methylation in early neoplasia. Cancer Biomarkers Sect A Dis Markers 2010;9:101–11.

[9] Weber M, Davies JJ, Wittig D, Oakeley EJ, Haase M, Lam WL, et al. Chromosome-wide and promoter-specific analyses identify sites of differential DNA methylation in normal and transformed human cells. Nat Genet 2005;37:853–62.

[10] Li Y, Tollefsbol TO. DNA methylation detection: bisulfite genomic sequencing analysis. Methods Mol Biol 2011;791:11–21.

[11] Feinberg AP, Vogelstein B. Hypomethylation distinguishes genes of some human cancers from their normal counterparts. Nature 1983;301:89–92.

[12] Esteller M. Epigenetics in cancer. N Engl J Med 2008;358:1148–59.

[13] Qu D, Sureban SM, Houchen CW. Epigenetic variants and biomarkers for colon cancer. Am J Pathol 2012;180:2205–7.

[14] Visvanathan K, Sukumar S, Davidson NE. Epigenetic biomarkers and breast cancer: cause for optimism. Clin Cancer Res Off J Am Assoc Cancer Res 2006;12:6591–3.

[15] Liloglou T, Bediaga NG, Brown BR, Field JK, Davies MP. Epigenetic biomarkers in lung cancer. Cancer Lett 2014;342:200–12.

[16] Jeronimo C, Henrique R. Epigenetic biomarkers in urological tumors: a systematic review. Cancer Lett 2014;342:264–74.

[17] Drummond EM, Gibney ER. Epigenetic regulation in obesity. Curr Opin Clin Nutr Metab Care 2013;16:392–7.

[18] Zaina S. Unraveling the DNA methylome of atherosclerosis. Curr Opin Lipidol 2014;25:148–53.

[19] Zhang Y, Zhao M, Sawalha AH, Richardson B, Lu Q. Impaired DNA methylation and its mechanisms in CD4(+)T cells of systemic lupus erythematosus. J Autoimmun 2013;41:92–9.

[20] Zhou Y, Qiu X, Luo Y, Yuan J, Li Y, Zhong Q, et al. Histone modifications and methyl-CpG-binding domain protein levels at the TNFSF7 (CD70) promoter in SLE CD4[+] T cells. Lupus 2011;20:1365–71.

[21] Akirav EM, Lebastchi J, Galvan EM, Henegariu O, Akirav M, Ablamunits V, et al. Detection of beta cell death in diabetes using differentially methylated circulating DNA. Proc Natl Acad Sci USA 2011;108:19018–23.

[22] Smolarek I, Wyszko E, Barciszewska AM, Nowak S, Gawronska I, Jablecka A, et al. Global DNA methylation changes in blood of patients with essential hypertension. Med Sci Monit Int Med J Exp Clin Res 2010;16:CR149–155.

[23] Breton CV, Byun HM, Wenten M, Pan F, Yang A, Gilliland FD. Prenatal tobacco smoke exposure affects global and gene-specific DNA methylation. Am J Respir Crit Care Med 2009;180:462–7.

[24] Tobi EW, Lumey LH, Talens RP, Kremer D, Putter H, Stein AD, et al. DNA methylation differences after exposure to prenatal famine are common and timing- and sex-specific. Hum Mol Genet 2009;18:4046–53.

[25] Calin GA, Dumitru CD, Shimizu M, Bichi R, Zupo S, Noch E, et al. Frequent deletions and down-regulation of micro- RNA genes miR15 and miR16 at 13q14 in chronic lymphocytic leukemia. Proc Natl Acad Sci USA 2002;99:15524–9.

[26] Hake SB, Xiao A, Allis CD. Linking the epigenetic 'language' of covalent histone modifications to cancer. Br J cancer 2004;90:761–9.

[27] Hu N, Qiu X, Luo Y, Yuan J, Li Y, Lei W, et al. Abnormal histone modification patterns in lupus CD4[+] T cells. J Rheumatol 2008;35:804–10.

[28] Ausio J, de Paz AM, Esteller M. MeCP2: the long trip from a chromatin protein to neurological disorders. Trends Mol Med 2014;20:487–98.

[29] Seligson DB, Horvath S, Shi T, Yu H, Tze S, Grunstein M, et al. Global histone modification patterns predict risk of prostate cancer recurrence. Nature 2005;435:1262–6.

[30] Seligson DB, Horvath S, McBrian MA, Mah V, Yu H, Tze S, et al. Global levels of histone modifications predict prognosis in different cancers. Am J Pathol 2009;174:1619–28.

[31] Ulivi P, Zoli W. miRNAs as non-invasive biomarkers for lung cancer diagnosis. Molecules 2014;19:8220–37.

[32] Lu J, Getz G, Miska EA, Alvarez-Saavedra E, Lamb J, Peck D, et al. MicroRNA expression profiles classify human cancers. Nature 2005;435:834–8.

[33] Liu CG, Calin GA, Meloon B, Gamliel N, Sevignani C, Ferracin M, et al. An oligonucleotide microchip for genome-wide microRNA profiling in human and mouse tissues. Proc Natl Acad Sci USA 2004;101:9740–4.

[34] Tijsen AJ, Creemers EE, Moerland PD, de Windt LJ, van der Wal AC, Kok WE, et al. MiR423-5p as a circulating biomarker for heart failure. Circ Res 2010;106:1035–9.

[35] Fichtlscherer S, De Rosa S, Fox H, Schwietz T, Fischer A, Liebetrau C, et al. Circulating microRNAs in patients with coronary artery disease. Circ Res 2010;107:677–84.

[36] Zandberga E, Kozirovskis V, Abols A, Andrejeva D, Purkalne G, Line A. Cell-free microRNAs as diagnostic, prognostic, and predictive biomarkers for lung cancer. Genes Chromosomes Cancer 2013;52:356–69.

[37] Vendetti FP, Rudin CM. Epigenetic therapy in non-small-cell lung cancer: targeting DNA methyltransferases and histone deacetylases. Expert Opin Biol Ther 2013;13:1273–85.

[38] Li H, Chiappinelli KB, Guzzetta AA, Easwaran H, Yen RW, Vatapalli R, et al. Immune regulation by low doses of the DNA methyltransferase inhibitor 5-azacitidine in common human epithelial cancers. Oncotarget 2014;5:587–98.

[39] Rius M, Lyko F. Epigenetic cancer therapy: rationales, targets and drugs. Oncogene 2012;31:4257–65.

[40] Foulks JM, Parnell KM, Nix RN, Chau S, Swierczek K, Saunders M, et al. Epigenetic drug discovery: targeting DNA methyltransferases. J Biomol Screen 2012;17:2–17.

[41] Kowluru RA, Santos JM, Mishra M. Epigenetic modifications and diabetic retinopathy. BioMed Res Int 2013;2013:635284.

[42] Prebet T, Vey N. Vorinostat in acute myeloid leukemia and myelodysplastic syndromes. Expert Opin Invest Drugs 2011;20:287–95.

[43] Vigna E, Recchia AG, Madeo A, Gentile M, Bossio S, Mazzone C, et al. Epigenetic regulation in myelodysplastic syndromes: implications for therapy. Expert Opin Invest Drugs 2011;20:465–93.

[44] Tsokos GC. Systemic lupus erythematosus. N Engl J Med 2011;365:2110–21.

[45] Reilly CM, Thomas M, Gogal Jr R, Olgun S, Santo A, Sodhi R, et al. The histone deacetylase inhibitor trichostatin A upregulates regulatory T cells and modulates autoimmunity in NZB/W F1 mice. J Autoimmun 2008;31:123–30.

[46] Lim SJ, Tan TW, Tong JC. Computational epigenetics: the new scientific paradigm. Bioinformation 2010;4:331–7.

[47] Frommer M, McDonald LE, Millar DS, Collis CM, Watt F, Grigg GW, et al. A genomic sequencing protocol that yields a positive display of 5-methylcytosine residues in individual DNA strands. Proc Natl Acad Sci USA 1992;89:1827–31.

[48] Hajkova P, el-Maarri O, Engemann S, Oswald J, Olek A, Walter J. DNA-methylation analysis by the bisulfite-assisted genomic sequencing method. Methods Mol Biol 2002;200:143–54.

[49] Anbazhagan R, Herman JG, Enika K, Gabrielson E. Spreadsheet-based program for the analysis of DNA methylation. BioTechniques 2001;30:110–4.

[50] Bock C, Reither S, Mikeska T, Paulsen M, Walter J, Lengauer T. BiQ Analyzer: visualization and quality control for DNA methylation data from bisulfite sequencing. Bioinformatics 2005;21:4067–8.

[51] Yauk CL, Rowan-Carroll A, Stead JD, Williams A. Cross-platform analysis of global microRNA expression technologies. BMC Genomics 2010;11:330.

[52] Sturm M, Hackenberg M, Langenberger D, Frishman D. TargetSpy: a supervised machine learning approach for microRNA target prediction. BMC Bioinf 2010;11:292.

[53] Chandrasekharan S, Cook-Deegan R. Gene patents and personalized medicine–what lies ahead? Genome Med 2009;1:92.

Index

Arabidopsis accessions, 25
ARE. *See* Antioxidant response element
Arsenic (As), 231–232, 264–265
Arsenic methyltransferase (AS3MT), 294
Aryl-hydrocarbon receptor repressor
  (AHRR), 303
AS3MT. *See* Arsenic methyltransferase
ASD. *See* Autism spectrum disorder
ASFMR1. See Antisense Fragile X mental
  retardation 1
ASM. *See* Allele-specific methylation
ASO. *See* Antisense oligonucleotide
Asymmetric DNA methylation, 125–128
"Atom-to-cell" approach, 374
Atrial natriuretic peptide (ANP), 292–293,
  527
Atrial vasopressin (AVP), 292–293
Autism spectrum disorder (ASD), 205–206,
  398
Autoimmune diseases (AIDs), 207, 477,
    554, 563. *See also* Autoimmunity
  big data, 494–495
  epigenetic drugs, 497
  epigenetics, 494–495
  gene expression heat map, 496f
  HDACs, 481t
  lessons from oncology, 481–484
  miRNAs, 482f
  pathogenesis of diseases, 495–496
  personalized epigenetic therapy in, 484,
    498t–499t
Autoimmune encephalomyelitis (EAE), 486
Autoimmunity
  environmental agents in, 491
    minerals, 493–494
    pollutants, 492–493
    viruses, 491
    vitamins, 493–494
  epigenetic agents in
    DNA-methylation agents, 486–488
    HDAC inhibitors, 484–486
    MicroRNAs, 488–491
  epigenetic basis for, 479–481
  evolution of treatment, 477–479
Autosomal-recessive diseases, 102–104
AVP. *See* Atrial vasopressin
a^{vy}. See agouti viable yellow allele
A^{vy} mouse model. *See Agouti* variable
  yellow mouse model
axin fused (axin^{fu}), 27
5-Aza-2′-deoxycytidine (5-Aza-CdR), 91
5-Aza-CdR. *See* 5-aza-2′-deoxycytidine

5-Aza-CR. *See* 5-azacytidine
5-Aza-deoxycytidine (DAC), 329
5-Azacytidine (5-Aza-CR), 91, 487–488

B
B-cell lymphoma 2 (Bcl-2), 331–332
B-cell lymphoma-6 (BCL-6), 91
BACE1. *See* β-site amyloid precursor
  protein cleaving enzyme 1
BAT. *See* Brown adipose tissue
Bayesian methods, 173
Bayesian mixture model, 171
BC. *See* Black carbon
Bcl-2. *See* B-cell lymphoma 2
BDNF. *See* Brain-derived neurotrophic
  factor
Beckwith-Wiedemann syndromes, 33–34
Behavioral disorders, 205–207
Benzamides, 484–485
Big data, 494–495, 563
Bio2RDF approach, 167–168
Bioactive compounds, 321
Bioactive food
  components, 381
  compounds, 326–333
Bioinformatics, 6, 154, 157–158
Biological age estimation, 228–231
Biomarker, 557
BisChIP-seq. *See* Bisulfite sequencing of
  chromatin-immunoprecipitated
Bisphenol A (BPA), 269
Bisulfite sequencing, 261–262
Bisulfite sequencing of chromatin-
  immunoprecipitated (BisChIP-seq),
  290
Black carbon (BC), 290–291
Bladder cancer (BlCa), 201
Blood, 229–230
Blood cellular heterogeneity and
  methylation variation, 134–136
BMI. *See* Body mass index
BNP. *See* Brain natriuretic peptide
Body fluid/tissue identification system,
  225
Body mass index (BMI), 334–335, 380, 430
BPA. *See* Bisphenol A
Brain cellular heterogeneity and
  methylation variation, 136–138. *See
  also* Stem cellular heterogeneity and
  methylation variation
Brain natriuretic peptide (BNP), 527
Brain tumors, 137

Printed in the United States
By Bookmasters